Legumes: Physiology and Molecular Biology of Abiotic Stress Tolerance

Prakash Muthu Arjuna Samy •
Anandan Ramasamy •
Viswanathan Chinnusamy •
B. Sunil Kumar
Editors

Legumes: Physiology and Molecular Biology of Abiotic Stress Tolerance

Springer

Editors
Prakash Muthu Arjuna Samy
Department of Genetics and
Plant Breeding
Annamalai University
Annamalai Nagar, Tamil Nadu, India

Anandan Ramasamy
Department of Genetics and Plant Breeding
Annamalai University
Annamalai Nagar, Tamil Nadu, India

Viswanathan Chinnusamy
Division of Plant Physiology
Indian Agricultural Research Institute
New Delhi, India

B. Sunil Kumar
Department of Genetics and Plant Breeding
Annamalai University
Annamalai Nagar, Tamil Nadu, India

ISBN 978-981-19-5816-8 ISBN 978-981-19-5817-5 (eBook)
https://doi.org/10.1007/978-981-19-5817-5

This Springer imprint is published by the registered company Springer Nature Singapore Pte Ltd.
The registered company address is: 152 Beach Road, #21-01/04 Gateway East, Singapore 189721, Singapore

Preface

Legumes are commonly referred to as "poor man's meat" and play a significant role in balanced diet. They constitute the primary source of protein, oil, fiber, and micronutrients for both human beings and livestock. They also have the ability to fix atmospheric nitrogen, which is crucial for crop production. Legumes occupy third place after cereals and oilseeds in terms of global production. They have a significant impact on the environment, agriculture, animal and human nutrition, and health. Legumes are susceptible to a wide range of abiotic stresses, including cold, drought, ultraviolet light, high temperatures, mineral toxicity and deficiency, salinity, and alkalinity. Hence, the yield of legume is drastically affected by these abiotic stresses. These limitations could be facilitated by using recent biotechnological techniques like molecular marker-assisted breeding, gene pyramiding, transgenic breeding, somaclonal variation, in vitro mutagenesis, in vitro selection, transgenomics, transcriptomics, and proteomics. It is essential to comprehend how plants react to these abiotic stresses in order to create new crop types that are more suited to harsh environmental circumstances.

This book offers a thorough overview of the agronomical, physiological, and molecular basis of plant responses to abiotic stresses. This book includes 15 chapters covering most of the legume crops and gives a complete description of plant responses to various environmental stresses. It is intended for academics, technologists, policy makers, and undergraduate and postgraduate students interested in plant physiology and molecular biology for sustainable agricultural production. This book is an important contribution to agricultural college and university libraries and research centers at state and national levels, where plant physiology and agricultural and horticultural science are being taught. We wholeheartedly thank every author who contributed their valuable chapters for the excellent outcome of this book. We are incredibly thankful to the publisher, Springer Nature, for assisting us to publish globally.

Annamalai Nagar, Tamil Nadu, India — Prakash Muthu Arjuna Samy
Annamalai Nagar, Tamil Nadu, India — Anandan Ramasamy
New Delhi, India — Viswanathan Chinnusamy
Annamalai Nagar, Tamil Nadu, India — B. Sunil Kumar

Contents

Editors and Contributors

About the Editors

Prakash Muthu Arjuna Samy, born in 1966, served as the Head of the Department of Genetics and Plant Breeding from 2006 to 2012 and is presently serving as the Controller of Examinations, Annamalai University. He has put in 29 years of teaching, research, and extension service. He has guided 8 Ph.D. scholars and 25 PG scholars. He has published more than 150 research papers, 10 books, and 15 book chapters. He has organized 3 international and 15 national seminars/conferences and completed 5 research projects. He has earned the status of Departmental Research Support of University Grants Commission (UGC-SAP-DRS) with a financial assistance of Rs. 50 lakhs at phase I (2009–2014) and Rs. 1.025 crore at phase II (2015–2020) level as Programme Coordinator.

He was awarded with the National Merit Scholarship, Government of India, during 1981 1988; ICAR Junior Fellowship during 1988–1990; J.J. Chinoy Gold Medal Award in 2017; Dr. B.P. Pal Memorial NABS Best Scientist Award in 2017; Best Researcher (Publication) Prize in 2016–2017, 2018–2019, and 2021; Fellow of Indian Society of Plant Physiology, New Delhi, in 2015; Fellow of National Academy of Biological Sciences, Chennai, in 2015; and Fellow of National Environmental Science Academy, New Delhi, in 2016.

Anandan Ramasamy is presently working as the Assistant Professor of Biotechnology, Department of Genetics and Plant Breeding, Faculty of Agriculture, Annamalai University. He has 15 years of teaching and research experience in plant molecular biology and biotechnology. His areas of teaching and research include plant tissue culture, molecular marker, molecular biology, gene cloning, plant genetic engineering, and bioinformatics. He has implemented three research projects funded by the TNSCST, DST, and UGC-GDA-XII Plan Innovative Research (Co-PI). He has published more than 35 research papers in national and international journals, book chapters, and proceedings of seminars and organized two national level workshops on techniques in plant molecular biology.

He has supervised many students at the M.Sc. (Ag.) level for their dissertations. He has deposited several partially amplified sequences of DNA barcode in the NCBI database. He has also qualified ARS/CSIR NET examination in the year 2009. He has served as session chairman/rapporteur and invited speaker for several national conferences/seminars. He has also received best researcher award by the Annamalai University in 2007, 2012, 2016, and 2019 and several other awards to his credit. He was instrumental in establishing the molecular marker laboratory from the grant sanctioned by the UGC-SAP, DST-FIST, PURSE, and RUSA during 2012–2019.

Viswanathan Chinnusamy is the Principal Scientist and Head, Division of Plant Physiology, at the ICAR-Indian Agricultural Research Institute (IARI), New Delhi. He completed his B.Sc. Agriculture from AC&RI (TNAU), Killikulam, and M.Sc. and Ph.D. from the Indian Agricultural Research Institute, New Delhi. He carried out his postdoctoral research at the University of Arizona, Tucson, USA; University of California Riverside, USA; and Harvard University, Boston, USA.

He joined Agricultural Research Service in 1996 at the IARI, New Delhi. He has 26 years of research and teaching experience in plant physiology. He has published about 175 peer-reviewed publications with total citations >17,000. He has contributed to the state-of-the-art research facilities, namely Nanaji Deshmukh Plant Phenomics Centre and Discovery Centre at the IARI, New Delhi. His current research interest includes phenomics, genetic engineering, and genome editing for deciphering the mechanisms of abiotic stress tolerance and developing climate-resilient crops.

Dr. Chinnusamy is a Fellow of the National Academy of Agricultural Sciences and the Indian Society for Plant Physiology. He is also the Honorary Secretary of the Indian Society for Plant Physiology and has been the Executive Editor of Plant Physiology Reports from 2019.

B. Sunil Kumar serving as the Associate Professor in the Department of Genetics and Plant Breeding, Faculty of Agriculture, Annamalai University, Chidambaram, Tamil Nadu, obtained M.Sc. (Ag.) in Genetics and Plant Breeding and Ph.D. in Agricultural Botany from the Annamalai University. He had cleared Agricultural Research Service (ARS)/National Eligibility Test (NET) in 2001. He is actively involved in cutting-edge research in biotic stress management in legume crops, especially in mung bean and urdbean at biometric, biochemical, biophysical, and molecular levels.

He had operated 3 national level projects and published more than 60 research publications in peer-reviewed journals. He has 4 books and 15 book chapters to his credit and has received 6 awards including Best Researcher Award, Outstanding Scientist Award, etc. He has also organized many international/national conferences and workshops. He is a life member and serving as a subject matter specialist/committee member in various journals and academic boards.

Contributors

Neha Anand Mineral Nutrition Lab, Division of Plant Physiology, ICAR-Indian Agricultural Research Institute, New Delhi, India

R. Anandan Department of Genetics and Plant Breeding, Faculty of Agriculture, Annamalai University, Annamalai Nagar, Tamil Nadu, India

Muraleedhar S. Aski Division of Genetics, ICAR-Indian Agricultural Research Institute, New Delhi, India

Ruchi Bansal Plant Physiology, ICAR-Indian Agricultural Research Institute, New Delhi, India

Anjali Bhardwaj Department of Botany, Punjab University, Chandigarh, India

HanumanthaRao Bindumadahva Dr. Marri Channa Reddy Foundation (MCRF), Hyderabad, Telangana, India

Shikha Chaudhary Department of Botany, Punjab University, Chandigarh, India

Manoj Choudhary ICAR-Central Arid Zone Research Institute, Jodhpur, Rajasthan, India

Alok Das Division of Plant Biotechnology, ICAR-Indian Institute of Pulses Research, Kanpur, Uttar Pradesh, India

Jyoti Devi ICAR-Indian Institute of Vegetable Research, Varanasi, India

Poonam Devi Department of Botany, Punjab University, Chandigarh, India

Harsh K. Dikshit Division of Genetics, ICAR-Indian Agricultural Research Institute, New Delhi, India

M. Djanaguiraman Department of Crop Physiology, Tamil Nadu Agricultural University, Coimbatore, Tamil Nadu, India

Kumar Durgesh ICAR-Indian Agricultural Research Institute, New Delhi, India

Gayacharan Germplasm Evaluation Division, ICAR-National Bureau of Plant Genetic resources, New Delhi, India

V. Geethalakshmi Tamil Nadu Agricultural University, Coimbatore, Tamil Nadu, India

J. Gokulakrishnan Department of Genetics and Plant Breeding, Faculty of Agriculture, Annamalai University, Annamalai Nagar, Tamil Nadu, India

Mir Asif Iquebal Centre for Agricultural Bioinformatics, ICAR-Indian Agricultural Statistics Research Institute, New Delhi, India

E. Jamuna Agricultural College and Research Institute, Tamil Nadu Agricultural University, Tiruvannamalai, Tamil Nadu, India

Souframanien Jegadeesan Nuclear Agriculture and Biotechnology Division, Bhabha Atomic Research Centre, Mumbai, Maharashtra, India

P. Jeyakumar Department of Crop Physiology, Tamil Nadu Agricultural University, Coimbatore, Tamil Nadu, India

Uday Jha Crop Improvement Division, Indian Institute of Pulses Research, Kanpur, India

R. K. Kakani ICAR-Central Arid Zone Research Institute, Jodhpur, Rajasthan, India

M. K. Kalarani Directorate of Crop Management, Tamil Nadu Agricultural University, Coimbatore, Tamil Nadu, India

Rajwant K. Kalia ICAR-Central Arid Zone Research Institute, Jodhpur, Rajasthan, India

A. Karthikeyan Department of Biotechnology, Centre of Excellence in Innovation, Agricultural College and Research Institute, Tamil Nadu Agricultural University, Madurai, Tamil Nadu, India
Subtropical Horticulture Research Institute, Jeju National University, Jeju, South Korea

A. Krishnaveni Agricultural College and Research Institute, Tamil Nadu Agricultural University, Tiruvannamalai, Tamil Nadu, India

Shiv Kumar Biodiversity and Crop Improvement Program, International Center for Agriculture Research in the Dry Areas (ICARDA), Rabat, Morocco
International Center for Agricultural Research in the Dry Areas (ICARDA), New Delhi, India

Sudhir Kumar Division of Plant Biotechnology, ICAR-Indian Institute of Pulses Research, Kanpur, Uttar Pradesh, India

Gyan P. Mishra Division of Genetics, ICAR-Indian Agricultural Research Institute, New Delhi, India

Biswajit Mondal Division of Crop Improvement, ICAR-Indian Institute of Pulses Research, Kanpur, Uttar Pradesh, India

Harsh Nayyar Department of Botany, Punjab University, Chandigarh, India

Kandiah Pakeerathan Department of Agricultural Biology, Faculty of Agriculture, University of Jaffna, Kilinochchi, Sri Lanka

Madan Pal Division of Plant Physiology, ICAR-Indian Agricultural Research Institute, New Delhi, India

Pooja Panchariya ICAR-Central Arid Zone Research Institute, Jodhpur, Rajasthan, India

Renu Pandey Mineral Nutrition Lab, Division of Plant Physiology, ICAR-Indian Agricultural Research Institute, New Delhi, India

M. Pandiyan Agricultural College and Research Institute, Tamil Nadu Agricultural University, Vazhavachanur, Tiruvannamalai, Tamil Nadu, India

Rakesh Pathak ICAR-Central Arid Zone Research Institute, Jodhpur, Rajasthan, India

M. Prakash Department of Genetics and Plant Breeding, Faculty of Agriculture, Annamalai University, Annamalai Nagar, Tamil Nadu, India

Aditya Pratap Crop Improvement Division, Indian Institute of Pulses Research, Kanpur, India

Manu Priya Department of Botany, Punjab University, Chandigarh, India

S. Punitha Directorate of Crop Management, Tamil Nadu Agricultural University, Coimbatore, Tamil Nadu, India

Dhanasekar Punniyamoorthy Nuclear Agriculture and Biotechnology Division, Bhabha Atomic Research Centre, Mumbai, Maharashtra, India

B. Rakavi Department of Crop Physiology, Tamil Nadu Agricultural University, Coimbatore, Tamil Nadu, India

Reena Rani ICAR-Central Arid Zone Research Institute, Jodhpur, Rajasthan, India

Meenal Rathore Division of Plant Biotechnology, ICAR-Indian Institute of Pulses Research, Kanpur, Uttar Pradesh, India

V. G. Renganathan Department of Biotechnology, Centre of Excellence in Innovation, Agricultural College and Research Institute, Tamil Nadu Agricultural University, Madurai, Tamil Nadu, India

K. R. Saravanan Department of Genetics and Plant Breeding, Faculty of Agriculture, Annamalai University, Annamalai Nagar, Tamil Nadu, India

Sarika Centre for Agricultural Bioinformatics, ICAR-Indian Agricultural Statistics Research Institute, New Delhi, India

G. Sathiyanarayanan Department of Genetics and Plant Breeding, Faculty of Agriculture, Annamalai University, Annamalai Nagar, Tamil Nadu, India

A. Senthil Directorate of Crop Management, Tamil Nadu Agricultural University, Coimbatore, Tamil Nadu, India

N. Senthil Department of Biotechnology, Centre of Excellence in Innovation, Agricultural College and Research Institute, Tamil Nadu Agricultural University, Madurai, Tamil Nadu, India
Department of Plant Molecular Biology and Bioinformatics, Center for Plant Molecular Biology and Biotechnology, Tamil Nadu Agricultural University, Coimbatore, Tamil Nadu, India

Sandeep Sharma Mineral Nutrition Lab, Division of Plant Physiology, ICAR-Indian Agricultural Research Institute, New Delhi, India

Kadambot H. M. Siddique Department of Agronomy, Kansas State University, Manhattan, KS, USA
The UWA Institute of Agriculture, The University of Western Australia, Perth, WA, Australia

Akansha Singh Amity Institute of Organic Agriculture, Amity University, Noida, Uttar Pradesh, India

Chandan Kumar Singh Division of Genetics, ICAR-Indian Agricultural Research Institute, New Delhi, India

Deepti Singh Department of Botany, Meerut College, Meerut, India

Dharmendra Singh Division of Genetics, ICAR-Indian Agricultural Research Institute, New Delhi, India

Inderjit Singh Department of Plant Breeding and Genetics, Punjab Agricultural University, Ludhiana, Punjab, India

Prateek Singh Division of Plant Biotechnology, ICAR-Indian Institute of Pulses Research, Kanpur, Uttar Pradesh, India

Sarvjeet Singh Department of Plant Breeding and Genetics, Punjab Agricultural University, Ludhiana, Punjab, India

M. Sivaji Agricultural College and Research Institute, Tamil Nadu Agricultural University, Tiruvannamalai, Tamil Nadu, India

C. Sivakumar Agricultural College and Research Institute, Tamil Nadu Agricultural University, Tiruvannamalai, Tamil Nadu, India

Kantilal Solanki ICAR-Central Arid Zone Research Institute, Jodhpur, Rajasthan, India

S. Sowmyapriya Directorate of Crop Management, Tamil Nadu Agricultural University, Coimbatore, Tamil Nadu, India

P. Sudhakar Department of Agronomy, Faculty of Agriculture, Annamalai University, Annamalai Nagar, Tamil Nadu, India

B. Sunil Kumar Department of Genetics and Plant Breeding, Faculty of Agriculture, Annamalai University, Annamalai Nagar, Tamil Nadu, India

Jyoti Taunk Division of Plant Physiology, ICAR-Indian Agricultural Research Institute, New Delhi, India
Department of Biotechnology, University Centre for Research and Development (UCRD), Chandigarh University, Mohali, India

Shallu Thakur Division of Plant Biotechnology, ICAR-Indian Institute of Pulses Research, Kanpur, Uttar Pradesh, India

Ram Sewak Singh Tomar College of Horticulture and Forestry, Rani Lakshmi Bai Central Agricultural University, Jhansi, India

Sripad Udupa International Center for Agricultural Research in the Dry Areas (ICARDA), Rabat, Morocco

M. Umapathi Department of Crop Physiology, Tamil Nadu Agricultural University, Coimbatore, Tamil Nadu, India

P. V. Vara Prasad Department of Agronomy, Kansas State University, Manhattan, KS, USA

Krishnapriya Vengavasi Plant Physiology Section, Division of Crop Production, ICAR-Sugarcane Breeding Institute, Coimbatore, Tamil Nadu, India

C. Viswanathan Nanaji Deshmukh Plant Phenomics Centre, Division of Plant Physiology, Indian Agricultural Research Institute, New Delhi, India

Prachi S. Yadav Division of Genetics, ICAR-Indian Agricultural Research Institute, New Delhi, India

M. Yuvaraj Agricultural College and Research Institute, Tamil Nadu Agricultural University, Tiruvannamalai, Tamil Nadu, India

Physiology and Molecular Biology of Abiotic Stress Tolerance in Legumes

1

R. Anandan, B. Sunil Kumar, M. Prakash, and C. Viswanathan

Abstract

Agricultural productivity in legumes is hampered due to several abiotic stresses, including extreme temperatures, salinity, flood, drought, heavy metals, ultraviolet radiation, and nutrient deficiencies. Generally, it is empathized that legumes are sensitive to abiotic stresses, and abiotic stresses negatively influence the plant survival and agricultural productivity. Over a decade, advances in crop physiology and genetics and scientific developments in omics such as genomics, transcriptomics, proteomics, lipidomics, metabolomics, and epigenomics have substantially enhanced our understanding of crop response to these stresses. To explore the underlying complex multilayered abiotic tolerance mechanism, a comprehensive understanding of abiotic stress, especially molecular-physiological strategies, is essential for breeding involving abiotic stress tolerance. This chapter addresses the diverse abiotic stresses and their management to increase the agricultural productivity.

Keywords

Abiotic stress · Legumes · Omics · Drought tolerance · Molecular breeding · Phenotyping

R. Anandan (✉) · B. Sunil Kumar · M. Prakash
Department of Genetics & Plant Breeding, Faculty of Agriculture, Annamalai University, Annamalai Nagar, Tamil Nadu, India

C. Viswanathan
Nanaji Deshmukh Plant Phenomics Centre, Division of Plant Physiology, Indian Agricultural Research Institute, New Delhi, India

P. Muthu Arjuna Samy et al. (eds.), *Legumes: Physiology and Molecular Biology of Abiotic Stress Tolerance*, https://doi.org/10.1007/978-981-19-5817-5_1

1.1 Introduction

Around 87% of the area under pulses is rainfed and predominantly restricted to marginal and submarginal soils, and abiotic stressors are the key impediments to attaining the yield potential. Losses in pulses owing to biotic and abiotic stressors range from 30% to 100%, depending on the degree of the stress (Rana et al. 2016). Due to the restricted availability of breeding lines/materials obtained from crossings between landraces and wild progenitors, grain legume breeding is time expensive and results in relatively poor yield gains when compared to cereal crops (Abdelrahman et al. 2017). Legumes including soybean (*Glycine max*), pea (*Pisum sativum*), chickpea (*Cicer arietinum*), lentil (*Lens culinaris*), and faba bean (*Vicia faba*) are high in protein, starch, fiber, vitamins, and minerals (Clemente and Olias 2017).

Genomics-based techniques give access to agronomically desirable alleles present at quantitative trait loci (QTLs) that influence such responses, allowing us to more effectively increase abiotic stress tolerance and yield of crops grown in stressed situations (Tuberosa and Salvi 2006). As a result, a variety of complex adaptations evolved, some of which are unique in the biological world (Ceccon 2008). By using a number of methodologies, interdisciplinary scientists have attempted to understand and dissect the processes of plant resistance to various stresses; nevertheless, progress has been limited (Mir et al. 2012). Molecular approaches have been utilized to better understand the processes by which plants sense environmental cues and transmit them to cellular machinery to trigger adaptive responses in the previous decade (Osakabe and Osakabe 2012).

This review discusses about recent advances in plant physiology for precision phenotyping of abiotic stress response, which is a prerequisite for implementing genetic and molecular-physiological strategies to unveil the multilayered drought tolerance mechanism and further exploration using molecular breeding approaches for crop improvement.

1.2 Abiotic Stress

Abiotic stress is a severe threat to life on Earth, especially for crops whose growth and yield are harmed. Plants have developed a variety of physiological, biochemical, and metabolic strategies to deal with abiotic stressors. Normally, it is harder to envision the complex signaling pathways that are activated and deactivated in response to various abiotic stresses (Chawla et al. 2011). According to recent findings, molecular techniques play a crucial part in abiotic stress stimulation in several crops. Farm revenues and agricultural advantages are reduced as a result of abiotic stressors (Waraich et al. 2022). Reactive oxygen species are produced in response to abiotic stresses, causing detrimental effects on carbohydrates, nucleic acids, lipids, and proteins. Plant growth is harmed as a result of oxidative stress (Zhu-Salzman et al. 2004). Furthermore, agricultural plant transpiration, stomatal conductance, and photosynthesis might be harmed by water deficit and heat stress (Varshney et al. 2014).

1.3 Drought-Stress Response and Signaling

Drought is among the most major abiotic stress factor that affects productivity in many regions of the world, and it has proven to be the most resistant to standard breeding methods (Tuberosa and Salvi 2006). Due to the complexity of the water-limiting environment and climatic change, drought stress has become one of the major constraints on worldwide agricultural productivity (Umezawa et al. 2010). It is important to note that water constraint alone is responsible for 70% of global agricultural productivity losses. Drought stress has a negative impact on the phenological stages of legumes (Maqbool et al. 2017). Also, when it comes to drought, desiccation has been identified as the most severe kind of drought that results in protoplasmic water loss. Furthermore, lack of water hinders the most important biological function of photosynthesis as well as other plant metabolic activities (Chaves et al. 2003; Pinheiro and Chaves 2011). To maximize legume yield under drought stress, it is critical to first understand the mechanisms of tolerance (Nadeem et al. 2019). Drought escape is one of the most important adaptive mechanisms. Legumes can withstand drought by shortening their life cycle to prevent stress, keeping higher tissue water potential, and minimizing water loss (Siddique et al. 1993). When phenological development is successfully connected to phases of soil moisture availability, drought escape occurs, especially when the growing season is shorter and terminal drought stress is more prevalent (Farooq et al. 2014).

Legume crops with an uncertain growth habit (like common bean and cowpea) may help to lower the negative effects of short-term drought stress by growing new organs throughout the stress recovery phase. Plants with a deep rooting system and a perpetual growth habit can endure stress better than annuals with shallow root systems (Chowdhury et al. 2016). If drought strikes early on, drought escape crops can use the succulent approach or a more progressive drought tolerance mechanism such as osmolyte synthesis and high water-use efficiency to gradually transition to drought avoidance (Amede et al. 2004). Drought-stressed plants produce ROS, which functions as a signal to activate defense systems in the plant (Choudhury et al. 2017). ROS works as a signaling molecule at low concentrations, triggering a variety of reactions in response to dryness. When the level of ROS exceeds the defense system, it causes oxidative stress to proteins, lipids, and nucleic acids, resulting in changes in biomolecule inherent properties and cell death (Kurutas 2016; Salmanoglu and Kurutas 2020). The defense mechanism of ROS in cell is regulated by both enzymatic and nonenzymatic components and by maintaining a larger concentration of antioxidants or antioxidant enzymes that has been shown to be a responsive adaptation under drought stress (Al Hassan et al. 2017; Sahitya et al. 2018). The activities of superoxide dismutase, ascorbate peroxide, glutathione reductase, dehydroascorbate reductase, hydrogen peroxide, glutathione peroxidase, and POD activities increased under drought stress, in resistant cultivars of green bean (Chakrabarty et al. 2016), pea (Noctor et al. 2000; Zoz et al. 2013), soybean (Yasar et al. 2013), chickpea (Osman 2015), common bean (Guler and Pehlivan 2016; Saglam et al. 2011), and horse gram (Jyoti and Yadav 2012; Patel et al. 2011). Moreover, elevated antioxidant activity in legumes might aid drought tolerance by guarding against oxidative stress.

Gibberellins, cytokinins, auxins, ABA, and ethylene are phytohormones that govern and control all factors of plant growth and development. Drought tolerance is aided by these plant hormones (Ullah et al. 2018). For example, a surge in cytokinin levels in xylem sap during a water shortage induces stomatal opening by decreasing ABA sensitivity (Bielach et al. 2017). When there is a lack of water, gibberellin, cytokinin, and auxin levels drop, while ethylene and ABA levels rise (Weyers and Paterson 2001). In kidney beans, decreased stomatal conductance was associated to a rise in ABA concentration produced by rewatering (Miyashita et al. 2005). Jasmonic acid induced drought resistance in plants by a variety of mechanisms, including stomatal closure, ROS scavenging, and root growth (Ullah et al. 2018). Jasmonates like ABA participated in the control of stomatal closure and stomatal modulation in response to drought stress, according to several studies (Kazan 2015; Munemasa et al. 2011; Riemann et al. 2015; Savchenko et al. 2014).

Methyl jasmonate promotes drought tolerance and plant development in soybean (Mohamed and Latif 2017). The formation of vast roots was thought to be a key characteristic for drought resistance in chickpeas, since it provided more coverage for more water intake and increased yields by avoiding terminal drought (Kashiwagi et al. 2006). Water uptake, water use, and temporal aspects should be prioritized over the roots themselves (Vadez and Ratnakumar 2016). Drought tolerance is a multifaceted quantitative characteristic governed by a number of small-effect genes, or QTLs. Understanding the physiological and genetic underpinnings of plant drought responses is critical for addressing the complexity of these responses (Passioura 2012). Drought tolerance QTLs have been discovered in various significant and essential crop species, such as soybean (Mian et al. 1996, 1998; Specht et al. 2001; Bhatnagar et al. 2005; Monteros et al. 2006) and common bean (Blair et al. 2012). Root features including root depth and root proliferation have been recognized as the most promising qualities in chickpea for terminal drought resistance because they aid in the extraction of available soil moisture (Varshney et al. 2017).

The accumulation of better alleles by marker-assisted recurrent selection is used to improve drought tolerance (Varshney et al. 2017). Stress has an effect on gene expression as well. Drought stress causes the expression of many genes to be either upregulated or downregulated. Copper-related genes are targeted by miR408, a conserved miRNA found in terrestrial plants. Despite the fact that numerous environmental conditions, such as drought stress, alter miR408 expression, the biological activity of miR408 remains unknown. To investigate the role of miR408 in chickpeas under drought stress, transgenic lines overexpressing the miR408 gene were created. Plants with increased miR408 expression showed induced tolerance after a 17-day lack of water (Hajyzadeh et al. 2015). Plantacyanin transcript suppression caused by overexpression resulted in DREB and other drought-sensitive genes being regulated (Maqbool et al. 2017). CarNAC3, a member of the NAP family, is involved in plant growth and abiotic stress responses. Under stress condition, CarNAC3 expression produced increases in proline and photosynthetic pigment levels, as well as antioxidant enzyme activity (Movahedi et al. 2015). Additionally, as compared to the wild-type control, CarNAC3 expression reduced malondialdehyde levels. According to phylogenetic research, CarNAC3 is a member of the NAP subgroup of the NAC protein family (Peng et al. 2009). When compared to control plants, the transgenic OX-amiR408 cowpea lines showed improved drought and salinity tolerance, with greater chlorophyll, relative water content, and proline content, and reduced cellular H_2O_2 concentration under drought stress (Mishra et al. 2022). The identification and characterization of this plant's drought-responsive miRNAs might be a precious genetic resource for understanding the molecular pathways of drought tolerance in plants. By sequencing short RNA libraries from normal and moisture-stressed leaf tissues, Bhat et al. found 143 and 128 conserved miRNAs, respectively (Bhat et al. 2020). *Pseudomonas putida* strain MTCC5279 is a plant-growth-promoting rhizobacterium that increases plant growth and development by colonizing the root surface and provides drought-stress resistance in chickpeas (Jatan et al. 2019).

Water stress has a negative effect on several aspects of plant physiology, particularly photosynthetic potential. Plant development and productivity are drastically affected if the stress is extended (Osakabe et al. 2014). Plants have developed complex biochemical, molecular, physiological, morphological, and cellular responses to survive with different stresses and evolved a range of molecular processes to limit resource use and modify growth in response to changing environmental conditions (Ha et al. 2012). One of the important contributors in the EL stress is water stress, which causes a leaf water potential reduction and stomatal opening, resulting in downregulation of photosynthesis-related genes and decreased CO_2 availability (Osakabe et al. 2011). Stress responses include a variety of molecular networks, including signal transduction (Nishiyama et al. 2013). Environmental stresses can influence stomatal function, which can affect CO_2 absorption and consequently photosynthesis and plant development. Under drought, endogenous ABA is rapidly released, causing a series of physiological responses, including stomatal closure, which is controlled by a signal transduction system (Endo et al. 2008). The majority of ABA biosynthesis genes have been discovered, and they are mostly expressed in leaf vascular tissues. NCED is a major enzyme in ABA biosynthesis, and one of the products of such genes as *AtNCED3* is mainly induced by dehydration in *Arabidopsis* (Behnam et al. 2013). However, increased seed dormancy was one of the detrimental consequences of overexpression of NCED in tomato employing constitutive promoters (Tung et al. 2008). At drought condition, OsbZIP16 expression was significantly increased. Transgenic rice plants overexpressing *OsbZIP16* showed considerably increased drought tolerance at both the seedling and tillering phases, which was positively linked with *OsbZIP16* expression levels (Chen et al. 2012). Drought resistance is controlled by hundreds of genes and small-effect loci that govern physiological and morphological responses to drought (Hu and Xiong 2014).

Drought-responsive gene expression is linked to major physiological processes, according to transcriptomic and proteomic analyses in various species such as soyabean (Alam et al. 2010), *Ammopiptanthus mongolicus* (Zhou et al. 2012), banana (Muthusamy et al. 2016), *Agropyron mongolicum* (Zhao et al. 2018), *Arachis duranensis* (Carmo et al. 2019), sesame (You et al. 2019), maize (Jin et al. 2019; Liu et al. 2020; Zenda et al. 2019), *Vitis champinii* cv. Ramsey (Cochetel et al. 2020), *Arachis hypogaea* (Jiang et al. 2021), opium poppy (Kundrátová et al. 2021), and *Phoebe zhennan* (Xie et al. 2022). Several types of kinases such as CPKs (Campo et al. 2014; Cieśla et al. 2016; Bundó and Coca 2017; Huang et al. 2018), CIPK (Xunji Chen et al. 2021; Ketehouli et al. 2021; Lu et al. 2021; Luo et al. 2017; Wang et al. 2018c), MAPKs (Li et al. 2017; Muhammad et al. 2019; Lin et al. 2021; Zhu et al. 2021), and SnRK2 (SNF1-related kinase 2) (Maszkowska et al. 2019; Kamiyama et al. 2021) have been reported to be involved in drought response.

RSA, which is comprised of structural features such as root length, spread, number, and length of lateral roots, among others, shows a lot of flexibility in response to environmental factors and might be crucial in producing more efficient roots (Hu and Xiong 2014). The potential of roots to penetrate and the degree of drought resistance have been proven to be positively related (Aslam et al. 2015; Wasaya et al. 2018). Woody plant seedlings in arid environments have vertical roots that are ten times longer than the height aboveground. Plants may sustain a higher water potential and a longer duration of transpiration during drought situations because of their large root system and rooting depth, which gives additional benefits to their growth and development. Plants dynamically adapt and adjust their root system morphology in response to soil water deficits by modifying their root development in a variety of ways, depending on the species (Tardieu 2012; Monneveux et al. 2013; del Pozo and Ramirez-Parra 2014; Lynch 2018). Drought causes changes in leaf morphology and ultrastructure in most of the plant species. Alterations in leaf size, stomata submersion in succulent plants and xerophytes, thickening of leaf cell walls, cutinization of leaf surface, underdevelopment of the conductive system but a rise in the number of big xylem vessels, leaf rolling in cereals, and induction of early senescence are all examples of changes (Anjum et al. 2011, 2017). Glaucousness is another trait that conserves moisture content by minimizing transpiration when there is a water shortage (Kaur and Asthir 2017).

1.4 Temperature Stress

In legumes, temperature has a significant impact on seed yield and quality (Christophe et al. 2011; Ruelland and Zachowski 2010). Heat stress in plants is defined as an increase in air temperature of one degree over a threshold level (Teixeira et al. 2013). Heat stress has different effects depending on the severity, duration, and degree of the high temperature. Extreme temperature changes, both high and low, may negatively impact plant development by affecting plant growth and function (Wahid et al. 2007). Physiological processes of plants can be disrupted by heat stress, leading to shortened vegetative and pod-filling phases (Adnane et al.

2015), poor crop stands and consequently lower yield, photosynthetic inhibition, reduced reproductive development, stigma receptivity, pollen viability, nitrogen anabolism, ovule size, ovule viability, fertilization, seed composition, grain filling, seed/fruit set, seed quality, higher protein catabolism, and reduced radicle and plumule growth, as reported in legumes (Prasad et al. 2002; Chakraborty and Pradhan 2011; Kumar et al. 2013; Kaushal et al. 2013; Tzudir et al. 2014; HanumanthaRao et al. 2016).

After dry bean and field pea, chickpea is the world's third most popular cool-season grain legume crop. Chickpea production is harmed by freezing and below-zero temperature in many of its growing locations. At the cellular, molecular, canopy, and whole-plant levels, response mechanisms to chilling and freezing are studied. A plant's freezing tolerance varies substantially across different tissues, such as top and lower leaves of the canopy, meristems, stems, and roots (Croser et al. 2003).

Among various environmental conditions, low temperature is one of the most critical factors limiting plant productivity and distribution (Theocharis et al. 2012). Many physiological and biochemical cell activities have been linked to observable symptoms as a result of exposure to low temperatures (necrosis, chlorosis, or wilting) (Ruelland and Zachowski 2010). Alterations in the expression of certain genes producing proteins that give enhanced cold tolerance are required for plant acclimatization to low temperatures (Doherty et al. 2009). Plants with higher antioxidative enzyme activity in chickpea were shown to be more cold tolerant (Kumar et al. 2012b). Modifying preexisting proteins and up- or downregulating gene expression and protein synthesis are all part of the process. The activity of cold/chilling-induced genes has been linked to the metabolic alterations that provide low-temperature tolerance in several studies (Doherty et al. 2009; Thomashow 2010).

Cold stress, which includes chilling (0–15 °C) and freezing (<0 °C), is an abiotic stress that has a negative impact on plant development and productivity (Guo et al. 2018; Liu et al. 2018). Cold is an important factor that affects agriculture productivity, among the several abiotic stressors. Low temperatures have an impact on the growth and development of agricultural species all over the world (Pearce 2001). A series of complex physiological and metabolic changes occur throughout this process. Many chemicals or protective proteins, such as soluble sugars, proline, and cold-resistance proteins, are generated at the physiological level in plants (Kaplan et al. 2007). Chilling stress impacts plant cell membrane rigidification, which is thought to be the fundamental process that causes plant cold-stress responses (Orvar et al. 2000; Cano-Ramirez et al. 2021; Fan et al. 2015). Freezing stress lowers the activity of enzymes such as ROS scavenging enzymes and disrupts the stability of proteins or protein complexes. Photoinhibition and reduced photosynthesis are the results of these processes, as well as significant membrane damage (Siddiqui and Cavicchioli 2006; Ruelland et al. 2009; Shi et al. 2015; Ding et al. 2019). Chilling stress has also an impact on gene expression and protein synthesis because it promotes the development of secondary RNA structures (Rajkowitsch et al. 2007; Ruelland et al. 2009).

Several components of cold-stress signaling pathways have been found over the last two decades, including protein kinases and phosphatases, messenger molecules, and transcription factors (Ding et al. 2019). The *CBF-COR* signaling pathway is the most well studied of them. *CBE/DREB1* genes are strongly activated by cold stress and play an important role in plant cold adaptation (Wu et al. 2015; Jan et al. 2017). Frost tolerance was significantly improved in both *TaDREB2* and *TaDREB3* transgenic plants with constitutive overexpression of the wheat transgene. Higher expression of *TaDREB2* and *TaDREB3* led to increased expression of ten additional *CBF/DREB* genes in transgenic wheat (Morran et al. 2011), *Lolium perenne* (Li et al. 2011), *Moso Bamboo* (Wu et al. 2015), *Arabidopsis thaliana* (Hua 2016; Hu et al. 2018), barley (Yunfei Yang et al. 2020), and *Kandelia obovata* (Peng et al. 2020).

Successful cold-stress acclimation pathways are controlled by a number of TFs and proteins. The *ICE-CBF-COR* signaling system in plants governs how plants acclimate to cold stress. Cold stress triggers signal transmission, resulting in the activation and regulation of *ICE* genes, which upregulate the transcription and expression of *CBF* genes. The *CBF* protein activates transcription by binding to the *CRT/DRE*, a homeopathic element of the *COR* promoter. These activities result in a high level of cold-stress tolerance (Hwarari et al. 2022). Exogenous melatonin induces cold tolerance on strawberry seedlings via the *DREB/CBF-COR* pathway (Hayat et al. 2022). Compared to wild-type plants, overexpressed *PvC3H72* driven by the maize ubiquitin promoter showed significantly improved chilling tolerance at 4 °C, as evidenced by less electrolyte leakage and higher relative water content, as well as a considerably higher survival rate after freezing treatment at −5 °C. *PvC3H72* transgenic lines with improved cold tolerance have considerably upregulated the expression of the *ICE1-CBF-COR* regulon and ABA-responsive genes under cold condition (Xie et al. 2019).

Lee and Seo (2015) reported that an R2R3-type *MYB* transcription factor, *MYB96*, integrates the ABA and cold signaling pathways. MYB96 is activated by cold stress in an ABA-independent way and hence stimulates freezing tolerance. Large-scale alterations in the transcriptome are linked to this process, which are influenced by a collection of tandemly duplicated *CBF* transcription factors found at the *Fr-2* gene (Pearce et al. 2013). During the day/night cycle, a plastid signal helps to regulate *CBF* expression and downstream expression of cold-responsive genes (Norén et al. 2016). *BpERF13* overexpression lines were more tolerant to subfreezing and exhibited lower levels of ROS in *B. platyphylla*, which shows that the transcription factor *BpERF13* affects physiological processes in woody plants that promote cold tolerance (Lv et al. 2020). Many of the molecular responses to cold stress were reported in various crops (Kim et al. 2015; Liu et al. 2018; Guo et al. 2018; Song et al. 2021; Wang et al. 2018b; Zhang et al. 2020; Bielsa et al. 2021). Cold stress also leads to changes of some protected enzymes such as POD, SOD, and CAT. The activities of SOD, POD, and CAT increased under low temperature in *Avena nuda* L. seedlings (Liu et al. 2013).

High-temperature stress is a key environmental stress that restricts plant growth, metabolism, and production. Biochemical and physiological responses to heat stress have been hot topics in recent days, and molecular techniques are being used to develop high-temperature tolerance in plants (Hasanuzzaman et al. 2013). Heat stress reduces photosynthetic efficiency significantly. Melatonin is a bio-stimulator that regulates abiotic stress tolerance in a variety of ways. The fundamental processes of melatonin-mediated photosynthesis in heat-stressed plants, on the other hand, are still largely unknown (Jahan et al. 2021). The reproductive period is very vulnerable to environmental challenges, particularly high temperatures (HT), which drastically limit commercial crop yields (Almeida et al. 2021). Molecular techniques that discover the response and tolerance mechanisms will pave the road for creating plants that can tolerate HT and might serve as the foundation for developing crop varieties that can provide economic yields in HT-affected environments (Źróbek-Sokolnik 2012). Plants' genomes include multiple heat stress-sensitive genes, and DNA is the starting point for all molecular data linked to heat-stress tolerance (Yeh et al. 2012). Chakraborty and Pradhan (2011) exposed 5-day-old thermotolerant genotype, namely BPR-542-6, and thermosusceptible genotype, namely NPJ-119, of *B. juncea* to HT (45.0 ± 0.5 °C) stress. Kumar et al. (2012a) explored the antioxidant defense system and the comparative response of HT in *O. sativa* and *Z. mays* plants under stress. When compared to the susceptible variety, the tolerant types were able to sustain higher levels of activity at HT. HT stress has a variety of effects on different crop species, such as *Glycine max* (Djanaguiraman et al. 2011), *Nicotiana tabacum* (Tan et al. 2011), *Triticum aestivum* (Zhang et al. 2013), *Abelmoschus esculentus* (Gunawardhana and De Silva 2011), and *Zea mays* (Edreira and Otegui 2012).

Heat stress may be minimized by applying different genetic engineering and transgenic techniques to generate agricultural plants with enhanced thermal tolerance. Grover et al. (2013) suggested that transgenic plants could be used to develop HT stress tolerance by overexpressing *HSP* genes or altering levels of *HSFs* that regulate the expression of heat-shock and non-heat-shock genes, as well as overexpression of other trans-acting factors such as *DREB2A*, *bZIP28*, and *WRKY* proteins. Under high-temperature stress, the activities of SOD, POX, CAT, APX, and GR raised, although the rise was substantially larger in the tolerant genotype (Rani et al. 2013). HSPs/chaperones are regulated by a variety of heat-shock factors, which are activated in response to stress (Jacob et al. 2017). When miR398 is downregulated in response to oxidative stress, one of its target genes, *CSD2*, is upregulated, which helps plants cope with oxidative stress. Heat stress increases the expression of miR398 and decreases the expression of its target genes *CSD1*, *CSD2*, and *CCS* (Guan et al. 2013).

1.5 Heavy Metal Tolerance

Since heavy metals accumulate in many parts of agricultural plants, they limit plant growth/productivity and pose serious health risks to humans (Rai et al. 2021). Various metals and metalloids, such as mercury (Hg), cadmium (Cd), cobalt (Co), chromium (Cr), lead (Pb), zinc (Zn), aluminum (Al), arsenic (As), and nickel (Ni), cause significant toxicity when they reach the soil agroecosystem by anthropogenic or natural processes (Neilson and Rajakaruna 2015). Elevated heavy metal bioaccumulation over the threshold level has been found to have a severe influence on the natural food chain and microbial flora and is now being viewed as a serious danger to the ecosystem and environment (Singh and Kumar 2017). Heavy metals, when present at low quantities, stimulate plant growth and development by serving as cofactors for a variety of enzymes engaged in numerous physiological and metabolic pathways (Mohammed et al. 2011; Luo et al. 2015). *Lablab purpureus* L., often known as the hyacinth bean or Indian bean, has been shown to be resistant to heavy metals such as Cd, Hg, Pb, Zn, P, and Cr (Ruthrof et al. 2018). Heavy metal toxicity has been related to a variety of processes that occur at the same time, posing severe metal-induced toxic effects via the production of oxidative stress (Pottier et al. 2015).

Plant tolerance to a specific heavy metal (HM) is controlled by a complex network of physiological and molecular processes, and knowing these mechanisms and their genetic base is critical for developing plants as phytoremediation agents (DalCorso et al. 2010; Hossain et al. 2010). In trying to adapt with stress signals, plants must coordinate complex biochemical and physiological processes, changes in metabolite compositions, protein modifications, and gene expression culminating in proper stress signal recognition and tolerance (Urano et al. 2010). To figure out what causes HM buildup, tolerance, and adaptive responses to HM stress, scientists are still using physiological, biochemical, and molecular approaches (Hossain et al. 2012).

Zinc is used as a cofactor by about 200 transcription factors and 300 enzymes involved in auxin metabolism, membrane integrity, and reproduction (Ricachenevsky et al. 2013). Several heavy metals, including Cr, Al, Cd, Hg, Pb, and others, are exceedingly harmful even at very low quantities, despite the fact that they are nonessential and play no physiological role (Garzón et al. 2011; Hayat et al. 2012; Shahid et al. 2012; Gill et al. 2013; Chong-qing Wang et al. 2013). Excessive levels of heavy metals cause inactivation and denaturation of enzymes and proteins, blockage of functional groups of metabolically important molecules, displacement or substitution of critical metal ions from biomolecules, conformational abnormalities, and disruption of membranosomes, to name a few toxicity symptoms (Villiers et al. 2011). Brassicaceae members make for 25% of all known metal-hyperaccumulation species and might be employed in phytoremediation (Gall and Rajakaruna 2013). Plants tackle HM stress by overexpressing a variety of stress-related proteins, glutathione-mediated tolerance pathways, and signaling proteins involved in a variety of stress regulatory networks (Thapa et al. 2012).

Though each HM response may differ and be more particular in terms of stress network regulation, plant growth, biomass, and photosynthetic pigments were found to be increased with rising metal concentrations in soil up to 1.0 mM and then declined as metal levels climbed (Tauqeer et al. 2016). *Suaeda glauca* and *Arabidopsis thaliana* plants' growth and physiological responses were studied along with the soil conditions under various amounts of Cd, Pb, and Mn, and it was found that *S. glauca* showed better tolerance capacity for Mn, Cd, and Pb, when compared with *Arabidopsis* (Zhang et al. 2018b).

1.6 Saline/Salt Tolerance

Salt stress is a persistent threat to agricultural production, especially in nations where agriculture is irrigated. Efforts to increase salt tolerance in agricultural plants are critical for ensuring future food supply by ensuring sustainable crop production on marginal areas (Farooq et al. 2017). According to the FAO, salinity affects roughly 800 million hectares of soil worldwide. Grain legumes are vulnerable to salt stress, which results in decreased yield (Flexas et al. 2004), nutritional imbalances, hormonal disturbances, specific ion and osmotic effects (El Sayed and El Sayed 2011), delayed blooming, and lower flower numbers and pod set (Chowdhury et al. 2016). Salt-stress tolerance in legumes is linked to alterations in various physiological, molecular, and biochemical processes, including Na^+ sequestration, antioxidative stress induction, and osmoprotectant accumulation (Adnane et al. 2015; Zhang et al. 2017).

Salt-tolerant plants are equipped with a diverse array of antioxidant enzymes, such as GPX, DHAR, MADHAR, CAT, SOD, GR, APX, GPX, and GST, and certain nonenzymatic antioxidants, such as glutathione, carotenoids, tocopherols, ascorbic acid, flavonoids, and flavones (Hernandez et al. 1999; Kukreja et al. 2005). Rehabilitation of salt-degraded soils is dependent not only on salt-tolerant legumes, but also on rhizobia survival under saline environments (Coba de la Pena and Pueyo 2012; Ventorino et al. 2012; Bruning et al. 2015). A comparative ecophysiological examination of salt-stress tolerance in wild (*Glycine soja*) and cultivated soybean was conducted in another research, and it was noted that wild soybean was able to sustain a greater relative water content, accumulate more osmolytes such as proline and glycine betaine, and enhance K^+ inflow and Na^+ efflux to maintain a higher K^+/Na^+ ratio by increasing K^+ influx and Na^+ efflux (Hasegawa et al. 2000; Li et al. 2006; Waheed et al. 2006; Phang et al. 2008; Turner et al. 2013; Farooq et al. 2017). *SOS1* overexpression enhances plant salt tolerance. *SOS1* is an antiporter that helps to remove excess Na^+ from roots and is involved in long-distance Na^+ transport in the xylem (Shi et al. 2000). It was revealed that wild soybeans had alternative tolerance mechanisms or varying amounts of the same mechanisms, allowing them to tolerate salinity better than cultivated soybeans (Wu et al. 2014). Exogenous proline can increase plant tolerance to salt stress by regulating endogenous proline metabolism, which is done partly by differential expression of particular proline-related genes (*P5CS*) (de Freitas et al. 2019). Adding exogenous proline inhibited *P5CS* function in both stressed and unstressed plants, but only in unstressed plants did it enhance PDH activity (Zheng et al. 2015). Beans have been researched under drought stress with the addition of ABA or miRNA accumulation, but their function under saline circumstances has not been addressed (Covarrubias and Reyes 2010). Upregulation of miRNAs, on the other hand, is critical for soybean salinity stress management (Dong et al. 2013). Furthermore, the *PR10a* gene was important in reducing faba bean salt tolerance (Hanafy et al. 2013). Similarly, transgenic lines of chickpea with the AP2-type TFs, *CAP2*, improve SS tolerance (Frugier et al. 2000). The altered root system of overexpressing plants was able to maintain growth under high salinity, while roots with *MtNAC969* downregulation grew better under salt stress. As a result, production of salt stress markers was reduced or increased in *MtNAC969*-overexpressing or RNAi roots, suggesting that this transcription factor has a regulatory role in the salt-stress response. Methyl jasmonate, glucose, mannitol, and NaCl treatment dramatically elevated the expression of the two soybean flavone synthase genes, *GmFNSII-1* and *GmFNSII-2* (Yan et al. 2014). The above examples provided an insight into mechanisms that enable various legumes to adapt to salt stress.

Therefore, understanding the processes of salt tolerance is critical in order to develop plants that respond better to this abiotic stress (Hernández 2019). Salt tolerance is a valuable economic trait for crops grown in both irrigated and nonirrigated regions. Salt tolerance is a multigene-controlled characteristic that includes a variety of biochemical and physiological pathways (Zhang and Shi 2013). Complex metabolic processes, physiological features, and molecular or gene networks are all involved in plant adaptation or tolerance to salt stress (Gupta and Huang 2014).

The response of salt-tolerant and salt-sensitive Populus species to salinity injury (photosynthesis, plant growth) and primary salt tolerance mechanisms (accumulation of soluble osmolytes, ion homeostasis), as well as reactive oxygen and nitrogen species (ROS) metabolism and signaling networks induced by salinity, were studied, and candidate genes for improving salt tolerance were discovered (Zhang et al. 2019). Transcriptome sequencing might give a functional understanding of plant salt stress resistance pathways, and Wang et al. (2018a) investigated the transcriptome of the algae *Chlamydomonas reinhardtii* after a short-term (24-h) adaptation to salt stress (200 mM NaCl). The role of the *bZIP* transcription factors in response to salt stress was investigated using *C. reinhardtii* as an experimental organism (Ji et al. 2018). Six *CrebZIP* genes were found to be involved in stress response and lipid accumulation after qRT-PCR expression profiling of *CrebZIP* genes.

Wu et al. (2019) sequenced the *Linum usitatissimum* L. transcriptome to identify DEUs under NaCl stress. The transcriptome profile under abiotic stressors has been widely revealed and compared using next-generation sequencing technology based on high-throughput RNA-Seq technology (Haider et al. 2017), which provides large-scale data to identify and characterize the differentially expressed genes DEGs. Miao et al. (2018) identified a new ROP gene from banana (*MaROP5g*) that boosted salt tolerance in transgenic *Arabidopsis thaliana* plants when it was overexpressed. A transgenic plum line (J8-1) with four copies of the pea cytosolic ascorbate peroxidase gene (cytapx) responds to salt stress (Bernal-Vicente et al. 2018).

Overexpression of MAX2 from *Sapium sebiferum* (*SsMAX2*) in *Arabidopsis* plants considerably improved resilience to abiotic stressors such as osmotic, drought, and salinity (Wang et al. 2019). Overexpression of an SKn-type dehydrin from *Capsicum annuum* L. (*CaDHN5*) in *Arabidopsis* plants resulted in greater tolerance to salt and osmotic stresses, suggesting that *CaDHN5* plays an essential role in response to the abiotic stressors indicated (Luo et al. 2019). The CPK12-RNAi mutant was more susceptible to salinity than wild-type plants in terms of seedling development, demonstrating the role of CDPKs in *Arabidopsis* adaptation to salt stress (Zhang et al. 2018a). Overexpression of genes in plants has been engineered.

Antiporter encoding gene has been established as a viable strategy for producing salt-tolerant plants such as vacuolar Na^+/H^+ antiporter Ms NHX1 from *Arabidopsis* (Bao-Yan et al. 2008), rice (Verma et al. 2007), and vacuolar Na^+/H^+ antiporter Pg NHX1 from tobacco (Zhang et al. 2008). *Brassica juncea* and *Brassica campestris* are two examples of increased antiporter gene expression in response to salt stress (Chakraborty et al. 2012). Various saline-tolerant genes are reported, *SOS1*, *SOS2*, *SOS3*, *AtNHX* genes, *OsPRP*, *SAG*, *HSPC025* (Roshandel and Flowers 2009), *OsHSP23.7*, *OsHSP71.1*, *OsHSP80* (Zou et al. 2009), *OsHsp17.0*, *OsHsp23.7* (Zou et al. 2012), *Arabidopsis thaliana AtSKIP* (Lim et al. 2010), *JcDREB* (Tang et al. 2011), and *DcHsp17.7* in carrot (Song and Ahn 2011).

In the plant cell, ionic homeostasis is maintained. Without salt stress under normal conditions, 14-3-3 and GI proteins interact with and repress SOS2 kinase activity, and the activity of plasma membrane H^+-ATPase is inhibited by SCaBP1/CBL2 (calcium-binding protein) and PKS5/SOS2-like SCaBP8 inhibits the action of AKT1. A calcium signal activates the SOS pathway during salt stress, and SCaBP8 is phosphorylated by SOS2, which may dissociate from AKT1, a potassium channel. SOS1 sodium transport necessitates the creation of a proton gradient by PMH^+-ATPases, whose activity is stimulated by DnaJhomolog3 (J3; heat-shock protein 40-like) suppression of PKS5 kinase activity. The vacuolar Na^+/H^+ exchanger (a vacuolar Na^+/H^+ antiporter) is a vacuolar Na^+/H^+ antiporter that is driven by the proton gradient generated by vacuolar H^+-ATPases and H^+-pyrophosphatases The activity of the vacuolar H^+-ATPase and Na^+/H^+ exchanger can be activated by SOS2 (Yang and Guo 2018).

1.7 Flood Tolerance

The establishment of fine roots in the surface aerobic layers of a flooded soil can therefore attain a degree of flood tolerance, and while the usable soil volume may be limited and shoot growth may be slowed, the intake of phototoxins can be reduced (Laan et al. 1989). Higher levels of decreased ascorbate play a significant role in plant defense against flooding damage, according to studies (Kawano et al. 2002; Das et al. 2004). Due to a scarcity of tolerant germplasm and possible target genes, traditional breeding to generate tolerant cultivars is limited (Tyagi et al. 2022). The overexpression of enzymes involved in ascorbic acid production aids the plant's capacity to withstand stress (Hasanuzzaman et al. 2012). During flooding, it interacts with ROS in both photosystems I and II via ASC-GSH, as well as the xanthophyll cycle (Damanik et al. 2010). A high overexpression of tyrosine protein kinase and a downregulation of linoleate 9S lipoxygenase 5, a fat metabolism gene, suggested an energy-saving approach in Kaspa. In NL2, however, the upregulation of a subtilase family protein and peroxisomal adenine nucleotide carrier 2, a fat-metabolizing gene, suggested a quicker energy consumption approach (Zaman et al. 2019). Oram et al. (2021) discovered that grasses were more resistant to floods than legumes, and that legumes recovered more quickly. The resistance of resource-conserving grasses was stronger, but resource-acquiring grasses recovered faster.

N_2O emissions were reduced by resilient grass and legume species. Grasses with lower intrinsic leaf and root ^{13}C (as well as legumes with lower root ^{13}C) produced less N_2O during and after the flood. NO and H_2S are known to affect essential physiological pathways such as leaf senescence, stomatal closure, and regulation of various stress signaling pathways, and NO is also involved in the production of adventitious roots under waterlogging (Tyagi et al. 2022).

1.8 Conclusion

Abiotic stress is one of the key factors limiting crop development and productivity in the surrounding environment. In abiotic environment, crop growth, development, and yield were all hampered. We outlined how plants respond to osmotic, ion, drought, flood, temperature, salt, and oxidative stresses in this work and compiled a large number of research advances on the effect of abiotic stresses on plants. Thorough study on plants' physiological and biochemical adaptation to abiotic stress, along with genetic engineering, will help to clarify the plant abiotic tolerance mechanism and give enough theoretical direction for the future production of abiotic-resistant crops. Plant abiotic tolerance has to be improved further, and there are a lot of abiotic-tolerant plant species that need to be studied.

References

Abdelrahman M, El-Sayed M, Jogaiah S, Burritt DJ, Tran L-SP (2017) The "STAY-GREEN" trait and phytohormone signaling networks in plants under heat stress. Plant Cell Rep 36(7): 1009–1025

Adnane B, Mainassara ZA, Mohamed F, Mohamed L, Jean-Jacques D, Rim TM, Georg C (2015) Physiological and molecular aspects of tolerance to environmental constraints in grain and forage legumes. Int J Mol Sci 16(8):18976–19008

Al Hassan M, Chaura J, Donat-Torres MP, Boscaiu M, Vicente O (2017) Antioxidant responses under salinity and drought in three closely related wild monocots with different ecological optima. AoB Plants 9(2):plx009

Alam I, Sharmin SA, Kim K-H, Yang JK, Choi MS, Lee B-H (2010) Proteome analysis of soybean roots subjected to short-term drought stress. Plant Soil 333(1):491–505

Almeida J, Perez-Fons L, Fraser PD (2021) A transcriptomic, metabolomic and cellular approach to the physiological adaptation of tomato fruit to high temperature. Plant Cell Environ 44(7): 2211–2229

Amede T, Schubert S, Stahr K (2004) Mechanisms of drought resistance in grain legumes I: Osmotic adjustment. SINET Ethiopian J Sci 26(1):37–46

Anjum SA, Xie X, Wang L, Saleem MF, Man C, Lei W (2011) Morphological, physiological and biochemical responses of plants to drought stress. Afr J Agric Res 6(9):2026–2032

Anjum SA, Ashraf U, Zohaib A, Tanveer M, Naeem M, Ali I, Tabassum T, Nazir U (2017) Growth and development responses of crop plants under drought stress: a review. Zemdirbyste 104(3): 267–276

Aslam M, Maqbool MA, Cengiz R (2015) Mechanisms of drought resistance. In: Drought stress in maize (*Zea mays L.*). Springer, New York, pp 19–36

Bao-Yan AN, Yan LUO, Jia-Rui LI, Wei-Hua Q, Zhang X-S, Xin-Qi GAO (2008) Expression of a vacuolar Na^+/H^+ antiporter gene of alfalfa enhances salinity tolerance in transgenic Arabidopsis. Acta Agron Sin 34(4):557–564

Behnam B, Iuchi S, Fujita M, Fujita Y, Takasaki H, Osakabe Y, Yamaguchi-Shinozaki K, Kobayashi M, Shinozaki K (2013) Characterization of the promoter region of an Arabidopsis gene for 9-cis-epoxycarotenoid dioxygenase involved in dehydration-inducible transcription. DNA Res 20(4):315–324

Bernal-Vicente A, Cantabella D, Petri C, Hernández JA, Diaz-Vivancos P (2018) The salt-stress response of the transgenic plum line J8-1 and its interaction with the salicylic acid biosynthetic pathway from mandelonitrile. Int J Mol Sci 19(11):3519

Bhat KV, Mondal TK, Gaikwad AB, Kole PR, Chandel G, Mohapatra T (2020) Genome-wide identification of drought-responsive miRNAs in grass pea (*Lathyrus sativus* L.). Plant Gene 21:100210

Bhatnagar S, King C, Purcell L, Ray J (2005) Identification and mapping of quantitative trait loci associated with crop responses to water-deficit stress in soybean [*Glycine max* (L.) Merr.]. The ASA-CSSA-SSSA international annual meeting poster abstract, 6–10

Bielach A, Hrtyan M, Tognetti VB (2017) Plants under stress: involvement of auxin and cytokinin. Int J Mol Sci 18(7):1427

Bielsa B, Ávila-Alonso JI, Fernández i Martí Á, Grimplet J, Rubio-Cabetas MJ (2021) Gene expression analysis in cold stress conditions reveals BBX20 and CLO as potential biomarkers for cold tolerance in almond. Horticulturae 7(12):527

Blair MW, Galeano CH, Tovar E, Muñoz Torres MC, Castrillón AV, Beebe SE, Rao IM (2012) Development of a Mesoamerican intra-genepool genetic map for quantitative trait loci detection in a drought tolerant × susceptible common bean (*Phaseolus vulgaris* L.) cross. Mol Breed 29 (1):71–88

Bruning B, van Logtestijn R, Brockman R, de Vos A, González AP, Rozema J (2015) Growth and nitrogen fixation of legumes at increased salinity under field conditions: implications for the use of green manures in saline environments. AoB Plants 7:plv010

Bundó M, Coca M (2017) Calcium dependent protein kinase OsCPK10 mediates both drought tolerance and blast disease resistance in rice plants. J Exp Bot 68(11):2963–2975

Campo S, Baldrich P, Messeguer J, Lalanne E, Coca M, San Segundo B (2014) Overexpression of a calcium-dependent protein kinase confers salt and drought tolerance in rice by preventing membrane lipid peroxidation. Plant Physiol 165(2):688–704

Cano-Ramirez DL, Carmona-Salazar L, Morales-Cedillo F, Ramírez-Salcedo J, Cahoon EB, Gavilanes-Ruíz M (2021) Plasma membrane fluidity: an environment thermal detector in plants. Cell 10(10):2778

Carmo LST, Martins ACQ, Martins CCC, Passos MAS, Silva LP, Araujo ACG, Brasileiro ACM, Miller RNG, Guimarães PM, Mehta A (2019) Comparative proteomics and gene expression analysis in Arachis duranensis reveal stress response proteins associated to drought tolerance. J Proteomics 192:299–310

Ceccon E (2008) Plant productivity and environment. New For 35(3):443–448

Chakrabarty A, Aditya M, Dey N, Banik N, Bhattacharjee S (2016) Antioxidant signaling and redox regulation in drought-and salinity-stressed plants. In: Hossain MA et al (eds) Drought stress tolerance in plants, vol 1. Springer, New York, pp 465–498

Chakraborty U, Pradhan D (2011) High temperature-induced oxidative stress in Lens culinaris, role of antioxidants and amelioration of stress by chemical pre-treatments. J Plant Interact 6(1): 43–52

Chakraborty K, Sairam RK, Bhattacharya RC (2012) Differential expression of salt overly sensitive pathway genes determines salinity stress tolerance in Brassica genotypes. Plant Physiol Biochem 51:90–101

Chaves MM, Maroco JP, Pereira JS (2003) Understanding plant responses to drought - from genes to the whole plant. Funct Plant Biol 30(3):239–264

Chawla K, Barah P, Kuiper M, Bones AM (2011) Systems biology: a promising tool to study abiotic stress responses. In: Tuteja N, Gill SS, Tuteja R (eds) Omics and plant abiotic stress tolerance. Bentham Publishers, Sharjah, pp 163–172

Chen H, Chen W, Zhou J, He H, Chen L, Chen H, Deng XW (2012) Basic leucine zipper transcription factor OsbZIP16 positively regulates drought resistance in rice. Plant Sci 193–194:8–17

Chen X, Chen G, Li J, Hao X, Tuerxun Z, Chang X, Gao S, Huang Q (2021) A maize calcineurin B-like interacting protein kinase ZmCIPK42 confers salt stress tolerance. Physiol Plant 171(1): 161–172

Choudhury FK, Rivero RM, Blumwald E, Mittler R (2017) Reactive oxygen species, abiotic stress and stress combination. Plant J 90(5):856–867

Chowdhury J, Karim M, Khaliq Q, Ahmed A, Khan M (2016) Effect of drought stress on gas exchange characteristics of four soybean genotypes. Bangladesh J Agric Res 41(2):195–205

Christophe S, Jean-Christophe A, Annabelle L, Alain O, Marion P, Anne-Sophie V (2011) Plant N fluxes and modulation by nitrogen, heat and water stresses: a review based on comparison of legumes and non legume plants. In: Shanker A, Venkateswarlu B (eds) Abiotic stress in plants–mechanisms and adaptations. Intech Open Access Publisher, Rijeka, pp 79–118

Cieśla A, Mituła F, Misztal L, Fedorowicz-Strońska O, Janicka S, Tajdel-Zielińska M, Marczak M, Janicki M, Ludwików A, Sadowski J (2016) A role for barley calcium-dependent protein kinase CPK2a in the response to drought. Front Plant Sci 7:1550

Clemente A, Olias R (2017) Beneficial effects of legumes in gut health. Curr Opin Food Sci 14:32–36

Coba de la Pena T, Pueyo JJ (2012) Legumes in the reclamation of marginal soils, from cultivar and inoculant selection to transgenic approaches. Agron Sustain Dev 32(1):65–91

Cochetel N, Ghan R, Toups HS, Degu A, Tillett RL, Schlauch KA, Cramer GR (2020) Drought tolerance of the grapevine, Vitis champinii cv. Ramsey, is associated with higher photosynthesis and greater transcriptomic responsiveness of abscisic acid biosynthesis and signaling. BMC Plant Biol 20(1):1–25

Covarrubias AA, Reyes JL (2010) Post-transcriptional gene regulation of salinity and drought responses by plant microRNAs. Plant Cell Environ 33(4):481–489

Croser JS, Clarke HJ, Siddique KHM, Khan TN (2003) Low-temperature stress: implications for chickpea (Cicer arietinum L.) improvement. Crit Rev Plant Sci 22(2):185–219

DalCorso G, Farinati S, Furini A (2010) Regulatory networks of cadmium stress in plants. Plant Signal Behav 5(6):663–667

Damanik RI, Maziah M, Ismail MR, Ahmad S, Zain AM (2010) Responses of the antioxidative enzymes in Malaysian rice (*Oryza sativa* L.) cultivars under submergence condition. Acta Physiol Plant 32(4):739–747

Das KK, Panda D, Nagaraju M, Sharma SG, Sarkar RK (2004) Antioxidant enzymes and aldehyde releasing capacity of rice cultivars (*Oryza sativa* L.) as determinants of anaerobic seedling establishment capacity. Bulgar J Plant Physiol 30(1–2):34–44

de Freitas PAF, de Carvalho HH, Costa JH, Miranda RS, Saraiva KDC, de Oliveira FDB, Coelho DG, Prisco JT, Gomes-Filho E (2019) Salt acclimation in sorghum plants by exogenous proline: physiological and biochemical changes and regulation of proline metabolism. Plant Cell Rep 38(3):403–416

del Pozo JC, Ramirez-Parra E (2014) Deciphering the molecular bases for drought tolerance in Arabidopsis autotetraploids. Plant Cell Environ 37(12):2722–2737

Ding Y, Shi Y, Yang S (2019) Advances and challenges in uncovering cold tolerance regulatory mechanisms in plants. New Phytol 222(4):1690–1704

Djanaguiraman M, Prasad PVV, Al-Khatib K (2011) Ethylene perception inhibitor 1-MCP decreases oxidative damage of leaves through enhanced antioxidant defense mechanisms in soybean plants grown under high temperature stress. Environ Exp Bot 71(2):215–223

Doherty CJ, Van Buskirk HA, Myers SJ, Thomashow MF (2009) Roles for Arabidopsis CAMTA transcription factors in cold-regulated gene expression and freezing tolerance. Plant Cell 21(3): 972–984

Dong Z, Shi L, Wang Y, Chen L, Cai Z, Wang Y, Jin J, Li X (2013) Identification and dynamic regulation of microRNAs involved in salt stress responses in functional soybean nodules by high-throughput sequencing. Int J Mol Sci 14(2):2717–2738

Edreira JIR, Otegui ME (2012) Heat stress in temperate and tropical maize hybrids: differences in crop growth, biomass partitioning and reserves use. Field Crop Res 130:87–98

El Sayed H, El Sayed A (2011) Influence of NaCl and Na_2SO_4 treatments on growth development of broad bean (*Vicia Faba*, L.) plant. J Life Sci 5(7):513–523

Endo A, Sawada Y, Takahashi H, Okamoto M, Ikegami K, Koiwai H, Seo M, Toyomasu T, Mitsuhashi W, Shinozaki K (2008) Drought induction of Arabidopsis 9-cis-epoxycarotenoid dioxygenase occurs in vascular parenchyma cells. Plant Physiol 147(4):1984–1993

Fan TT, Ni JJ, Dong WC, An LZ, Xiang Y, Cao SQ (2015) Effect of low temperature on profilins and ADFs transcription and actin cytoskeleton reorganization in Arabidopsis. Biol Plant 59(4): 793–796

Farooq M, Hussain M, Siddique KHM (2014) Drought stress in wheat during flowering and grain-filling periods. Crit Rev Plant Sci 33(4):331–349

Farooq M, Gogoi N, Hussain M, Barthakur S, Paul S, Bharadwaj N, Migdadi HM, Alghamdi SS, Siddique KHM (2017) Effects, tolerance mechanisms and management of salt stress in grain legumes. Plant Physiol Biochem 118:199–217

Flexas J, Bota J, Loreto F, Cornic G, Sharkey TD (2004) Diffusive and metabolic limitations to photosynthesis under drought and salinity in C3 plants. Plant Biol 6(03):269–279

Frugier F, Poirier S, Satiat-Jeunemaître B, Kondorosi A, Crespi M (2000) A Krüppel-like zinc finger protein is involved in nitrogen-fixing root nodule organogenesis. Genes Dev 14(4): 475–482

Gall JE, Rajakaruna N (2013) The physiology, functional genomics, and applied ecology of heavy metal-tolerant Brassicaceae. In: Lang M (ed) Brassicaceae: characterization, functional genomics and health benefits. Nova Science, New York, pp 121–148

Garzón T, Gunsé B, Moreno AR, Tomos AD, Barceló J, Poschenrieder C (2011) Aluminium-induced alteration of ion homeostasis in root tip vacuoles of two maize varieties differing in Al tolerance. Plant Sci 180(5):709–715

Gill SS, Hasanuzzaman M, Nahar K, Macovei A, Tuteja N (2013) Importance of nitric oxide in cadmium stress tolerance in crop plants. Plant Physiol Biochem 63:254–261

Grover A, Mittal D, Negi M, Lavania D (2013) Generating high temperature tolerant transgenic plants: achievements and challenges. Plant Sci 205:38–47

Guan Q, Lu X, Zeng H, Zhang Y, Zhu J (2013) Heat stress induction of miR398 triggers a regulatory loop that is critical for thermotolerance in Arabidopsis. Plant J 74(5):840–851

Guler NS, Pehlivan N (2016) Exogenous low-dose hydrogen peroxide enhances drought tolerance of soybean (*Glycine max* L.) through inducing antioxidant system. Acta Biol Hung 67(2):169–183

Gunawardhana MDM, De Silva CS (2011) Impact of temperature and water stress on growth yield and related biochemical parameters of okra. Trop Agric Res 23(1):77–83

Guo X, Liu D, Chong K (2018) Cold signaling in plants: insights into mechanisms and regulation. J Integr Plant Biol 60(9):745–756

Gupta B, Huang B (2014) Mechanism of salinity tolerance in plants: physiological, biochemical, and molecular characterization. Int J Genomics 2014:701596

Ha S, Vankova R, Yamaguchi-Shinozaki K, Shinozaki K, Tran L-SP (2012) Cytokinins: metabolism and function in plant adaptation to environmental stresses. Trends Plant Sci 17(3):172–179

Haider MS, Kurjogi MM, Khalil-Ur-Rehman M, Fiaz M, Pervaiz T, Jiu S, Haifeng J, Chen W, Fang J (2017) Grapevine immune signaling network in response to drought stress as revealed by transcriptomic analysis. Plant Physiol Biochem 121:187–195

Hajyzadeh M, Turktas M, Khawar KM, Unver T (2015) MiR408 overexpression causes increased drought tolerance in chickpea. Gene 555(2):186–193

Hanafy MS, El-Banna A, Schumacher HM, Jacobsen H-J, Hassan FS (2013) Enhanced tolerance to drought and salt stresses in transgenic faba bean (*Vicia faba* L.) plants by heterologous expression of the PR10a gene from potato. Plant Cell Rep 32(5):663–674

HanumanthaRao B, Nair RM, Nayyar H (2016) Salinity and high temperature tolerance in mungbean [*Vigna radiata* (L.) Wilczek] from a physiological perspective. Front Plant Sci 7:957

Hasanuzzaman M, Hossain MA, Silva JA, Fujita M (2012) Plant response and tolerance to abiotic oxidative stress: antioxidant defense is a key factor. In: Crop stress and its management: perspectives and strategies. Springer, Dordrecht, pp 261–315

Hasanuzzaman M, Nahar K, Alam M, Roychowdhury R, Fujita M (2013) Physiological, biochemical, and molecular mechanisms of heat stress tolerance in plants. Int J Mol Sci 14(5):9643–9684

Hasegawa PM, Bressan RA, Zhu J-K, Bohnert HJ (2000) Plant cellular and molecular responses to high salinity. Annu Rev Plant Biol 51(1):463–499

Hayat S, Khalique G, Irfan M, Wani AS, Tripathi BN, Ahmad A (2012) Physiological changes induced by chromium stress in plants: an overview. Protoplasma 249(3):599–611

Hayat F, Sun Z, Ni Z, Iqbal S, Xu W, Gao Z, Qiao Y, Tufail MA, Jahan MS, Khan U (2022) Exogenous melatonin improves cold tolerance of strawberry (Fragaria × ananassa Duch.) through modulation of DREB/CBF-COR pathway and antioxidant defense system. Horticulturae 8(3):194

Hernández JA (2019) Salinity tolerance in plants: trends and perspectives. Int J Mol Sci 20(10): 2408

Hernandez JA, Campillo A, Jimenez A, Alarcon JJ, Sevilla F (1999) Response of antioxidant systems and leaf water relations to NaCl stress in pea plants. New Phytol 141(2):241–251

Hossain MA, Hasanuzzaman M, Fujita M (2010) Up-regulation of antioxidant and glyoxalase systems by exogenous glycinebetaine and proline in mung bean confer tolerance to cadmium stress. Physiol Mol Biol Plants 16(3):259–272

Hossain MA, Piyatida P, da Silva JAT, Fujita M (2012) Molecular mechanism of heavy metal toxicity and tolerance in plants: central role of glutathione in detoxification of reactive oxygen species and methylglyoxal and in heavy metal chelation. J Bot 2012:872875

Hu H, Xiong L (2014) Genetic engineering and breeding of drought-resistant crops. Annu Rev Plant Biol 65:715–741

Hu X, Xu X, Li C (2018) Ectopic expression of the LoERF017 transcription factor from Larix olgensis Henry enhances salt and osmotic-stress tolerance in Arabidopsis thaliana. Plant Biotechnol Rep 12(2):93–104

Hua J (2016) Defining roles of tandemly arrayed CBF genes in freezing tolerance with new genome editing tools. New Phytol 212(2):301–302

Huang K, Peng L, Liu Y, Yao R, Liu Z, Li X, Yang Y, Wang J (2018) Arabidopsis calcium-dependent protein kinase AtCPK1 plays a positive role in salt/drought-stress response. Biochem Biophys Res Commun 498(1):92–98

Hwarari D, Guan Y, Ahmad B, Movahedi A, Min T, Hao Z, Lu Y, Chen J, Yang L (2022) ICE-CBF-COR signaling cascade and its regulation in plants responding to cold stress. Int J Mol Sci 23(3):1549

Jacob P, Hirt H, Bendahmane A (2017) The heat-shock protein/chaperone network and multiple stress resistance. Plant Biotechnol J 15(4):405–414

Jahan MS, Guo S, Sun J, Shu S, Wang Y, Abou El-Yazied A, Alabdallah NM, Hikal M, Mohamed MHM, Ibrahim MFM (2021) Melatonin-mediated photosynthetic performance of tomato seedlings under high-temperature stress. Plant Physiol Biochem 167:309–320

Jan AU, Hadi F, Midrarullah AA, Rahman K (2017) Role of CBF/DREB gene expression in abiotic stress tolerance. A review. Int J Hortic Agric 2:1–12

Jatan R, Tiwari S, Asif MH, Lata C (2019) Genome-wide profiling reveals extensive alterations in Pseudomonas putida-mediated miRNAs expression during drought stress in chickpea (*Cicer arietinum* L.). Environ Exp Bot 157:217–227

Ji C, Mao X, Hao J, Wang X, Xue J, Cui H, Li R (2018) Analysis of bZIP transcription factor family and their expressions under salt stress in *Chlamydomonas reinhardtii*. Int J Mol Sci 19(9):2800

Jiang C, Li X, Zou J, Ren J, Jin C, Zhang H, Yu H, Jin H (2021) Comparative transcriptome analysis of genes involved in the drought stress response of two peanut (*Arachis hypogaea* L.) varieties. BMC Plant Biol 21(1):1–14

Jin H, Liu S, Zenda T, Wang X, Liu G, Duan H (2019) Maize leaves drought-responsive genes revealed by comparative transcriptome of two cultivars during the filling stage. PLoS One 14(10):e0223786

Jyoti B, Yadav SK (2012) Comparative study on biochemical parameters and antioxidant enzymes in a drought tolerant and a sensitive variety of horsegram (*Macrotyloma uniflorum*) under drought stress. Am J Plant Physiol 7(1):17–29

Kamiyama Y, Katagiri S, Umezawa T (2021) Growth promotion or osmotic stress response: how SNF1-related protein kinase 2 (SnRK2) kinases are activated and manage intracellular signaling in plants. Plants 10(7):1443

Kaplan F, Kopka J, Sung DY, Zhao W, Popp M, Porat R, Guy CL (2007) Transcript and metabolite profiling during cold acclimation of Arabidopsis reveals an intricate relationship of cold-regulated gene expression with modifications in metabolite content. Plant J 50(6):967–981

Kashiwagi J, Krishnamurthy L, Crouch JH, Serraj R (2006) Variability of root length density and its contributions to seed yield in chickpea (*Cicer arietinum* L.) under terminal drought stress. Field Crop Res 95(2):171–181

Kaur G, Asthir B (2017) Molecular responses to drought stress in plants. Biol Plant 61(2):201–209

Kaushal N, Awasthi R, Gupta K, Gaur P, Siddique KHM, Nayyar H (2013) Heat-stress-induced reproductive failures in chickpea (*Cicer arietinum*) are associated with impaired sucrose metabolism in leaves and anthers. Funct Plant Biol 40(12):1334–1349

Kawano N, Ella E, Ito O, Yamauchi Y, Tanaka K (2002) Metabolic changes in rice seedlings with different submergence tolerance after desubmergence. Environ Exp Bot 47(3):195–203

Kazan K (2015) Diverse roles of jasmonates and ethylene in abiotic stress tolerance. Trends Plant Sci 20(4):219–229

Ketchouli T, Zhou Y-G, Dai S-Y, Carther KFI, Sun D-Q, Li Y, Nguyen QVH, Xu H, Wang F-W, Liu W-C (2021) A soybean calcineurin B-like protein-interacting protein kinase, GmPKS4, regulates plant responses to salt and alkali stresses. J Plant Physiol 256:153331

Kim YS, Lee M, Lee J-H, Lee H-J, Park C-M (2015) The unified ICE–CBF pathway provides a transcriptional feedback control of freezing tolerance during cold acclimation in Arabidopsis. Plant Mol Biol 89(1):187–201

Kukreja S, Nandwal AS, Kumar N, Sharma SK, Unvi V, Sharma PK (2005) Plant water status, H_2O_2 scavenging enzymes, ethylene evolution and membrane integrity of Cicer arietinum roots as affected by salinity. Biol Plant 49(2):305–308

Kumar S, Gupta D, Nayyar H (2012a) Comparative response of maize and rice genotypes to heat stress: status of oxidative stress and antioxidants. Acta Physiol Plant 34(1):75–86

Kumar S, Kaushal N, Nayyar H, Gaur P (2012b) Abscisic acid induces heat tolerance in chickpea (*Cicer arietinum* L.) seedlings by facilitated accumulation of osmoprotectants. Acta Physiol Plant 34(5):1651–1658

Kumar S, Thakur P, Kaushal N, Malik JA, Gaur P, Nayyar H (2013) Effect of varying high temperatures during reproductive growth on reproductive function, oxidative stress and seed yield in chickpea genotypes differing in heat sensitivity. Arch Agron Soil Sci 59(6):823–843

Kundrátová K, Bartas M, Pečinka P, Hejna O, Rychlá A, Čurn V, Červeň J (2021) Transcriptomic and proteomic analysis of drought stress response in opium poppy plants during the first week of germination. Plants 10(9):1878

Kurutas EB (2016) The importance of antioxidants which play the role in cellular response against oxidative/nitrosative stress: current state. Nutr J 15(1):1–22. https://doi.org/10.1186/s12937-016-0186-5

Laan P, Smolders A, Blom C, Armstrong W (1989) The relative roles of internal aeration, radial oxygen losses, iron exclusion and nutrient balances in flood-tolerance of Rumex species. Acta Bot Neerlandica 38(2):131–145

Lee HG, Seo PJ (2015) The MYB 96–HHP module integrates cold and abscisic acid signaling to activate the CBF–COR pathway in Arabidopsis. Plant J 82(6):962–977

Li WF, Wong F, Tsai S, Phang T, Shao G, Lam H (2006) Tonoplast-located GmCLC1 and GmNHX1 from soybean enhance NaCl tolerance in transgenic bright yellow (BY)-2 cells. Plant Cell Environ 29(6):1122–1137

Li X, Cheng X, Liu J, Zeng H, Han L, Tang W (2011) Heterologous expression of the Arabidopsis DREB1A/CBF3 gene enhances drought and freezing tolerance in transgenic Lolium perenne plants. Plant Biotechnol Rep 5(1):61–69

Li F, Li M, Wang P, Cox KL Jr, Duan L, Dever JK, Shan L, Li Z, He P (2017) Regulation of cotton (*Gossypium hirsutum*) drought responses by mitogen-activated protein (MAP) kinase cascade-mediated phosphorylation of Gh WRKY 59. New Phytol 215(4):1462–1475

Lim G, Zhang X, Chung M, Lee DJ, Woo Y, Cheong H, Kim CS (2010) A putative novel transcription factor, AtSKIP, is involved in abscisic acid signalling and confers salt and osmotic tolerance in Arabidopsis. New Phytol 185(1):103–113

Lin L, Wu J, Jiang M, Wang Y (2021) Plant mitogen-activated protein kinase cascades in environmental stresses. Int J Mol Sci 22(4):1543

Liu W, Yu K, He T, Li F, Zhang D, Liu J (2013) The low temperature induced physiological responses of *Avena nuda* L., a cold-tolerant plant species. Sci World J 2013:658793

Liu J, Shi Y, Yang S (2018) Insights into the regulation of C-repeat binding factors in plant cold signaling. J Integr Plant Biol 60(9):780–795

Liu S, Zenda T, Li J, Wang Y, Liu X, Duan H (2020) Comparative transcriptomic analysis of contrasting hybrid cultivars reveal key drought-responsive genes and metabolic pathways regulating drought stress tolerance in maize at various stages. PLoS One 15(10):e0240468

Lu L, Chen X, Wang P, Lu Y, Zhang J, Yang X, Cheng T, Shi J, Chen J (2021) CIPK11: a calcineurin B-like protein-interacting protein kinase from Nitraria tangutorum, confers tolerance to salt and drought in Arabidopsis. BMC Plant Biol 21(1):1–16

Luo X-S, Xue Y, Wang Y-L, Cang L, Xu B, Ding J (2015) Source identification and apportionment of heavy metals in urban soil profiles. Chemosphere 127:152–157

Luo Q, Wei Q, Wang R, Zhang Y, Zhang F, He Y, Zhou S, Feng J, Yang G, He G (2017) BdCIPK31, a calcineurin B-like protein-interacting protein kinase, regulates plant response to drought and salt stress. Front Plant Sci 8:1184

Luo D, Hou X, Zhang Y, Meng Y, Zhang H, Liu S, Wang X, Chen R (2019) CaDHN5, a dehydrin gene from pepper, plays an important role in salt and osmotic stress responses. Int J Mol Sci 20(8):1989

Lv K, Li J, Zhao K, Chen S, Nie J, Zhang W, Liu G, Wei H (2020) Overexpression of an AP2/ERF family gene, BpERF13, in birch enhances cold tolerance through upregulating CBF genes and mitigating reactive oxygen species. Plant Sci 292:110375

Lynch JP (2018) Rightsizing root phenotypes for drought resistance. J Exp Bot 69(13):3279–3292

Maqbool MA, Aslam M, Ali H (2017) Breeding for improved drought tolerance in Chickpea (*Cicer arietinum* L.). Plant Breed 136(3):300–318

Maszkowska J, Dębski J, Kulik A, Kistowski M, Bucholc M, Lichocka M, Klimecka M, Sztatelman O, Szymańska KP, Dadlez M (2019) Phosphoproteomic analysis reveals that dehydrins ERD10 and ERD14 are phosphorylated by SNF1-related protein kinase 2.10 in response to osmotic stress. Plant Cell Environ 42(3):931–946

Mian MAR, Bailey MA, Ashley DA, Wells R, Carter TE Jr, Parrott WA, Boerma HR (1996) Molecular markers associated with water use efficiency and leaf ash in soybean. Crop Sci 36(5): 1252–1257

Mian MAR, Ashley DA, Boerma HR (1998) An additional QTL for water use efficiency in soybean. Crop Sci 38(2):390–393

Miao H, Sun P, Liu J, Wang J, Xu B, Jin Z (2018) Overexpression of a novel ROP gene from the banana (MaROP5g) confers increased salt stress tolerance. Int J Mol Sci 19(10):3108

Mir RR, Zaman-Allah M, Sreenivasulu N, Trethowan R, Varshney RK (2012) Integrated genomics, physiology and breeding approaches for improving drought tolerance in crops. Theor Appl Genet 125(4):625–645. https://doi.org/10.1007/s00122-012-1904-9

Mishra S, Sahu G, Shaw BP (2022) Integrative small RNA and transcriptome analysis provides insight into key role of miR408 towards drought tolerance response in cowpea. Plant Cell Rep 41(1):75–94

Miyashita K, Tanakamaru S, Maitani T, Kimura K (2005) Recovery responses of photosynthesis, transpiration, and stomatal conductance in kidney bean following drought stress. Environ Exp Bot 53(2):205–214

Mohamed HI, Latif HH (2017) Improvement of drought tolerance of soybean plants by using methyl jasmonate. Physiol Mol Biol Plants 23(3):545–556

Mohammed AS, Kapri A, Goel R (2011) Heavy metal pollution: source, impact, and remedies. In: Khan MS, Zaidi A, Goel R, Musarrat J (eds) Biomanagement of metal-contaminated soils. Springer, Netherlands, pp 1–28

Monneveux P, Ramírez DA, Pino M-T (2013) Drought tolerance in potato (*S. tuberosum* L.): can we learn from drought tolerance research in cereals? Plant Sci 205:76–86

Monteros M, Lee G, Missaoui A, Carter T, Boerma H (2006) Identification and confirmation of QTL conditioning drought tolerance in Nepalese soybean PI471938. The 11th Biennial conference on the molecular and cellular biology of the soybean

Morran S, Eini O, Pyvovarenko T, Parent B, Singh R, Ismagul A, Eliby S, Shirley N, Langridge P, Lopato S (2011) Improvement of stress tolerance of wheat and barley by modulation of expression of DREB/CBF factors. Plant Biotechnol J 9(2):230–249

Movahedi A, Zhang J, Gao P, Yang Y, Wang L, Yin T, Kadkhodaei S, Ebrahimi M, Zhuge Q (2015) Expression of the chickpea CarNAC3 gene enhances salinity and drought tolerance in transgenic poplars. Plant Cell Tissue Organ Cult 120(1):141–154

Muhammad T, Zhang J, Ma Y, Li Y, Zhang F, Zhang Y, Liang Y (2019) Overexpression of a mitogen-activated protein kinase SlMAPK3 positively regulates tomato tolerance to cadmium and drought stress. Molecules 24(3):556

Munemasa S, Mori IC, Murata Y (2011) Methyl jasmonate signaling and signal crosstalk between methyl jasmonate and abscisic acid in guard cells. Plant Signal Behav 6(7):939–941

Muthusamy M, Uma S, Backiyarani S, Saraswathi MS, Chandrasekar A (2016) Transcriptomic changes of drought-tolerant and sensitive banana cultivars exposed to drought stress. Front Plant Sci 7:1609

Nadeem M, Li J, Yahya M, Sher A, Ma C, Wang X, Qiu L (2019) Research progress and perspective on drought stress in legumes: a review. Int J Mol Sci 20(10):2541

Neilson S, Rajakaruna N (2015) Phytoremediation of agricultural soils: using plants to clean metal-contaminated arable land. In: Phytoremediation. Springer, New York, pp 159–168

Nishiyama R, Watanabe Y, Leyva-Gonzalez MA, Van Ha C, Fujita Y, Tanaka M, Seki M, Yamaguchi-Shinozaki K, Shinozaki K, Herrera-Estrella L (2013) Arabidopsis AHP2, AHP3, and AHP5 histidine phosphotransfer proteins function as redundant negative regulators of drought stress response. Proc Natl Acad Sci 110(12):4840–4845

Noctor G, Veljovic-Jovanovic S, Foyer CH (2000) Peroxide processing in photosynthesis: antioxidant coupling and redox signalling. Philos Trans R Soc Lond B Biol Sci 355(1402):1465–1475

Norén L, Kindgren P, Stachula P, Rühl M, Eriksson ME, Hurry V, Strand Å (2016) Circadian and plastid signaling pathways are integrated to ensure correct expression of the CBF and COR genes during photoperiodic growth. Plant Physiol 171(2):1392–1406

Oram NJ, Sun Y, Abalos D, van Groenigen JW, Hartley S, De Deyn GB (2021) Plant traits of grass and legume species for flood resilience and N_2O mitigation. Funct Ecol 35(10):2205–2218

Orvar BL, Sangwan V, Omann F, Dhindsa RS (2000) Early steps in cold sensing by plant cells: the role of actin cytoskeleton and membrane fluidity. Plant J 23(6):785–794

Osakabe Y, Osakabe K (2012) Abiotic stress responses in plants: present and future. In: Ahmad P, Prasad MNV (eds) Abiotic Stress: new research. Springer, New York, pp 171–180

Osakabe Y, Kajita S, Osakabe K (2011) Genetic engineering of woody plants: current and future targets in a stressful environment. Physiol Plant 142(2):105–117

Osakabe Y, Osakabe K, Shinozaki K, Tran L-S (2014) Response of plants to water stress. Front Plant Sci 5:1–8

Osman HS (2015) Enhancing antioxidant–yield relationship of pea plant under drought at different growth stages by exogenously applied glycine betaine and proline. Ann Agric Sci 60(2): 389–402

Passioura JB (2012) Phenotyping for drought tolerance in grain crops: when is it useful to breeders? Funct Plant Biol 39(11):851–859

Patel PK, Hemantaranjan A, Sarma BK, Radha S (2011) Growth and antioxidant system under drought stress in chickpea (*Cicer arietinum* L.) as sustained by salicylic acid. J Stress Physiol Biochem 7(4):130–144

Pearce RS (2001) Plant freezing and damage. Ann Bot 87(4):417–424

Pearce S, Zhu J, Boldizsár Á, Vágújfalvi A, Burke A, Garland-Campbell K, Galiba G, Dubcovsky J (2013) Large deletions in the CBF gene cluster at the Fr-B2 locus are associated with reduced frost tolerance in wheat. Theor Appl Genet 126(11):2683–2697

Peng H, Cheng H-Y, Chen C, Yu X-W, Yang J-N, Gao W-R, Shi Q-H, Zhang H, Li J-G, Ma H (2009) A NAC transcription factor gene of Chickpea (*Cicer arietinum*), CarNAC3, is involved in drought stress response and various developmental processes. J Plant Physiol 166(17):1934–1945

Peng Y-L, Wang Y-S, Fei J, Sun C-C (2020) Isolation and expression analysis of two novel C-repeat binding factor (CBF) genes involved in plant growth and abiotic stress response in mangrove Kandelia obovata. Ecotoxicology 29(6):718–725

Phang T, Shao G, Lam H (2008) Salt tolerance in soybean. J Integr Plant Biol 50(10):1196–1212

Pinheiro C, Chaves MM (2011) Photosynthesis and drought: can we make metabolic connections from available data? J Exp Bot 62(3):869–882

Pottier M, Oomen R, Picco C, Giraudat J, Scholz-Starke J, Richaud P, Carpaneto A, Thomine S (2015) Identification of mutations allowing Natural Resistance Associated Macrophage Proteins (NRAMP) to discriminate against cadmium. Plant J 83(4):625–637

Prasad PVV, Boote KJ, Allen LH Jr, Thomas JMG (2002) Effects of elevated temperature and carbon dioxide on seed-set and yield of kidney bean (*Phaseolus vulgaris* L.). Glob Chang Biol 8(8):710–721

Rai KK, Pandey N, Meena RP, Rai SP (2021) Biotechnological strategies for enhancing heavy metal tolerance in neglected and underutilized legume crops: a comprehensive review. Ecotoxicol Environ Saf 208:111750

Rajkowitsch L, Chen D, Stampfl S, Semrad K, Waldsich C, Mayer O, Jantsch MF, Konrat R, Bläsi U, Schroeder R (2007) RNA chaperones, RNA annealers and RNA helicases. RNA Biol 4(3):118–130

Rana D, Dass A, Rajanna G, Kaur R (2016) Biotic and abiotic stress management in pulses. J Agron 61:238–248

Rani B, Dhawan K, Jain V, Chhabra ML, Singh D (2013) High temperature induced changes in antioxidative enzymes in *Brassica juncea* (L) Czern & Coss. Recuperado El:12

Ricachenevsky FK, Menguer PK, Sperotto RA, Williams LE, Fett JP (2013) Roles of plant metal tolerance proteins (MTP) in metal storage and potential use in biofortification strategies. Front Plant Sci 4:144

Riemann M, Dhakarey R, Hazman M, Miro B, Kohli A, Nick P (2015) Exploring jasmonates in the hormonal network of drought and salinity responses. Front Plant Sci 6:1077

Roshandel P, Flowers T (2009) The ionic effects of NaCl on physiology and gene expression in rice genotypes differing in salt tolerance. Plant Soil 315(1):135–147

Ruelland E, Zachowski A (2010) How plants sense temperature. Environ Exp Bot 69(3):225–232

Ruelland E, Vaultier M-N, Zachowski A, Hurry V (2009) Cold signalling and cold acclimation in plants. Adv Bot Res 49:35–150
Ruthrof KX, Fontaine JB, Hopkins AJM, McHenry MP, O'Hara G, McComb J, Hardy GESJ, Howieson J (2018) Potassium amendment increases biomass and reduces heavy metal concentrations in Lablab purpureus after phosphate mining. Land Degrad Dev 29(3):398–407
Saglam A, Saruhan N, Terzi R, Kadioglu A (2011) The relations between antioxidant enzymes and chlorophyll fluorescence parameters in common bean cultivars differing in sensitivity to drought stress. Russ J Plant Physiol 58(1):60–68
Sahitya UL, Krishna MSR, Deepthi R, Prasad GS, Kasim D (2018) Seed antioxidants interplay with drought stress tolerance indices in chilli (*Capsicum annuum* L) seedlings. Biomed Res Int 2018:1605096
Salmanoglu E, Kurutas EB (2020) The levels of oxidative and nitrosative stress in patients who had 99mTc-MIBI myocardial perfusion scintigraphy and 99mTc-DMSA, 99mTc-MAG-3 renal scintigraphy. Nucl Med Rev 23(2):89–96
Savchenko T, Kolla VA, Wang C-Q, Nasafi Z, Hicks DR, Phadungchob B, Chehab WE, Brandizzi F, Froehlich J, Dehesh K (2014) Functional convergence of oxylipin and abscisic acid pathways controls stomatal closure in response to drought. Plant Physiol 164(3): 1151–1160
Shahid M, Pinelli E, Dumat C (2012) Review of Pb availability and toxicity to plants in relation with metal speciation; role of synthetic and natural organic ligands. J Hazard Mater 219:1–12
Shi H, Ishitani M, Kim C, Zhu J-K (2000) The Arabidopsis thaliana salt tolerance gene SOS1 encodes a putative Na^+/H^+ antiporter. Proc Natl Acad Sci 97(12):6896–6901
Shi Y, Ding Y, Yang S (2015) Cold signal transduction and its interplay with phytohormones during cold acclimation. Plant Cell Physiol 56(1):7–15
Siddique KHM, Walton GH, Seymour M (1993) A comparison of seed yields of winter grain legumes in Western Australia. Aust J Exp Agric 33(7):915–922
Siddiqui KS, Cavicchioli R (2006) Cold-adapted enzymes. Annu Rev Biochem 75(1):403–433
Singh UK, Kumar B (2017) Pathways of heavy metals contamination and associated human health risk in Ajay River basin, India. Chemosphere 174:183–199
Song N-H, Ahn Y-J (2011) DcHsp17. 7, a small heat shock protein in carrot, is tissue-specifically expressed under salt stress and confers tolerance to salinity. New Biotechnol 28(6):698–704
Song Q, Wang X, Li J, Chen THH, Liu Y, Yang X (2021) CBF1 and CBF4 in Solanum tuberosum L. differ in their effect on low-temperature tolerance and development. Environ Exp Bot 185: 104416
Specht JE, Chase K, Macrander M, Graef GL, Chung J, Markwell JP, Germann M, Orf JH, Lark KG (2001) Soybean response to water: a QTL analysis of drought tolerance. Crop Sci 41(2): 493–509
Tan W, Wei Meng Q, Brestic M, Olsovska K, Yang X (2011) Photosynthesis is improved by exogenous calcium in heat-stressed tobacco plants. J Plant Physiol 168(17):2063–2071
Tang M, Liu X, Deng H, Shen S (2011) Over-expression of JcDREB, a putative AP2/EREBP domain-containing transcription factor gene in woody biodiesel plant Jatropha curcas, enhances salt and freezing tolerance in transgenic Arabidopsis thaliana. Plant Sci 181(6):623–631
Tardieu F (2012) Any trait or trait-related allele can confer drought tolerance: just design the right drought scenario. J Exp Bot 63(1):25–31
Tauqeer HM, Ali S, Rizwan M, Ali Q, Saeed R, Iftikhar U, Ahmad R, Farid M, Abbasi GH (2016) Phytoremediation of heavy metals by Alternanthera bettzickiana: growth and physiological response. Ecotoxicol Environ Saf 126:138–146
Teixeira E, Fischer G, Velthuizen H, Walter C, Ewert F (2013) Global hot-spots of heat stress on agricultural crops due to climate change. Agric For Meteorol 170:206–215
Thapa G, Sadhukhan A, Panda SK, Sahoo L (2012) Molecular mechanistic model of plant heavy metal tolerance. Biometals 25(3):489–505
Theocharis A, Clément C, Barka EA (2012) Physiological and molecular changes in plants grown at low temperatures. Planta 235(6):1091–1105

Thomashow MF (2010) Molecular basis of plant cold acclimation: insights gained from studying the CBF cold response pathway. Plant Physiol 154(2):571–577

Tuberosa R, Salvi S (2006) Genomics-based approaches to improve drought tolerance of crops. Trends Plant Sci 11(8):405–412

Tung SA, Smeeton R, White CA, Black CR, Taylor IB, Hilton HW, Thompson AJ (2008) Over-expression of LeNCED1 in tomato (*Solanum lycopersicum* L.) with the rbcS3C promoter allows recovery of lines that accumulate very high levels of abscisic acid and exhibit severe phenotypes. Plant Cell Environ 31(7):968–981

Turner NC, Colmer TD, Quealy J, Pushpavalli R, Krishnamurthy L, Kaur J, Singh G, Siddique KHM, Vadez V (2013) Salinity tolerance and ion accumulation in chickpea (*Cicer arietinum* L.) subjected to salt stress. Plant Soil 365(1):347–361

Tyagi A, Sharma S, Ali S, Gaikwad K (2022) Crosstalk between H_2S and NO: an emerging signalling pathway during waterlogging stress in legume crops. Plant Biol 24(4):576–586

Tzudir L, Bera PS, Chakraborty PK (2014) Impact of temperature on the reproductive development in mungbean (*Vigna radiata*) varieties under different dates of sowing. Int J Bioresour Stress Manag 5(2):194–199

Ullah A, Manghwar H, Shaban M, Khan AH, Akbar A, Ali U, Ali E, Fahad S (2018) Phytohormones enhanced drought tolerance in plants: a coping strategy. Environ Sci Pollut Res 25(33):33103–33118

Umezawa T, Nakashima K, Miyakawa T, Kuromori T, Tanokura M, Shinozaki K, Yamaguchi-Shinozaki K (2010) Molecular basis of the core regulatory network in ABA responses: sensing, signaling and transport. Plant Cell Physiol 51(11):1821–1839

Urano K, Kurihara Y, Seki M, Shinozaki K (2010) 'Omics' analyses of regulatory networks in plant abiotic stress responses. Curr Opin Plant Biol 13(2):132–138

Vadez V, Ratnakumar P (2016) High transpiration efficiency increases pod yield under intermittent drought in dry and hot atmospheric conditions but less so under wetter and cooler conditions in groundnut (*Arachis hypogaea* L.). Field Crop Res 193:16–23

Varshney RK, Thudi M, Nayak SN, Gaur PM, Kashiwagi J, Krishnamurthy L, Jaganathan D, Koppolu J, Bohra A, Tripathi S, Rathore A, Jukanti AK, Jayalakshmi V, Vemula A, Singh SJ, Yasin M, Sheshshayee MS, Viswanatha KP (2014) Genetic dissection of drought tolerance in chickpea (*Cicer arietinum* L.). Theor Appl Genet 127(2):445–462

Varshney RK, Thudi M, Muehlbauer FJ (2017) Future prospects for chickpea research. In: The chickpea genome. Springer, New York, pp 135–142

Ventorino V, Caputo R, De Pascale S, Fagnano M, Pepe O, Moschetti G (2012) Response to salinity stress of Rhizobium leguminosarum bv. viciae strains in the presence of different legume host plants. Ann Microbiol 62(2):811–823

Verma D, Singla-Pareek SL, Rajagopal D, Reddy MK, Sopory SK (2007) Functional validation of a novel isoform of Na^+/H^+ antiporter from Pennisetum glaucum for enhancing salinity tolerance in rice. J Biosci 32(3):621–628

Villiers F, Ducruix C, Hugouvieux V, Jarno N, Ezan E, Garin J, Junot C, Bourguignon J (2011) Investigating the plant response to cadmium exposure by proteomic and metabolomic approaches. Proteomics 11(9):1650–1663

Waheed A, Hafiz IA, Qadir G, Murtaza G, Mahmood T, Ashraf M (2006) Effect of salinity on germination, growth, yield, ionic balance and solute composition of Pigeon pea (*Cajanus cajan* (L.) Millsp). Pak J Bot 38(4):1103

Wahid A, Gelani S, Ashraf M, Foolad MR (2007) Heat tolerance in plants: an overview. Environ Exp Bot 61(3):199–223

Wang C-q, Tao W, Ping MU, Li Z, Ling Y (2013) Quantitative trait loci for mercury tolerance in rice seedlings. Rice Sci 20(3):238–242

Wang N, Qian Z, Luo M, Fan S, Zhang X, Zhang L (2018a) Identification of salt stress responding genes using transcriptome analysis in green alga Chlamydomonas reinhardtii. Int J Mol Sci 19(11):3359

Wang X, Wang L, Wang Y, Liu H, Hu D, Zhang N, Zhang S, Cao H, Cao Q, Zhang Z (2018b) Arabidopsis PCaP2 plays an important role in chilling tolerance and ABA response by activating CBF- and SnRK2-mediated transcriptional regulatory network. Front Plant Sci 9:215

Wang Y, Li T, John SJ, Chen M, Chang J, Yang G, He G (2018c) A CBL-interacting protein kinase TaCIPK27 confers drought tolerance and exogenous ABA sensitivity in transgenic Arabidopsis. Plant Physiol Biochem 123:103–113

Wang Q, Ni J, Shah F, Liu W, Wang D, Yao Y, Hu H, Huang S, Hou J, Fu S (2019) Overexpression of the stress-inducible SsMAX2 promotes drought and salt resistance via the regulation of redox homeostasis in Arabidopsis. Int J Mol Sci 20(4):837

Waraich EA, Ahmad R, Ashraf MY (2022) Role of mineral nutrition in alleviation of drought stress in plants. Aust J Crop Sci 5(6):764–777

Wasaya A, Zhang X, Fang Q, Yan Z (2018) Root phenotyping for drought tolerance: a review. Agronomy 8(11):241

Weyers JDB, Paterson NW (2001) Plant hormones and the control of physiological processes. New Phytol 152(3):375–407

Wu G, Zhou Z, Chen P, Tang X, Shao H, Wang H (2014) Comparative ecophysiological study of salt stress for wild and cultivated soybean species from the Yellow River Delta, China. ScientificWorldJournal 2014:651745

Wu H, Lv H, Li L, Liu J, Mu S, Li X, Gao J (2015) Genome-wide analysis of the AP2/ERF transcription factors family and the expression patterns of DREB genes in Moso Bamboo (*Phyllostachys edulis*). PLoS One 10(5):e0126657

Wu J, Zhao Q, Wu G, Yuan H, Ma Y, Lin H, Pan L, Li S, Sun D (2019) Comprehensive analysis of differentially expressed unigenes under nacl stress in flax (*Linum usitatissimum* L.) using RNA-Seq. Int J Mol Sci 20(2):369

Xie Z, Lin W, Yu G, Cheng Q, Xu B, Huang B (2019) Improved cold tolerance in switchgrass by a novel CCCH-type zinc finger transcription factor gene, PvC3H72, associated with ICE1–CBF–COR regulon and ABA-responsive genes. Biotechnol Biofuels 12(1):224

Xie N, Li B, Yu J, Shi R, Zeng Q, Jiang Y, Zhao D (2022) Transcriptomic and proteomic analyses uncover the drought adaption landscape of Phoebe zhennan. BMC Plant Biol 22(1):1–16

Yan J, Wang B, Jiang Y, Cheng L, Wu T (2014) GmFNSII-controlled soybean flavone metabolism responds to abiotic stresses and regulates plant salt tolerance. Plant Cell Physiol 55(1):74–86

Yang Y, Guo Y (2018) Elucidating the molecular mechanisms mediating plant salt-stress responses. New Phytol 217(2):523–539

Yang Y, Al-Baidhani HHJ, Harris J, Riboni M, Li Y, Mazonka I, Bazanova N, Chirkova L, Sarfraz Hussain S, Hrmova M (2020) DREB/CBF expression in wheat and barley using the stress-inducible promoters of HD-Zip I genes: impact on plant development, stress tolerance and yield. Plant Biotechnol J 18(3):829–844

Yasar F, Uzal O, Yasar TO (2013) Investigation of the relationship between the tolerance to drought stress levels and antioxidant enzyme activities in green bean (*Phaseolus Vulgaris* L.) genotypes. Afr J Agric Res 8:5759–5763

Yeh C-H, Kaplinsky NJ, Hu C, Charng Y (2012) Some like it hot, some like it warm: phenotyping to explore thermotolerance diversity. Plant Sci 195:10–23

You J, Zhang Y, Liu A, Li D, Wang X, Dossa K, Zhou R, Yu J, Zhang Y, Wang L (2019) Transcriptomic and metabolomic profiling of drought-tolerant and susceptible sesame genotypes in response to drought stress. BMC Plant Biol 19(1):1–16

Zaman MSU, Malik AI, Erskine W, Kaur P (2019) Changes in gene expression during germination reveal pea genotypes with either "quiescence" or "escape" mechanisms of waterlogging tolerance. Plant Cell Environ 42(1):245–258

Zenda T, Liu S, Wang X, Liu G, Jin H, Dong A, Yang Y, Duan H (2019) Key maize drought-responsive genes and pathways revealed by comparative transcriptome and physiological analyses of contrasting inbred lines. Int J Mol Sci 20(6):1268

Zhang J-L, Shi H (2013) Physiological and molecular mechanisms of plant salt tolerance. Photosynth Res 115(1):1–22

Zhang G-H, Su Q, An L-J, Wu S (2008) Characterization and expression of a vacuolar Na^+/H^+ antiporter gene from the monocot halophyte Aeluropus littoralis. Plant Physiol Biochem 46(2): 117–126

Zhang X, Cai J, Wollenweber B, Liu F, Dai T, Cao W, Jiang D (2013) Multiple heat and drought events affect grain yield and accumulations of high molecular weight glutenin subunits and glutenin macropolymers in wheat. J Cereal Sci 57(1):134–140

Zhang D, Kumar M, Xu L, Wan Q, Huang Y, Xu Z-L, He X-L, Ma J-B, Pandey GK, Shao H-B (2017) Genome-wide identification of major intrinsic proteins in Glycine soja and characterization of GmTIP2; 1 function under salt and water stress. Sci Rep 7(1):1–12

Zhang H, Zhang Y, Deng C, Deng S, Li N, Zhao C, Zhao R, Liang S, Chen S (2018a) The Arabidopsis Ca^{2+}-dependent protein kinase CPK12 is involved in plant response to salt stress. Int J Mol Sci 19(12):4062

Zhang X, Li M, Yang H, Li X, Cui Z (2018b) Physiological responses of Suaeda glauca and Arabidopsis thaliana in phytoremediation of heavy metals. J Environ Manag 223:132–139

Zhang X, Liu L, Chen B, Qin Z, Xiao Y, Zhang Y, Yao R, Liu H, Yang H (2019) Progress in understanding the physiological and molecular responses of Populus to salt stress. Int J Mol Sci 20(6):1312

Zhang L, Jiang X, Liu Q, Ahammed GJ, Lin R, Wang L, Shao S, Yu J, Zhou Y (2020) The HY5 and MYB15 transcription factors positively regulate cold tolerance in tomato via the CBF pathway. Plant Cell Environ 43(11):2712–2726

Zhao Y, Gao C, Shi F, Yun L, Jia Y, Wen J (2018) Transcriptomic and proteomic analyses of drought responsive genes and proteins in Agropyron mongolicum Keng. Curr Plant Biol 14:19–29

Zheng J-L, Zhao L-Y, Wu C-W, Shen B, Zhu A-Y (2015) Exogenous proline reduces NaCl-induced damage by mediating ionic and osmotic adjustment and enhancing antioxidant defense in *Eurya emarginata*. Acta Physiol Plant 37(9):1–10

Zhou Y, Gao F, Liu R, Feng J, Li H (2012) De novo sequencing and analysis of root transcriptome using 454 pyrosequencing to discover putative genes associated with drought tolerance in Ammopiptanthus mongolicus. BMC Genomics 13(1):1–13

Zhu X, Zhang N, Liu X, Li S, Yang J, Hong X, Wang F, Si H (2021) Mitogen-activated protein kinase 11 (MAPK11) maintains growth and photosynthesis of potato plant under drought condition. Plant Cell Rep 40(3):491–506

Zhu-Salzman K, Salzman RA, Ahn J-E, Koiwa H (2004) Transcriptional regulation of sorghum defense determinants against a phloem-feeding aphid. Plant Physiol 134(1):420–431

Zou J, Liu A, Chen X, Zhou X, Gao G, Wang W, Zhang X (2009) Expression analysis of nine rice heat shock protein genes under abiotic stresses and ABA treatment. J Plant Physiol 166(8): 851–861

Zou J, Liu C, Liu A, Zou D, Chen X (2012) Overexpression of OsHsp17.0 and OsHsp23.7 enhances drought and salt tolerance in rice. J Plant Physiol 169(6):628–635

Zoz T, Steiner F, Guimaraes VF, Castagnara DD, Meinerz CC, Fey R (2013) Peroxidase activity as an indicator of water deficit tolerance in soybean cultivars. Biosci J 29:1664–1677

Źróbek-Sokolnik A (2012) Temperature stress and responses of plants. In: Environmental adaptations and stress tolerance of plants in the era of climate change. Springer, New York, pp 113–134

Harnessing Genetic Variation in Physiological and Molecular Traits to Improve Heat Tolerance in Food Legumes

2

Poonam Devi, Shikha Chaudhary, Anjali Bhardwaj, Manu Priya, Uday Jha, Aditya Pratap, Shiv Kumar, HanumanthaRao Bindumadahva, Inderjit Singh, Sarvjeet Singh, P. V. Vara Prasad, Kadambot H. M. Siddique, and Harsh Nayyar

Abstract

Plant genetic variations provide opportunity to develop new and improved cultivars with desired characteristics, hence gaining major attention from the scientists and breeders all over the world. Harnessing genetic variability is the key factor in the adaptation of plants to ever-rising temperature. Nowadays, such characteristic traits among the population can be used to develop various heat-resilient crop varieties and have a profound effect on restoring the balance between climate change and agriculture. Genetic variations in physiological and

P. Devi · S. Chaudhary · A. Bhardwaj · M. Priya · H. Nayyar (✉)
Department of Botany, Punjab University, Chandigarh, India

U. Jha · A. Pratap
Crop Improvement Division, Indian Institute of Pulses Research, Kanpur, India

S. Kumar
Biodiversity and Crop Improvement Program, International Center for Agriculture Research in the Dry Areas (ICARDA), Rabat, Morocco

H. Bindumadahva
Dr. Marri Channa Reddy Foundation (MCRF), Hyderabad, Telangana, India

I. Singh · S. Singh
Department of Plant Breeding and Genetics, Punjab Agricultural University, Ludhiana, Punjab, India

P. V. Vara Prasad
Department of Agronomy, Kansas State University, Manhattam, KS, USA

K. H. M. Siddique
Department of Agronomy, Kansas State University, Manhattam, KS, USA

The UWA Institute of Agriculture, The University of Western Australia, Perth, WA, Australia

P. Muthu Arjuna Samy et al. (eds.), *Legumes: Physiology and Molecular Biology of Abiotic Stress Tolerance*, https://doi.org/10.1007/978-981-19-5817-5_2

molecular traits proved to be the major components for breeding programs to augment the gene pool. With genetic variations, it is possible to identify the phenotypic variations governed either by a single gene or by many genes that will be helpful for mapping associated quantitative trait loci. Genetic variations can also be traced by examining various physiological traits of a crop plant like growth traits (biomass, plant height, and root growth), leaf traits (stomatal conductance, chlorophyll content, chlorophyll fluorescence, photosynthetic rate, membrane stability, sucrose content, and canopy temperature depression), and floral traits (mainly associated with male gametophyte). Yield traits can also display enormous variation, making it highly useful/reliable for screening purposes. Further, genetic variation at the biochemical level can be assessed by measuring the expression of enzymes (related to oxidative stress and antioxidants) and metabolites (both primary and secondary). Evaluating how genetic variation influences phenotype is the ultimate objective of genetics, and using omics approaches can improve the understanding of heat tolerance-governing mechanisms. Further, collecting molecular data at different levels of plant growth and development will help to accelerate our understanding of the mechanisms linking genotype to phenotype.

Keywords

Genetic variations · Physiological and molecular traits · Metabolites · Phenotype · Heat tolerance · Omics approaches

2.1 Introduction

The Earth's rising average surface temperature, possibly due to global warming, poses a significant threat to the production potential of plants (Bita and Gerats 2013). Temperature is one of the main factors affecting plant phenology and plays a significant role in plant species distribution around the globe (Li et al. 2018). All plant species have a threshold temperature for growth to reach their yield potential; temperatures beyond the threshold are stressful at all plant growth stages, affecting overall performance (Wahid et al. 2007). Heat stress is supraoptimal temperatures that cause irreversible damage to plants (Hasanuzzaman et al. 2013). The impact of heat stress depends on species, specific growth stage, and intensity and duration of the stress (Farooq et al. 2017; Li et al. 2018).

Heat stress affects all stages of plant growth, viz., (1) seed germination (decreases seed germination rate and seedling root and shoot lengths), (2) vegetative growth (decreases plant height, biomass production, and root growth), (3) leaf structure and function (damages membrane structure, increases canopy temperature, decreases stomatal conductance, chlorophyll fluorescence, photosynthetic rate, and sucrose metabolism), (4) reproductive traits (mainly male gametophyte), (5) cellular homeostasis (elevated reactive oxygen species production), and (6) yield (reduced seed number, seed weight, and seed-filling rate). The reproductive stage is much more

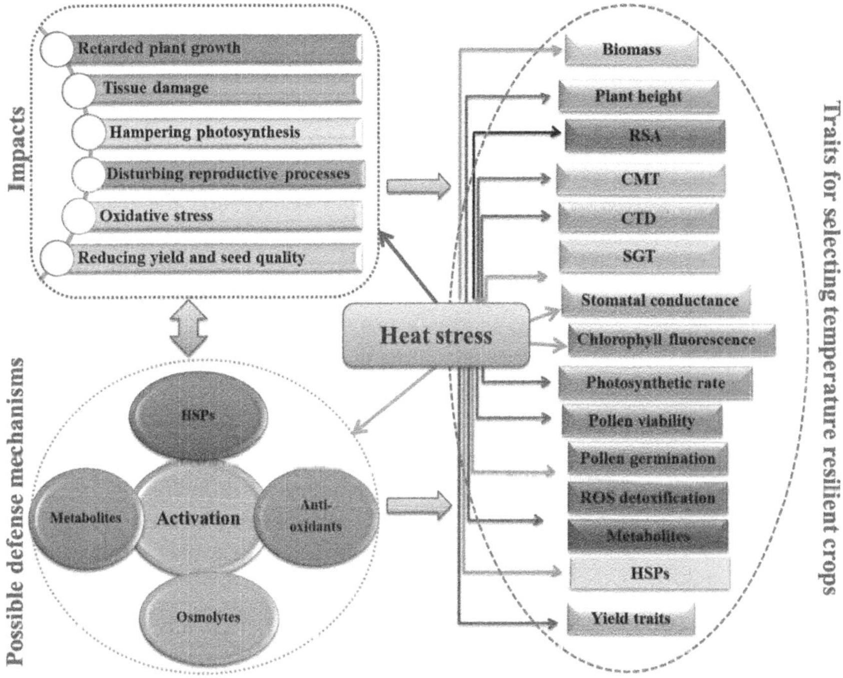

Fig. 2.1 Impacts, defense mechanisms against heat stress, and possible screening traits used for selecting temperature-resilient crops. High temperature adversely affects plant growth, causes tissue damage, and impairs vital processes such as photosynthesis, respiration, and reproduction. The injuries caused by heat stress lead to oxidative stress due to the production of reactive oxygen species, reducing crop yields. Plants implement various mechanisms to cope with heat stress, including antioxidant and metabolite production, accumulation and adjustment of compatible solutes, and most importantly chaperone (heat-stress proteins, HSPs) signaling and transcriptional activation. These mechanisms, regulated at the molecular level, enable plants to thrive under heat stress. Various growth traits [e.g., plant biomass, plant height, root system architecture (RSA)], leaf traits [e.g., cell membrane thermostability (CMT), canopy temperature depression (CTD), stay-green trait (SGT), stomatal conductance, chlorophyll fluorescence, photosynthetic rate], reproductive traits (e.g., pollen viability, pollen germination), biochemical traits [e.g., reactive oxygen species (ROS) detoxification, various metabolites, HSP levels], and yield traits have been explored as heat-tolerance indicators for screening and breeding for heat tolerance

sensitive to heat stress than the vegetative stages, leading to lower seed weights and thus yield (Farooq et al. 2017). Plants are sessile organisms that can develop various adaptive mechanisms to endure heat waves, such as antioxidant production, synthesis of low-molecular-weight secondary metabolites, increasing heat-shock proteins (HSPs), and upregulating various transcription factors (Fig. 2.1). These endurance mechanisms vary between crop species, growth stage, and growth traits (Bita and Gerats 2013; Prasad et al. 2017).

2.2 Heat Stress and Legumes

Food legumes are an indispensable part of the human diet in developing countries. The major food legumes consumed worldwide are pea (*Pisum sativum* L.), chickpea (*Cicer arietinum* L.), common bean (*Phaseolus vulgaris* L.), lentil (*Lens culinaris* Medik.), mung bean/green gram (*Vigna radiata* L.), urdbean/black gram (*Vigna mungo* L.), and cowpea [*Vigna unguiculata* (L.) Walp.], and the major oilseed legumes include peanut (*Arachis hypogaea* L.) and soybean (*Glycine max* L.) (Maphosa and Jideani 2017). Due to their high nutritional value, legumes are ranked second after cereals. They are rich in protein (20–45%), carbohydrates (60%), dietary fiber (5–37%), and mineral matter (calcium, iron, potassium, phosphorus, copper, and zinc) with no cholesterol and low fat (Iqbal et al. 2006). Environmental factors, mainly rising temperatures, are major constraints on the growth and yield of food legumes. Heat stress adversely affects physiological and reproductive stages, resulting in poor seed yield and quality (Sita et al. 2017). Table 2.1 shows the threshold temperatures for commonly grown legumes in different regions of the world. Various studies have reported the impact of heat stress on seed germination, including poor emergence, germination percentage and radicle and plumule growth, and abnormal seedling vigor. For instance, chickpea germinated well at temperatures from 15 to 35 °C but poorly at temperatures above 40 °C (Kumari et al. 2018). Temperature beyond the threshold range showed lethal effects on the chickpea seedlings (Kumari et al. 2018). Similarly, a 50 °C heat treatment for 30 min significantly reduced seed germination, seed vigor, and seedling growth of dry black gram (Piramila et al. 2012).

Heat stress affects early vegetative growth, decreasing biomass accumulation and root growth and stunting plant height (Huang and Xu 2008; Kaushal et al. 2013).

Table 2.1 Threshold temperatures of few selective food legumes

Food legumes	Threshold temperature (°C)	References
Pulses		
Chickpea (*Cicer arietinum* L.)	16–27	Devasirvatham et al. (2013)
Common bean (*Phaseolus vulgaris* L.)	27–30	Rainey and Griffiths (2005)
Cowpea (*Vigna unguiculata* L.)	18–28	Craufurd et al. (1998)
Faba bean (*Vicia faba* L.)	22–23	Lavania et al. (2015)
Lentil (*Lens culinaris* Medik.)	18–30	Sita et al. (2017)
Mung bean (*Vigna radiata* L.)	28–30	Kaur et al. (2015)
Pea (*Pisum sativum* L.)	18–24	Jiang et al. (2015)
Urdbean/black gram (*Vigna mungo* L.)	30–35	Anitha et al. (2016)
Oilseeds		
Peanut (*Arachis hypogaea* L.)	22–28	Prasad et al. (1999)
Soybean (*Glycine max* L.)	20–26	Nahar et al. (2016)

Various studies have reported that heat stress inhibits physiological processes and cellular response activation, including decreased cellular membrane thermostability (Xu et al. 2006). Heat stress dramatically affects the photosynthetic process by disrupting chloroplast structures (thylakoid leakiness and grana stacking) and damaging the D1 protein of PSII due to the accumulation of reactive oxygen species (ROS) (Allakhverdiev et al. 2008; Sharkey 2005). Deactivation of the RuBisCo enzyme even at moderate–high temperatures further hampers photosynthesis (Allakhverdiev et al. 2008).

High temperatures significantly affect the reproductive phase, as reported in various food legumes, including mung bean (Kaur et al. 2015), chickpea (Kaushal et al. 2013), lentil (Bhandari et al. 2016; Sita et al. 2017), and peanut (Prasad et al. 1999). The main reproductive events affected by heat stress are male gametophyte development (meiosis in microspore mother cell, tapetum development in viable pollen, reduced pollen germination, reduced pollen tube growth), female gametophyte development (meiosis in the megaspore mother cell, tapetum development in viable eggs, altered stigmatic and style positions, reduced stigma receptivity), and fertilization (double fertilization and triple fusion) (Farooq et al. 2017; Prasad et al. 2017). Heat stress accelerates seed filling, inhibiting the accumulation of reserves in developing seeds, resulting in poor-quality seeds (Calderini et al. 2006) and reduced seed yields in food legumes such as chickpea (Awasthi et al. 2014) and lentil (Sehgal et al. 2018).

Understanding the impact of heat stress and the related mechanisms will help improve crop genotypes under heat stress. Therefore, identifying traits through extensive screening experiments related to heat tolerance is important for selecting better performing heat-tolerant genotypes of food legumes. This chapter identifies various traits in genotypes of various food legumes with different heat sensitivity/tolerance levels (Fig. 2.1) and offers insight into the overall traits and mechanisms used to select heat-tolerant genotypes.

2.3 Growth-Based Studies

High temperature adversely affects the growth and development of various legumes, restricting the growth cycle from emergence to seed set (Sehgal et al. 2018). Seed germination and seedling establishment, including root and shoot lengths and seedling vigor, are highly sensitive to high temperature. For instance, mung bean seedlings exposed to 45/35 °C had reduced growth (Kumar et al. 2011), and chickpea seedlings exposed to 40 °C for 96 h died (Kumari et al. 2018). Heat stress accelerates early vegetative growth, decreasing leaf number and dry matter accumulation (Tahir et al. 2008). Even moderate heat stress leads to rapid growth and development, resulting in shorter crop duration and less carbon assimilation over the plant's life cycle (Driedonks et al. 2016; Hatfield and Prueger 2015). Many studies have shown that disturbances in fundamental physiological processes, such as photosynthesis, respiration, water status, membrane stability, primary and secondary metabolites, and ROS generation, due to metabolic disparity resulted in fewer and malformed plant parts (Wahid et al. 2007). Reduced vegetative growth also results from various anatomical and structural changes in cellular organelles, leading to necrosis, chlorosis, sunburn, senescence, and abscission of leaves, twigs, branches, and stems. Further, heat stress negatively affects plant architecture, including branching pattern, leaf area, internode elongation, and leaf/branch angles (Sabagh et al. 2020). The above studies indicate that several processes and molecules are involved in heat stress, reducing plant growth. Many studies have reported reduced vegetative growth in legumes, suggesting an interaction between potential yield and vegetative growth traits, for instance, in chickpea (Awasthi et al. 2014), common bean (Soltani et al. 2019; Yoldas and Esiyok 2009), faba bean (Siddiqui et al. 2015), lentil (Sita et al. 2017), mung bean (Kumar et al. 2011; Sharma et al. 2016), and soybean (Sabagh et al. 2020). Thus, the impact of heat stress on plant growth can be evaluated by assessing traits such as plant height, biomass, and root system architecture. Studies on contrasting genotypes revealed genetic variation in these traits in response to heat stress, which will help identify the mechanisms associated with heat tolerance in legumes.

2.3.1 Biomass

Biomass is an indicator of dry matter accumulation during plant growth, which is adversely affected by heat stress in various legumes (Sabagh et al. 2020). Several studies have revealed genetic variations in biomass accumulation in legumes under high temperatures. Thus, chickpea under heat stress (>32/20 °C) in a greenhouse had 22–30% less biomass than control plants (Kaushal et al. 2013). High temperature decreased biomass more in heat-sensitive chickpea genotypes (ICC5912, ICC10685) than heat-tolerant genotypes (ICC15614 and ICCV92944) (Kaushal et al. 2013). In another greenhouse study, heat stress (38/35 °C) decreased alfalfa (*Medicago sativa*) biomass, more so in heat-sensitive Wl712 than heat-tolerant Bara310SC, compared to the control (25 °C) (Wassie et al. 2019). In the field, heat stress (>32/20 °C) significantly decreased lentil biomass (Sita et al. 2017). Genotypes IG3263, IG2507, IG3297, IG3312, IGG3327, IG3330, IG3546, IG3745, IG4258, and FLIP2009 retained the most biomass and were considered heat tolerant, while genotypes IG2519, IG2802, IG2506, IG2849, IG2821, IG2878, IG3326, IG3290, IG3973, IG3964, IG4242, DPL15, DP315, IG4221, and IG3568 were considered heat sensitive. High temperature (>40/28 °C) in the field significantly reduced (76%) plant biomass in 45 mung bean accessions from the World Vegetable Center, compared to control conditions (34/16 °C)—genotypes EC693357, EC693358, EC693369, Harsha, and ML 1299 retained the most biomass under heat stress and were considered heat tolerant, while genotypes EC693363, EC693361, KPS1, EC693370, and IPM02-3 retained the least biomass and were considered heat sensitive (Sharma et al. 2016).

2.3.2 Plant Height

Heat stress suppresses the overall vegetative growth of plants by affecting various growth-related mechanisms involving hormones and enzymes (Siddiqui et al. 2015). Plant height at different growth stages is a vital indicator of plant growth under stress situations and has been correlated with heat stress sensitivity (Prasad et al. 2008). A field study was undertaken to screen 12 Kabuli chickpea lines through delayed sowing for heat exposure (39.4 °C) (Mishra and Babbar 2014). Four chickpea lines—KAK2, JGK2, ICCV07311, and ICCV06301—were selected as heat tolerant based on plant height and other yield traits, with positive correlations between phenological traits (days to flowering, days to 50% flowering, maturity days, number of secondary branches, plant height) and yield traits (Mishra and Babbar 2014). Soybean genotypes (64) exposed to heat stress (40/32 °C; seedling stage for 20 days) varied in plant height—IREANE, CZ4898RY, CZ5242LL, CZ5375, ELLIS, 5N393R2, CZ4181, and 45A46 were categorized as heat tolerant, and 5115LL, S47-K5, S45-W9, 483C, 38R10, R01-416F, JTN-5110, S48RS53, and DG4825RR2/STS as heat sensitive, with the remainder categorized as moderately heat tolerant or moderately heat sensitive (Alsajri et al. 2019). Similarly, high temperature imposed on four common bean genotypes (Gima, Volare, Amboto, Nassan) by delaying normal sowing (late-sown) significantly reduced yields, relative to normal-sown plants, due to a shorter vegetative cycle, and genotypes Gima and Volare maintained taller plants than Amboto and Nassan (Yoldas and Esiyok 2009). In a greenhouse study, ten faba bean genotypes raised under high temperatures (HT1: 31 °C and HT2: 37 °C) had markedly reduced plant height compared to the control plants. Genotype C5 produced the tallest plants (heat tolerant), while Espan produced the shortest plants (heat sensitive) (Siddiqui et al. 2015).

2.3.3 Root System Architecture

Root system architecture (RSA) is the structure and spatial and temporal configuration of plant root systems (de Dorlodot et al. 2007). On a macroscale, RSA can determine the organization of the primary and secondary roots (Smith and De Smet 2012). On a microscale, RSA can determine root microstructures, such as fine root hairs and root tips and their interactions with soil and soil microorganisms responsible for water and mineral uptake (Wu et al. 2018). The spatial and temporal distribution of roots determines the crop's ability to exploit heterogeneously distributed soil resources (Brussaard et al. 2007). Heat stress directly affects plant roots by restricting carbohydrate transport from shoots to roots (Huang and Xu 2008). A comprehensive understanding of RSA helps us understand the effect of environmental conditions and management practices on crops, decreasing the deviation between potential and actual average yields (Garnett et al. 2009; Judd et al. 2015; Ryan et al. 2016). RSA plays an important role in plant–soil–microbe interactions and resolving the cross talk with beneficial soil microbes in the rhizosphere (Ryan et al. 2016).

Root architecture adapts to fluctuating environments. Therefore, we can improve crop performance by increasing root traits, such as root development allocation, and morphological, anatomical, or developmental plasticity (Sultan 2000). Thus, understanding the genetic and molecular mechanisms determining root phenotypic plasticity is necessary for effective selection and crop breeding efforts. Direct relationships between individual root architectural plasticity and yield have been reported across changing environments in various species (Niones et al. 2013; Sadras 2009). Root branching is important for improving soil anchorage and root surface area, enabling plants to reach more distant water reserves. In plants, high- and low-temperature stress generally reduces primary root length, lateral root density (number of lateral roots per unit primary root length), and emergence angle of lateral roots from the primary root, but does not affect the average lateral root length (McMichael and Quisenberry 1993; Nagel et al. 2009). Heat stress affects nutrient uptake due to a decline in root biomass and root hair surface area. In mung bean, high temperatures of 40/30 °C and 45/35 °C inhibited root growth by 13% and 23%, respectively (Kumar et al. 2011).

Root growth has lower optimal growth temperatures and is more sensitive to high temperatures than shoot growth (Huang and Gao 2000; Xu and Huang 2000). Some plant roots synthesize heat-shock proteins (HSPs) by ameliorating their working efficiency (Nieto-Sotelo et al. 2002). Root phenotyping of 577 common bean genotypes in variable heat environments revealed significant relationships between seed yield and seedling basal root number, seedling adventitious root abundance, and seedling taproot length (Strock et al. 2019). The Mesoamerican genotypes yielded higher than the Andean genotypes under heat stress (Strock et al. 2019). In another study, five chickpea genotypes were assessed for thermotolerance at 30, 35, and 40 °C using root length and root branching as criteria, which identified CSJD 884 and RSG 895 as heat tolerant and C 235 as heat sensitive (Kumari et al. 2018). The 40 °C treatment for 96 h negatively affected root branching in chickpea (Kumari et al. 2018).

Similarly, screening 48 lentil genotypes in a growth chamber at 34 °C using root length as one of the selection criteria identified Ranjan, Moitree, 14-4-1, IC 201710, and IC 208329 as heat tolerant (Choudhury et al. 2012). In another lentil study, heat-tolerant genotypes (IG2507, IG3263, IG3745, IG4258, and FLIP2009) had 1.8–22-fold more root nodulation than heat-sensitive genotypes (IG2821, IG2849, IG4242, IG3973, IG3964) under heat stress (>32/20 °C) (Sita et al. 2017).

2.4 Yield-Based Traits

Heat stress negatively impacts reproductive efficiencies and seed development stages, reducing crop yield and quality (Sehgal et al. 2018). Various studies have shown that the relative performance of plants in terms of yield under heat stress is useful for selecting genotypes for crop improvement programs (detailed below). Heat stress severely affects seed development and seed filling in many crop species, resulting in abnormal and shriveled seeds (Egli 1998). The direct effect of heat stress on the sink potential of maturing seeds (Commuri and Jones 1999) disrupts cell division in the endosperm, decreases the number of starch granules, and reduces starch accumulation. Many screening studies under heat stress have included yield traits, such as seed number, seed weight, seed-filling rate, and duration (Farooq et al. 2017).

2.4.1 Seed Number

Heat stress disrupts pollination and fertilization events that directly curtail seed number. For instance, high temperature (45/32 °C) reduced seed number in mung bean genotypes relative to the control (34/16 °C), more so in heat-sensitive genotypes (EC693363, EC693361, KPS1, EC693370, and IPM02-3) than heat-tolerant genotypes (EC693357, EC693358, EC693369, Harsha, and ML 1299) (Sharma et al. 2016). Similarly, in a greenhouse study, the 33/30 °C treatment reduced pod number and seed number per pod the most in 24 common bean genotypes exposed to varying temperatures (24/21 °C, 27/24 °C, 30/27 °C, 33/30 °C), more so in heat-sensitive genotypes (−66%; A55, Labrador, Majestic, IJR) than heat-tolerant genotypes (−31%; Brio, Carson, G122, HB1880, HT38, Venture) (Rainey and Griffiths 2005). In another study, heat stress (36/27 °C) reduced seed number per pod in all but two cowpea lines (heat tolerant B89-600 and TN88-63) evaluated for heat tolerance in a greenhouse (Ehlers and Hall 1998). In another greenhouse study, high temperature (38 °C) during the reproductive stage of 211 pea genotypes revealed HUDP-25, IPF-400, HFP-4, and DDR-56 as heat tolerant and VL-40, KPMR-615, DDR-61, and KPMR-557 as heat susceptible based on yield parameters; for example, heat-tolerant genotypes had more seeds per plant (35–197) than heat-sensitive genotypes (1–58) (Mohapatra et al. 2020).

2.4.2 Seed Weight

Seed weight is one of the major traits governing crop yield and is thus used as a screening trait in many studies to select heat-tolerant varieties. For example, chickpea exposed to different temperatures (35/25 °C, 40/30 °C, and 45/35 °C) in a growth chamber decreased seed weight at 40/30 °C by 37–45% in sensitive genotypes (ICC14183, ICC5912) relative to tolerant genotypes (ICCV07110, ICCV92944). However, higher temperature (45/35 °C) had a more severe effect, with fewer seeds in tolerant genotypes and no pod set in sensitive genotypes (Kumar et al. 2013). Similar findings were recorded in mung bean when high temperatures (45/32 °C) coincided with reproductive growth; seed weights declined by 48.3% in the sensitive genotype (SML668) and 35.1% in the tolerant genotype (SML832), relative to the control (Kaur et al. 2015). Likewise, seed weight of lentil plants exposed to high temperature (>32/20 °C) in the field declined, relative to control plants (Bhandari et al. 2016), more so in the heat-sensitive genotypes (−50%; LL699 and LL1122) than the heat-tolerant genotype (−33%; LL931).

In common bean, a high temperature of 33/30 °C was adequate for selecting heat-tolerant (Carson, G122, Brio, HB1880, HT38, Venture) and heat-sensitive genotypes (Labrador, A55, Majestic, IJR), based on seed weight trait in the field; seed weights decreased by 88% in heat-sensitive genotypes compared with 35% in heat-tolerant genotypes (Rainey and Griffiths 2005). Different location-based yield trials—Coachella (USA; 41/25 °C) and Riverside (USA; 36/17 °C)—were used to screen three groups of cowpea genotypes differing in heat sensitivity (Ismail and Hall 1999). Yield parameters, mainly seed weight, and seeds/pod, decreased significantly as the temperature increased. Tolerant genotypes (H36, H8-9, DLS99) retained more seed weight (193 mg/seed) at higher temperature (41/25 °C) than heat-sensitive genotypes (168 mg/seed; CB5, CB3, DLS127). Mohapatra et al. (2020) reported that heat stress reduced 25-seed weight in pea in heat-susceptible genotypes (VL-40, KPMR-615, DDR-61, KPMR-557) to a mean value of 4.13 g, while heat-tolerant genotypes (HUDP-25, IPF-400, HFP-4, DDR-56) had higher seed weights (4.60 g).

Heat stress accelerates the seed-filling rate but decreases the seed-filling duration. In cowpea, increasing the temperature from 15.5 to 26.6 °C increased the seed-filling duration by 14–21 days (Nielsen and Hall 1985). During seed development, heat stress (>32/20 °C) increased the seed-filling rate in six chickpea genotypes relative to the optimum temperature, and shortened the seed-filling duration, more so in heat-sensitive (ICC4567) than heat-tolerant (ICC1356, ICC15614) genotypes (Awasthi et al. 2014). Thus, reduced seed weight due to heat stress could be related to a decline in seed-filling processes (Sehgal et al. 2017).

2.5 Pollen Grain Traits

Pollen grains are sensitive to extreme temperatures from early pollen development to fertilization, including meiosis I and meiosis II of the microspore mother cell, early dissolution of the tapetum layer, anther dehiscence, pollen shedding, pollen viability, pollen germination, pollen tube growth, and fertilization (Barnabas et al. 2008; Hedhly 2011; Kumar et al. 2013). Observations on heat stress-induced arrest of male gametophyte development revealed the importance of starch accumulation during pollen development because it gives rise to carbohydrates at maturity (Raja et al. 2019). Heat stress prevents starch accumulation during pollen development, which possibly contributes to reduced pollen viability (Pressman et al. 2002). High temperature during anthesis leads to yield losses due to poor pollen traits such as pollen viability, pollen production, and pollen tube length in crop plants, including chickpea (Devasirvatham et al. 2012; Kaushal et al. 2013), common bean (Suzuki et al. 2001), mung bean (Kaur et al. 2015), lentil (Kumar et al. 2016; Sita et al. 2017), and soybean (Salem et al. 2007). Heat-tolerant and heat-sensitive common bean genotypes were identified based on pollen stainability—exposure to high temperature (>28 °C) for 8–11 days before anthesis decreased pollen stainability and increased flower abortion, reducing pod yield (Suzuki et al. 2001). Heat-sensitive genotypes (Kentucky Wonder, Oregon, and Okinawa Local) had <20% pollen stainability, while the heat-tolerant genotype (Haibushi) had 60% pollen stainability under heat-stress conditions. Heat stress (43/30 °C and 45/32 °C) in mung bean adversely affected reproductive components, reducing pollen viability, pollen germination, and pollen tube length (Kaur et al. 2015), compared to the controls (>40/25 °C). Moreover, high temperature during microsporogenesis reduced pollen number and produced shriveled pollen grains, more so in the heat-sensitive genotype than the heat-tolerant genotype. Another field study exposed 45 mung bean genotypes to high temperature (42 °C) during the flowering stage (Sharma et al. 2016).

An in vitro pollen study revealed that heat-tolerant mung bean genotypes (C693357, EC693358, EC693369, Harsha, ML1299) had better pollen viability and pollen germination than sensitive genotypes (KPS1, EC693361, EC693363, EC693370, IPM02-3) (Sharma et al. 2016). Other pollen traits (pollen germination and pollen load) were used to screen chickpea, identifying heat-tolerant (ICC15614, ICCV92944) and -sensitive (ICC10685, ICC5912) genotypes (Kaushal et al. 2013). Another study identified tolerant and sensitive chickpea genotypes using pollen traits (Devasirvatham et al. 2013) under heat stress (≥35 °C); pollen grains were more sensitive to high temperature than stigmas, with genotype ICC1205 identified as heat tolerant and ICC4567 as heat sensitive. Kumar et al. (2016) screened 334 lentil accessions for heat tolerance under field conditions (>35/25 °C) and selected heat-tolerant genotypes (FLIP2009-55L, IG2507, and IG4258) based on pollen traits. Sita et al. (2017) revealed that high temperature (>32/20 °C) in the field reduced pollen viability to a greater extent than control (<32/20 °C), with higher pollen germination in heat-tolerant genotypes (48–50%; IG2507, IG3263, IG3745, IG4258, and FLIP2009) than heat-sensitive genotypes (28–33%).

Sixteen pea accessions were screened for heat tolerance by exposing plants to 45 °C for 2 h; the Ran1 line was selected as heat tolerant and R–Af-1, C–Af-2, and Cs–Af–3 as heat sensitive based on pollen traits (pollen viability, pollen germination, pollen tube growth) (Petkova et al. 2009). In another study, two pea cultivars were tested for their differential sensitivity to high temperature (27/18 °C, 30/18 °C, 33/18 °C, and 36/18 °C) based on in vitro pollen germination, pollen tube length, pollen surface morphology, and pollen wall structure; as a result, CDC Sage was classified as tolerant and CDC Golden as sensitive genotype based on its higher pollen germination and stable lipid composition in pollen than the heat-sensitive genotype at 36 °C (Jiang et al. 2015).

Pollen-based traits were also used to screen 44 soybean genotypes for heat tolerance at 38/30 °C (Salem et al. 2007). The total stress response index based on reproductive traits such as pollen germination and pollen tube length was used to categorize the genotypes. Three of these genotypes, heat tolerant (DG 5630RR), heat intermediate (PI 471938), and heat sensitive (Stewart III), were selected for pollen grain morphology; the heat-sensitive genotype had deformed pollen with reduced aperture. Based on the studies mentioned above, pollen grain structure and function could be used as a screening tool for heat tolerance in soybean (Salem et al. 2007).

2.6 Leaf-Based Parameters

2.6.1 Stomatal Conductance

Stomatal conductance is a measure of stomatal opening or the rate of passage of CO_2 entering and water vapor releasing through leaf stomata. Stomatal conductance is affected by many environmental factors, including high temperature. Stomatal conductance increases with increasing temperature to increase photosynthesis, which can help plants endure short heat waves (Urban et al. 2017). Moreover, plants acclimatize to high temperatures by evaporating more water, keeping their canopies cool despite the presence of fewer stomata (Crawford et al. 2012). Therefore, regulating stomatal conductance under high temperatures is a useful trait for screening contrasting genotypes. Stomatal conductance can be recorded with a leaf porometer and expressed in mmol m^{-2} s^{-1} (Priya et al. 2018). Heat-tolerant chickpea genotypes (ICC15614, ICCV92944) had higher stomatal conductance (265–271 mmol H_2O m^{-2} s^{-1}) than heat-sensitive genotypes (ICC5912, ICC10685; 187–210 mmol H_2O m^{-2} s^{-1}) under high temperatures (>32/20 °C) imposed by late sowing (Kaushal et al. 2013). Similarly, for late-sown mung bean genotypes, the heat-tolerant genotype (SML 868) had higher stomatal conductance (99 mmol m^{-2} s^{-1}) than the heat-sensitive genotype (SML 668; 90 mmol m^{-2} s^{-1}) (Kaur et al. 2015). In another study, five common bean genotypes (SB761, SB776, SB781, Jaguar, TB1) were screened in the greenhouse at three temperature regimes (35/30 °C, 40/35 °C, 45/40 °C); stomatal conductance in all genotypes increased with increasing temperature until 40/35 °C but declined at 45/40 °C except in genotype TB1, which was identified as heat tolerant (Traub et al. 2018). Similarly, Sita et al. (2017) identified heat-tolerant (IG2507, IG3263, IG3745, IG4258, FLIP2009) and heat-sensitive (IG2821, IG2849, IG4242, IG3973, IG3964) lentil genotypes based on stomatal conductance—the heat-tolerant genotypes had higher stomatal conductance values (390–497 mmol m^{-2} s^{-1}) than heat-sensitive genotype (205–313 mmol m^{-2} s^{-1}) in a late-sown environment.

2.6.2 Stay-Green Trait

Heat stress negatively affects photosynthesis by decreasing leaf pigment content and damaging leaf ultrastructure in heat-sensitive genotypes. Chloroplasts play a vital role in photosynthesis as one of the most heat-sensitive organelles (Abdelmageed and Gruda 2009; Krause and Santarius 1975). Decreased total chlorophyll content and changes in the chlorophyll a/b ratio have been correlated with reduced photosynthesis during heat stress due to reduced "antenna (pigment unit)" size that reduces light harvesting (Blum 1986; Harding et al. 1990; Shanmugam et al. 2013). Chlorophyll retention (chlorophyll content) is an integrative trait and is considered a good criterion for screening heat-stress tolerance in legume crops. For example, high-temperature (38/28 °C) stress for 14 days at the flowering stage in a growth chamber caused anatomical and structural changes, including damaged plasma membrane, chloroplast membrane, and thylakoid membranes and reduced leaf photosynthetic rate, in the leaves of soybean genotype K 03-2897. Plant chlorophyll maintenance, also known as the stay-green (SGR) trait, is affected by high temperature. Understanding the physiological and molecular mechanisms of the stay-green trait is important for controlling photosynthetic ability (Abdelrahman et al. 2017). The SGR trait, or delayed leaf senescence (DLS), allows plants to retain leaves in an active photosynthetic state under high temperatures to maintain assimilation and increase crop yield (Gregersen et al. 2013; Kumari et al. 2013). Stay-green genotypes can carry out photosynthesis for longer than senescent types, often with yield benefits (Borrell et al. 2014). The development of contrasting F6 and F7 recombinant-inbred lines of cowpea for the DLS trait under heat stress revealed that the DLS trait increased plant survival and seed size under heat stress (Ismail et al. 2000). Of ten common bean genotypes, only BRS Expedito, FT-Taruma, and BAF071 had the stay-green trait, with higher initial chlorophyll *a* contents, less chlorophyll degradation, and higher grain yields under heat stress than the other genotypes (Schmit et al. 2019).

A field experiment screening 58 chickpea genotypes for high-temperature tolerance (25–40 °C) during the reproductive phase identified eight genotypes—Pusa 1103, Pusa 1003, KWR 108, BGM 408, BG 240, PG 95333, JG 14, and BG 1077—as heat tolerant, with higher chlorophyll contents than the heat-sensitive genotypes (ICC1882, PUSA 332, PUSA 112, RSG 803) (Kumar et al. 2017). Two heat-tolerant chickpea genotypes (ICC1356, ICC15614) maintained higher chlorophyll contents under heat stress (>32 °C/20 °C) in the field than two heat-sensitive genotypes (ICC4567, ICC5912) (Awasthi et al. 2017). In another study, chickpea genotypes were grown in the greenhouse to flowering (42 and 46 DAS) and then in a growth chamber under increasing temperatures (by 2 °C per day from 27/18 °C to 42/25 °C; day/night) for 8 days (anthesis), which revealed that genotype JG14 (heat tolerant) had higher total leaf chlorophyll content than genotype ICC16374 (heat sensitive) (Parankusam et al. 2017). Similarly, heat-tolerant chickpea genotypes Pusa-1103 and BGD-72 had significantly higher chlorophyll contents than heat-sensitive genotypes Pusa-256 and RSG-991 under high temperatures (25/35 °C) in wooden polyethylene chambers (Singh et al. 2018). Likewise, Kaushal et al. (2013) identified two heat-tolerant (ICC15614, ICCV92944) and two heat-sensitive (ICC10685, ICC5912) chickpea genotypes based on the chlorophyll content, after exposure to heat stress (>32/20 °C) in the field during reproductive development. A field study on lentils measured the stay-green trait as the loss of total chlorophyll (Chl) in leaves under high temperature (>32/20 °C) during the reproductive phase; heat-stressed plants had lower total chlorophyll concentrations than the control plants, and the heat-tolerant genotype (IG3263) retained more Chl than the heat-sensitive genotype (IG4242) (Sita et al. 2017). Similarly, lentil genotypes LL699 and LL931 (heat tolerant) retained more chlorophyll than genotype LL1122 (heat sensitive) in outdoor conditions (>32/23 °C), which was confirmed in a controlled environment with plants subjected to 33/15 °C or 35/20 °C during reproductive growth (Bhandari et al. 2016). Heat stress in the field (>30/20 °C) during reproductive growth and seed filling revealed two lentil heat-tolerant genotypes (1G 2507 and 1G 4258) with high leaf chlorophyll concentrations and two heat-sensitive genotypes (1G 3973 and 1G 3964) with lower chlorophyll concentrations (Sehgal et al. 2017). In another study, common bean genotypes exposed to 32/25 °C at the V4 developmental stage identified two genotypes (Sacramento and NY-105) with high chlorophyll contents, indicating their high thermotolerance, relative to the thermosensitive genotype Redhawk with low chlorophyll content (Soltani et al. 2019). Likewise, in a heat-sensitive mung bean genotype (SML668), chlorophyll content declined, relative to the heat-tolerant genotype (SML832), grown under heat stress (43/30 °C and 45/32 °C) in outdoor late-sown conditions, contributing to an increase in leaf temperature (Kaur et al. 2015). Mung bean genotypes EC693357, EC693358, EC693369, Harsha, and ML 1299 produced more chlorophyll content under heat stress than genotypes EC693363, EC693361, KPS1, EC693370, and IPM02-3 (Sharma et al. 2016). Screening of ten faba bean genotypes for heat-stress tolerance (37 °C) revealed that genotype C5 tolerated high temperature by retaining more chlorophyll, while genotype Espan had less chlorophyll and was relatively more sensitive to heat stress (Siddiqui et al. 2015). In a recent study, 4-week-old seedlings of 15 alfalfa cultivars were exposed to heat treatment (38/35 °C) for 7 days in a growth incubator; genotypes Gibraltar, WL354HQ, Golden Queen, Siriver, WL712, and Sanditi had significantly lower Chl contents (heat sensitive) than genotypes Bara310SC, WL363HQ, WL656HQ, and Magna995 (heat tolerant) (Wassie et al. 2019).

2.6.3 Chlorophyll Fluorescence

Chlorophyll (Chl) fluorescence (F_v/F_m ratio) is used as an indicator of functional changes in photosynthetic apparatus under abiotic or biotic stress (Yamada et al. 1996). The relationships between essential photosynthetic responses and chlorophyll fluorescence are pivotal as they provide information on the plant's photosynthetic ability and acclimation limit under stress conditions (Kalaji et al. 2018; Lichtenthaler 1987). Chlorophyll fluorescence is a fast, nondestructive, and effective common tool for determining heat-stress responses as it can reveal damage before visible stress symptoms appear (Baker 2008; Méthy et al. 1994; Wilson and Greaves 1990). Of the photosynthetic apparatus, photosystem II (PSII) is the most heat-labile cell structure (Vacha et al. 2007). Since damage to PSII is often the first response of plants subjected to thermal stress (Mathur et al. 2011), measuring chlorophyll *a* fluorescence is an effective and noninvasive technique for identifying damage to PSII efficiency (Baker 2008; Baker and Rosenqvist 2004). The ratio between variable fluorescence (F_v) and maximum fluorescence (F_m), or F_v/F_m, reflects the maximum quantum efficiency of PSII (Butler 1978). When plants are exposed to abiotic stress, including thermal stress, F_v/F_m often declines (Molina-Bravo et al. 2011; Sharma et al. 2012; Willits and Peet 2001). Screening methodologies have used chlorophyll fluorescence to detect and quantify damage in PSII and thylakoid membranes in several legume crops under heat stress, including chickpea, groundnut, pigeon pea, and soybean (Herzog and Chai-Arree 2012; Srinivasan et al. 1996). Recent study assessed the response of four chickpea genotypes to a natural temperature gradient during the reproductive stage in the field and a climate chamber using chlorophyll fluorescence. Field experiments were conducted over two winter seasons; two genotypes (Acc#RR-3, Acc#7) showed tolerance (F_v/F_m 0.83–0.85) and two (Acc#2, Acc#8) showed sensitivity (F_v/F_m 0.78–0.80) to heat stress. The field results were validated in the climate chamber experiment, where F_v/F_m declined more in the heat-sensitive (0.74–0.75 at 35/30 °C) than heat-tolerant (0.78–0.81 at 35/30 °C) genotypes when exposed to short-term heat treatments (30/25 °C and 35/30 °C) (Makonya et al. 2019). In another chickpea study, heat stress (>30 °C) in the field during the reproductive stage reduced F_v/F_m more in two heat-sensitive genotypes ICC10685 and ICC5912 (0.48, 0.41) than in two heat-tolerant genotypes ICC15614 and ICCV92944 (0.64, 0.60) (Awasthi et al. 2014; Kaushal et al. 2013). A similar study, where four contrasting chickpea genotypes—two heat tolerant (ICC1356, ICC15614) and two heat sensitive (ICC4567, ICC5912)—were analyzed in the field, revealed that the tolerant genotypes maintained higher chlorophyll fluorescence (F_v/F_m 0.60) on exposure to heat stress (>32/20 °C) than the sensitive genotypes (F_v/F_m 0.50) (Awasthi et al. 2017). In lentils, photosynthetic efficiency was measured as PSII function (F_v/F_m ratio) in the field by exposing plants to heat stress (>32/20 °C) during the reproductive stage. Heat-tolerant genotypes—IG2507, IG3263, IG3297, IG3312, IG3327, IG3546, IG3330, IG3745, IG4258, and FLIP2009—maintained higher chlorophyll fluorescence (F_v/F_m 0.71) under stress than heat-sensitive genotypes IG2821, IG2849, IG4242, IG3973, and IG3964 (F_v/F_m 0.58) (Sita et al. 2017). Similarly, two heat-tolerant lentil genotypes (1G 2507 and 1G 4258) exposed to heat stress (>25 °C) during reproductive growth and seed filling in the field had higher chlorophyll fluorescence (F_v/F_m 0.67) than two heat-sensitive genotypes (1G 3973 and 1G 3964; F_v/F_m 0.57) (Sehgal et al. 2017). Likewise, the screening of 41 mung bean lines grown outdoors and exposed to high temperatures (>40/28 °C) during the reproductive stage revealed several promising heat-tolerant lines (EC693358, EC693357, EC693369, Harsha, ML1299) with high F_v/F_m ratios (0.73–0.75) compared to sensitive lines (0.61–0.67), which could serve as useful donor/s for breeding programs and as a suitable base plant source to gain insight into heat stress-induced effects in cell metabolism (Sharma et al. 2016). Nine common bean lines were evaluated for changes in chlorophyll fluorescence under heat stress during flowering (45 °C for 2 h) in a greenhouse; thermotolerant lines 83201007 and RRR46 had higher F_v/F_m values under heat stress than the heat-sensitive line Secuntsa (Petkova et al. 2009). In another study, 12 varieties and lines of common bean were exposed to 42 °C in the field during the reproductive period; two genotypes (Ranit and Nerine) maintained their F_v/F_m values at 42 °C, relative to the controls at 26 °C, and were considered heat tolerant. These two genotypes also showed good productivity and quality and can be used as parental lines in bean breeding programs (Petkova et al. 2007). Screening of 15 alfalfa genotypes by exposing seedlings to 38/35 °C day/night for 7 days in a growth chamber identified Bara310SC (F_v/F_m 0.79) and WL712 (F_v/F_m <0.79) as heat-tolerant and heat-sensitive cultivars, respectively (Wassie et al. 2019), showing that F_v/F_m is an effective tool for phenotyping contrasting genotypes for heat tolerance.

2.6.4 Photosynthetic Rate

Heat stress affects the stay-green trait, chlorophyll content, and chlorophyll fluorescence, which affects RuBisCo activation, decreasing the photosynthetic rate (Salvucci Michael and Crafts-Brandner 2004; Sharkey 2005). Hence, photosynthetic rate can be used as a screening parameter for selecting heat-tolerant genotypes. Variation in photosynthetic rate among plant species in response to heat stress has been well documented. For example, the response of four chickpea genotypes to a natural temperature gradient in the field at the flowering stage identified two heat-tolerant genotypes (Acc#RR-3, Acc#7) with high P_n and two heat-sensitive genotypes (Acc#2, Acc#8) with lower P_n; these results were validated in a climate chamber experiment set at 30/25 °C and 35/30 °C (Makonya et al. 2019). In another study, 56 chickpea genotypes were exposed to high temperatures in the field from flowering to crop maturity (maximum temperatures 25–40 °C)—the tolerant genotypes (PUSA1103, PUSA1003, KWR108, BGM408, BG240, PG95333, JG14, BG) had higher P_n than the sensitive genotypes (ICC1882, PUSA372, PUSA2024) (Kumar et al. 2017). In a similar study in lentil, two heat-tolerant (1G 2507 and 1G 4258) genotypes had higher photosynthetic rate (P_n) than two heat-sensitive (1G 3973 and 1G 3964) genotypes exposed to heat stress (>25 °C) in the field during reproductive growth and seed filling (Sehgal et al. 2017).

Soybean cultivars IA3023 and KS4694 and PI lines PI393540 and PI588026A expressed heat tolerance and susceptibility with high and low P_n, respectively (Djanaguiraman et al. 2019). The two cultivars had less thylakoid membrane damage than the PI lines. In an earlier study on soybean, genotype K 03-2897, exposed to high temperature (38/28 °C) in a growth chamber for 14 days at the flowering stage, significantly decreased P_n due to anatomical and structural changes (increased thickness of palisade and spongy layers and lower epidermis) in cells and cell organelles, particularly damage to chloroplasts and mitochondria (Djanaguiraman and Prasad 2010).

2.6.5 Sucrose

Leaf photosynthates are transported to sink organs primarily as sucrose, and sucrose synthase (SS) is a key enzyme for sucrose to enter various metabolic pathways (Calderini et al. 2006). Downregulation of SS indirectly inhibits carbohydrate production, eventually reducing yield and quality. Maintaining sucrose levels is vital during stressed conditions, which depend on its synthesis and hydrolysis. Heat-stressed plants had significantly lower activities of key enzymes—sucrose phosphate synthase (SPS) and SS—involved in sucrose synthesis than non-stressed plants. Sucrose availability to reproductive organs is crucial for sustaining their function (Kaushal et al. 2013). Heat-tolerant genotypes can stabilize the photosynthetic process better than heat-sensitive genotypes. Heat stress disturbs sucrose production in leaves and impairs its transportation to reproductive organs (Kaushal et al. 2013; Li et al. 2012). Limitations in sucrose supply to reproductive organs, particularly under thermal stress, restrict flower development and function and pod and seed filling, reducing crop yield (Kaushal et al. 2013; Li et al. 2012). Measuring sucrose concentrations reveals the photosynthetic status of plants under heat stress (Awasthi et al. 2014). Sucrose synthase is strongly associated with heat tolerance in chickpea; heat-sensitive genotypes produced far less leaf sucrose than heat-tolerant genotypes, which impaired its supply to developing reproductive organs (flowers, pods, and seeds) in chickpea (Kaushal et al. 2013). Screening a large core collection of chickpea genotypes for heat tolerance (32/20 °C) in field condition identified two heat-tolerant (ICC15614, ICCV92944) and two heat-sensitive (ICC10685, ICC5912) genotypes. The heat-sensitive genotypes had significantly greater inhibition of RuBisCo (carbon-fixing enzyme), SPS, and SS than the heat-tolerant genotypes and thus produced less sucrose than the tolerant genotypes (Kaushal et al. 2013). Heat-sensitive (ICC16374) and heat-tolerant (JG14) chickpea genotypes exposed to gradually increasing temperatures (2 °C per day from 27/18 °C to 42/25 °C; day/night) for 8 days at anthesis in a growth chamber revealed greater sucrose synthase expression in JG14 than ICC16374 (Parankusam et al. 2017). Two tolerant chickpea genotypes (Acc#7 and Acc#RR-3) had higher starch contents and were relatively unaffected by heat-stress exposure compared to two heat-sensitive genotypes (Acc#2, Acc#8) at high temperature (35/30 °C) in a control chamber (Makonya et al. 2019). Therefore, an increased abundance of sucrose synthase in the tolerant genotype reasserted its potential role during heat-stress tolerance; this may ensure successful fertilization due to sustained pollen viability under heat stress, enhancing pod set and yield, as reported earlier for the tolerant genotype (ICC15614) (Krishnamurthy et al. 2011).

In lentil, sucrose production is vital for leaf and anther function and has been correlated with SPS activity in natural high-temperature environments (>32/20 °C). Heat-tolerant lentil genotypes (IG2507, IG3263, IG3297, IG3312, IG3327, IG3546, IG3330, IG3745, IG4258, FLIP2009) produced more sucrose in leaves (65–73%) and anthers (35–78%) than heat-sensitive genotypes (IG2821, IG2849, IG4242, IG3973, IG3964), which was associated with superior reproductive function and nodulation in tolerant genotypes (Sita et al. 2017). Limitations in sucrose supply may disrupt the development and function of reproductive organs (Prasad and Djanaguiraman 2011; Snider et al. 2011). In a similar study, two heat-tolerant (1G 2507 and 1G 4258) lentil genotypes exposed to heat stress (>25 °C) in the field had higher SS activity and thus higher sucrose contents in leaves and seeds than two heat-sensitive (1G 3973 and 1G 3964) genotypes (Sehgal et al. 2017). Thus, sucrose synthase in seeds and leaves is strongly correlated with seed yield; therefore, reductions in seed size and weight are attributed mainly to reductions in sucrose content.

Mung bean genotypes tested under heat stress (>40/25 °C day/night) during flowering and podding outdoors and in a controlled environment showed that two heat-tolerant genotypes (SML832 and SML668) had more sucrose than the heat-susceptible genotype (SML832). Thus, sucrose concentrations in leaves and anthers and SS and SPS activities declined significantly in sensitive genotypes under heat stress (Kaur et al. 2015). Exposure of common bean genotypes at the V4 developmental stage to heat treatment (32/25 °C) in a growth chamber significantly reduced leaf sucrose concentration in genotype Redhawk (most heat-sensitive genotype) and increased sugar contents in Sacramento (58%) and NY-105 (most heat tolerant) (Soltani et al. 2019).

2.6.6 Cell Membrane Thermostability

Under heat stress, protein denaturation, lipid liquefaction, and loss of membrane integrity are some of the chief physiological, biochemical, and molecular changes in plant metabolism (Gulen and Eris 2004). Most of the changes that appear during acclimation to heat stress are reversible, but death can occur if the stress is too intense (Saelim and Zwiazek 2000). Cell membranes are the principal target of environmental stresses, including heat stress (Chen et al. 2014; Sita et al. 2017). Protein denaturation and increased membrane fluidity, enzyme inactivation, decreased protein synthesis, protein degradation, and alterations in membrane integrity are documented injuries under heat stress (Howarth 2005). By accelerating the kinetic energy and movement of molecules across membranes, heat stress releases chemical bonds within the molecules of biological membranes, resulting in membrane fluidity by protein denaturation or increased unsaturated fatty acids (Savchenko et al. 2002). Decreased cell membrane thermostability or increased ionic leakage caused by the alteration of membrane protein structure is an important indicator of heat stress. The increased membrane fluidity caused by protein denaturation and increased unsaturated fatty acids in the membrane under high temperatures affect membrane structure and function (Wahid et al. 2007), causing symptoms, such as photooxidation of chlorophyll pigments, impaired electron flow, inhibited carbon fixation, and water loss from leaves (Prasad et al. 2017; Sharifi et al. 2012; Sita et al. 2017). The relationship between cell membrane thermostability (CMT) and crop yield changes from plant to plant under high temperatures. Ion leakage from plant tissues has been used as a membrane damage indicator in plants exposed to heat stress. Thus, CMT is an indirect indicator of heat-stress tolerance in legumes, such as soybean (Martineau et al. 1979), lentil (Sita et al. 2017), chickpea (Kaushal et al. 2013), and mung bean (Sharma et al. 2016). Membrane damage occurs under heat and cold stress, more so under heat stress, as reported for *Medicago* (Mo et al. 2011). Cell membrane thermostability (CMT) tends to decline during the late developmental phase of plants (Ahmad and Prasad 2011).

In addition to conventional breeding techniques, noticeable variations in membrane thermostability among genotypes, combined with biochemical and physiological screening methods, could be used to improve the selection for breeding objectives (Hemantaranjan et al. 2014). Membrane thermostability has been used to assess thermotolerance in many food crops worldwide. Depending on the growing season, electrolyte leakage in plants varies among tissues, organs, and growth stages and is affected by plant/tissue age, sampling organ, developmental stage, growing season, degree of hardening, and plant species. A significant positive relationship between CMT and yield was reported in sorghum (Sullivan and Ross 1979). In crop plants such as barley (*Hordeum vulgare* L.), cotton (*Gossypium* spp.), sorghum, and cowpea, increased electrolyte leakage decreased membrane thermostability (Wahid et al. 2007; Wahid and Shabbir 2005). In leguminous crops, electrolyte leakage has been used to assess thermotolerance. For example, heat stress at 34 °C in lentil revealed genotypes Ranjan, Moitree, 14-4-1, IC201710, and IC208329 as heat tolerant and genotypes ICC201655, ICC201661, ICC201662, ICC201670, ICC201675, ICC201681, ICC201698, ICC201743, ICC201794, ICC248959, Asha, Sagardeep Local, and UP local as heat sensitive, based on cell membrane stability in field and growth chamber studies (Choudhury et al. 2012). In another study, lentil genotypes exposed to high temperature (45 °C) at the flowering stage revealed Qazvin and B4400 as heat-tolerant and -sensitive genotypes, with 98.13% and 33.19% CMT, respectively (Barghi et al. 2013). At 38/28 °C and 40/30 °C in a controlled environment, heat-tolerant lentil genotypes IG2507, IG3263, IG3745, IG4258, and FLIP2009 had less membrane damage (<20% electrolyte leakage) than heat-sensitive genotypes IG2821, IG2849, IG4242, IG3973, and IG3964 (>30%) (Sita et al. 2017).

Among various legumes (pigeon pea, peanut, chickpeas, and soybean), chickpea was the most sensitive to high temperature based on CMT (Devasirvatham et al. 2012). Heat-tolerant chickpea genotypes ICCV07110 and ICCV92944 had less membrane damage (22.6% and 20.6%) than heat-sensitive genotypes ICC14183 and ICC5912 (30.4% and 33.3%) under high temperatures of 40/30 °C and 45/35 °C (Kumar et al. 2013). In another study, high temperature (>32/20 °C) during the reproductive stage caused the most membrane damage in heat-sensitive chickpea genotypes ICC10685 (28.3%) and ICC5912 (26.3%) and the least membrane damage in heat-tolerant genotypes ICC15614 (17.3%) and ICCV 92944 (19.6%) (Kaushal et al. 2013). A gradual rise in temperature (42/25 °C) at anthesis for 8 days increased electrolyte leakage (EL) by 20–25% greater in heat-sensitive chickpea genotype ICC16374 compared to heat-tolerant genotype ICCV92944 (Parankusam et al. 2017). At 37/27 °C, electrolyte leakage increased by a maximum of 16–25% in chickpea genotypes (Pareek et al. 2019), with ICC1205 identified as heat tolerant (13–14%). Similarly, Dua et al. (2001) reported ICCV88, ICC512, and ICC513 as heat-tolerant chickpea genotypes under heat stress. Another study on six chickpea genotypes revealed DG36 (EL: 36.7%) and Pusa 372 (EL: 50.7%) as heat-tolerant and heat-sensitive genotypes, respectively, when exposed to high temperature (>38 °C) under field conditions, based on EL (Singh et al. 2004). Of 115 chickpea genotypes screened at high temperature (36.5 °C) in the field, GNG 663 and Pusa 244 were selected as heat tolerant and heat sensitive, with electrolyte leakage values of 23% and 50%, respectively (Kumar et al. 2012). Among 30 chickpea genotypes screened for heat tolerance (>30 °C), Pusa 240 and GG2 genotypes were identified as heat-tolerant and -sensitive genotypes, respectively, with minimum (45%) and maximum (69%) cell membrane injury (Kumar et al. 2013).

Screening of nine cowpea genotypes exposed to heat stress (33/20 °C) during flowering and pod revealed less leaf electrolyte leakage in heat-tolerant genotypes H36, H8-9, and DLS99 (35.8–36.7%) than heat-susceptible genotypes CB5, CB3, and DLS127 (66.2–79.0%) (Ismail and Hall 1999). In another study at high temperature (38/30 °C), cell membrane injury was negatively corelated with yield in heat-tolerant (CB 27, Prima, UCR 193) and heat-sensitive genotypes (CB 5, CB 46) (Singh et al. 2010), with less membrane damage in heat-tolerant genotypes.

Screening of 15 *Medicago* cultivars at high temperature (38/35 °C) using membrane damage revealed "Bara310SC" and "WL712" as heat-tolerant and heat-sensitive genotypes with 24.07% and 53.2% electrolyte leakage, respectively (Wassie et al. 2019). Similarly, screening studies on 116 green gram genotypes at high temperature (45/25 °C) identified EC 3398889 and LGG460 as heat tolerant and heat sensitive, with minimum and maximum cell membrane damage, respectively (Basu et al. 2019). Gradual exposure to high temperature (35–50 °C) of 4-week-old three common bean genotype seedlings in a growth chamber revealed "local genotype" and "Ferasetsiz" as heat-sensitive genotypes, while "Balkız" was a relatively heat-sensitive genotype (Tokyol and Turhan 2019). Gross and Kigel (1994) used electrolyte leakage as a criterion for assessing heat tolerance at 32/28 °C during the reproductive stage and reported PI 271998 and BBL 47 as heat-tolerant and heat-sensitive genotypes in common bean, respectively. High-temperature studies (>40/28 °C) at the reproductive stage in mung bean showed high electrolyte leakage (21.8–23.6%) in heat-sensitive lines (EC 693363, EC 693361, EC 693370, KPS1, IPM02-3) compared to heat-tolerant lines (16.8–20.4%; EC693357, EC693358, EC693369, Harsha, ML1299) (Sharma et al. 2016). Another study on mung bean at high temperature (>35 °C) identified genotype MH 421 as heat tolerant and Basanti as heat sensitive, with low (34.88%) and high (41.34%) electrolyte leakage, respectively (Jha et al. 2015). Screening of ten faba bean genotypes exposed to heat stress (37 °C) 60 days after sowing revealed C5 as heat tolerant and Espan as heat sensitive, based on low (57.67%) and high (76%) membrane damage, respectively (Siddiqui et al. 2015).

2.6.7 Canopy Temperature Depression

Canopy temperature depression (CTD) is the plant canopy temperature deviation from the ambient temperature (Balota et al. 2007). At the whole-crop level, leaf temperature decreases below air temperature when water evaporates. CTD acts as an indirect measure of transpiration (Reynolds et al. 2001) and plant water status (Araus et al. 2003) and indicates the relative metabolic fitness of genotypes in a given environment (Reynolds 1997). CTD is a key trait for assessing the response of genotypes to low water usage, high temperature, and other stresses (Balota et al. 2007). At high temperatures, transpiration increases for some time, with plants using more water during growth due to more open stomata and lower CTD. A positive CTD value [i.e., difference between air temperature (T_a) and canopy temperature (T_c)] occurs when the canopy is cooler than the air ($CTD = T_a - T_c$) (Balota et al. 2008).

Canopy temperature depression is heritable and can be measured on cloudless days using an infrared thermometer (Reynolds et al. 1997). To maintain canopy temperature at a metabolically comfortable range, plants transpire through open stomata. Plants close stomata during stress acclimation, increasing the canopy temperature (Kashiwagi et al. 2008). Canopy temperature can be affected by biological and environmental factors, such as soil water status, wind, evapotranspiration, cloudiness, conduction systems, plant metabolism, air temperature, relative humidity, and continuous radiations (Reynolds et al. 2001). Canopy temperature is an indicator of plant water status or the equilibrium between root water uptake and shoot transpiration (Berger et al. 2010). CTD can act as a desirable criterion for selecting heat-tolerant genotypes based on phenotypic variation (Mason and Singh 2014). It can be used to determine yield potential and metabolic fitness of crop plants under specific environmental conditions (Kumari et al. 2013). It acts as a mechanism of heat escape and is strongly correlated with yield (Reynolds et al. 2001); affected by many physiological factors, it is a strong trait for determining genotype fitness.

Epicuticular leaf wax QTL and CTD are strongly interlinked, with wax load affecting plant canopy temperature (Awika et al. 2017). Stay-green genotypes have high CTD values and thus low canopy temperature due to transpirational cooling under heat stress (Fischer et al. 1998; Reynolds et al. 1994). In chickpea, CTD is negatively correlated with water potential, osmotic pressure, relative leaf water content, and seed yield (Sharma et al. 2015). Differences in canopy temperature are not detectable in high-humidity environments because the effect of evaporative leaf cooling is negligible (de Souza et al. 2012). CTD has been successfully used to select for heat tolerance in various crop species, including legumes. For example, heat-tolerant chickpea genotypes ICCVs 95311, 98902, 07109, and 92944 had higher CTD values than sensitive genotypes ICCVs 07116, 07117, and 14592, which had negative CTD values (Devasirvatham et al. 2015). Another study screened 30 chickpea genotypes exposed to temperature >30 °C to reveal Pusa 240 as a heat-tolerant genotype due to its cooler canopy than other genotypes (Kumar et al. 2013). Similarly, screening chickpea genotypes subjected to 36.5 °C identified GNG 663 and Vaibhavaas as heat tolerant and heat sensitive, respectively, with CTD values of 4.8 °C (maximum) and 1.8 °C (minimum) (Kumar et al. 2012). In a screening study of 56 chickpea genotypes for heat tolerance (40 °C), CTD values ranged from 5.0 to 7.5 °C; eight genotypes (Pusa 1103, Pusa 1003, KWR 108, BGM 408, BG 240, PG 95333, JG 14, BG 1077) were identified as heat tolerant, with maximum CTD values compared to other genotypes (Kumar et al. 2017). In mung bean, seed yield positively correlated with CTD, while canopy temperature negatively correlated with root traits, such as the number of lateral branches and dry root weight (Raina et al. 2019). In another study, mung bean genotype MH 421 (CTD 5.78 °C) was selected as heat tolerant compared to Basanti (CTD 4.37 °C) when tested at high temperature (>35 °C) (Jha et al. 2015). In pea, CTD is affected by canopy structure, and increased pod number and pod-to-node ratio associated with CTD (Tafesse et al. 2019).

2.7 Biochemical Traits

2.7.1 Oxidative Stress and Antioxidants

Heat stress is a major environmental factor affecting vital metabolic processes in plants, hampering proper growth and development. Disturbances in these metabolic processes lead to ROS generation, such as hydrogen peroxide, hydroxyl radicals, and superoxides (Chakraborty and Pradhan 2011). ROS production damages cellular activity by inactivating enzymes, denaturing proteins, and damaging membranes and DNA. Plants shield such injuries by activating cascades of enzymatic activities, such as superoxide dismutase (SOD), catalase (CAT), peroxidase (POD), ascorbate peroxidase (APX), and glutathione reductase (GR), and nonenzymatic activities, such as glutathione (GSH) and ascorbic acid (ASC) (Suzuki et al. 2012). The selection of contrasting genotypes based on the expression level of these antioxidants is effective in leguminous plants (Kumar et al. 2013). For example, chickpea genotypes raised under natural conditions until 50% flowering and then in a growth chamber for heat treatment (30/20 °C, 35/25 °C, 40/30 °C, and 45/35 °C) revealed that heat-tolerant genotypes (ICCV92944, ICCV07110) had lower H_2O_2 and MDA concentrations than sensitive genotypes (ICC5912, ICC14183). Tolerant genotypes face fewer injuries due to greater expression of antioxidants, such as APX and GR (Kumar et al. 2013). Similarly, 41 mung bean genotypes were screened, and contrasting genotypes were selected based on oxidative stress damage and antioxidant activity. Heat-tolerant genotypes (EC693357, EC693358, EC693369, Harsha, ML1299) experienced less oxidative damage (1.52–2.0-fold increase in MDA; 1.59–1.96-fold increase in H_2O_2) than sensitive genotypes (2.2–2.4-fold increase in MDA; 2.21–2.93-fold increase in H_2O_2) (Sharma et al. 2016). Moreover, heat-tolerant genotypes increased APX activity (by 1.48–1.77-fold) more than sensitive genotypes (1.27–1.37-fold). Likewise, of 38 lentil genotypes screened for heat tolerance (>35/20 °C) during the reproductive phase, heat-tolerant genotypes (IG2507, IG3263, IG3745, IG4258, FLIP2009) had less oxidative damage (MDA and H_2O_2 contents increased) and higher SOD, CAT, APX, and GR activities than heat-sensitive genotypes (IG2821, IG2849, IG4242, IG3973, IG3964) (Sita et al. 2017). In another study on lentil exposed to heat stress (30, 35, 40, 45, and 50 °C for 4 h) in plant growth chambers, SOD, CAT, and APOX activities initially increased in four heat-tolerant lentil varieties (IPL 81, IPL 406, Asha, Subrata) at 35, 40, and 45 °C but decreased at 50 °C, and decreased in heat-sensitive genotypes (Sehore and Lv) at all temperatures, except 30 °C (Chakraborty and Pradhan 2011). Further accumulation of carotenoids and ascorbate followed a similar trend, indicating the association of heat sensitivity with antioxidant expression.

2.7.2 Metabolites

Metabolite detection and quantification are an effective and powerful tool for selecting genotypes in response to environmental stresses (Bueno and Lopes

2020). Metabolites include low-molecular-weight compounds, including precursors and intermediate metabolic pathways, which are an indispensable part of plant metabolism, regulating vital biological processes and involved in stress tolerance (Wahid et al. 2007). The primary metabolites upregulated during abiotic stress are amino acids (proline), carbohydrates (sucrose, hexoses, polyhydric alcohols), polyamines (spermidine, spermine, putrescine), and glycine betaine. Correspondingly, secondary metabolites include terpenoids (saponins, tocopherols), phenolic compounds (flavonoids, isoflavonoids, anthocyanins), and nitrogen-containing metabolites (alkaloids and glucosinolates) (Rodziewicz et al. 2014). About one million specific metabolites varying in chemical structures, polarity, and physiochemical properties are present in the plant kingdom and can be analyzed through metabolomics profiling and metabolic fingerprinting. Due to heat stress, plants reshuffle their metabolites to sustain plant growth (Serrano et al. 2019). Metabolite production is regulated by genes; thus, the activation of heat-shock factors, mainly HSFA2 and HSFA3, increases metabolite content, such as galactinol (Song et al. 2016). Knowledge on metabolite production is important for developing metabolite markers to select heat-tolerant varieties.

Chebrolu et al. (2016) raised heat-tolerant (04025-1-1-4-1-1) and heat-sensitive (DT97-4290) soybean genotypes in a growth chamber, which were maintained under control conditions (28/22 °C) until flowering. Heat stress [moderate (36/24 °C) and severe (42/26 °C)] was imposed from flowering to maturity, with metabolite profiling undertaken on harvested seeds. The seeds of genotypes collected at 42/26 °C were highly abnormal and small and had high nitrogen levels compared with the sensitive genotype. Two hundred and seventy-five metabolites were traced and compared for 36/24 °C and 28/22 °C; 83 metabolites (48 downregulated and 35 upregulated were differentially altered in tolerant than sensitive genotypes) significantly differed between genotypes at 36/24 °C, compared to 61 metabolites (−30 and +31 in tolerant than sensitive genotypes) at 28/22 °C. Most traced compounds were antioxidants belonging to tocopherol, terpenoid, and flavonoid precursors. The tolerant genotype had more gulono-1,4-lactones (precursor for ascorbic acid) than the sensitive genotype, which was attributed to its higher tolerance to heat stress and positively correlated with seed vigor, seed germination, seed weight, and oil content.

Proline is a multifunctional amino acid involved in plant growth and development that acts as a compatible osmolyte and ROS scavenger to regulate plant function in stressed environments (Szepesi and Szőllősi 2018). Under stress, proline has diverse roles, such as stabilizing membranes, proteins, subcellular structures, and energy sources, thus maintaining cellular homeostasis. Therefore, an increase in compatible solutes such as proline under stressful conditions is valuable for plants (Kaur and Asthir 2015). Leaf proline concentrations were measured in four chickpea genotypes varying in their sensitivity to high temperature (4.5 °C higher than the ambient temperature for 15 days); heat-treated genotypes had significant higher proline concentrations than the control, more so in Pusa 1103 and BGD-72 (tolerant genotypes) than Pusa 256 and Pusa 261 (sensitive genotypes) (Arunkumar et al. 2012). Similarly, a high-temperature treatment (45 °C for 8 h) on 6-day-old common bean seedlings increased proline content compared to control plants (25 °C) (Babu and Devaraj 2008).

2.7.3 Heat-Shock Proteins

Heat-shock proteins are specific proteins accumulated during rapid heat stress. Heat-shock genes are upregulated for plant survival under heat stress and responsible for encoding HSPs (Chang et al. 2007). A sudden change in temperature increases HSP production (Wahid et al. 2007). In all organisms, HSP expression is a general response to high temperature (Vierling 1991). HSP90, HSP70, and low-molecular-weight proteins are three classes of proteins according to molecular weight. Under stress conditions, HSPs perform chaperone-like functions in protein synthesis, maturation, targeting, renaturation, and membrane stabilization (Reddy et al. 2010, 2016). HSPs also play a role in protein translation and translocation, perform proteolysis and protein folding, and reactivate denatured proteins (Zhang et al. 2005). Under heat stress, the expression of HSPs protects the machinery of protein biosynthesis (Miroshnichenko et al. 2005). Membrane lipid composition, membrane integrity osmoprotectants, and HSPs play important roles in heat tolerance (Blum 2018). HSPs are located mainly in the cytoplasm, nucleus, mitochondria, chloroplast, and endoplasmic reticulum (Waters et al. 1996). In plant species such as potato, maize, soybean, and barley, specific HSPs have been identified in mitochondria in response to high temperature (Neumann et al. 1994). HSPs maintain membrane stability and protect PSII from oxidative stress (Barua et al. 2003). In *Medicago truncatula*, the role of HSPs was determined by cloning and characterization (Li et al. 2016). The roots of some plants also synthesize HSPs to cope with heat stress (Nieto-Sotelo et al. 2002). The expression profiles of HSPs have been compared in plant species/genotypes contrasting in heat sensitivity. In a comparative study on cowpea and eight common bean varieties at 40 °C, cowpea showed more HSP expression than common bean and was thus more tolerant to high temperature. IPA 7 had the highest HSP expression of the eight common bean genotypes (Simões-Araújo et al. 2003).

In chickpea exposed to high temperature (42/25 °C) at anthesis, the levels of HSPs increased in genotype JG14 compared to ICC16374 (Parankusam et al. 2017). In another study, five chickpea genotypes were assessed for thermotolerance at 30, 35, and 40 °C, with CSJD 884 and RSG 895 identified as heat tolerant and C 235 as heat sensitive (Kumari et al. 2018). In peanut genotypes exposed to 50 °C for 30 min, ICGS 76, COC038, COC050, COC041, and COC068 were identified as heat tolerant and COC812, COC166, COC115, COC277, COC227, Tamrun OL 02, and Spanco as heat sensitive (Selvaraj et al. 2011). Heat-tolerant peanut genotype ICGS 44 had higher HSP expression than heat-sensitive genotypes AK 159 and DRG 1 under heat stress (45 °C) (Chakraborty et al. 2018). The level of thermotolerance positively correlated with HSP accumulation. Thirty varieties of pea seedlings exposed to high temperature (46–49 °C) in growth chambers for different time intervals (1–3 h) identified Acc#623 and Acc#476 as heat-tolerant and heat-sensitive varieties, respectively, with Acc#623 having higher levels of HSP70, HSP90, and HSP104 than Acc#476 (Srikanthbabu et al. 2002). In soybean under 38/30 °C, cultivar PI 471938 had higher HSP expression (especially HSP70), conferring heat tolerance, than R95-1705 (Katam et al. 2020).

2.8 Genes for Heat Tolerance

Diverse genes have been identified using omics analyses (transcriptomics, genomics, and proteomics) in various plant species for heat resilience mechanisms; these genes are essential for developing stable cultivars (Singh et al. 2019). A lentil population was developed by crossing heat-tolerant (PDL-1 and PDL-2) and heat-sensitive (JL-3 and E-153) genotypes for molecular mapping and genetics studies (Singh et al. 2017). For this purpose, simple sequence repeat (SSR) marker analysis and QTL analysis were performed, using 495 SSR markers, which detected seven SSR markers and two QTLs—qHt_ss and qHt_ps were closely linked with SSR markers (PBA_LC_1507, PLC_105, PBA_LC_1288, LC_03, PBA_LC_1684, PBA_LC_1752, PBA_LC_1480). Further, SSR marker PBA_LC_1507 was closely linked to pod set and seedling survival trait. Another lentil study revealed genetic diversity for heat tolerance among 119 genotypes using SSR markers (Zhang et al. 2005). High-temperature stress was applied at the seedling (35/33 °C) and anthesis (35/20 °C) stages to study the effects on morphophysiological and reproductive traits of non-stressed and stressed plants in the field. A set of 209 alleles were identified using 35 SSR markers. Genotypes were clustered into nine groups based on SSR markers. Clusters 1 and 6 had significant variation, which could help produce better segregants for heat tolerance. The genotypes in clusters 2, 3, 4, 5, 7, 8, and 9 were moderately tolerant or moderately sensitive to heat stress. Significant differences among clusters were observed for seedling survivability, heat tolerance scores, membrane stability index, pollen viability, pollen germination, pod and seed set, and seed yield. The finding suggests that identifying the genetic distances between clusters will maximize their use for breeding heat-tolerant lentils. Results from the RT-PCR confirmed differential gene expression in heat-sensitive fescue genotype PI283316 and heat-tolerant genotype PI297901 (Zhang et al. 2005).

Similarly, in chickpea, phenotyping of RILs developed from a cross between ICC4567 (heat-sensitive) and ICC156614 (heat-tolerant) genotypes exhibited two genomic regions (CaLG05 and CaLG06) with four QTLs for the number of filled pods, seed number, grain yield, and pod set. Further, 25 genes responsible for heat tolerance were reported in these two genomic regions—five encoding HSPs and heat-shock transcription factors, three responsible for detoxifying ROS, five encoding proteins like farnesylated protein 6 and ethylene-responsive transcription factors, and all these genes collectively upregulating other genes like MYB4, AKH3, and RAN1 that are involved in the mitigation of heat stress in chickpea (Paul et al. 2018). Molecular characterization in mung bean genotype VC1973A revealed 24 *VrHsf* genes responsible for the synthesis of heat-shock transcription factors that mediate plant responses under heat stress, suggesting their potential role in investigating mechanisms related to heat tolerance (Liu et al. 2019). Similarly, in a soybean study, 26 *GmHsf* genes coded for heat-shock transcription factors, with *GmHsf12*, *GmHsf28*, *GmHsf34*, *GmHsf35*, *and GmHsf47*, highly upregulated during heat stress (Chung et al. 2013).

2.9 Scope of Harnessing Germplasm for Designing Heat Tolerance

Harnessing crop germplasm variability is one of the cheapest and most environmentally friendly approaches for developing abiotic stress, including heat stress tolerance (Jha et al. 2014). Like other crops, substantial genetic variation has been harnessed to develop grain legumes that tolerate heat stress (Craufurd et al. 2003; Jha et al. 2017; Krishnamurthy et al. 2011). Several breeder-friendly techniques, such as field-based screening of grain legumes in targeted heat-stress environments, enabled the selection of potential heat-tolerant grain legumes in chickpea, soybean, common bean, pea, lentil, and cowpea. Based on the early phenology, an important heat stress, some important chickpea genotypes, viz., ICC 14346, ACC 316, and ACC 317, showing heat stress escape mechanisms have been reported (Canci and Toker 2009; Upadhyaya et al. 2011). Selection relying on yield and yield-related traits, such as high pod and seed set, low grain yield reduction, and maintaining high biomass, has been used to directly identify heat-tolerant lines, including ICC1205, ICC15614, BG256, and Vaibhav in chickpea (Devasirvatham et al. 2013; Gaur et al. 2012; Jha et al. 2015; Jumrani et al. 2018); G122, PI 163120, PI 271998, G122, A55, and Cornell 503 in common bean (Miklas et al. 2000; Rainey and Griffiths 2005; Shonnard and Gepts 1994); TN88-63, Tvu 4552, and Prima in cowpea (Nielsen and Hall 1985; Warrag and Hall 1983); 55-437, 796, 796, 55-437, ICG 1236, ICGV 86021, ICGV 87281, and ICGV 92121 in groundnut (Craufurd et al. 2003; Ntare et al. 2001); 72578, 70548, 71457, and 73838 in lentil (Delahunty et al. 2015); Dieng, IA3023, and KS4694 in soybean (Djanaguiraman et al. 2019; Puteh et al. 2013); C.52/1/1/1 and C.42 in faba bean (Abdelmula and Abuanja 2007); and JP-625, IARI-2877, PMR-38 II, EC-318760, EC-328758, and IARI-2904 in pea (Mohapatra et al. 2020). Similar studies based on various physiological parameters, including cell membrane stability, identified heat-tolerant ILC 482, Annegiri, and ICCV 10 in chickpea (Srinivasan et al. 1996), PI 271998 in common bean (Marsh et al. 1985), and SPT 06-07 in groundnut (Singh et al. 2016), and studies based on pollen germination and fertilization under heat stress identified heat-tolerant ICC 15614, ICCV 92944, and ICC1205 in chickpea (Devasirvatham et al. 2010; Kaushal et al. 2013), 55-437, ICG 1236, TMV 2, and ICGS 11 in groundnut (Kakani et al. 2002), DG 5630RR, NRC 7, and EC 538828 in soybean (Jumrani et al. 2018; Salem et al. 2007), and Haibushi in common bean (Tsukaguchi et al. 2003). In addition, studies based on superior yield performance and genotype × genotype × environment biplot analysis identified heat-tolerant ICC 4958, RVG 203, RVG 202, JAKI 9218, and JG 130 in chickpea (Jha et al. 2018, 2019), and studies based on several heat-stress tolerance indices identified heat-tolerant lines in soybean (Sapra and Anaele

1991), chickpea (Jha et al. 2018), and common bean (Porch 2006). Harnessing existing genetic variability in crop wild relatives and landraces should be considered to broaden the genetic base of grain legumes for higher heat tolerance in the future.

2.10 Genetics of Heat Tolerance

Classical genetics and quantitative genetics approaches, such as generation mean analysis and diallel analysis, provided preliminary information on heat-stress tolerance in chickpea (Jha et al. 2019), cowpea (Marfo and Hall 1992; Patel and Hall 1988), and common bean (Miklas et al. 2000; Rainey and Griffiths 2005) based on yield and yield-related traits under heat stress. However, this genetic information does not provide a complete picture of heat tolerance in these grain legumes, as this trait is governed by multigenes and highly influenced by G × E interactions (Upadhyaya et al. 2011).

2.11 Genomic Resources for Heat Tolerance

Unprecedented advances in genomic resource development have enabled the precise mapping of various traits of breeding importance, including heat-stress tolerance in various grain legume crops (Jha et al. 2021; Paul et al. 2018; Pottorff et al. 2014; Varshney et al. 2019). In parallel, the availability of reference genome sequences for major grain legumes has enriched the genomics resources in legume crops. Using a biparental mapping approach, several QTLs controlling heat-stress tolerance have been elucidated in chickpea (Jha et al. 2019; Paul et al. 2018), cowpea (Lucas et al. 2013; Pottorff et al. 2014), lentil (Singh et al. 2017), and pea (Huang et al. 2017). In chickpea, four important QTLs related to yield traits were identified on CaLG05 and CaLG06 from an ICC15614 × ICC4567 RIL population under heat stress (Paul et al. 2018). Jha et al. (2021) reported that 37 major QTLs related to heat tolerance in chickpea were discovered. Five QTLs were elucidated in cowpea under heat stress (Lucas et al. 2013). Similarly, an evaluation of IT93K-503-1 × CB46 and IT84S-2246 × TVu14676 RIL populations identified three QTLs (*Hbs-1*, *Hbs-2*, and *Hbs-3*) contributing to heat tolerance in cowpea (Pottorff et al. 2014). Many QTLs contribute to phenological traits, such as days to flowering, with yield-related QTLs reported in pea under heat stress (Huang et al. 2017).

The availability of high-throughput SNP markers elucidated genomic regions controlling heat tolerance across the whole genome in a large set of chickpea germplasm using a genome-wide association mapping approach (Tafesse et al. 2020; Varshney et al. 2019). In this context, several marker-trait associations (MTAs) for various heat-stress traits have been deciphered in chickpea (Thudi et al. 2014; Varshney et al. 2019), pea (Tafesse et al. 2020), and common bean (López-Hernández and Cortés 2019). In whole genome resequencing derived SNP markers based GWAS analysis involving a large panel of chickpea germplasm, several significant MTAs for various physiological and yield traits were unveiled under heat stress (Varshney et al., 2019). Likewise, Tafesse et al. (2020) identified several significant MTAs for chlorophyll content, photochemical reflectance index, canopy temperature, and pod number in pea under heat stress. In common bean, GWAS in 78 "geo-referenced" wild common bean accessions revealed several candidate genes (e.g., *MED23*, *MED25*, *HSFB1*, *HSP40*, *HSP20*, *phospholipase C*, *MBD9*, *PAP*) related to heat-stress tolerance (López-Hernández and Cortés 2019). These MTAs could be important in marker-assisted breeding for developing heat-tolerant grain legumes.

2.12 Transcriptomics for Unfolding Candidate Genes for Heat Tolerance

In the past decade, technical interventions in functional genomics, especially next-generation sequencing-based RNA-seq facility, have offered great insights into gaining function of candidate gene(s) controlling various complex traits, including heat stress in various grain legumes (Agarwal et al. 2016; Singh et al. 2019; Wang et al. 2018). Using the RNA-seq technique, *Ca_25811*, *Ca_23016*, *Ca_09743*, *Ca_17680*, and *Ca_25602* candidate genes were deciphered from heat-treated reproductive tissues of heat-tolerant and heat-sensitive chickpea genotypes (Agarwal et al. 2016). In soybean, RNA-seq analysis of contrasting genotypes treated with combined drought and heat stress revealed several differentially expressed genes, primarily involved in the defense response, photosynthesis, and metabolic processes (Wang et al. 2018). RNA-seq analysis of heat-treated soybean leaf tissue at the reproductive stage revealed a plethora of up- and down-regulatory differentially expressed genes and unearthed genes involved in flowering, oxidative stress, osmoregulation, HSPs, and ethylene biosynthesis (Xu et al. 2020). Transcriptional analysis of heat-treated soybean root tissue revealed numerous differentially expressed genes involved in regulating the heat-stress response (Valdés-López et al. 2016). In lentil, transcriptome analysis of contrasting heat-tolerant and heat-sensitive genotypes (PDL-2 and JL-3) revealed several genes encoding a WRKY transcription factor, DnaJ homolog subfamily B member 13, and 17.1 kDa class II heat-shock protein and cell wall (Singh et al. 2019). However, higher expression of NAC and WRKY transcription factor genes conferred heat tolerance in the PDL-2 genotype.

2.13 Proteomics and Metabolomics Resolving Gene Networks for Heat Tolerance in Grain Legumes

A proteomics approach could endow us with the whole landscape of proteins responding to various biotic and abiotic stresses (Ramalingam et al. 2015). A series of proteins contributing to switching on various complex signal transduction mechanisms and intricate gene networks associated with adapting the plant response to heat stress have been investigated (Rathi et al. 2016). However, the role of proteomics in mediating heat-stress tolerance remains limited in grain legumes.

Various types of HSPs, such as ClpB/HSP100 and VfHsp17.9-CII (Kumar et al. 2015), EF-Tu protein (Das et al. 2016), tissue-specific proteins (Ahsan et al. 2010), and early response to dehydration (ERD)-related proteins (ERD10 and ERD14) (Kovacs et al. 2008), act as chaperones, protecting cells from heat stress-related injuries. Similarly, heat stress increased HSP expression in chickpea genotype JG14 (Parankusam et al. 2017) and groundnut genotype ICGS 44 (Chakraborty et al. 2018). Further, Das et al. (2016) reported 25 proteins contributing to various cellular metabolic activities under heat stress in soybean. Furthermore, the participatory role of dehydrin-like proteins recovered from mitochondria and their plausible role in safeguarding mitochondrial membrane in yellow lupin under heat stress are worth noting (Rurek 2010). Valdés-López et al. (2016) reported 30 commonly up- and downregulated heat stress-responsive proteins involved in cell wall formation, amino acid and lipid biosynthesis, and ROS reduction in soybean.

Like proteomics, metabolomics is a robust approach for enriching our understanding of various primary and secondary metabolites produced in response to abiotic stresses, including heat stress (Janni et al. 2020; Ramalingam et al. 2015). Among the various metabolites, tocopherol and its isoforms, ascorbate, flavonoids, phenolic compounds, proline, polyamines, and glycine betaine help plants adjust to heat stress (Chebrolu et al. 2016; Kaplan et al. 2004). For example, a heat-tolerant soybean genotype had a higher abundance of flavonoids and tocopherols acting as antioxidants than a heat-sensitive genotype (Chebrolu et al. 2016). Further technical innovations and bioinformatic analysis of metabolomics-derived data could shed light on the complex gene network of heat-stress adaptation in grain legumes.

2.14 Conclusions

Increasing episodes of heat stress are becoming a serious issue worldwide, challenging the yield potential of various crops, including grain legumes. Harnessing genetic resources could be an important approach for sustaining legumes under rising temperatures. In addition to yield traits, incorporating various physiological traits could enable plants to adapt and sustain grain yield under heat stress (Reynolds and Langridge 2016).

As crop wild relatives are the reservoir of novel gene(s)/QTLs for various stress tolerance including heat-stress tolerance, introgression of heat-tolerance genomic region into elite legume cultivars using a pre-breeding approach could sustain legume yields under rising global temperatures (Chaudhary et al. 2020). Likewise, capitalizing on the various adaptive traits conferring heat tolerance from legume landraces could assist in developing grain legumes that tolerate heat stress. Furthermore, advances in grain legume genomics, especially molecular markers, and availability of grain legume genome assemblies have helped pinpoint heat-tolerance genomic regions in various legumes. Whole-genome resequencing efforts have also enabled the discovery of novel haplotypes controlling heat tolerance (Varshney et al. 2019). In parallel, progress in functional genomics, including RNA-seq-based transcriptomics, has enabled the discovery of underlying candidate gene

(s) involved in heat tolerance and putative functions (Agarwal et al. 2016; Singh et al. 2019; Wang et al. 2018). Additionally, advances in proteomics and metabolomics have uncovered various participatory proteins, especially HSPs and heat stress-responsive metabolites, and various novel signaling molecules in legumes (Chebrolu et al. 2016; Parankusam et al. 2017). Therefore, leveraging various breeding, physiological, and "omics" approaches combined with emerging "speed breeding," genomic selection, and genome editing technology could help develop climate-resilient grain legumes to meet the increasing demand for plant-based dietary protein.

References

Abdelmageed AHA, Gruda N (2009) Influence of high temperatures on gas exchange rate and growth of eight tomato cultivars under controlled heat stress conditions. Eur J Hortic Sci 74: 152–159

Abdelmula AA, Abuanja IK (2007) Genotypic responses, yield stability, and association between characters among some of Sudanese faba bean (*Vicia faba* L.) genotypes under heat stress. In: Conference on international agricultural research for development, University of Kassel-Witzenhausen and University of Göttingen, Germany, 9–11 October 2007

Abdelrahman M, El-Sayed M, Jogaiah S, Burritt DJ, Tran LSP (2017) The "STAY-GREEN" trait and phytohormone signalling networks in plants under heat stress. Plant Cell Rep 36:1009–1025. https://doi.org/10.1007/s00299-017-2119-y

Agarwal G, Garg V, Kudapa H, Doddamani D, Pazhamala LT, Khan AW, Thudi M, Lee S, Varshney RK (2016) Genome-wide dissection of AP2/ERF and HSP90 gene families in five legumes and expression profiles in chickpea and pigeonpea. Plant Biotechnol J 14:1563–1577. https://doi.org/10.1111/pbi.12520

Ahmad P, Prasad MNV (2011) Abiotic stress responses in plants: metabolism, productivity and sustainability. Springer, New York

Ahsan N, Donnart T, Nouri MZ, Komatsu S (2010) Tissue-specific defense and thermo-adaptive mechanisms of soybean seedlings under heat stress revealed by proteomic approach. J Proteome Res 9:4189–4204. https://doi.org/10.1021/pr100504j

Allakhverdiev SI, Kreslavski VD, Klimov VV, Los DA, Carpentier R, Mohanty P (2008) Heat stress: an overview of molecular responses in photosynthesis. Photosynth Res 98:541–550. https://doi.org/10.1007/s11120-008-9331-0

Alsajri FA, Singh B, Wijewardana C, Irby J, Gao W, Reddy KR (2019) Evaluating soybean cultivars for low- and high-temperature tolerance during the seedling growth stage. Agronomy 9:1–20. https://doi.org/10.3390/agronomy9010013

Anitha Y, Vanaja M, Kumar VG (2016) Identification of attributes contributing to high temperature tolerance in blackgram (*Vigna mungo* L. Hepper) genotypes. Int J Sci Res 5:1021–1025

Araus JL, Bort J, Steduto P, Villegas D, Royo C (2003) Breeding cereals for Mediterranean conditions: ecophysiological clues for biotechnology application. Ann Appl Biol 142:129–141. https://doi.org/10.1111/j.1744-7348.2003.tb00238.x

Arunkumar R, Sairam RK, Deshmukh PS, Pal M, Khetarpal S, Pandey SK, Kushwaha SR, Singh TP (2012) High temperature stress and accumulation of compatible solutes in chickpea (*Cicer arietinum* L.). Indian J Plant Physiol 17:145–150

Awasthi R, Kaushal N, Vadez V, Turner NC, Berger J, Siddique KH, Nayyar H (2014) Individual and combined effects of transient drought and heat stress on carbon assimilation and seed filling in chickpea. Funct Plant Biol 41:1148–1167. https://doi.org/10.1071/FP13340

Awasthi R, Gaur P, Turner NC, Vadez V, Siddique KHM, Nayyar H (2017) Effects of individual and combined heat and drought stress during seed filling on the oxidative metabolism and yield of chickpea (*Cicer arietinum*) genotypes differing in heat and drought tolerance. Crop Pasture Sci 68:823–841. https://doi.org/10.1071/CP17028

Awika HO, Hays DB, Mullet JE, Rooney WL, Weers BD (2017) QTL mapping and loci dissection for leaf epicuticular wax load and canopy temperature depression and their association with QTL for staygreen in *Sorghum bicolor* under stress. Euphytica 213:1–22. https://doi.org/10.1007/s10681-017-1990-5

Babu NR, Devaraj VR (2008) High temperature and salt stress response in French bean (*Phaseolus vulgaris*). Aust J Crop Sci 2:40–48

Baker NR (2008) Chlorophyll fluorescence: a probe of photosynthesis in vivo. Annu Rev Plant Biol 59:89–113. https://doi.org/10.1146/annurev.arplant.59.032607.092759

Baker NR, Rosenqvist E (2004) Applications of chlorophyll fluorescence can improve crop production strategies: an examination of future possibilities. J Exp Bot 55:1607–1621. https://doi.org/10.1093/jxb/erh196

Balota M, Payne WA, Evett SR, Lazar MD (2007) Canopy temperature depression sampling to assess grain yield and genotypic differentiation in winter wheat. Crop Sci 47:1518–1529. https://doi.org/10.2135/cropsci2006.06.0383

Balota M, Payne WA, Evett SR, Peters TR (2008) Morphological and physiological traits associated with canopy temperature depression in three closely related wheat lines. Crop Sci 48:1897–1910. https://doi.org/10.2135/cropsci2007.06.0317

Barghi SS, Mostafaii H, Peighami F, Zakaria RA, Nejhad RF (2013) Response of in vitro pollen germination and cell membrane thermostability of lentil genotypes to high temperature. Int J Agric Res Rev 3:13–20

Barnabas B, Jager K, Feher A (2008) The effect of drought and heat stress on reproductive processes in cereals. Plant Cell Environ 31:11–38. https://doi.org/10.1111/j.1365-3040.2007.01727.x

Barua D, Downs CA, Heckathorn SA (2003) Variation in chloroplast small heat-shock protein function is a major determinant of variation in thermotolerance of photosynthetic electron transport among ecotypes of *Chenopodium album*. Funct Plant Biol 30:1071–1079. https://doi.org/10.1071/FP03106

Basu PS, Pratap A, Gupta S, Sharma K, Tomar R, Singh NP (2019) Physiological traits for shortening crop duration and improving productivity of greengram (*Vigna radiata* L. Wilczek) under high temperature. Front Plant Sci 10:1–18. https://doi.org/10.3389/fpls.2019.01508

Berger B, Parent B, Tester M (2010) High-throughput shoot imaging to study drought responses. J Exp Bot 61:3519–3528. https://doi.org/10.1093/jxb/erq201

Bhandari K, Siddique KHM, Turner NC, Kaur J, Singh S, Agrawal SK, Nayyar H (2016) Heat stress at reproductive stage disrupts leaf carbohydrate metabolism, impairs reproductive function, and severely reduces seed yield in lentil. J Crop Improv 30:118–151. https://doi.org/10.1080/15427528.2015.1134744

Bita CE, Gerats T (2013) Plant tolerance to high temperature in a changing environment: scientific fundamentals and production of heat stress-tolerant crops. Front Plant Sci 4:1–18. https://doi.org/10.3389/fpls.2013.00273

Blum A (1986) The effect of heat stress on wheat leaf and ear photosynthesis. J Exp Bot 37:111–118. https://doi.org/10.1093/jxb/37.1.111

Blum A (2018) Plant breeding for stress environments. CRC Press, Boca Raton, FL

Borrell AK, van Oosterom EJ, Mullet JE, George-Jaeggli B, Jordan DR, Klein PE, Hammer GL (2014) Stay-green alleles individually enhance grain yield in sorghum under drought by modifying canopy development and water uptake patterns. New Phytol 203:817–830. https://doi.org/10.1111/nph.12869

Brussaard L, De Ruiter PC, Brown GG (2007) Soil biodiversity for agricultural sustainability. Agric Ecosyst Environ 121:233–244. https://doi.org/10.1016/j.agee.2006.12.013

Bueno PC, Lopes NP (2020) Metabolomics to characterize adaptive and signaling responses in legume crops under abiotic stresses. ACS Omega 5:1752–1763. https://doi.org/10.1021/acsomega.9b03668

Butler WL (1978) Energy distribution in the photochemical apparatus of photosynthesis. Annu Rev Plant Physiol 29:345–378. https://doi.org/10.1146/annurev.pp.29.060178.002021

Calderini DF, Reynolds MP, Slafer GA (2006) Source–sink effects on grain weight of bread wheat, durum wheat, and triticale at different locations. Aust J Agric Res 57:227–233. https://doi.org/10.1071/AR05107

Canci H, Toker C (2009) Evaluation of yield criteria for drought and heat resistance in chickpea (*Cicer arietinum* L.). J Agron Crop Sci 195:47–54. https://doi.org/10.1111/j.1439-037X.2008.00345.x

Chakraborty U, Pradhan D (2011) High temperature-induced oxidative stress in *Lens culinaris*, role of antioxidants and amelioration of stress by chemical pre-treatments. J Plant Interact 6:43–52. https://doi.org/10.1111/jac.12260

Chakraborty K, Bishi SK, Singh AL, Zala PV, Mahatma MK, Kalariya KA, Jat RA (2018) Rapid induction of small heat shock proteins improves physiological adaptation to high temperature stress in peanut. J Agron Crop Sci 204:285–297. https://doi.org/10.1111/jac.12260

Chang PFL, Jinn TL, Huang WK, Chen Y, Chang HM, Wang CW (2007) Induction of a cDNA clone from rice encoding a class II small heat shock protein by heat stress, mechanical injury, and salicylic acid. Plant Sci 172:64–75. https://doi.org/10.1016/j.plantsci.2006.07.017

Chaudhary S, Devi P, Bhardwaj A, Jha UC, Sharma KD, Prasad PVV, HanumanthaRao B, Kumar S, Nayyar H (2020) Identification and characterization of contrasting genotypes/cultivars for developing heat tolerance in agricultural crops: current status and prospects. Front Plant Sci 11:1–34. https://doi.org/10.3389/fpls.2020.587264

Chebrolu KK, Fritschi FB, Ye S, Krishnan HB, Smith JR, Gillman JD (2016) Impact of heat stress during seed development on soybean seed metabolome. Metabolomics 12:1–14. https://doi.org/10.1007/s11306-015-0941-1

Chen X, Lin S, Liu Q, Huang J, Zhang W, Lin J, Wang Y, Ke Y, He H (2014) Expression and interaction of small heat shock proteins (sHsps) in rice in response to heat stress. Biochim Biophys Acta 1844:818–828. https://doi.org/10.1016/j.bbapap.2014.02.010

Choudhury DR, Tarafdar S, Das M, Kundagrami S (2012) Screening lentil (*Lens culinaris* Medik.) germplasms for heat tolerance. Trends Biosci 5:143–146

Chung E, Kim KM, Lee JH (2013) Genome-wide analysis and molecular characterization of heat shock transcription factor family in *Glycine max*. J Genet Genomics 40:127–135. https://doi.org/10.1016/j.jgg.2012.12.002

Commuri PD, Jones RJ (1999) Ultrastructural characterization of maize (*Zea mays* L.) kernels exposed to high temperature during endosperm cell division. Plant Cell Environ 22:375–385. https://doi.org/10.1046/j.1365-3040.1999.00424.x

Craufurd PQ, Bojang M, Wheeler TR, Summerfield RJ (1998) Heat tolerance in cowpea: effect of timing and duration of heat stress. Ann Appl Biol 133:257–267. https://doi.org/10.1111/j.1744-7348.1998.tb05826.x

Craufurd PQ, Prasad PV, Kakani VG, Wheeler TR, Nigam SN (2003) Heat tolerance in groundnut. Field Crops Res 80:63–77. https://doi.org/10.1016/S0378-4290(02)001557

Crawford AJ, McLachlan DH, Hetherington AM, Franklin KA (2012) High temperature exposure increases plant cooling capacity. Curr Biol 22:R396–R397. https://doi.org/10.1016/j.cub.2012.03.044

Das A, Eldakak M, Paudel B, Kim DW, Hemmati H, Basu C, Rohila JS (2016) Leaf proteome analysis reveals prospective drought and heat stress response mechanisms in soybean. Biomed Res Int 2016:1–23. https://doi.org/10.1155/2016/6021047

de Dorlodot S, Forster B, Pages L, Price A, Tuberosa R, Draye X (2007) Root system architecture: opportunities and constraints for genetic improvement of crops. Trends Plant Sci 12:474–481. https://doi.org/10.1016/j.tplants.2007.08.012

de Souza MA, Pimentel AJB, Ribeiro G (2012) Breeding for heat-stress tolerance. In: Fritsche-Neto R, Borem A (eds) (1) Plant breeding for abiotic stress tolerance. Springer, Berlin, pp 137–156

Delahunty A, Nuttall J, Nicolas M, Brand J (2015) Genotypic heat tolerance in lentil. Paper presented in the Proceedings of the 17th ASA Conference, Hobart, Australia, 20–24 September 2018

Devasirvatham V, Tan DKY, Trethowan RM, Gaur PM, Mallikarjuna N (2010) Impact of high temperature on the reproductive stage of chickpea. Paper presented at the Food security from sustainable agriculture proceedings of the 15th Australian Society of Agronomy Conference, New Zealand, Lincoln, 15–18 November 2010

Devasirvatham V, Tan DKY, Gaur PM, Raju TN, Trethowan RM (2012) High temperature tolerance in chickpea and its implications for plant improvement. Crop Pasture Sci 63:419–428. https://doi.org/10.1071/CP11218

Devasirvatham V, Gaur PM, Mallikarjuna N, Raju TN, Trethowan RM, Tan DKY (2013) Reproductive biology of chickpea response to heat stress in the field is associated with the performance in controlled environments. Field Crops Res 142:9–19. https://doi.org/10.1016/j.fcr.2012.11.011

Devasirvatham V, Gaur PM, Raju TN, Trethowan RM, Tan DKY (2015) Field response of chickpea (*Cicer arietinum* L.) to high temperature. Field Crops Res 172:59–71. https://doi.org/10.1016/j.fcr.2014.11.017

Djanaguiraman M, Prasad PVV (2010) Ethylene production under high temperature stress causes premature leaf senescence in soybean. Funct Plant Biol 37:1071–1084. https://doi.org/10.1071/FP10089

Djanaguiraman M, Schapaugh W, Fritschi F, Nguyen H, Prasad PV (2019) Reproductive success of soybean (*Glycine max* L. Merrill) cultivars and exotic lines under high daytime temperature. Plant Cell Environ 42:321–336. https://doi.org/10.1111/pce.13421

Driedonks N, Rieu I, Vriezen WH (2016) Breeding for plant heat tolerance at vegetative and reproductive stages. Plant Reprod 29:67–79. https://doi.org/10.1007/s00497-016-0275-9

Dua RP, Chaturvedi SK, Sewak S (2001) Reference varieties of chickpea for IPR regime. Indian Institute of Pulses Research, Kanpur duration genotypes in chickpea (*Cicer arietinum* L.). Legume Res 21:121–124

Egli DB (1998) Seed biology and the yield of grain crops. CAB International, Oxford, UK, p 178

Ehlers JD, Hall AE (1998) Heat tolerance of contrasting cowpea lines in short and long days. Field Crops Res 55:11–21. https://doi.org/10.1016/S0378-4290(97)00055-5

Farooq M, Nadeem F, Gogoi N, Ullah A, Alghamdi SS, Nayyar H, Siddique KHM (2017) Heat stress in grain legumes during reproductive and grain-filling phases. Crop Pasture Sci 68:985–1005. https://doi.org/10.1071/CP17012

Fischer RA, Rees D, Sayre KD, Lu ZM, Condon AG, Saavedra AL (1998) Wheat yield progress associated with higher stomatal conductance and photosynthetic rate, and cooler canopies. Crop Sci 38:1467–1475. https://doi.org/10.2135/cropsci1998.0011183X003800060011x

Garnett T, Conn V, Kaiser BN (2009) Root based approaches to improving nitrogen use efficiency in plants. Plant Cell Environ 32:1272–1283. https://doi.org/10.1111/j.1365-3040.2009.02011.x

Gaur PM, Jukanti AK, Varshney RK (2012) Impact of genomic technologies on chickpea breeding strategies. Agron J 2:199–221. https://doi.org/10.3390/agronomy2030199

Gregersen PL, Culetic A, Boschian L, Krupinska K (2013) Plant senescence and crop productivity. Plant Mol Biol 82:603–622. https://doi.org/10.1007/s11103-013-0013-8

Gross Y, Kigel J (1994) Differential sensitivity to high temperature of stages in the reproductive development of common bean (*Phaseolus vulgaris* L.). Field Crops Res 36:201–212. https://doi.org/10.1016/0378-4290(94)90112-0

Gulen H, Eris A (2004) Effect of heat stress on peroxidase activity and total protein content in strawberry plants. Plant Sci 166:739–744. https://doi.org/10.1016/j.plantsci.2003.11.014

Harding SA, Guikema JA, Paulsen GM (1990) Photosynthetic decline from high temperature stress during maturation of wheat: I. Interaction with senescence processes. Plant Physiol 92:648–653. https://doi.org/10.1104/pp.92.3.648

Hasanuzzaman M, Nahar K, Alam M, Roychowdhury R, Fujita M (2013) Physiological, biochemical, and molecular mechanisms of heat stress tolerance in plants. Int J Mol Sci 14:9643–9684. https://doi.org/10.3390/ijms14059643

Hatfield JL, Prueger JH (2015) Temperature extremes: effect on plant growth and development. Weather Clim Extremes 10:4–10. https://doi.org/10.1016/j.wace.2015.08.001

Hedhly A (2011) Sensitivity of flowering plant gametophytes to temperature fluctuations. Environ Exp Bot 74:9–16. https://doi.org/10.1016/j.envexpbot.2011.03.016

Hemantaranjan A, Bhanu AN, Singh MN, Yadav DK, Patel PK, Singh R, Katiyar D (2014) Heat stress responses and thermotolerance. Adv Plants Agric Res 1:1–10

Herzog H, Chai-Arree W (2012) Gas exchange of five warm-season grain legumes and their susceptibility to heat stress. J Agron Crop Sci 198:466–474. https://doi.org/10.1111/j.1439-037X.2012.00517.x

Howarth CJ (2005) Genetic improvements of tolerance to high temperature. In: Ashraf M, Harris P (eds) Abiotic stresses: plant resistance through breeding and molecular approaches, 1st edn. CRC Press, Boca Raton, FL, pp 277–300

Huang B, Gao H (2000) Growth and carbohydrate metabolism of creeping bentgrass cultivars in response to increasing temperatures. Crop Sci 40:1115–1120. https://doi.org/10.2135/cropsci2000.4041115x

Huang B, Xu C (2008) Identification and characterization of proteins associated with plant tolerance to heat stress. J Integr Plant Biol 50:1230–1237. https://doi.org/10.1111/j.1744-7909.2008.00735.x

Huang S, Gali KK, Tar'an B, Warkentin TD, Bueckert RA (2017) Pea phenology: crop potential in a warming environment. Crop Sci 57:1540–1551. https://doi.org/10.2135/cropsci2016.12.0974

Iqbal A, Khalil IA, Ateeq N, Khan MS (2006) Nutritional quality of important food legumes. Food Chem 97:331–335. https://doi.org/10.1016/j.foodchem.2005.05.011

Ismail AM, Hall AE (1999) Reproductive-stage heat tolerance, leaf membrane thermostability and plant morphology in cowpea. Crop Sci 39:1762–1768. https://doi.org/10.2135/cropsci1999.3961762x

Ismail AM, Hall AE, Ehlers JD (2000) Delayed-leaf-senescence and heat-tolerance traits mainly are independently expressed in cowpea. Crop Sci 40:1049–1055. https://doi.org/10.2135/cropsci2000.4041049x

Janni M, Gullì M, Maestri E, Marmiroli M, Valliyodan B, Nguyen HT, Marmiroli N (2020) Molecular and genetic bases of heat stress responses in crop plants and breeding for increased resilience and productivity. J Exp Bot 71:3780–3802. https://doi.org/10.1093/jxb/eraa034

Jha UC, Bohra A, Singh NP (2014) Heat stress in crop plants: its nature, impacts and integrated breeding strategies to improve heat tolerance. Plant Breed 133:679–701. https://doi.org/10.1111/pbr.12217

Jha UC, Basu P, Singh D (2015) Genetic variation and diversity analysis of chickpea genotypes based on quantitative traits under high temperature stress. Int J Stress Manag 6:700–706. https://doi.org/10.5958/0976-4038.2015.00108.6

Jha UC, Bohra A, Jha R, Parida S (2017) Integrated 'omics' approaches to sustain major global grain legume productivity under heat stress. Plant Breed 136:437–459. https://doi.org/10.1111/pbr.12489

Jha UC, Jha R, Singh NP, Shil S, Kole PC (2018) Heat tolerance indices and their role in selection of heat stress tolerant chickpea (*Cicer arietinum*) genotypes. Indian J Agric Sci 88:260–267

Jha UC, Kole PC, Singh NP, Shil S, Kumar H (2019) GGE bi-plot analysis for grain yield in chickpea (*Cicer arietinum*) under normal and heat stress conditions. Indian J Agric Sci 89:721–725

Jha UC, Palakurthi R, Nayyar H, Jha R, Valluri VK, Bajaj P, Chitikineni A, Singh NP, Varshney RK, Thudi M (2021) Major QTLs and potential candidate genes for heat stress tolerance identified in chickpea (*Cicer arietinum* L.). Front Plant Sci 12:1–16. https://doi.org/10.3389/fpls.2021.655103

Jiang Y, Lahlali R, Karunakaran C, Kumar S, Davis AR, Bueckert RA (2015) Seed set, pollen morphology and pollen surface composition response to heat stress in field pea. Plant Cell Environ 38:2387–2397. https://doi.org/10.1111/pce.12589

Judd LA, Jackson BE, Fonteno WC (2015) Advancements in root growth measurement technologies and observation capabilities for container-grown plants. Plants 4:369–392. https://doi.org/10.3390/plants4030369

Jumrani K, Bhatia VS, Pandey GP (2018) Screening soybean genotypes for high temperature tolerance by in vitro pollen germination, pollen tube length, reproductive efficiency and seed yield. Indian J Plant Physiol 23:77–90. https://doi.org/10.1007/s40502-018-0360-1

Kakani VG, Prasad PVV, Craufurd PQ, Wheeler TR (2002) Response of in vitro pollen germination and pollen tube growth of groundnut (*Arachis hypogaea* L.) genotypes to temperature. Plant Cell Environ 25:1651–1661. https://doi.org/10.1046/j.1365-3040.2002.00943.x

Kalaji HM, Rastogi A, Živčák M, Brestic M, Daszkowska-Golec A, Sitko K, Cetner MD (2018) Prompt chlorophyll fluorescence as a tool for crop phenotyping: an example of barley landraces exposed to various abiotic stress factors. Photosynthetica 56:953–961. https://doi.org/10.1007/s11099-018-0766-z

Kaplan F, Kopka J, Haskell DW, Zhao W, Schiller KC, Gatzke N, Guy CL (2004) Exploring the temperature-stress metabolome of Arabidopsis. Plant Physiol 136:4159–4168. https://doi.org/10.1104/pp.104.052142

Kashiwagi J, Krishnamurthy L, Upadhyaya HD, Gaur PM (2008) Rapid screening technique for canopy temperature status and its relevance to drought tolerance improvement in chickpea. J Agric Res 6:105–104

Katam R, Shokri S, Murthy N, Singh SK, Suravajhala P, Khan MN, Reddy KR (2020) Proteomics, physiological, and biochemical analysis of cross tolerance mechanisms in response to heat and water stresses in soybean. PLoS One 15:1–29. https://doi.org/10.1371/journal.pone.0233905

Kaur G, Asthir BJBP (2015) Proline: a key player in plant abiotic stress tolerance. Biol Plant 59: 609–619. https://doi.org/10.1007/s10535-015-0549-3

Kaur R, Bains TS, Bindumadhava H, Nayyar H (2015) Responses of mungbean (*Vigna radiata* L.) genotypes to heat stress: effects on reproductive biology, leaf function and yield traits. Sci Hortic 197:527–541. https://doi.org/10.1016/j.scienta.2015.10.015

Kaushal N, Awasthi R, Gupta K, Gaur P, Siddique KHM, Nayyar H (2013) Heat-stress-induced reproductive failures in chickpea (*Cicer arietinum*) are associated with impaired sucrose metabolism in leaves and anthers. Funct Plant Biol 40:1334–1349. https://doi.org/10.1071/FP13082

Kovacs D, Kalmar E, Torok Z, Tompa P (2008) Chaperone activity of ERD10 and ERD14, two disordered stress-related plant proteins. Plant Physiol 147:381–390. https://doi.org/10.1104/pp.108.118208

Krause GH, Santarius KA (1975) Relative thermostability of the chloroplast envelope. Planta 127: 285–299. https://doi.org/10.1007/BF00380726

Krishnamurthy L, Gaur PM, Basu PS, Chaturvedi SK, Tripathi S, Vadez V (2011) Large genetic variation for heat tolerance in the reference collection of chickpea (*Cicer arietinum* L.) germplasm. Plant Genet Resour 9:59–69. https://doi.org/10.1017/S1479262110000

Kumar S, Kaur R, Kaur N, Bhandhari K, Kaushal N, Gupta K, Nayyar H (2011) Heat-stress induced inhibition in growth and chlorosis in mungbean (*Phaseolus aureus* Roxb.) is partly mitigated by ascorbic acid application and is related to reduction in oxidative stress. Acta Physiol Plant 33: 2091–2101. https://doi.org/10.1007/s11738-011-0748-2

Kumar S, Gupta D, Nayyar H (2012) Comparative response of maize and rice genotypes to heat stress: status of oxidative stress and antioxidants. Acta Physiol Plant 34:75–86. https://doi.org/10.1007/s11738-011-0806-9

Kumar S, Thakur P, Kaushal N, Malik JA, Gaur P, Nayyar H (2013) Effect of varying high temperatures during reproductive growth on reproductive function, oxidative stress and seed yield in chickpea genotypes differing in heat sensitivity. Arch Agron Soil Sci 59:823–843. https://doi.org/10.1080/03650340.2012.683424

Kumar R, Lavania D, Singh AK, Negi M, Siddiqui MH, Al-Whaibi MH, Grover A (2015) Identification and characterization of a small heat shock protein 17.9-CII gene from faba bean (*Vicia faba L.*). Acta Physiol Plant 37:1–13. https://doi.org/10.1007/s11738-015-1943-3

Kumar J, Kant R, Kumar S, Basu PS, Sarker A, Singh NP (2016) Heat tolerance in lentil under field conditions. Legume Genomics Genet 7:1–11. https://doi.org/10.5376/lgg.2016.07.0001

Kumar P, Shah D, Singh MP (2017) Evaluation of chickpea (*Cicer arietinum* L.) genotypes for heat tolerance: a physiological assessment. Indian J Plant Physiol 22:164–177. https://doi.org/10.1007/s40502-017-0301-4

Kumari M, Pudake RN, Singh VP, Joshi AK (2013) Association of stay green trait with canopy temperature depression and yield traits under terminal heat stress in wheat (*Triticum aestivum* L.). Euphytica 190:87–97. https://doi.org/10.1007/s10681-012-0780-3

Kumari P, Singh S, Yadav S (2018) Analysis of thermotolerance behaviour of five chickpea genotypes at early growth stages. Rev Bras Bot 41:551–565. https://doi.org/10.1007/s40415-018-0484-6

Lavania D, Siddiqui MH, Al-Whaibi MH, Singh AK, Kumar R, Grover A (2015) Genetic approaches for breeding heat stress tolerance in faba bean (*Vicia faba* L.). Acta Physiol Plant 37:1–9. https://doi.org/10.1007/s11738-014-1737-z

Li Z, Palmer WM, Martin AP, Wang R, Rainsford F, Jin Y (2012) High invertase activity in tomato reproductive organs correlates with enhanced sucrose import into, and heat tolerance of, young fruit. J Exp Bot 63:1155–1166. https://doi.org/10.1093/jxb/err329

Li Y, Chen X, Chen Z, Cai R, Zhang H, Xiang Y (2016) Identification and expression analysis of BURP domain-containing genes in *Medicago truncatula*. Front Plant Sci 7:1–16. https://doi.org/10.3389/fpls.2016.00485

Li B, Gao K, Ren H, Tang W (2018) Molecular mechanisms governing plant responses to high temperatures. J Integr Plant Biol 60:757–779. https://doi.org/10.1111/jipb.12701

Lichtenthaler HK (1987) Chlorophylls and carotenoids: pigments of photosynthetic biomembranes. Methods Enzymol 148:350–382. https://doi.org/10.1016/0076-6879(87)48036-1

Liu Y, Li J, Zhu Y, Jones A, Rose RJ, Song Y (2019) Heat stress in legume seed setting: effects, causes, and future prospects. Front Plant Sci 10:1–12. https://doi.org/10.3389/fpls.2019.00938

López-Hernández F, Cortés AJ (2019) Last-generation genome–environment associations reveal the genetic basis of heat tolerance in common bean (*Phaseolus vulgaris* L.). Front Genet 10:1–22. https://doi.org/10.3389/fgene.2019.00954

Lucas MR, Ehlers JD, Huynh BL, Diop NN, Roberts PA, Close TJ (2013) Markers for breeding heat-tolerant cowpea. Mol Breed 31:529–536. https://doi.org/10.1007/s11032-012-9810-z

Makonya GM, Ogola JB, Muasya AM, Crespo O, Maseko S, Valentine AJ, Chimphango SB (2019) Chlorophyll fluorescence and carbohydrate concentration as field selection traits for heat tolerant chickpea genotypes. Plant Physiol Biochem 141:172–182. https://doi.org/10.1016/j.plaphy.2019.05.031

Maphosa Y, Jideani VA (2017) The role of legumes in human nutrition. In: Hueda MC (ed) Functional food-improve health through adequate food. Intech Open, Rijeka, p 1, 13. https://doi.org/10.5772/intechopen.69127

Marfo KO, Hall AE (1992) Inheritance of heat tolerance during pod set in cowpea. Crop Sci 32:912–918. https://doi.org/10.2135/cropsci1992.0011183X003200040015x

Marsh LE, Davis DW, Li PH (1985) Selection and inheritance of heat tolerance in the common bean by use of conductivity. J Am Soc Hortic Sci 110:680–683

Martineau JR, Specht JE, Williams JH, Sullivan CY (1979) Temperature tolerance in soybeans. I. Evaluation of a technique for assessing cellular membrane thermostability. Crop Sci 19:75–78. https://doi.org/10.2135/cropsci1979.0011183X001900010017x

Mason RE, Singh RP (2014) Considerations when deploying canopy temperature to select high yielding wheat breeding lines under drought and heat stress. Agronomy 4:191–201. https://doi.org/10.3390/agronomy4020191

Mathur S, Jajoo A, Mehta P, Bharti S (2011) Analysis of elevated temperature-induced inhibition of photosystem II using chlorophyll a fluorescence induction kinetics in wheat leaves (*Triticum aestivum*). Plant Biol 13:1–6. https://doi.org/10.1111/j.1438-8677.2009.00319.x

McMichael BL, Quisenberry JE (1993) The impact of the soil environment on the growth of root systems. Environ Exp Bot 33:53–61. https://doi.org/10.1016/0098-8472(93)90055-K

Méthy M, Olioso A, Trabaud L (1994) Chlorophyll fluorescence as a tool for management of plant resources. Remote Sens Environ 47:2–9. https://doi.org/10.1016/0034-4257(94)90121-X

Miklas PN, Hannan R, Smith JR, Beaver JS, Riley R, Antonius S (2000) Transferring heat tolerance and indeterminancy from Indeterminate Jamaica Red (PI 163122) to kidney bean. Annu Rep Bean Improv Crop 43:68–69

Miroshnichenko S, Tripp J, Nieden UZ, Neumann D, Conrad U, Manteuffel R (2005) Immunomodulation of function of small heat shock proteins prevents their assembly into heat stress granules and results in cell death at sublethal temperatures. Plant J 41:269–281. https://doi.org/10.1111/j.1365-313X.2004.02290.x

Mishra S, Babbar A (2014) Selection strategies to assess the promising kabuli chickpea promising lines under normal and heat stress environments. Electron J Plant Breed 5:260–267

Mo Y, Liang G, Shi W, Xie J (2011) Metabolic responses of alfalfa (*Medicago sativa* L.) leaves to low and high temperature induced stresses. Afr J Biotechnol 10:1117–1124. https://doi.org/10.5897/AJB10.1433

Mohapatra C, Chand R, Tiwari JK, Singh AK (2020) Effect of heat stress during flowering and pod formation in pea (*Pisum sativum* L.). Physiol Mol Biol Plant 26:1119–1125. https://doi.org/10.1007/s12298-020-00803-4

Molina-Bravo R, Arellano C, Sosinski BR, Fernandez GE (2011) A protocol to assess heat tolerance in a segregating population of raspberry using chlorophyll fluorescence. Sci Hortic 130:524–530. https://doi.org/10.1016/j.scienta.2011.07.022

Nagel KA, Kastenholz B, Jahnke S, Van Dusschoten D, Aach T, Mühlich M, Schurr U (2009) Temperature responses of roots: impact on growth, root system architecture and implications for phenotyping. Funct Plant Biol 36:947–959. https://doi.org/10.1071/FP09184

Nahar K, Hasanuzzaman M, Fujita M (2016) Heat stress responses and thermotolerance in soybean. In: Miransari M (ed) Abiotic and biotic stresses in soybean production, 1. Academic, San Diego, pp 261–284. https://doi.org/10.1016/B978-0-12-801536-0.00012-8

Neumann D, Lichtenberger O, Günther D, Tschiersch K, Nover L (1994) Heat-shock proteins induce heavy-metal tolerance in higher plants. Planta 194:360–367. https://doi.org/10.1007/BF00197536

Nielsen CL, Hall AE (1985) Responses of cowpea (*Vigna unguiculata* (L.) Walp.) in the field to high night air temperature during flowering. I. Thermal regimes of production regions and field experimental system. Field Crops Res 10:167–179. https://doi.org/10.1016/0378-4290(85)90024-3

Nieto-Sotelo J, Martínez LM, Ponce G, Cassab GI, Alagón A, Meeley RB, Yang R (2002) Maize HSP101 plays important roles in both induced and basal thermotolerance and primary root growth. Plant Cell 14:1621–1633. https://doi.org/10.1105/tpc.010487

Niones JM, Suralta RR, Inukai Y, Yamauchi A (2013) Roles of root aerenchyma development and its associated QTL in dry matter production under transient moisture stress in rice. Plant Prod Sci 16:205–216. https://doi.org/10.1626/pps.16.205

Ntare BR, Williams JH, Fatondji D (2001) Evaluation of groundnut genotypes for heat tolerance under field conditions in a Sahelian environment using a simple physiological model for yield. J Agric Sci 136:81–88. https://doi.org/10.1017/S0021859600008583

Parankusam S, Bhatnagar-Mathur P, Sharma KK (2017) Heat responsive proteome changes reveal molecular mechanisms underlying heat tolerance in chickpea. Environ Exp Bot 141:132–144. https://doi.org/10.1016/j.envexpbot.2017.07.007

Pareek A, Rathi D, Mishra D, Chakraborty S, Chakraborty N (2019) Physiological plasticity to high temperature stress in chickpea: adaptive responses and variable tolerance. Plant Sci 289:110258. https://doi.org/10.1016/j.plantsci.2019.110258

Patel PN, Hall AE (1988) Inheritance of heat-induced brown discoloration in seed coats of cowpea. Crop Sci 28:929–932. https://doi.org/10.2135/cropsci1988.0011183X002800060011x

Paul PJ, Samineni S, Thudi M, Sajja SB, Rathore A, Das RR, Gaur PM (2018) Molecular mapping of QTLs for heat tolerance in chickpea. Int J Mol Sci 19:1–20. https://doi.org/10.3390/ijms19082166

Petkova V, Denev ID, Cholakov D, Porjazov I (2007) Field screening for heat tolerant common bean cultivars (*Phaseolus vulgaris* L.) by measuring of chlorophyll fluorescence induction parameters. Sci Hortic 111:101–106. https://doi.org/10.1016/j.scienta.2006.10.005

Petkova V, Denev I, Stefanov D (2009) Resistance to high temperature stress of various bean (*Phaseolus vulgaris* L.) cultivars and lines. Gen Appl Plant Physiol 35:117–121

Piramila BHM, Prabha AL, Nandagopalan V, Stanley AL (2012) Effect of heat treatment on germination, seedling growth and some biochemical parameters of dry seeds of black gram. Int J Pharm Phytopharmacol Res 1:194–202

Porch TG (2006) Application of stress indices for heat tolerance screening of common bean. J Agron Crop Sci 192:390–394. https://doi.org/10.1111/j.1439-037X.2006.00229.x

Pottorff M, Roberts PA, Close TJ, Lonardi S, Wanamaker S, Ehlers JD (2014) Identification of candidate genes and molecular markers for heat-induced brown discoloration of seed coats in cowpea [*Vigna unguiculata* (L.) Walp]. BMC Genomics 15:1–11. https://doi.org/10.1186/1471-2164-15-328

Prasad PV, Djanaguiraman M (2011) High night temperature decreases leaf photosynthesis and pollen function in grain sorghum. Funct Plant Biol 38:993–1003. https://doi.org/10.1071/FP11035

Prasad PV, Craufurd PQ, Summerfield RJ (1999) Sensitivity of peanut to timing of heat stress during reproductive development. Crop Sci 39:1352–1357. https://doi.org/10.2135/cropsci1999.3951352x

Prasad PVV, Staggenborg SA, Ristic Z (2008) Impacts of drought and/or heat stress on physiological, developmental, growth, and yield processes of crop plants. In: Ahuja LR, Reddy VR, Saseendran SA, Yu Q (eds) Response of crops to limited water: understanding and modeling water stress effects on plant growth processes. ASA-CSSA-SSSA, Baltimore, MD, pp 301–356

Prasad PV, Bheemanahalli R, Jagadish SK (2017) Field crops and the fear of heat stress—opportunities, challenges and future directions. Field Crop Res 200:114–121. https://doi.org/10.1016/j.fcr.2016.09.024

Pressman E, Peet MM, Pharr DM (2002) The effect of heat stress on tomato pollen characteristics is associated with changes in carbohydrate concentration in the developing anthers. Ann Bot 90: 631–636. https://doi.org/10.1093/aob/mcf240

Priya M, Siddique KHM, Dhankhar OP, Prasad PV, Rao BH, Nair RM, Nayyar H (2018) Molecular breeding approaches involving physiological and reproductive traits for heat tolerance in food crops. Indian J Plant Physiol 23:697–720. https://doi.org/10.1007/s40502-018-0427-z

Puteh AB, Thu Zar M, Mondal MMA, Abdullah APB, Halim MRA (2013) Soybean [*Glycine max* (L.) Merrill] seed yield response to high temperature stress during reproductive growth stages. Aust J Crop Sci 7:1472–1479. https://doi.org/10.3316/informit.618691733672149

Raina SK, Rane J, Raskar N, Singh AK, Govindasamy V, Kumar M, Minhas PS (2019) Physiological traits reveal potential for identification of drought tolerant mungbean [*Vigna radiata* (L.) Wilczek] genotypes under moderate soil-moisture deficit. Indian J Genet 79:427–437. https://doi.org/10.31742/IJGPB.79.2.6

Rainey KM, Griffiths PD (2005) Differential response of common bean genotypes to high temperature. J Am Soc Hortic Sci 130:18–23. https://doi.org/10.21273/JASHS.130.1.18

Raja MM, Vijayalakshmi G, Naik ML, Basha PO, Sergeant K, Hausman JF, Khan PSSV (2019) Pollen development and function under heat stress: from effects to responses. Acta Physiol Plant 41:1–20. https://doi.org/10.1007/s11738-019-2835-8

Ramalingam A, Kudapa H, Pazhamala LT, Weckwerth W, Varsh-ney RK (2015) Proteomics and metabolomics: two emerging areas for legume improvement. Front Plant Sci 6:1–21. https://doi.org/10.3389/fpls.2015.01116

Rathi D, Gayen D, Gayali S, Chakraborty S, Chakraborty N (2016) Legume proteomics: progress, prospects, and challenges. Proteomics 16:310–327. https://doi.org/10.1002/pmic.201500257

Reddy PS, Mallikarjuna G, Kaul T, Chakradhar T, Mishra RN, Sopory SK, Reddy MK (2010) Molecular cloning and characterization of gene encoding for cytoplasmic Hsc70 from *Pennisetum glaucum* may play a protective role against abiotic stresses. Mol Genet Genomics 283:243–254. https://doi.org/10.1007/s00438-010-0518-7

Reddy PS, Chakradhar T, Reddy RA, Nitnavare RB, Mahanty S, Reddy MK (2016) Role of heat shock proteins in improving heat stress tolerance in crop plants. In: Asea AAA, Kaur P, Calderwood SK (eds) Heat shock proteins and plants. Springer, Cham, Switzerland p, pp 283–307

Reynolds MP (1997) Using canopy temperature depression to select for yield potential of wheat in heat-stressed environments, vol No. 42. CIMMYT, Mexico

Reynolds M, Langridge P (2016) Physiological breeding. Curr Opin Plant Biol 31:162–171. https://doi.org/10.1016/j.pbi.2016.04.005

Reynolds MP, Balota M, Delgado MIB, Amani I, Fischer RA (1994) Physiological and morphological traits associated with spring wheat yield under hot, irrigated conditions. Funct Plant Biol 21:717–730. https://doi.org/10.1071/PP9940717

Reynolds MP, Nagarajan S, Razzaque MA, Ageeb OAA (2001) Heat tolerance. In Reynolds MP, Ortiz-Monasterio JI, McNab A (eds) Application of physiology in wheat breeding CIMMYT, Mexico, p 124–135

Reynolds MP, Nagarayan S, Razzaue MA, Ageeb OAA (1997) Using canopy temperature depression to select for yield potential of wheat in heat-stressed environments. Wheat Special Rep. No. 42. CIMMYT, Mexico

Rodziewicz P, Swarcewicz B, Chmielewska K, Wojakowska A, Stobiecki M (2014) Influence of abiotic stresses on plant proteome and metabolome changes. Acta Physiol Plant 36:1–19. https://doi.org/10.1007/s11738-013-1402-y

Rurek M (2010) Diverse accumulation of several dehydrin-like proteins in cauliflower (*Brassica oleracea* var. botrytis), *Arabidopsis thaliana* and yellow lupin (*Lupinus luteus*) mitochondria under cold and heat stress. BMC Plant Biol 10:1–17. https://doi.org/10.1186/1471-2229-10-181

Ryan PR, Delhaize E, Watt M, Richardson AE (2016) Plant roots: understanding structure and function in an ocean of complexity. Ann Bot 118:555–559. https://doi.org/10.1093/aob/mcw192

Sabagh AE, Hossain A, Islam MS, Iqbal MA, Fahad S, Ratnasekera D et al (2020) Consequences and mitigation strategies of heat stress for sustainability of soybean (*Glycine max* L. Merr.) production under the changing climate. In: Plant stress physiology. Intech Open, London, UK. https://doi.org/10.5772/intechopen.92098

Sadras VO (2009) Does partial root-zone drying improve irrigation water productivity in the field? A meta-analysis. Irrig Sci 27:183–190. https://doi.org/10.1007/s00271-008-0141-0

Saelim S, Zwiazek JJ (2000) Preservation of thermal stability of cell membranes and gas exchange in high temperature acclimated *Xylia xylocarpa* seedlings. J Plant Physiol 156:380–385. https://doi.org/10.1016/S0176-1617(00)80077-2

Salem MA, Kakani VG, Koti S, Reddy KR (2007) Pollen-based screening of soybean genotypes for high temperatures. Crop Sci 47:219–231. https://doi.org/10.2135/cropsci2006.07.0443

Salvucci Michael E, Crafts-Brandner SJ (2004) Relationship between the heat tolerance of photosynthesis and the thermal stability of Rubisco activase in plants from contrasting thermal environments. Plant Physiol 134:1460–1470. https://doi.org/10.1104/pp.103.038323

Sapra VT, Anaele AO (1991) Screening soybean genotypes for drought and heat tolerance. J Agron Crop Sci 167:96–102. https://doi.org/10.1111/j.1439-037X.1991.tb00939.x

Savchenko GE, Klyuchareva EA, Abramchik LM, Serdyuchenko EV (2002) Effect of periodic heat shock on the inner membrane system of etioplasts. Russ J Plant Physiol 49:349–359. https://doi.org/10.1023/A:1015592902659

Schmit R, de Melo RC, Trevisani N, Guidolin AF, Coimbra JLM (2019) Screening and agronomic benefits of the stay-green trait in common bean genotypes. J Biosci 35:869–877. https://doi.org/10.14393/BJ-v35n3a2019-42022

Sehgal A, Sita K, Kumar J, Kumar S, Singh S, Siddique KH, Nayyar H (2017) Effects of drought, heat and their interaction on the growth, yield and photosynthetic function of lentil (*Lens culinaris* Medikus) genotypes varying in heat and drought sensitivity. Front Plant Sci 8:1–19. https://doi.org/10.3389/fpls.2017.01776

Sehgal A, Sita K, Siddique KHM, Kumar R, Bhogireddy S, Varshney RK et al (2018) Drought or/and heat-stress effects on seed filling in food crops: impacts on functional biochemistry, seed yields, and nutritional quality. Front Plant Sci 871:1–19. https://doi.org/10.3389/fpls.2018.01705

Selvaraj MG, Burow G, Burke JJ, Belamkar V, Puppala N, Burow MD (2011) Heat stress screening of peanut (*Arachis hypogaea* L.) seedlings for acquired thermotolerance. Plant Growth Regul 65:83–91. https://doi.org/10.1007/s10725-011-9577-y

Serrano N, Ling Y, Bahieldin A, Mahfouz MM (2019) Thermopriming reprograms metabolic homeostasis to confer heat tolerance. Sci Rep 9:1–14. https://doi.org/10.1038/s41598-018-36484-z

Shanmugam S, Kjær KH, Ottosen CO, Rosenqvist E, Kumari Sharma D, Wollen Weber B (2013) The alleviating effect of elevated CO_2 on heat stress susceptibility of two wheat (*Triticum aestivum* L.) cultivars. J Agron Crop Sci 199:340–350. https://doi.org/10.1111/jac.12023

Sharifi P, Amirnia R, Majidi E, Hadi H, Nakhoda B, Alipoor HM, Moradi F (2012) Relationship between drought stress and some antioxidant enzymes with cell membrane and chlorophyll stability in wheat lines. Afr J Microbiol Res 6:617–623. https://doi.org/10.5897/AJMR

Sharkey TD (2005) Effects of moderate heat stress on photosynthesis: importance of thylakoid reactions, rubisco deactivation, reactive oxygen species, and thermotolerance provided by isoprene. Plant Cell Environ 28:269–277. https://doi.org/10.1111/j.1365-3040.2005.01324.x

Sharma DK, Andersen SB, Ottosen CO, Rosenqvist E (2012) Phenotyping of wheat cultivars for heat tolerance using chlorophyll a fluorescence. Funct Plant Biol 39:936–947. https://doi.org/10.1071/FP12100

Sharma DK, Andersen SB, Ottosen CO, Rosenqvist E (2015) Wheat cultivars selected for high Fv/Fm under heat stress maintain high photosynthesis, total chlorophyll, stomatal conductance, transpiration and dry matter. Physiol Plant 153:284–298. https://doi.org/10.1111/ppl.12245

Sharma L, Priya M, Bindumadhava H, Nair RM, Nayyar H (2016) Influence of high temperature stress on growth, phenology and yield performance of mungbean [*Vigna radiata* (L.) Wilczek] under managed growth conditions. Sci Hortic 213:379–391. https://doi.org/10.1016/j.scienta.2016.10.033

Shonnard GC, Gepts P (1994) Genetics of heat tolerance during reproductive development in common bean. Crop Sci 34:1168–1175. https://doi.org/10.2135/cropsci1994.0011183X003400050005x

Siddiqui MH, Al-Khaishany MY, Al-Qutami MA, Al-Whaibi MH, Grover A, Ali HM, Al-Wahibi MS (2015) Morphological and physiological characterization of different genotypes of faba bean under heat stress. Saudi J Biol Sci 22:656–663. https://doi.org/10.1016/j.sjbs.2015.06.002

Simões-Araújo JL, Rumjanek NG, Margis-Pinheiro M (2003) Small heat shock proteins genes are differentially expressed in distinct varieties of common bean. Braz J Plant Physiol 15:33–41. https://doi.org/10.1590/S1677-04202003000100005

Singh N, Sandhu KS, Kaur M (2004) Characterization of starches separated from Indian chickpea (*Cicer arietinum* L.) cultivars. J Food Eng 63:441–449. https://doi.org/10.1016/j.jfoodeng.2003.09.003

Singh SK, Kakani VG, Surabhi GK, Reddy KR (2010) Cowpea (*Vigna unguiculata* [L.] Walp.) genotypes response to multiple abiotic stresses. J Photochem Photobiol B 100:135–146. https://doi.org/10.1016/j.jphotobiol.2010.05.013

Singh D, Singh CK, Tomar RSS, Chaturvedi AK, Shah D, Kumar A, Pal M (2016) Exploring genetic diversity for heat tolerance among lentil (*Lens culinaris* Medik.) genotypes of variant habitats by simple sequence repeat markers. Plant Breed 135:215–223. https://doi.org/10.1111/pbr.12341

Singh D, Singh CK, Singh Tomar RS, Pal M (2017) Genetics and molecular mapping of heat tolerance for seedling survival and pod set in lentil. Crop Sci 57:3059–3067. https://doi.org/10.2135/cropsci2017.05.0284

Singh AK, Ganapathysubramanian B, Sarkar S, Singh A (2018) Deep learning for plant stress phenotyping: trends and future perspectives. Trends Plant Sci 23:883–898. https://doi.org/10.1016/j.tplants.2018.07.004

Singh D, Singh CK, Taunk J, Jadon V, Pal M, Gaikwad K (2019) Genome wide transcriptome analysis reveals vital role of heat responsive genes in regulatory mechanisms of lentil (*Lens culinaris* Medikus). Sci Rep 9:1–19. https://doi.org/10.1038/s41598-019-49496-0

Sita K, Sehgal A, Kumar J, Kumar S, Singh S, Siddique KHM, Nayyar H (2017) Identification of high-temperature tolerant lentil (*Lens culinaris* Medik.) genotypes through leaf and pollen traits. Front Plant Sci 8:744. https://doi.org/10.3389/fpls.2017.00744

Smith S, De Smet I (2012) Root system architecture: insights from Arabidopsis and cereal crops. Phil Trans R Soc B 367:1441–1452. https://doi.org/10.1098/rstb.2011.0234

Snider JL, Oosterhuis DM, Loka DA, Kawakami EM (2011) High temperature limits in vivo pollen tube growth rates by altering diurnal carbohydrate balance in field-grown *Gossypium hirsutum* pistils. J Plant Physiol 168:1168–1175. https://doi.org/10.1016/j.jplph.2010.12.011

Soltani A, Weraduwage SM, Sharkey TD, Lowry DB (2019) Elevated temperatures cause loss of seed set in common bean (*Phaseolus vulgaris* L.) potentially through the disruption of source-sink relationships. BMC Genomics 20:1–18. https://doi.org/10.1186/s12864-019-5669-2

Song C, Chung WS, Lim CO (2016) Overexpression of heat shock factor gene HsfA3 increases galactinol levels and oxidative stress tolerance in Arabidopsis. Mol Cells 39:477–483. https://doi.org/10.14348/molcells.2016.0027

Srikanthbabu V, Krishnaprasad BT, Gopalakrishna R, Savitha M, Udayakumar M (2002) Identification of pea genotypes with enhanced thermotolerance using temperature induction response technique (TIR). J Plant Physiol 159:535–545. https://doi.org/10.1078/0176-1617-00650

Srinivasan A, Takeda H, Senboku T (1996) Heat tolerance in food legumes as evaluated by cell membrane thermostability and chlorophyll fluorescence techniques. Euphytica 88:35–45. https://doi.org/10.1007/BF00029263

Strock CF, Burridge J, Massas AS, Beaver J, Beebe S, Camilo SA, Fourie D, Jochua C, Miguel M, Miklas PN, Mndolwa E, Nchimbi-Msolla S, Polania J, Porch TG, Rosas JC, Trapp JJ, Lynch JP (2019) Seedling root architecture and its relationship with seed yield across diverse environments in *Phaseolus vulgaris*. Field Crops Res 237:53–64. https://doi.org/10.1016/j.fcr.2019.04.012

Sullivan CY, Ross WM (1979) Selecting for drought and heat resistance in grain sorghum. In: Stress physiology in crop plants. Wiley, New York, pp 263–281

Sultan SE (2000) Phenotypic plasticity for plant development, function and life history. Trends Plant Sci 5:537–542. https://doi.org/10.1016/S1360-1385(00)01797-0

Suzuki K, Tsukaguchi T, Takeda H, Egawa Y (2001) Decrease of pollen stainability of green bean at high temperatures and relationship to heat tolerance. J Am Soc Hortic Sci 126:571–574. https://doi.org/10.21273/jashs.126.5.571

Suzuki N, Koussevitzky S, Mittler R, Miller G (2012) ROS and redox signalling in the response of plants to abiotic stress. Plant Cell Environ 35:259–270. https://doi.org/10.1111/j.1365-3040.2011.02336.x

Szepesi Á, Szőllősi R (2018) Mechanism of proline biosynthesis and role of proline metabolism enzymes under environmental stress in plants. In: Plant metabolites and regulation under environmental stress. Academic, San Diego, pp 337–353. https://doi.org/10.1016/B978-0-12-812689-9.00017-0

Tafesse EG, Warkentin TD, Bueckert RA (2019) Canopy architecture and leaf type as traits of heat resistance in pea. Field Crops Res 241:1–11. https://doi.org/10.1016/j.fcr.2019.107561

Tafesse EG, Gali KK, Lachagari VB, Bueckert R, Warkentin TD (2020) Genome-wide association mapping for heat stress responsive traits in field pea. Int J Mol Sci 21:1–26. https://doi.org/10.3390/ijms21062043

Tahir ISA, Nakata N, Yamaguchi T, Nakano J, Ali AM (2008) Influence of high shoot and root-zone temperatures on growth of three wheat genotypes during early vegetative stages. J Agron Crop Sci 194:141–151. https://doi.org/10.1111/j.1439-037X.2008.00298.x

Thudi M, Upadhyaya HD, Rathore A, Gaur PM, Krishnamurthy L, Roorkiwal M et al (2014) Genetic dissection of drought and heat tolerance in chickpea through genome-wide and candidate gene-based association mapping approaches. PLoS One 9:1–12. https://doi.org/10.1371/journal.pone.0096758

Tokyol A, Turhan E (2019) Heat stress tolerance of some green bean (*Phaseolus vulgaris* L.) genotypes. Agron Ser Sci Res 61:472–479

Traub J, Porch T, Naeem M, Urrea CA, Austic G, Kelly JD, Loescher W (2018) Screening for heat tolerance in *Phaseolus* spp. using multiple methods. Crop Sci 58:2459–2469. https://doi.org/10.2135/cropsci2018.04.0275

Tsukaguchi T, Kawamitsu Y, Takeda H, Suzuki K, Egawa Y (2003) Water status of flower buds and leaves as affected by high temperature in heat-tolerant and heat-sensitive cultivars of snap bean (*Phaseolus vulgaris* L.). Plant Prod Sci 6:24–27. https://doi.org/10.1626/pps.6.24

Upadhyaya HD, Dronavalli N, Gowda CLL, Singh S (2011) Identification and evaluation of chickpea germplasm for tolerance to heat stress. Crop Sci 51:2079–2094. https://doi.org/10.2135/cropsci2011.01.0018

Urban J, Ingwers M, McGuire MA, Teskey RO (2017) Stomatal conductance increases with rising temperature. Plant Signal Behav 12:1–3. https://doi.org/10.1080/15592324.2017.1356534

Vacha F, Adamec F, Valenta J, Vacha M (2007) Spatial location of photosystem pigment–protein complexes in thylakoid membranes of chloroplasts of *Pisum sativum* studied by chlorophyll fluorescence. J Luminesc 122–123:301–303. https://doi.org/10.1016/j.jlumin.2006.01.148

Valdés-López O, Batek J, Gomez-Hernandez N, Nguyen CT, Isidra-Arellano MC, Zhang N et al (2016) Soybean roots grown under heat stress show global changes in their transcriptional and proteomic profiles. Front Plant Sci 7:1–12. https://doi.org/10.3389/fpls.2016.00517

Varshney RK, Thudi M, Roorkiwal M, He W, Upadhyaya HD, Yang W, Bajaj P, Cubry P, Rathore A, Jian J et al (2019) Resequencing of 429 chickpea accessions from 45 countries provides insights into genome diversity, domestication and agronomic traits. Nat Genet 51:857–864. https://doi.org/10.1038/s41588-019-0401-3

Vierling E (1991) The roles of heat shock proteins in plants. Annu Rev Plant Biol 42:579–620. https://doi.org/10.1146/annurev.pp.42.060191.003051

Wahid A, Shabbir A (2005) Induction of heat stress tolerance in barley seedlings by pre-sowing seed treatment with glycinebetaine. Plant Growth Regul 46:133–141. https://doi.org/10.1007/s10725-005-8379-5

Wahid A, Gelani S, Ashraf M, Foolad MR (2007) Heat tolerance in plants: an overview. Environ Exp Bot 61:199–223. https://doi.org/10.1016/j.envexpbot.2007.05.011

Wang L, Liu L, Ma Y, Li S, Dong S, Zu W (2018) Transcriptome profiling analysis characterized the gene expression patterns responded to combined drought and heat stresses in soybean. Comput Biol Chem 77:413–429. https://doi.org/10.1016/j.compbiolchem.2018.09.012

Warrag MOA, Hall AE (1983) Reproductive responses of cowpea to heat stress: genotypic differences in tolerance to heat at flowering. Crop Sci 23:1088–1092. https://doi.org/10.2135/cropsci1983.0011183X002300060016x

Wassie M, Zhang W, Zhang Q, Ji K, Chen L (2019) Effect of heat stress on growth and physiological traits of alfalfa (*Medicago sativa* L.) and a comprehensive evaluation for heat tolerance. Agronomy 9:1–20. https://doi.org/10.3390/agronomy9100597

Waters ER, Lee GJ, Vierling E (1996) Evolution, structure and function of the small heat shock proteins in plants. J Exp Bot 47:325–338. https://doi.org/10.1093/jxb/47.3.325

Willits DH, Peet MM (2001) Measurement of chlorophyll fluorescence as a heat stress indicator in tomato: laboratory and greenhouse comparisons. J Am Soc Hortic Sci 126:188–194. https://doi.org/10.21273/JASHS.126.2.188

Wilson JM, Greaves JA (1990) Assessment of chilling sensitivity by chlorophyll fluorescence analysis. In: Wang YC (ed) Chilling injury of horticultural crops. CRC Press, Boca Raton, FL, pp 129–141

Wu W, Ma BL, Whalen JK (2018) Enhancing rapeseed tolerance to heat and drought stresses in a changing climate: perspectives for stress adaptation from root system architecture. Adv Agron 151:87–157. https://doi.org/10.1016/bs.agron.2018.05.002

Xu Q, Huang B (2000) Effects of differential air and soil temperature on carbohydrate metabolism in creeping bentgrass. Crop Sci 40:1368–1374. https://doi.org/10.2135/cropsci2000.4051368x

Xu P, Zhao PX, Cai XT, Mao JL, Miao ZQ, Xiang CB (2020) Integration of jasmonic acid and ethylene into auxin signaling in root development. Front Plant Sci 11:271. https://doi.org/10.3389/fpls.2020.00271

Xu S, Li J, Zhang X, Wei H, Cui L (2006) Effects of heat acclimation pretreatment on changes of membrane lipid peroxidation, antioxidant metabolites, and ultrastructure of chloroplasts in two cool-season turfgrass species under heat stress. Environ Exp Bot 56:274–285. https://doi.org/10.1016/j.envexpbot.2005.03.002

Yamada M, Hidaka T, Fukamachi H (1996) Heat tolerance in leaves of tropical fruit crops as measured by chlorophyll fluorescence. Sci Hortic 67:39–48. https://doi.org/10.1016/S0304-4238(96)00931-4

Yoldas F, Esiyok D (2009) The influence of temperature on growth and yield of green beans for processing. Int J Agric Res 4:123–130

Zhang J, Huang W, Pan Q, Liu Y (2005) Improvement of chilling tolerance and accumulation of heat shock proteins in grape berries (*Vitis vinifera* cv. Jingxiu) by heat pretreatment. Postharvest Biol Technol 38:80–90. https://doi.org/10.1016/j.postharvbio.2005.05.008

Traits Associated with Drought and High-Temperature Stress and Its Associated Mechanisms in Legumes

3

M. Djanaguiraman, B. Rakavi, and P. Jeyakumar

Abstract

Developments in crop management practices and novel breeding methods are important to sustain crop productivity for the current and upcoming challenges caused by drought and high-temperature stresses because the occurrence of these stresses during crop growth stages is determinantal to crop yield. Direct selection for yield per se under any abiotic stress conditions is often ineffective because of the low heritability for yield. One of the ways to increase the selection efficiency for stress tolerance is to select for any secondary traits which are easy to measure, presenting high heritability, and correlate highly with grain yield under stress situations. In this chapter, the secondary traits like plant water status, green leaf area duration, limited transpiration, canopy temperature depression, root architecture, early morning flowering, membrane integrity, photochemical efficiency, stem carbohydrate mobilization, and yield-associated traits are discussed. The above plant traits can be quantified under both controlled and field environments. The possibility of converting these traits under controlled environments into a method of quantification at field scale depends on the advancements in allied sectors of sciences, like spectroscopy, remote sensing, aeronautics, and high-end computing facilities. The use of these traits as a selection tool in crop breeding will pave the way for the development of drought and high temperature stress-tolerant genotypes.

Keywords

Legumes · Drought · High-temperature stress · Phenotyping traits · Yield

M. Djanaguiraman (✉) · B. Rakavi · P. Jeyakumar
Department of Crop Physiology, Tamil Nadu Agricultural University, Coimbatore, Tamil Nadu, India
e-mail: jani@tnau.ac.in

P. Muthu Arjuna Samy et al. (eds.), *Legumes: Physiology and Molecular Biology of Abiotic Stress Tolerance*, https://doi.org/10.1007/978-981-19-5817-5_3

3.1 Introduction

Phenotype is defined as the observable effect of a genotype and its interaction with a given environment. Phenotyping is the application of protocols and methodologies to measure the traits specifically related to the structure of plant or function to facilitate the right selection in the breeding program or to complement genotypic data for the identification of associated genes. Under plant breeding, selection can be defined as the science of discriminating among the biological variants in a population to detect and pick the desirable recombinants. However, the identification of recombinants that leads to superior phenotype is challenging because of the interaction of genotype with the environment. The major challenge in phenotyping is that it involves a large workforce, is expensive, and is prone to error if the methodology was not meticulously followed. Ample literature is available on the mechanisms of tolerance to drought and HT. However, many of the traits that have been reported to be promising for stress tolerance are not feasible for screening genotypes in large-scale or high-throughput mode.

In this chapter, we consider the traits which can be recorded at the lab, controlled environment, and field level. More information about field-based high-throughput phenotyping systems can be found in the review article by Cobb et al. (2013). The innovative use of technology and cautious development of tools to automate the processes without sacrificing predictive power will be very critical in the phenotyping platforms. Standardized phenotyping structures are not feasible practically for all research-related questions, but with thorough consideration and defined objectives, several techniques can be harnessed to examine specific characters under high-throughput settings. Agronomically important traits that are observable at the canopy level can help in discriminating the genotypes based on their capability to capture and use the natural resources, and these traits serve as a proxy for important agronomic characters. If traits associated with stress tolerance are identified and validated in a wide range of crops, then methodologies could be developed to quantify those traits under high-throughput systems using technologies such as digital imagery, remote sensing, robotics, thermography, and farm machinery.

3.2 Traits Associated with Drought and High Temperature (HT) Stress Tolerance and Its Phenotyping Method

3.2.1 Green Leaf Area Duration

Genotypes exhibiting the extended green leaf duration area are referred to as stay-green phenotypes. In contrast to cosmetic stay green, functional stay-green leaves supply more photosynthates to developing grains and can thus significantly contribute to grain yield (Thomas and Howarth 2000). Stay-green genotypes in various cereals have been reported, and selection of stay green has been targeted to improve the crop yield under drought and HT stress. Stay green can be studied at basic cell level, leaf level, or whole-plant level. At the cell level, Western blot analysis of proteins from leaves will indicate the integrity of chloroplast-associated proteins. In stay-green genotypes, delayed degradation of proteins is correlated with the phenomenon of delayed senescence onset (Borrell et al. 2001).

At leaf level, chlorophyll index measured through SPAD meter or chlorophyll meter is most frequently used to assess the greenness of the leaves. The chlorophyll index obtained from the SPAD meter is a point estimate (not cumulative), and it represents a section of the leaf and can be an indicator of the leaf senescence process. Studies have indicated that there is a strong positive relationship between chlorophyll index with actual leaf chlorophyll content (Hebbar et al. 2016). These phenotyping tools can allow the frequent measurement to assess the rate of leaf senescence, which is otherwise difficult if conventional spectrophotometric measurements are employed.

At the plant level, the expression of a stay-green trait can be evaluated using a stay-green score, or measuring canopy reflectance through a green seeker, and a spectroradiometer. Spectral reflectance from canopies over the visible and near-infrared (NIR) regions is mainly influenced by the canopy structure, leaf pigmentation, protein contents, and leaf water (Homolova et al. 2013). A healthy green canopy absorbs most of the red spectrum and reflects most of the NIR spectrum because chlorophyll molecule absorbs blue and red color and mesophyll region reflects the near-infrared spectrum. The normalized difference vegetation index (NDVI) is widely used to quantify the leaf senescence process or greenness of the canopy. The NDVI can be obtained at ground level, and from high, low, and satellite altitudes. The portable NDVI sensor, namely green seeker, provides rapid information on leaf area index and green area index. NDVI $= (R_{NIR} - R_{Red})/(R_{NIR} + R_{Red})$, where R_{NIR} is the reflectance at the NIR region, and R_{Red} is the reflectance at the red region. The majority of the portable NDVI sensors have their own sources of light, which allows the measurements to be made at any time of the day and any light condition. For NDVI sensor without a light source, measurements can be made on a bright sunny day with negligible wind, because slight wind or breeze can significantly alter the plant canopy structure. The plant surface should be devoid of dew, irrigation, and rain. Measurements can be made at any of the developmental stages or regular intervals from the emergence to maturity stage depending on the objective of the study. If the objective of the experiment is to compare the genotypes, measurements during heading and anthesis can be avoided because differences in phenology will confound the result. To identify the abiotic stress-tolerant genotypes, measurement during and after the stress (recovery) is generally recommended to discriminate between susceptible and tolerant genotypes. To track the rate of leaf senescence, NDVI measurements from the anthesis period to physiological maturity are generally followed. The genotype that maintains greenness, canopy green area, and duration are mostly associated with a higher yield.

Green leaf area duration in terms of NDVI can also be assessed using unmanned aerial vehicles or systems (UAV or UAS). Aerial vehicle systems have the capacity of acquiring images with temporal and spatial resolutions. When compared with other remote sensing platforms such as manned aircraft and satellites, UAVs can be deployed effortlessly and even have lower operational costs. Regardless of the class of UAV used, a range of customizable cameras and sensors can be integrated for agricultural studies. The images should be collected around solar noon under clear-sky conditions. Flight path, speed, and sensor parameters like aperture, exposure time, frame rate, and sensitivity are important so that there will be adequate overlap between images for mosaicking. After acquiring the images, preprocessing operations like mosaicking and radiometric calibration should be adopted for better results. After preprocessing, the UAV imagery has to be converted to reflectance data for the extraction of different vegetation indices.

3.2.2 Plant Water Status

Plant water status can be quantified through either psychrometric methods or pressure chamber methods, but both these methods are time consuming, very laborious, and less suited for plant breeding or screening genotypes for stress tolerance (Jones 2007). It is observed that relative water content (RWC) is often a good surrogate for pressure potential, and predawn water potential is a surrogate for soil water potential (Jones 2007). RWC evaluates the existing water content of the sampled leaf tissue relative to the maximal water content it can hold at complete turgidity. A major disadvantage of the RWC technique is the considerable time lag between obtaining and sampling the result. Further, the four weighing operations are required, which is time consuming. To overcome this disadvantage, relative tissue weight (a ratio between tissue fresh weight to tissue turgid weight) is used because relative tissue weight is linearly related to RWC (Smart and Bingham 1974).

Loss of water through the transpiration process changes the pattern of canopy reflectance, which indicates a reduction in the absorption of light by leaf due to both the radiative properties of water and drought-related changes in the leaf morphological properties, and leaf physiological status (Ollinger 2011). The extent of the increase in canopy reflectance under drought conditions is related to the duration of stress as well as the response of genotype. Penuelas and Filella (1998) have used canopy reflectance at specific wavelength bands in the visible and NIR region for estimating the plant water status. Hyperspectral active or passive sensors provide measurements of wavelengths in the visible (~400–700 nm) ranges and NIR (~700–2500 nm) ranges, from which different indices are calculated. The important NIR-based index is the water index, WI = R_{970}/R_{900}, which is used to quantify relative leaf water content (Peñuelas et al. 1993). Based on the WI, Babar et al. (2006) and Prasad et al. (2007) have developed normalized water indices NWI-1 = $([R_{970} - R_{900}]/[R_{970} + R_{900}])$, NWI-2 = $([R_{970} - R_{850}]/[R_{970} + R_{850}])$, NWI-3 = $([R_{970} - R_{880}]/[R_{970} + R_{880}])$, and NWI-4 = $([R_{970} - R_{920}]/[R_{970} + R_{920}])$ to screen spring wheat genotypes for drought tolerance. These five indices are now widely used as a selection tool for grain yield under drought stress in wheat (Prasad et al. 2007). Apart from the above, the available water absorption bands at 1450, 1900, and 2100 nm, overtones at 750 and 1250 nm, and numerous spectral vegetation indices for drought have been established for the recovery of crops' status under drought stress (Claudio et al. 2006; Cohen 1991; Hunt Jr et al. 1987).

3.2.3 Canopy Temperature Depression

Among the various traits that are associated with plant water status, the easiest method is a measurement of canopy temperature depression (CTD), which shows good correlations with parameters associated with plant water relation parameters (Mahan et al. 2012). Canopy temperature depression is expressed as the difference between the air and canopy temperature (CTD $= T_{\text{air}} - T_{\text{canopy}}$). Under high-drought conditions and solar radiation, stomatal conductance gets decreased because the soil moisture is inadequate to meet the evapotranspiration demands, resulting in an increase in canopy temperature. Canopy architecture influences the canopy temperature through mutual shading or leaf angle.

Canopy temperature depression is a negative or positive value based on the air temperature and canopy temperature. The CTD is influenced by both environmental and biological factors (Bahar et al. 2008). However, studies revealed a significant correlation between CTD and leaf water potential (Cohen et al. 2005), stomatal conductance (Rebetzke et al. 2012), and grain yield (Balota et al. 2007; Reynolds et al. 1994) under drought-stress conditions; therefore, it is used as a criterion for drought tolerance selection. Genotypes with cooler canopies than other genotypes in the same environment indicate the drought tolerance ability. Higher transpiration indicates cooler canopy and higher conductance of stomata, favoring net photosynthesis, and a lower canopy temperature in crops under drought indicates a relatively higher capacity for consuming soil moisture or for upholding a healthier plant water status. However, the suitability of CTD as an indicator of yield must be evaluated for the individual environment and in particular plant species (Blum et al. 1989) as the genotypes which keep their canopy cooler by deeper root system in the medium to deep soils may not perform better when grown on shallow soils.

Apart from this, CTD has remained a good estimate for screening genotypes to HT stress tolerance; measurement by CTD using infrared thermometer has some common genetic base under both drought and HT stress (Pinto et al. 2010). Under HT stress, due to the high vapor pressure deficit (VPD), the plants that are well watered raise their transpiration rate to cool the canopy through the evaporative cooling process. The cool canopies are associated with an increased rate of stomatal conductance and root lengths. Increased root length can explore the deeper layer of soil, and in a situation with higher VPD, it can extract more moisture, and it will be used to cool the canopy through open stomata. These genotypes are usually referred to as HT escaper. If subsoil moisture is not available for enhanced transpiration, the stomata close, leading to yield penalty. Real HT-tolerant genotypes give high yield under HT stress and also have inherent high leaf temperatures. Mutava et al. (2011) have studied sorghum [*Sorghum bicolor* (L.) Moench.] genotypes exhibiting higher leaf temperature and higher yield in the sorghum diversity panel, which can be used for improving drought and HT stress tolerance in sorghum.

Infrared thermal imaging is a remote sensing technique commonly used for quantifying canopy temperature (Jones and Schofield 2008). In earlier days, thermocouples and mercury thermometers were commonly used to measure leaf temperature. It was cumbersome and does not represent the canopy temperature because it is a point measurement. Hence, principles of thermometry and thermal imaging have been translated into the most commonly used remote sensing tools such as infrared thermometer (IRT) and thermal imaging cameras for assessing the canopy temperature of crop plants. Infrared thermometry can report the subtle differences in canopy temperature in fields and controlled conditions (Winterhalter et al. 2011). Thermal imaging has mostly preferred to quantify plant water relations because of the fast data collection and nondestructive nature, and it can also include a huge number of individual plants in a single image for calculating the temperature measurements. For higher accuracy, such measurements should be preferred when the canopy covers the soil, which otherwise contributes to background noise. These techniques were successful in differentiating drought tolerance among genotypes when used for assessing the irrigated crops on windless and cloudless days with high VPD. One of the most important issues while applying thermal imaging is to filter out the background soil from the image and get more precise canopy measurements. Also, the use of thermal imaging is influenced by air temperature, solar radiation, humidity, and wind speed, which keep one fluctuating under natural conditions. Repeated measurements can take these influences into account when assessing the genotypes for tolerance to high temperatures or drought. Hence, recently Internet of Things (IoT) is being used to replace handheld IRTs and wired IRTs for monitoring canopy temperature in wireless mode. The sensors are installed at the center position of each plot/genotype where there are maximum ground cover, uniform growth, and 0.15 m above the plant canopy height. A base station unit will be established at the edge or corner of the field, which collects the data transmitted by the sensors. Every sensor collects data from a circular field of view (60°) with a 0.15 m diameter every minute, and this is auto-averaged to every 15 min and is reported wirelessly to the base station. The canopy temperature data collected in the base station will be transmitted to a computer system for archiving process and subsequent analysis. Such automation in monitoring and assessing the high-temperature responses of crop genotypes can accelerate selection processes aiming at climate-resilient cultivars.

3.2.4 Limited Transpiration

The limited transpiration trait is usually referred to as a slow-wilting trait. Considerable intraspecific variations in the stomatal response to a change in vapor pressure deficit (VPD) have been reported in soybean (Fletcher et al. 2007), peanut (*Arachis hypogea* L., Devi et al. 2010), sorghum (Gholipoor et al. 2010), pearl millet [*Pennisetum glaucum* (L.) R. Br., Kholova et al. 2010], and chickpea (*Cicer arietinum* L., Zaman-Allah et al. 2011). Measurement of restricted transpiration is a semi-high-throughput phenotypic technique because it is not quick, but this trait can be measured simultaneously in a large number of samples. Restricted transpiration under high VPD by partial closure of the stomata may be associated with the decreased hydraulic conductance of leaf and root in plants, which limits the flow of water from roots to leaf (Sadok and Sinclair 2010). It has been assumed that the hydraulic conductivity connected with limited transpiration trait is related to the transmembrane transport of water via aquaporins. Therefore, application of aquaporin inhibitors, namely cycloheximide, mercury, and silver ions, has the potential to evaluate the expression of restricted transpiration trait.

An indirect approach for the identification of limited transpiration traits in a set of genotypes is through looking at a delay in canopy wilting under water-limiting field conditions. However, it cannot be definite because delayed wilting could be associated with other reasons too. An effective method to evaluate delayed wilting is to measure the canopy temperature under irrigated field conditions. Genotypes with higher canopy temperature and high VPD under well-watered conditions could indicate partial closure of stomata and be associated with limited transpiration rate. However, the environmental conditions should be unique because the difference in leaf temperature can be from several possibilities like temperature, light, relative humidity, and nutrition (Sinclair et al. 2017).

In another case, the expression of restricted transpiration can be quantified by measuring the transpiration rate and weight of the pot at different VPD in the intact plants as detailed by Riar et al. (2015), and here more care should be taken to reduce the evaporation from the soil. An alternate method is to study the stomatal conductance of plants under field conditions during the natural daily variations in VPD (Shekoofa et al. 2014). However, it is limited by the weather conditions when the measurements were made and also the number of lines that are repeatedly measured throughout a day for stomatal conductance. The LeasyScan platform allows quick measurement of plant leaf area and pot weight, and by using this platform, pot weight can be measured every hour to arrive at the transpiration rate. Using 3D laser scanning, the leaf area can be estimated. By using leaf area and transpiration rate, the limited transpiration trait can be phenotyped in a large number of genotypes. Evidence of concept in observing the limited transpiration has been confirmed in corn (*Zea mays* L.), pearl millet, cowpea (*Vigna unguiculata* L.), sorghum, and peanut (Sinclair et al. 2017).

3.2.5 Root Architecture

Huge phenotypic plasticity of root characters in response to soil physical and chemical conditions was observed, and lack of cost-effective and high-throughput screening techniques makes root studies highly challenging. Root architecture denotes the spatial arrangement of root systems, which determines the plant anchorage, ability of roots to absorb nutrients and water, and intra- and interplant competition. The rooting system of the plant responds to environmental stimuli through appropriate adaptive changes in morphological, structural, and physiological processes, which is referred to as root plasticity, and thus exploiting this through breeding by integrating the physiological phenes and root architectural traits will guide in breeding the genotypes for drought tolerance (Kashiwagi et al. 2006; Lynch 2011; Osmont et al. 2007).

Measurement of root system architecture is hindered not only by various complexities (physical, chemical, biological) of soil medium, but also by lack of comprehensive information about the root system architecture and life span of the root system of a plant (Ahmadi et al. 2011; McCully 1995). A lot of phenotyping and sampling methods for roots in the field have been suggested, namely monoliths, soil profiling, rhizotron, nail plates, trenching, probes, shovelomics, visualization, and digitalization of roots in the field (Costa et al. 2014; Pierret et al. 2003; Trachsel et al. 2011; Wu et al. 2015), to obtain information on root length, dry matter, surface area, dry weight, diameter, diameter class, and structure.

Spatially dispersed monolith sampling can be used for assessing crop root system architecture under the field. However, commonly used auger core sampling might suffer large errors when illustrating spatial distributions of roots. Shovelomics is an alternate high-throughput method for phenotyping root system architecture in the field, which provides a rapid sampling and quantification of rooting depth but not fine details of the root system. The heterogeneity of soil structure and composition can cause a confounding effect on the root system architecture within the same field. Several software packages (RootScan, RootNav, DART, GiARoots, RootSystemAnalyzer, RootReader, IJ Rhizo, RootReader3D, and RooTrak) were developed for extracting quantitative data and imaging roots from the captured root images (Lobet and Draye 2013).

Direct measurements of root traits under field conditions can be made by removing the soil, which can cause the death of the plant, the loss of root material, and the loss of geometric information. Earlier, digging of a trench close to growing plants to visualize the whole root was employed. Even though the trench method permits for a precise in situ observation of roots grown in the field, it is very slow and laborious. Apart from trench methods, researchers are using excavation techniques like soil coring to study the root architecture. Soil coring is done by introducing a metal cylinder down into the soil to obtain a soil core. The soil core is usually divided into segments with the same length, and each segment is washed over a screen to collect all the roots. The collected roots are scanned on a flatbed scanner for measuring subsequent length. The roots can also be dried, weighed, and counted. The soil coring method is used to estimate the rooting depth and root length density. With the development of the tractor hydraulic system, now the soil coring method is well automated.

Another method of excavation technique practiced is root crown phenotyping or shovelomics because the root crown is considered as the backbone of the root system. Using a regular shovel, the root crown which is the upper part of the root structure attached to the shoot is excavated, washed, and analyzed for root architecture. This procedure provides information on root placement in the soil and the number of roots, and their lengths and angles. Shovelomics is also a destructive method.

Field rhizotrons are considered as an enhanced version of trenches; in this method, a trench is dug, a glass window is positioned tightly over the vertical cut plane, and a roof is installed over the pit. A customized camera is inserted into the tube to image the soil with the roots around the tube. Through this method, root initiation, growth, and turnover of individual roots over a period can be assessed. A major limitation of this technique is low throughput (numbers of samples are small per unit time) and is highly influenced by soil properties. To reduce the environmental variability and increase the throughput, researchers are using rhizobox, which works similarly to rhizotron. In most cases, the rhizoboxes are maintained at an angle that forces the root to grow along with the glass so that it can be monitored frequently. However, in this method, it will be difficult to distinguish between thin roots and soil. Now researchers are using clear media such as agar (in Petri dishes) to grow roots, making the roots visible, and the images of the Petri dishes with roots can be analyzed for obtaining information on root angle.

On the other hand, X-ray computed tomography and magnetic resonance imaging are the two technologies that can be used for imaging the root systems over some time without destroying the plant. With X-ray computed tomography, both roots and soil are imaged, and custom-made software tools are required to segment the roots. On the contrary, magnetic resonance imaging can be adapted to image only roots in such a way as to avoid segmentation. However, these two techniques are low throughput and expensive. In summary, though there are many ways to grow and observe roots, appropriate methods should be selected based on the experimental question and ease of use.

3.2.6 Membrane Stability

Maintaining cell membrane stability is one of the adaptation mechanisms under abiotic stress. Cell membrane stability can be assessed directly through electron microscopy and indirectly by lipid peroxidation, electrolyte leakage, and chlorophyll *a* fluorescence. The level of lipid peroxidation detected as malondialdehyde (MDA) is an indicator of free radical damage to the cell membranes because lipid peroxidation alters the physiological functions of cell membranes. The traditional technique to detect MDA content in plants is the thiobarbituric acid-reactive substance (TBARS) test using the spectrophotometry technique. The TBARS such as aldehydes and malondialdehyde react with thiobarbituric acid at low pH and form [TBA]-MDA adduct, which is a pink chromogen having a maximum absorbance at 532 nm; the formed adduct is quantified through a spectrophotometer. The TBARS test is a standard test and is sensitive for microsomal and liposomal membrane lipid peroxidation tests. It rarely measures the free MDA content of the lipid system. TBA reactivity depends on the lipid content of the sample (Bhattacharjee 2014).

Also, high-performance liquid chromatography (HPLC) was used to determine MDA in plants (Davey et al. 2005). However, the HPLC method requires lots of time, chemical, and complex sample preparation (utmost care has to be undertaken to ensure the loss of oxidized material and artificial peroxidation). Kong et al. (2016) have assessed the feasibility of hyperspectral imaging with 400–1000 nm to detect MDA content in crops after herbicide application. The result indicated that the extreme learning machine model achieved the optimal prediction performance with 23 wavelengths selected by competitive adaptive reweighted sampling.

Assessment of damage to the thylakoid membrane under stress is a reliable measure of a plant's susceptibility to HT stress (Ristic et al. 2008). Impairment in the thylakoid membrane can be estimated by determining chlorophyll *a* fluorescence trait and measuring the ratio of constant fluorescence (*O*) and the peak of variable fluorescence (*P*) (Ristic et al. 2008). An increase in the *O*/*P* ratio represents the damage in thylakoid membranes; the higher the increase, the greater the damage. Larcher (1995) has observed a good correlation between chlorophyll fluorescence and electrolyte leakage (an indicator of membrane damage).

With the recent developments in tracer techniques, fluorescent dyes and nucleic acid stain (Sytox green) were used as molecular probes to track the membrane damage. After the brief incubation period with the stain, the nucleic acids of dead cells will fluoresce bright green. This property makes the stain Sytox green a simple, quantitative single-step dead cell (compromised membrane) indicator for use with fluorescence microscopes (Prasad and Djanaguiraman 2011).

3.2.7 Photochemical Efficiency

Chlorophyll fluorescence displays the fate of excitation energy in the photosynthetic apparatus that has been used as an early, in vivo indicator of stress (Yamada et al. 1996). In most of the studies, dark-adapted chlorophyll *a* fluorescence parameters are used to understand the reactions of plants to environmental cues. However, it is challenging under field conditions, due to time constraints (dark-adaptation time) to perform the dark-adapted test if the study involves many genotypes or treatments. Dark-adaptation times vary with a crop from 10 to 60 min, and some researchers use pre-dawn values for the basal fluorescence (F_o). The F_o measurement and its light-adapted equivalent F_o' are fundamental to the analysis of fluorescence. F_o' is measured immediately after switching off the actinic light, but accurate measurement of F_o' is difficult. Many fluorometers have the ability to apply a weak far-red light to measure both F_o and F_o'. Application of saturating pulse to a dark-adapted leaf triggers a maximum value of fluorescence by closing the reaction centers. At this time, in a non-stressed healthy leaf, there is no non-photochemical quenching (NPQ) since the leaf is dark-adapted leading to a maximum value of fluorescence (F_m). The F_v/F_m ratio is an indicator of the maximum quantum yield of PSII photochemistry. The value of F_v/F_m ratio in an unstressed leaf will be $\geq$0.80, and the presence of stress will decrease this ratio through photoinhibition or inactivation of PSII (Long et al. 1994). Thus, measuring the F_v/F_m ratio after an appropriate dark adaptation is the most commonly used technique to quantify stress in leaves. To attain this precisely in a light-adapted leaf, we need to make sure that PSII is fully oxidized, and this can be succeeded using a pulse of far-red light. Precise measurements of basal fluorescence in the field are challenging due to the relaxation kinetics of the chlorophyll molecule under the dark-adapted state. When a leaf is dark-adapted, the movement of electrons in the thylakoid should stop almost immediately. However, NPQ "relaxes" more leisurely because the protective NPQ processes remain active. Therefore, to obtain the true maximum value of fluorescence, we must allow the leaf to remain in the dark for a span of time ample for these processes to complete (i.e., NPQ to become zero) (Murchie and Lawson 2013).

Due to the ease of measurement of the maximum efficiency of PSII photochemistry in the light, this is widely used as an indicator of the operating efficiency of PSII in the light. However, care should be taken in this measurement because PSI may contribute to fluorescence when measurements are made above 700 nm and the existence of "multiple turnovers" of PSII during the saturating pulse. From the value of F_v/F_m ratio, the rate of electron transport can be calculated using the photosynthetic active radiation (PAR) value and a fraction of light intercepted by PSII and PSI. It is difficult to measure the latter, and an assumption of equal absorption is made. Although these standard values are expected to be constants, they will differ between the leaves having different optical properties or the same leaf suffering from stress treatments. For instance, relating the electron transport rate (ETR) values between control (fully hydrated leaf) with a drought-stressed leaf (low turgor value) is not appropriate.

Similarly, leaf samples with different pigment contents or photosystem stoichiometry will vary for a light interception, causing inaccuracies in ETR calculation (Walters 2005). The measurement of F_o' can be open to error if the far-red light applied does not sufficiently oxidize Q_A and if the relaxation of NPQ causes F_o' to rise quickly after the actinic light has been switched off. Thus, care should be taken during measuring the PSII quantum yield to assess the impact of stress on the plants. Now, imaging chlorophyll fluorescence as a diagnostic tool is becoming increasingly popular for screening germplasm. Chlorophyll fluorescence imaging has been combined into many phenotyping platforms for high-throughput analysis. Imaging provides additional evidence on the spatial and temporal heterogeneity of measured parameters. Now, chlorophyll fluorescence imaging is integrated with infrared gas exchange techniques, thermography, and hyperspectral imaging to explore and integrate various traits to understand the stress response (Bauriegel et al. 2011).

3.2.8 Yield-Forming Traits

Grain yield is the final product of many processes, and in cereals, for example, it is primarily determined by yield-associated traits like the number of spikes/panicles per plant, number of grains per spike/panicle, and individual grain weight. Seed numbers are a function of seed set, and the decrease in seed set percentage under stress is linked with early or delayed flowering, asynchrony of male and female reproductive development, and impairments in parental tissues, in male and female gametes (Zinn et al. 2010). Estimation of seed set percent is a more accurate estimation of the response of gametes to HT stress. However, data from marked floret instead of whole spike/panicle will provide a good estimate because within a spike each floret will have a different developmental stage or day of anthesis (Aiqing et al. 2018). During estimating seed set percentage, researchers may consider ill-filled seeds/grains as a seed, because the formation of seed indicates the function of gametes. However, it will be wise to discard the ill-filled grains during calculating the seed set percentage because it will overestimate the seed yield potential of the genotype. A strong positive correlation between seed set and pollen availability, and seed set and seed yield, was established (Prasad et al. 2017, 2019). Limited gamete functions may be considered the most important factor for seed set under drought and HT stress environments. Gamete functions depend on its viability, which can be evaluated by viability assays like staining and in vitro and in vivo germination.

The choice of pollen viability method depends on the crop or species. Viability has been defined as having the ability to dwell, nurture, germinate, or develop, and the loss of viability is a constant variable. Thus, the viability of pollen grains has been used to define the ability of pollen grains to germinate on the stigmatic surface, germinating in vitro, picking up certain stains, and effective seed set following pollination. Viable pollen grains cannot germinate under in vitro or in vivo conditions if the circumstances are not favorable (environment and pistil response). Therefore, assessment of pollen viability based on seed set indicates the response of both male and female gametes. Methods of determining pollen viability are enormous, and the method of determining pollen fertility (ability to set the seed) is through quantifying seed set percentage.

All the methods of assessing pollen viability depend on factors like cytoplasm content, enzyme activity, plasmalemma integrity, and environmental conditions. However, none of the methods can be able to confirm that the pollen is inviable and unable to fertilize and set seed. Therefore, it gives a likelihood estimate. The approaches used to evaluate the pollen viability are measuring the respiration rate (very rarely used), staining techniques (vital stain to indicate membrane integrity, presence of the cytoplasm, respiration rate, enzyme activity, and starch content), in vitro germination, and capacity to set seeds. In vitro pollen germination method is rapid, fully quantitative, and reasonably simple, and it is highly correlated with seed set percent in many species (Prasad et al. 2019). The results depend on the time of pollen collection, the composition of the medium, the temperature of the growth medium, and the duration of the test. Low germination under in vitro conditions indicates that the pollen is still fertile and able to set seed. The in vivo method is more valid than the in vitro method. However, it must be accompanied by a stigma receptivity test. The major drawbacks are that vital stains tend to stain old and dead pollen, overestimation of viability, pollen with cytoplasm or starch is not necessarily fully fertile, and immature or aborted pollen grains also pick the stains (Dafni and Firmage 2000). The important elements that showed up during assessing pollen viability are (1) information about the test environment, (2) freshly collected pollen preferred for the assay, (3) testing in parallel the dead pollen as a negative control, (4) testing hydrated vs. dehydrated pollen to understand the effect of moisture level, and (5) running several tests simultaneously to understand which method is best for the test species.

Cardinal temperatures (T_{min}, T_{opt}, and T_{max}) for pollen grain germination are used to screen germplasm for HT stress tolerance, and this proved to be a good screening tool. Results from in vitro studies on peanut, sorghum, pearl millet, rice, and coconut (*Cocos nucifera* L.) presented that genotypes varied in response to temperature for cardinal temperatures, and the differences in cardinal temperatures were mainly responsible for the tolerance/susceptibility level of genotypes to HT stress (Djanaguiraman et al. 2014, 2018; Hebbar et al. 2018). The genotypes having higher ceiling temperature (T_{max}) for pollen germination values tend to be HT tolerant in most cases. However, research also indicated that there is no relationship between cardinal temperature and tolerance/susceptibility to HT stress because the cultivars which had a higher optimum temperature (T_{opt}) for pollen germination did not always have a T_{max} or vice versa (Kakani et al. 2002).

To evaluate the effect of HT on seed abortion (postfertilization stage), the number of seeds per pod and number of locules are used. The weight of individual grain is a product of the rate and duration of grain filling. Grain-filling duration can be expressed as the time between anthesis and physiological maturity; beyond this point, there will be no significant increase in grain dry matter. The average for grain-filling rate was calculated from the ratio of maximum grain weight to grain-filling duration, which was estimated from quadratic or cubic polynomial curves. Linear regression has been employed to find out the grain-filling rate, and the intersection of two regression lines has been used to attain the grain-filling duration. Among genotypes, genetic variability exists for grain-filling rate and duration and can be exploited for developing high-yielding cultivars.

3.3 Conclusion

Selection for grain yield per se under drought and HT stress is limited by the low heritability of grain yield under stressful conditions. This situation requires the identification of highly reliable secondary traits that are closely related to grain yield under stress and having high heritability. The traits like plant water status, canopy temperature depression, green leaf area duration, root architecture, membrane stability, gamete viability, and stem reserves are key traits among those listed in this chapter that are highly associated with stress tolerance. The relative importance of each trait under drought and HT stress was provided in Table 3.1. Regarding

Table 3.1 A subjective classification of the relative value of different trait measures for stress tolerance

Traits	Stress	
	Drought	High temperature
Green leaf area duration	+++	++
Plant water status	+++	+
Canopy temperature depression	++	+
Limited transpiration	++	+
Root architecture	+++	++
Membrane stability	+	+++
Photochemical efficiency	++	++
Early-morning flowering	–	+
Stem reserve mobilization	+++	+++
Yield-forming traits	++++	++++

More +s indicates greater value, while (–) indicates limited value quantification of the above traits, they are semi- to high throughput in nature. The advent of high-throughput genotyping technologies urges us to gather high-quality phenotypic data for marker-based selection in crop breeding. However, a collection of phenotypic data is laborious and highly influenced by the genotype, environment, and management. Therefore, the development of the high-throughput phenotypic platform is critical for accelerating the breeding programs. Field-based high throughput increases the accuracy of the estimation and reduces the time, leading to the selection of genotype and the identification of genetic loci with high precision. Despite major improvements in phenotyping, there are still large shortcomings, namely quality of data collection, management of digital data, image resolution, and accurate analysis. Robust computational skills will be needed to handle the phenotypic data collected from the phenotypic platform. Care should be taken to phenotype at the target environment with standardized protocols. The field phenotyping process must go hand in hand with the methodologies to characterize and control the field variations, and user-friendly data management. Imaging and spectroscopy techniques can provide nondestructive measures of traits like chlorophyll fluorescence and green leaf area duration, and this technique can be extended to quantify other traits through surrogate parameters. All these advances in phenotyping are likely to accelerate genomics application for enhancing crop productivity under drought and HT environments.

References

Ahmadi S, Plauborg F, Andersen M, Sepaskhah A, Jensen C, Hansen S (2011) Effects of irrigation strategies and soils on field grown potatoes: root distribution. Agric Water Manage 98:1280–1290

Aiqing S, Somayanda I, Sebastian SV, Singh K, Gill K, Prasad PVV, Jagadish SVK (2018) Heat stress during flowering affects time of day of flowering, seed set, and grain quality in spring wheat. Crop Sci 58(1):380–392

Babar MA, Reynolds MP, Van Ginkel M, Klatt AR, Raun WR, Stone ML (2006) Spectral reflectance indices as a potential indirect selection criteria for wheat yield under irrigation. Crop Sci 46(2):578–588

Bahar B, Yildirim M, Barutcular C, Ibrahim G (2008) Effect of canopy temperature depression on grain yield and yield components in bread and durum wheat. Notulae Botanicae Horti Agrobotanici Cluj-Napoca 36(1):34–37

Balota M, Payne WA, Evett SR, Lazar MD (2007) Canopy temperature depression sampling to assess grain yield and genotypic differentiation in winter wheat. Crop Sci 47(4):1518–1529

Bauriegel E, Giebel A, Geyer M, Schmidt U, Herppich W (2011) Early detection of Fusarium infection in wheat using hyper-spectral imaging. Comput Electron Agric 75:304–312. https://doi.org/10.1016/j.compag.2010.12.006

Bhattacharjee S (2014) Membrane lipid peroxidation and its conflict of interest: the two faces of oxidative stress. Curr Sci 107:1811–1823

Blum A, Shpiler L, Golan G, Mayer J (1989) Yield stability and canopy temperature of wheat genotypes under drought-stress. Field Crop Res 22(4):289–296

Borrell A, Hammer G, Van Oosterom E (2001) Stay-green: a consequence of the balance between supply and demand for nitrogen during grain filling? Ann Appl Biol 138(1):91–95

Claudio HC, Cheng Y, Fuentes DA, Gamon JA, Luo H, Oechel W, Qiu H-L, Rahman AF, Sims DA (2006) Monitoring drought effects on vegetation water content and fluxes in chaparral with the 970 nm water band index. Remote Sens Environ 103(3):304–311

Cobb JN, DeClerck G, Greenberg A, Clark R, McCouch S (2013) Next-generation phenotyping: requirements and strategies for enhancing our understanding of genotype–phenotype relationships and its relevance to crop improvement. Theor Appl Genet 126(4):867–887

Cohen W (1991) Response of vegetation indices to changes in three measures of leaf water stress. Photogramm Eng Remote Sens 57(2):195–202

Cohen Y, Alchanatis V, Meron M, Saranga Y, Tsipris J (2005) Estimation of leaf water potential by thermal imagery and spatial analysis. J Exp Bot 56(417):1843–1852

Costa MCG, Cunha IML, Jorge LAC, Araújo ICS (2014) Public-domain software for root image analysis. Rev Bras Ciênc Solo 38:1359–1366

Dafni A, Firmage D (2000) Pollen viability and longevity: practical, ecological and evolutionary implications. In: Pollen and pollination. Springer, New York, pp 113–132

Davey MW, Stals E, Panis B, Keulemans J, Swennen RL (2005) High-throughput determination of malondialdehyde in plant tissues. Anal Biochem 347(2):201–207

Devi MJ, Sinclair TR, Vadez V (2010) Genotypic variation in peanut for transpiration response to vapor pressure deficit. Crop Sci 50(1):191–196

Djanaguiraman M, Prasad PVV, Murugan M, Perumal R, Reddy UK (2014) Physiological differences among sorghum (*Sorghum bicolor* L. Moench) genotypes under high temperature stress. Environ Exp Bot 100:43–54

Djanaguiraman M, Perumal R, Ciampitti IA, Gupta SK, Prasad PVV (2018) Quantifying pearl millet response to high temperature stress: thresholds, sensitive stages, genetic variability and relative sensitivity of pollen and pistil. Plant Cell Environ 41(5):993–1007

Fletcher AL, Sinclair TR, Allen LH Jr (2007) Transpiration responses to vapor pressure deficit in well watered 'slow-wilting' and commercial soybean. Environ Exp Bot 61(2):145–151

Gholipoor M, Prasad PVV, Mutava RN, Sinclair TR (2010) Genetic variability of transpiration response to vapor pressure deficit among sorghum genotypes. Field Crop Res 119(1):85–90

Hebbar KB, Subramanian P, Sheena TL, Shwetha K, Sugatha P, Arivalagan M, Varaprasad PV (2016) Chlorophyll and nitrogen determination in coconut using a non-destructive method. J Plant Nutr 39(11):1610–1619

Hebbar KB, Rose HM, Nair AR, Kannan S, Niral V, Arivalagan M, Gupta A, Samsudeen K, Chandran KP, Chowdappa P (2018) Differences in in vitro pollen germination and pollen tube growth of coconut (*Cocos nucifera* L.) cultivars in response to high temperature stress. Environ Exp Bot 153:35–44

Homolova L, Malenovský Z, Clevers JGPW, García-Santos G, Schaepman ME (2013) Review of optical-based remote sensing for plant trait mapping. Ecol Complex 15:1–16

Hunt ER Jr, Rock BN, Nobel PS (1987) Measurement of leaf relative water content by infrared reflectance. Remote Sens Environ 22(3):429–435

Jones HG (2007) Monitoring plant and soil water status: established and novel methods revisited and their relevance to studies of drought tolerance. J Exp Bot 58(2):119–130

Jones HG, Schofield P (2008) Thermal and other remote sensing of plant stress. Gen Appl Plant Physiol 34(1–2):19–32

Kakani VG, Prasad PVV, Craufurd PQ, Wheeler TR (2002) Response of in vitro pollen germination and pollen tube growth of groundnut (*Arachis hypogaea* L.) genotypes to temperature. Plant Cell Environ 25(12):1651–1661

Kashiwagi J, Krishnamurthy L, Crouch JH, Serraj R (2006) Variability of root length density and its contributions to seed yield in chickpea (*Cicer arietinum* L.) under terminal drought stress. Field Crop Res 95(2):171–181. https://doi.org/10.1016/j.fcr.2005.02.012

Kholova J, Hash CT, Kumar PL, Yadav RS, Kočová M, Vadez V (2010) Terminal drought-tolerant pearl millet [*Pennisetum glaucum* (L.) R. Br.] have high leaf ABA and limit transpiration at high vapour pressure deficit. J Exp Bot 61(5):1431–1440

Kong W, Liu F, Zhang C, Zhang J, Feng H (2016) Non-destructive determination of Malondialdehyde (MDA) distribution in oilseed rape leaves by laboratory scale NIR hyperspectral imaging. Sci Rep 6(1):1–8

Larcher W (1995) Photosynthesis as a tool for indicating temperature stress events. In: Ecophysiology of photosynthesis. Springer, New York, pp 261–277

Lobet G, Draye X (2013) Novel scanning procedure enabling the vectorization of entire rhizotron-grown root systems. Plant Methods 9(1):1–11

Long SP, Humphries S, Falkowski PG (1994) Photoinhibition of photosynthesis in nature. Annu Rev Plant Biol 45(1):633–662

Lynch JP (2011) Root phenes for enhanced soil exploration and phosphorus acquisition: tools for future crops. Plant Physiol 156(3):1041–1049

Mahan JR, Young AW, Payton P (2012) Deficit irrigation in a production setting: canopy temperature as an adjunct to ET estimates. Irrig Sci 30(2):127–137

McCully M (1995) How do real roots work? (Some new views of root structure). Plant Physiol 109(1):1

Murchie EH, Lawson T (2013) Chlorophyll fluorescence analysis: a guide to good practice and understanding some new applications. J Exp Bot 64(13):3983–3998

Mutava RN, Prasad PVV, Tuinstra MR, Kofoid KD, Yu J (2011) Characterization of sorghum genotypes for traits related to drought tolerance. Field Crop Res 123(1):10–18

Ollinger SV (2011) Sources of variability in canopy reflectance and the convergent properties of plants. New Phytol 189(2):375–394

Osmont KS, Sibout R, Hardtke CS (2007) Hidden branches: developments in root system architecture. Annu Rev Plant Biol 58:93–113

Penuelas J, Filella I (1998) Visible and near-infrared reflectance techniques for diagnosing plant physiological status. Trends Plant Sci 3(4):151–156

Peñuelas J, Filella I, Biel C, Serrano L, Save R (1993) The reflectance at the 950–970 nm region as an indicator of plant water status. Int J Remote Sens 14(10):1887–1905

Pierret A, Doussan C, Garrigues E, McKirby J (2003) Observing plant roots in their environment: current imaging options and specific contribution of two-dimensional approaches. Agronomie 23(5–6):471–479

Pinto RS, Reynolds MP, Mathews KL, McIntyre CL, Olivares-Villegas J-J, Chapman SC (2010) Heat and drought adaptive QTL in a wheat population designed to minimize confounding agronomic effects. Theor Appl Genet 121(6):1001–1021

Prasad PVV, Djanaguiraman M (2011) High night temperature decreases leaf photosynthesis and pollen function in grain sorghum. Funct Plant Biol 38(12):993–1003

Prasad B, Carver BF, Stone ML, Babar MA, Raun WR, Klatt AR (2007) Potential use of spectral reflectance indices as a selection tool for grain yield in winter wheat under great plains conditions. Crop Sci 47(4):1426–1440

Prasad PVV, Bheemanahalli R, Jagadish SVK (2017) Field crops and the fear of heat stress—opportunities, challenges and future directions. Field Crop Res 200:114–121

Prasad PVV, Djanaguiraman M, Jagadish SVK, Ciampitti IA (2019) Drought and high temperature stress and traits associated with tolerance. In: Sorghum: a state of the art and future perspetives, vol 58. American Society of Agronomy, pp 241–265

Rebetzke GJ, Rattey AR, Farquhar GD, Richards RA, Condon ATG (2012) Genomic regions for canopy temperature and their genetic association with stomatal conductance and grain yield in wheat. Funct Plant Biol 40(1):14–33

Reynolds MP, Balota M, Delgado MIB, Amani I, Fischer RA (1994) Physiological and morphological traits associated with spring wheat yield under hot, irrigated conditions. Funct Plant Biol 21(6):717–730

Riar MK, Sinclair TR, Prasad PVV (2015) Persistence of limited-transpiration-rate trait in sorghum at high temperature. Environ Exp Bot 115:58–62

Ristic Z, Bukovnik U, Momčilović I, Fu J, Prasad PVV (2008) Heat-induced accumulation of chloroplast protein synthesis elongation factor, EF-Tu, in winter wheat. J Plant Physiol 165(2): 192–202

Sadok W, Sinclair TR (2010) Transpiration response of 'slow-wilting' and commercial soybean (*Glycine max* (L.) Merr.) genotypes to three aquaporin inhibitors. J Exp Bot 61(3):821–829

Shekoofa A, Balota M, Sinclair TR (2014) Limited-transpiration trait evaluated in growth chamber and field for sorghum genotypes. Environ Exp Bot 99:175–179

Sinclair TR, Devi J, Shekoofa A, Choudhary S, Sadok W, Vadez V, Riar M, Rufty T (2017) Limited-transpiration response to high vapor pressure deficit in crop species. Plant Sci 260:109–118

Smart RE, Bingham GE (1974) Rapid estimates of relative water content. Plant Physiol 53(2): 258–260

Thomas H, Howarth CJ (2000) Five ways to stay green. J Exp Bot 51:329–337

Trachsel S, Kaeppler SM, Brown KM, Lynch JP (2011) Shovelomics: high throughput phenotyping of maize (Zea mays L.) root architecture in the field. Plant Soil 341(1):75–87

Walters RG (2005) Towards an understanding of photosynthetic acclimation. J Exp Bot 56(411): 435–447

Winterhalter L, Mistele B, Jampatong S, Schmidhalter U (2011) High throughput phenotyping of canopy water mass and canopy temperature in well-watered and drought stressed tropical maize hybrids in the vegetative stage. Eur J Agron 35(1):22–32

Wu J, Wu Q, Yang B, Guo Y (2015) Three-dimensional architecture of axile roots of field-grown maize. Plant Soil 387(1):363–377

Yamada M, Hidaka T, Fukamachi H (1996) Heat tolerance in leaves of tropical fruit crops as measured by chlorophyll fluorescence. Sci Hortic 67(1–2):39–48

Zaman-Allah M, Jenkinson DM, Vadez V (2011) Chickpea genotypes contrasting for seed yield under terminal drought stress in the field differ for traits related to the control of water use. Funct Plant Biol 38(4):270–281

Zinn KE, Tunc-Ozdemir M, Harper JF (2010) Temperature stress and plant sexual reproduction: uncovering the weakest links. J Exp Bot 61(7):1959–1968

Epigenetics of Abiotic Stress Tolerance in Legumes

4

Gyan P. Mishra, Harsh K. Dikshit, Jyoti Devi, Muraleedhar S. Aski, and Kumar Durgesh

Abstract

Epigenetic modifications are known to alter the activation pattern of some genes and not the per se DNA sequence. Stress to the plant causes epigenetic alterations in the plant either as hyper- or as hypo-methylation of certain DNA sequences. To overcome or to counter the various abiotic stress conditions, the plant's defense machinery including cellular signaling pathways gets regulated by several stress-responsive genes, which in turn are regulated by various mechanisms including DNA and chromatin modifications, and also through different small RNA-based mechanisms. There is a sudden spurt in the epigenetic studies aiming to find their role in the imposition of various types of abiotic stress tolerance in different plant species, mainly due to the quick advancements in the high-throughput NGS technologies. Many reports associating the DNA methylation response with that of various abiotic stress adaptations are available in many legume species like soybean, chickpea, pigeon pea, Medicago, lotus, peanut, and common beans using these techniques. These legumes have shown tolerance to several abiotic stresses because of unique epigenetic variations, which are present in the natural populations. Understanding the epigenetic mechanism regulating the tolerance to the abiotic stresses will help plant breeders in the development of more resilient and climate-smart varieties, giving higher yields under varied abiotic stresses. This chapter covers the current status of a novel and promising field of epigenetics in legume crops, especially for the imposition of different abiotic stress tolerance.

G. P. Mishra (✉) · H. K. Dikshit · M. S. Aski · K. Durgesh
ICAR-Indian Agricultural Research Institute, New Delhi, India

J. Devi
ICAR-Indian Institute of Vegetable Research, Varanasi, India

P. Muthu Arjuna Samy et al. (eds.), *Legumes: Physiology and Molecular Biology of Abiotic Stress Tolerance*, https://doi.org/10.1007/978-981-19-5817-5_4

Keywords

Abiotic stresses · DNA methylation · Epigenetic mark · Epigenome editing · Methylation landscape

4.1 Introduction

Legumes are unique in the sense that they can enhance soil fertility through natural nitrogen fixation ability and thereby help in the overall agricultural sustainability. Several grain legumes are the staple food and the key protein source, especially for the poor residing in developing and underdeveloped countries (Mishra et al. 2021). Ironically, most of the legumes suffer heavily due to the negative impact of many abiotic stresses like heat, cold, drought (water-deficit stress), and salinity (Bhalani et al. 2019; Sarkar et al. 2014, 2016). Thus, to sustain such important crops, there is a need to opt for novel approaches like epigenetics to develop the climate-resilient and abiotic stress-tolerant varieties in the legumes. Next-generation sequencing (NGS) technologies are giving an option for rapid and cost-effective omics technologies in many legumes including chickpea, mung bean, lentils, pigeon pea, peas, soybean, Medicago, etc. for the identification of key genes and regulatory pathways involved with different abiotic stress tolerance (Dasgupta et al. 2021; Mishra et al. 2020; Nawade et al. 2018). Various studies have proved the associations between methylation levels and different abiotic stresses, suggesting the pronounced role of epigenetic mechanisms in plant adaptability (Malabarba et al. 2021; Windels et al. 2021). Thus, understanding their role in the abiotic stress tolerance mechanism is very important to have improved productivity (Ramu et al. 2016).

In general, epigenetics (meaning above genetics) is being referred to as any heritable alteration which is unable to modify the DNA sequence(s) or genetic code yet causes modified gene expression and altered phenotype. However, the concept of epigenetics is constantly changing, and its exact definition is always debated (Deans and Maggert 2015). Epigenetic change results in the modification of the chromatin structure, which in turn affects the transcription pattern of the cells. Epigenetic regulation mechanisms can be largely classified into three groups, viz., DNA methylation, histone modification, and RNA interference (RNAi) (Saraswat et al. 2017).

Plants being sessile in nature are exposed continuously to the environmental vagaries and experience stresses of different kinds such as availability of water and nutrient, temperature and light regimes, and salinity (Patel et al. 2016, 2017; Reddy et al. 2020). Adaptation to these stressors needs constant dynamic changes in the plants at both morphological and molecular levels. To overcome such environmental vagaries, plants have developed several strategies including epigenetic regulation for better survivability (Saraswat et al. 2017; Shanker and Venkateswarlu 2011). Several epigenetic mechanisms including abiotic stress responses were identified mainly from the model plants like *Arabidopsis* (Pecinka et al. 2020) and rice (La et al. 2011). The knowledge derived from these species is being used to understand the similar phenomenon in legumes too (Chinnusamy and Zhu 2009; Gutzat and Scheid 2012). Also, reports mentioning the changes in the DNA methylation pattern in different plant species are available for different stresses such as water-deficit stress (Kapazoglou et al. 2013), temperature stress (Naydenov et al. 2015), and continuous cropping (Liang et al. 2019).

The most deeply studied model legumes at the genomic levels include soybean, *Medicago truncatula*, and lotus (Cañas and Beltrán 2018; Mochida et al. 2010; Ramesh et al. 2019). Lately, a few other legumes like *Phaseolus vulgaris* (common bean), chickpea, and cowpea are also being deeply investigated at the genomic level (Lobaton et al. 2018; Mishra et al. 2021, 2022; Timko et al. 2008; Varshney et al. 2013). Recently, many legume crops like pigeon pea, lentil, mung bean, peas, beans, Medicago, lotus, peanut, chickpea, and soybean have been sequenced, and the amount of genomic sequence data is increasing with each passing year (Ahmad et al. 2020; Bosamia et al. 2020; Garg et al. 2014; Mishra et al. 2020). Further, due to the rapid increase in the relatively cheap genomic technologies (including epigenome analysis), such studies even in the non-model organism have also been possible.

Under abiotic stress, plants respond differently involving multiple mechanisms through massive differential gene expressions and nuclear organizations including epigenetic changes (Budak et al. 2015). Nevertheless, studies delineating the role of epigenetics in the imposition of abiotic stresses in legumes are not very deeply understood, to date (Niederhuth and Schmitz 2014). Epigenetic variations in the DNA have been reported in response to many abiotic stresses (Pandey et al. 2016). Yet, the precise role of various enzymes catalyzing the active DNA methylations or other modifications has not been thoroughly understood.

Legumes generally have large genome sizes, many TEs, repeat regions, and numerous high-copy-number genes, and to understand their functions, legume breeding should include novel -omics technologies including the use of epigenetic approaches while going for the development of new high-yielding and climate-resilient varieties (Bosamia et al. 2015; Mishra et al. 2015; Salgotra and Gupta 2019). With this backdrop, this chapter gives an in-depth overview of various epigenetic studies in different legume species with a detailed focus on the role of epigenetics in abiotic stress responses in legumes.

4.2 Epigenetics and Major DNA Methylation Mechanisms

The "epigenetic landscape" and "epigenetics" terms were coined by Conrad Waddington way back during the early 1940s. Gene expression can be regulated through various epigenetic mechanisms like chromatin modifications (e.g., histone acetylation, methylation, phosphorylation, and ubiquitylation) and DNA modifications (e.g., cytosine methylation) (Gibney and Nolan 2010). These epigenetic modifications are prompted by various developmental and/or environmental reasons, which then modify the chromatin architecture without changing the DNA

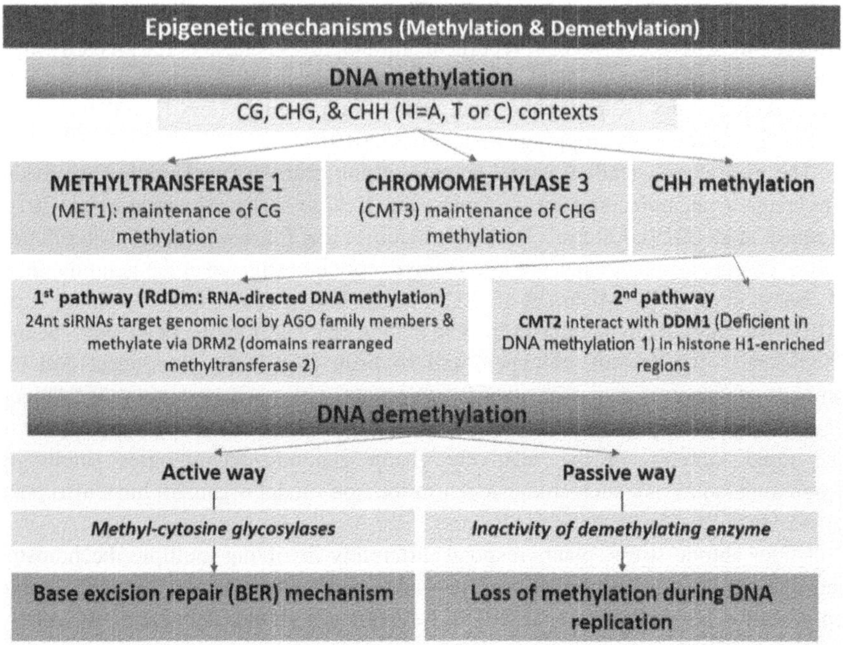

Fig. 4.1 Outline of DNA methylation and demethylation operating in the plants

sequence(s) (Chinnusamy and Zhu 2009). The detailed understanding of the role of epigenetic factors regulating various abiotic stresses is still very limited.

Among various epigenetic mechanisms, DNA methylation and posttranslational histone modifications (PHM) are the deeply studied DNA modification mechanisms. In the case of DNA methylation, a methyl group gets added from *S*-adenosyl-L-methionine to the fifth C of the cytosine ring, which results in the formation of 5-methylcytosine (5mC). In plant system, DNA methylation is reported to occur in three sequence contexts, viz., (1) symmetric CG, (2) symmetric CHG, and (3) asymmetric CHH, where H can be A, T, or C base or except G any other base (Malabarba et al. 2021). Process-wise DNA methylation can be of three types, viz., (1) de novo methylation, (2) maintenance of methylation, and (3) demethylation, which involves several enzymes (Fig. 4.1).

4.2.1 De Novo Methylation

The process uses domains rearranged methyltransferase-2 (DRM2), which gets controlled by RNA-directed DNA methylation (RdDM) pathway (Law and Jacobsen 2010; Matzke and Mosher 2014), wherein Pol IV (RNA pol IV) transcribes single-stranded RNAs (ssRNAs) which then form double-stranded RNA intermediates (dsRNAs) by RNA-dependent RNA polymerase 2 (RDR2). Afterward, DCL3

(RNase III-class DICER-LIKE 3) cleaves the dsRNAs to form 24-nt small interfering RNAs (siRNAs), which get incorporated into AGO4 (ARGONAUTE 4) and base-paired with Pol V and produce scaffold RNA and use DDR protein complex. This comprises proteins such as defective in RNA-directed DNA methylation 1 (DRD1), RNA-directed DNA methylation 1 (RDM1), and defective in meristem silencing 3 (DSM1) that stabilize the Pol V and chromatin interaction using MORC protein complex. Pol V then guides the AGO4 to the chromatin (Wierzbicki et al. 2009). These ultimately result in the DRM2 recruitment, which is followed by methylation of specific DNA base(s) (Matzke and Mosher 2014; Huiming Zhang and Zhu 2011). The precise function of the RdDM pathway indirect gene methylation and regulation is still not very clear. Rather, this pathway targets some repetitive sequences and transposable elements (TEs), which then controls the activation and repression of close-by gene(s) (Sigman and Slotkin 2016). The methylation of CHH context (de novo) of mostly heterochromatin regions especially that of TEs was regulated by the CMT2-dependent pathway (Zemach et al. 2013), which acts in a siRNA-independent way and is dependent on decreased in DNA methylation 1 (DDM1) chromatin remodeler.

4.2.2 Maintenance of Methylation

This is very much dependent on the sequence contexts; for example, methylation of CG context is reportedly maintained by methyltransferase 1 (MET1) and decrease in DNA methylation 1 (DDM1), whereas CHG by chromomethylase 2 and 3 (CMT2 and CMT3), and CHH by DRM2 and CMT2 (Chan et al. 2005).

4.2.3 DNA Demethylation

Demethylation can be through either active or passive ways, wherein passive demethylation denotes loss of methylation during DNA replication due to the inactivity of the demethylating enzyme (Zhu 2009). This process is being regulated by four bifunctional 5mC DNA glycosylases, viz., repressor of silencing 1 (ROS1), Demeter (DME), DME-like 2 (DML2), and DML3, which removes the 5mC using base excision repair (BER) pathway (Zhang and Zhu 2012). Due to the antagonistic effect of RdDM and ROS1 activity, some sort of coordination has been reported between DNA methylation and demethylation, which in turn stops the hypermethylation of certain loci (Tang et al. 2016). A 39-nt regulatory element or MEMS (DNA monitoring methylation sequence) is the ROS1 promoter and functions as a putative sensor of MET1 and RdDM pathway. Very high activity of MET1 and RdDM results in the hypermethylation of MEMS, which causes activation of ROS1 demethylase activity and regulates DNA methylation at the whole-genome level (Lei et al. 2015).

4.3 Methylation of Various Regions of the Gene

Gene expression also gets regulated by the methylation in the promotor region via inhibition of transcriptional activators/repressors. This may completely inhibit the tissue-based gene expression (Johnson et al. 2007; Zhang et al. 2006), or this may also regulate specific processes like gene imprinting during seed development or immune-responsive gene regulation (Matzke and Mosher 2014; Zhang et al. 2018a). However, the function of DNA methylation within the gene bodies is still not very clear, and two hypotheses have been proposed about their role, viz., (1) masking of the cryptic transcription sites, which will assist in its isoform splicing (Neri et al. 2017), and (2) reduction in the gene expression variations via exclusion of H2A.Z from the nucleosome (Zilberman et al. 2008). The function of methylation in the TE activity regulations is thoroughly studied, wherein it mainly functions as either TE silencing or a repressor of the transposition by hypermethylation of all the sequence contexts (Sigman and Slotkin 2016).

4.3.1 Histone Modifications

Chromatin accessibility in the gene's promotor region is observed through histone modifications, especially through methylation or acetylation (Berger 2007; Kouzarides 2007) (Fig. 4.2). A nucleosome consists of eight histone proteins (two copies each of H2A, H2B, H3, and H4 proteins), which are wrapped by 147 bp DNA (Peterson and Laniel 2004) and function through epigenetic modification of various genes by controlling the access and binding of regulatory elements (Berger 2007). Modification of the amino acids present at the N-terminal tails of histone proteins (H3 and H4) is reported, which can either activate the genes via acetylation, phosphorylation, and ubiquitination or repress the genes mainly via methylation (with some exceptions) (Zhao et al. 2019). Even though histones are considered as highly conserved proteins, plants do possess structurally and functionally discrete classes of H2A (H2A.X, H2A.Z) and H3 (H3.3) forms (Deal and Henikoff 2011). Increased H3K9ac (in the heterochromatic chromatin knobs) was found to be associated with an increase in the transcription, while increased H3K9me2 was found to be correlated with a decrease in the transcription of certain stress-responsive genes (Yong Hu et al. 2012). The stress-responsive genes in the plants show transient modifications in the histones under varied stress conditions (Zong et al. 2013).

4.3.2 Noncoding RNAs and Epigenetic Regulation Under Abiotic Stress

Noncoding RNAs (long ncRNA or small ncRNAs) regulate the opening and closing of the chromatin and are associated with both gene silencing and activation at both transcriptional and posttranscriptional levels. The ncRNAs which are associated

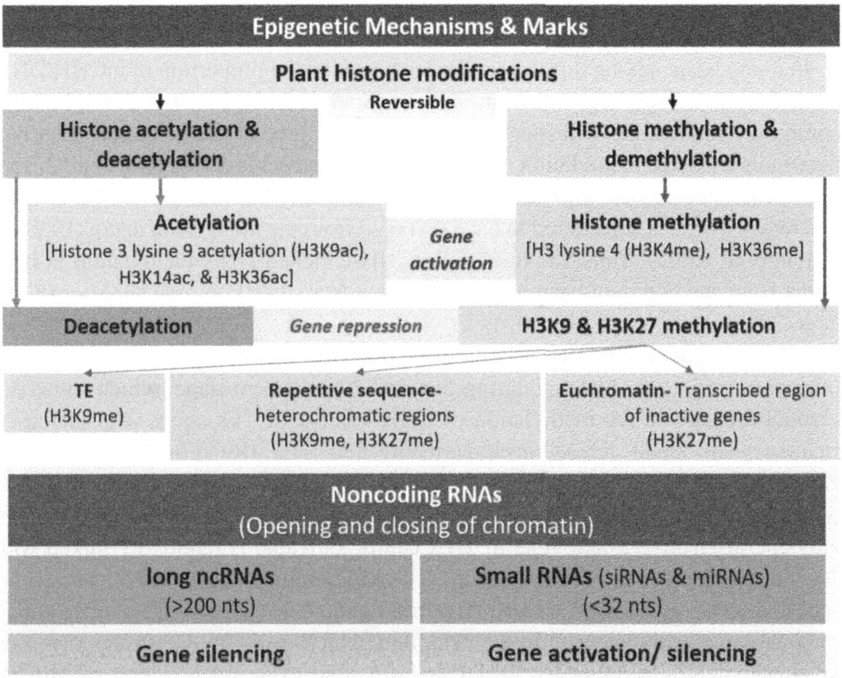

Fig. 4.2 Schematic representation of epigenetic mechanism and marks operating in plants

with epigenetic regulation (i.e., heterochromatin formation, histone modification, DNA methylation, and gene silencing) are of two major types, viz., the short ncRNAs (<30 nts) and the long ncRNAs (>200 nts). The short ncRNAs are grouped into three major types: (1) microRNAs (miRNAs), (2) short interfering RNAs (siRNAs), and (3) piwi-interacting RNAs (piRNAs) (Salgotra and Gupta 2019). Sequence-specific methylation is known to be caused by ds-RNA and RdDM (Law and Jacobsen 2010), while RNAi was reportedly associated with the RdDM causing cytosine methylation (Meister and Tuschl 2004; Wassenegger et al. 1994). In plants, miRNAs are known to cause RNA silencing and posttranscriptional gene regulation (Lee et al. 1993; Maxwell et al. 2012). Various classes of siRNAs (i.e., antisense siRNAs, heterochromatic siRNAs, trans-acting siRNAs) were known to enable gene silencing through methylation of histone and RdDM (Mosher et al. 2008; Xu et al. 2013).

Several reports have proved the role of ncRNAs in the regulation of gene expression under abiotic stress conditions. Zeller et al. (2009) reported the accumulation of various novel antisense transcripts under abiotic stress situations in the plants, which were the source of siRNAs. Downregulation of certain siRNAs such as Hc-siRNAs (heterochromatic siRNAs), siR441, and siR446 has been reported under abiotic stress situations, which seems essential for the gene regulation, especially under stress conditions (Yan et al. 2011). The miRNAs are known to have a major role under different abiotic stress-like conditions, especially under cold, heat, salinity, etc. (Salgotra and Gupta 2019).

In mung bean, among different sequence contexts, the proportion of mCHH DNA methylation was very high, while in soybean (Schmitz et al. 2013), mCG is the most commonly methylated sequence context. Both mung bean and common bean had maximum cytosine methylation in the CHH sequence context (Do Kim et al. 2015). These two species were known to have diverged nearly 8.0 Ma (million years ago), whereas soybean is considered to have diverged from the mung bean nearly 19.2 Ma (Lavin et al. 2005). Thus, the traces of mCHH-enriched DNA methylation in both mung bean and common bean might have come from their common ancestor (Kang et al. 2017).

The role of the RdDM pathway has been observed regulating the seed parameters (size and weight) in chickpea during the seed development stage, which showed a gradual increase in the methylation of CHH context of TEs along with increased frequency of small RNAs in hypermethylated TEs (Rajkumar et al. 2020). Kurdyukov et al. (2014) have studied a 2HA seed line of *Medicago truncatula* which was highly embryogenic, and microarray showed downregulation of an ethylene insensitive 3-like gene in 2HA callus. Ethylene is reportedly linked with several developmental processes such as somatic embryogenesis (SE) and various types of stress responses. The *Medicago truncatula EIL1* gene (MtEIL1) was found to be epigenetically silenced in the 2HA line, which could be due to the increased level of miRNA targeting its 3′UTR and also due to the methylation of *MtEIL1* (Kurdyukov et al. 2014). A plant-specific gene *MutS HOMOLOG1* (*MSH1*) has been used for the RNAi suppression in several plant species including soybean for the production of developmental changes including abiotic stress response along with methylome repatterning. Therefore, RNAi can be a direct means of exploitation of epigenetic variations associated with abiotic stress tolerance in plants (Raju et al. 2018).

4.4 Epigenetics and Abiotic Stress Tolerance in Legumes

4.4.1 Temperature-Stress Tolerance

Climate change induced by various factors including excessive greenhouse gas (GHG) emission has caused many negative impacts on crop plants primarily in the form of heat stress due to global warming. The mean temperature of the earth is reportedly increased to the tune of 0.35 °C from 1979 to 2003 (Venterea 2014; Walther et al. 2002). An increase in the temperature or heat stress is known to cause early flowering (Peñuelas et al. 2009), modified plant architecture (Wahid et al. 2007), poor seed development, decreased dormancy, reduced size, and poor grain yield (Folsom et al. 2014; Long and Ort 2010), and shifting of plant establishment to higher altitudes (Pauli et al. 2012). For the transcriptional machinery, chromatin conformation plays a major role by allowing the access of DNA sequences (Li et al. 2008). Several studies have proved the role of histone acetylation and methylation in the plant's response to abiotic stresses (Stępiński 2012). In the root meristem of the soybean, changes were recorded for the DNA methylation, histone methylation, and histone acetylation when grown under different temperature regimes.

4.4.1.1 Heat Stress

In response to abiotic stress like heat, the heat-stress factors (HSF) and secondary metabolites have been found to have a great role in managing such stress response (McClung and Davis 2010). In addition, even by the transcriptional reprogramming like by upregulation of kinases and various transcription factors (TF) and downregulation of growth-related genes, such stresses were also managed (Popova et al. 2013). ROS1 was reported to be demethylating all the DNA methylation contexts as in the ros1 mutants all the DNA methylation contexts get hypermethylated (Tang et al. 2016). Heat stress is also known to affect DNA methylation in different plant species as increased global methylation and homologous recombination frequency (Boyko et al. 2010).

Various methylation process proteins [e.g., DRM2, nuclear RNA polymerase D1 (NRPD1)] also show upregulation upon high-temperature stress and increased DNA methylation (Naydenov et al. 2015). The RdDM pathway is also reported to partially regulate the transcriptional response to high-temperature stress (Popova et al. 2013). Interestingly, global DNA methylation induced by heat stress is species and tissue specific. In addition, the result of the impact of heat stress on transgenerational memory was recorded as phenotypic changes over generations by different researchers (Migicovsky et al. 2014; Suter and Widmer 2013). During repeated heat stress conditions, methylation of histone H3K4 was found to be associated with the persistent expression and hyper-induction of high temperature-responsive genes (Lämke et al. 2016).

In the changing environmental scenario, an increase in the temperature is a major focus of various studies that are involved in unfolding the temperature-dependent genes and pathways (Arya et al. 2017). Temperature is a key factor affecting significantly the flowering in the plant system. Genome annotation has identified four copies of the *PIF4* gene in soybean, which were found expressing abundantly under short-day conditions (Wong et al. 2013). Also, eight copies of the *SVP* gene were identified in soybean, of which high expression of *Glyma01G02880*.1 and *Glyma02G04710*.1 has been recorded in shoot apical meristem during the floral transition phase (Wong et al. 2013). In soybean, miRNA 156 and miRNA 172 have shown upregulation during the floral transition stage, when plants were exposed to high temperatures (Li et al. 2015). H3K56ac and H3K4me3 methylation marks are usually recorded in class 3 SDGs, which showed differential expression in shoot apical meristem (SAM) in soybean, during the floral transition (Liew et al. 2013). Besides, several RNAi genes like RNA pol, dsRNA binding (DRB), Dicer-like, AGO, DNA methyltransferase, and DNA glycosylase were upregulated in SAM during the floral transition (Liew et al. 2013).

4.4.1.2 Chilling Stress

More of 5mC and H3K9me2 were recorded through immunostaining in the heterochromatin region of the soybean genome during chilling stress than during the recovery phase. However, indicators of permissive chromatin (i.e., H3K9ac, H4K12ac, and H3K4me) showed weak labeling in the euchromatin region (under stress) over recovered plants (Stępiński 2012). Ivashuta et al. (2002) reported transcriptional activation of specific retrotransposons in *Medicago sativa* under cold-stress conditions. Hypermethylation was recorded in *Cannabis sativa* genotypes when exposed to different levels of cold acclimation (Mayer et al. 2015). Epigenetic factors are known to regulate cold-stress tolerance in hemp. Under cold-stress conditions, soluble sugars are found to accumulate and be maintained in higher concentrations in the tolerant hemp genotypes. These genotypes expressed more of *COR* gene transcripts, which is associated with the de novo DNA methylation. Also, significantly higher methylcytosine levels at *COR* gene loci were recorded in the tolerant genotypes when deacclimated, thereby confirming the function of locus-specific DNA methylation (Mayer et al. 2015). Relatively more methylation over demethylation was recorded in the cold stress-tolerant genotypes over susceptible chickpea genotype. This could be due to the comparatively higher activation of cold stress-responsive genes in the tolerant genotypes (Rakei et al. 2016).

4.4.2 Drought-Stress Tolerance

Overall increased DNA methylation or hypermethylation has been recorded in both tolerant and sensitive genotypes of faba bean (Abid et al. 2017) and pea (Labra et al. 2002) under drought or water-deficit stress situations. In groundnut, the regulation of the *AhDREB1* gene (of AP2/ERF family TF) through acetylation of H3 helped in the positive regulation of water-deficit stress tolerance genes when exposed to the PEG-induced water-deficit stress conditions. Higher *AhDREB1* gene expression was recorded when an inhibitor of histone deacetylase (HDAC), i.e., trichostatin (TSA), was used, which then showed water-deficit stress tolerance in the peanut plant (Zhang et al. 2018b). In chickpea, drought stress was able to induce the *CaHDZ12* (an HD-Zip TF) activation, and its expression was found to be correlated with that of H3K9ac acetylation in the promoter region (Sen et al. 2017). Cytosine hypomethylation has been reported in the soybean roots under heat-stress conditions (Hossain et al. 2017). Reduced metabolic activity has been recorded in pea plants due to the drought-induced hyper-methylation of some key genes (Labra et al. 2002; Salgotra and Gupta 2019), while higher methylation was recorded in the drought-sensitive horse gram (*Macrotyloma uniflorum*) genotype than the tolerant genotype (Bhardwaj et al. 2013).

Epigenetic changes such as histone modifications, sRNAs, methylation of DNA, and lncRNAs were known to be associated with gene regulation in faba beans (Meyer 2015). A high degree of association between DNA methylation and gene expression under drought conditions in faba bean suggests the involvement of DNA methylation in the imposition of drought-stress tolerance (Abid et al. 2017). Several water-deficit stress-associated differentially methylated regions (DMRs) were identified by Abid et al. (2017), which became the basis for further understanding the epigenetic regulation of drought stress in faba bean.

In addition, a number of drought-responsive miRNAs are identified in various legumes (Mantri et al. 2013). Under drought and salinity stress, 259 differentially expressed miRNAs have been identified from the root tip tissue of chickpea. Many of these were found to have auxin and other abiotic stress-responsive cis-elements in their promoter region, which in turn imparted abiotic stress tolerance through various phytohormone syntheses (Khandal et al. 2017). In chickpea, under water-deficit stress, *miR408* was found to be accumulated (Hajyzadeh et al. 2015), while similar results were also reported for Medicago (Trindade et al. 2010). Interestingly, 24 novel miRNA families have been reported in water-deficit stress-tolerant and -sensitive cowpea genotype (Barrera-Figueroa et al. 2012), and of these, 22 families were reported from soybean too (Kulcheski et al. 2011). Most of the iso-miRNAs showed upregulation during water-deficit stress in sensitive genotypes, while downregulation in tolerant genotypes. miRNAs were known to play a key role in the abiotic stress tolerance in cowpea, and much water-deficit stress-associated microRNAs have been identified (Barrera-Figueroa et al. 2011). Inconsistent miRNA expression was reported in the studied cowpea genotypes and nine recorded predominant or exclusive expression in one of the two studied genotypes, while a few were found regulated under water-deficit stress in only one genotype (Barrera-Figueroa et al. 2011).

The genome sequencing of common bean has generated a lot of information that can provide the much-needed evidence about future PTGS studies (Vlasova et al. 2016). Posttranscriptional regulation in common bean under water-deficit stress is reportedly regulated via a legume-specific miR1514a, which targets the NAC family of TF and generates secondary phasiRNAs (Sosa-Valencia et al. 2017b). The miR1514a showed differential expression levels in the roots of the common bean when exposed to water-deficit stress conditions. In addition, an RNA-seq study has also identified the role of NAC 700 TF in the water-deficit stress, while degradome analysis revealed the two NAC TFs (Phvul.010g121000 and Phvul.010g120700) as the target of miR1514a. In addition, small RNA-seq data indicate the role of only Phvul.010g120700 in the generation and accumulation of phasiRNAs under water-deficit stress conditions (Sosa-Valencia et al. 2017b).

4.4.3 Salinity-Stress Tolerance

In the pigeon pea shoot tissues, a global decline in the methylation levels of DNA has been reported under salinity-stress conditions (Awana et al. 2019). Similarly, imposition of continuous stress for a relatively longer duration has resulted in an increase in the overall DNA demethylation in soybean, mainly in the tolerant genotypes, and this increase corresponded well with that of increased expression of various DNA demethylases (e.g., DML and ROS1). Further, different demethylation studies could identify that the regulatory genes having CG and CHG contexts were more crucial than the CHH for their adaptation to salinity stress (Liang et al. 2019). The salinity stress in *Medicago truncatula* revealed variations to the tune of 77% in CHH, while CHG and CG showed 9.1% and 13.9% changes, respectively.

Interestingly, no such correlation has been recorded between DNA methylation pattern and level of transcripts for other salinity-stress tolerance-associated key genes, which means that these genes might be regulated by other epigenetics processes (Yaish et al. 2018). On the contrary, four TFs showed induced response under salinity stress in soybean, and of these three showed demethylations in CG and non-CG contexts and also active enrichment of histone marks (H3K4me3 and H3K9ac) along with a reduction in the repressive mark H3K9me2. Thus, the possible interplay was recorded between methylation of DNA and histone modifications when exposed to salinity stress (Song et al. 2012). In the plants of salinity-stressed soybean, cross talk has been reported between histone methylation and acetylation (Liu et al. 2010; Stępiński 2012). The soybean plants under salt-stress condition showed the binding of GmPHD5 (a homeodomain TF) with H3K4me2 marks (salt-induced), which then recruits a complex associated with gene activation having GmISWI (a type of nonhistone proteins and a chromatin remodeling factor) and GmGNAT1 (an acetyltransferase), which selectively acetylates H3K14 for the activated expression of salinity-induced genes (Wu et al. 2011).

The methylation-sensitive amplified polymorphism (MSAP) and enzyme-linked immunosorbent assay (ELISA) showed significantly more methylation in mCG under salinity stress in *M. truncatula* (Al-Lawati et al. 2016; Yaish et al. 2014). A comparative whole-genome bisulfite sequencing (WGBS) on the DNA isolated from the root tissues of *Medicago truncatula* under salinity stress and control has revealed higher methylation levels in all sequence contexts, ranging from 3.8% to 10.2%. However, qPCR-based gene expression studies did not find any stable association between mCG methylation levels and transcript abundance of some key genes involved in the imposition of salinity tolerance. Thus, it seems that some other epigenetic controllers are regulating the gene expression under salinity stress (Yaish et al. 2018).

The role of MTases regulating different aspects of plant development including different abiotic stress responses has been unfolded in different legume species using expression analysis. Garg et al. (2014) have identified 16 members of the DNMT2 family in several legume species, and increased expression of DNMT (CaDNMT2) was observed in chickpea shoots under both salt- and drought-stress conditions, suggesting the role of DNMT in abiotic stress response. Overall, under abiotic stress, more transcript was recorded for CMT and DRM genes, signifying the role of stress-induced methylation in chickpea (Garg et al. 2014). In chickpea, salinity stress was found inducing the *CaHDZ12*, which also showed a correlation with that of H3K9ac acetylation in the promoter region (Sen et al. 2017). Preferential transmission of salinity tolerance and DNA methylation was reported through the female germline. However, paternal *dme* mutants recorded restoration of paternal memory, indicating that the active DNA methylation in male gametes is essential for the inhibition of paternal inheritance of hyperosmotic priming response (Wibowo et al. 2016). The details of epigenetic response in various legumes during abiotic stress-associated processes are presented in Table 4.1.

4.4.4 Abiotic Stress Tolerance and DNA Demethylation

Several reports are aimed to analyze the changes which occur in the DNA demethylase gene when exposed to various abiotic stresses, and only some mentioning the detailed analysis involving loss-of-function mutations are available (Parrilla-Doblas et al. 2019). Interestingly, some recent studies showed the function of active demethylation of DNA in the inter-generational transmission of "stress memory" helping rapid adaptation to short-term environmental variations called "priming" (Parrilla-Doblas et al. 2019). In addition, the response of a plant to abscisic acid (ABA) has also shown active demethylation of DNA under abiotic stress conditions. Still, various factors regulating such demethylation during abiotic stress response are superficially known to the scientific community. Additionally, miRNAs are now found to have some role in the active demethylation of DNA of certain genes (Parrilla-Doblas et al. 2019). This has been generally observed during gametophyte development (Slotkin et al. 2009). However, active demethylation of DNA considers enzyme-based elimination of methylated cytosine, by a family of DNA glycosylases (such as DME, ROS1, DML2, and DML3), which was followed by the base excision repair (BER) dependent process (Penterman et al. 2007; Zhu 2009). This does not only alter genome-wide epigenetics, but also regulate locus-specific genes with abiotic stress tolerance (Hsieh et al. 2009).

4.4.5 Abiotic Stress Tolerance and Epigenetics-Based Breeding Strategies in Legumes

Till now, we have compiled several ways that can be used for the enhancement of abiotic stress tolerance in various legumes. The use of epigenetics and epigenomics in improving the adaptation to abiotic stresses needs a combination of technical and biological innovations so that the breeders can go for the targeted gene-specific modifications of the epigenome for the desired trait improvement. Besides positive impact, stress-based memory may also have a negative impact on yield (Chinnusamy and Zhu 2009). Thus, care must be taken while going for an epigenetic-based approach for the abiotic stress improvement of the crops. We can use the impact prediction models for the epigenetic variations on a plant's phenotype and performance (Colicchio et al. 2015; Yaodong Hu et al. 2015). Identification of epialleles having an impact on the abiotic stress tolerance traits can result in epigenetic-based breeding of crop plants like the use of mutant lines, recurrent epi-selection, epigenomic selection, and editing (Greaves et al. 2014; Hauben et al. 2009; Lämke and Bäurle 2017; Oakey et al. 2016; Yang et al. 2015).

Table 4.1 Epigenetic response during various abiotic stress-associated processes in legumes

S. No.	Abiotic stress	Gene(s)/tissue/outcome	Crop species	Epigenetic response	Reference
1.	Cold	Reduced yield	*Cannabis sativa*	Methylome variation (locus-specific methylation and deacclimation)	Mayer et al. (2015)
2.	Drought	Reduced yield	Soybean	miR1514a modulation of a NAC TF transcript	Sosa-Valencia et al. (2017a, b)
3.	Drought	Reduced yield	Soybean	Up-regulation of isomiRNAs	Kulcheski et al. (2011)
4.	Drought	Reduced yield	Pea	Hypermethylation of cytosine residues	Labra et al. (2002)
5.	Drought	Reduced yield	Chickpea	Accumulation of miR408 transcripts; BHLH23 ERF/AP2	Hajyzadeh et al. (2015)
6.	Drought	Root	Chickpea	Accumulation of miRNAs at root apex	Khandal et al. (2017)
7.	Drought	Reduced yield	Cowpea	Very low vun-miR5021 and vun-miR156b-3p expression, while higher P5CS transcript level	Shui et al. (2013)
8.	Drought	Reduced yield	Faba bean (*Vicia faba*)	miR398a and miR2119 in dicistronic arrangement	De la Rosa et al. (2019)
9.	Drought	Reduced yield	Faba bean	Higher demethylation of *LOX*, *CDPK*, *ABC*, *GH*, and *PEPC* genes	Abid et al. (2017)
10.	Drought	Reduced yield	*Glycine max*	GmNFYA3	Ni et al. (2013)
11.	Drought	Reduced yield	*Phaseolus vulgaris*	NAF TF ARF10	Sosa-Valencia et al. (2017b)
12.	Drought	Reduced yield	*Vigna unguiculata*	Transferase family protein leucine repeat-rich transmembrane protein	Barrera-Figueroa et al. (2011)
13.	Drought	Reduced yield	*Macrotyloma uniflorum*	DNA methylation dynamics	Bhardwaj et al. (2013)
14.	Drought	DNMT	Chickpea	Increased expression of DNMT (CaDNMT2)	Garg et al. (2014)
15.	Heat	Roots	*Glycine max*	Hypomethylation of cytosine	Hossain et al. (2017)

16.	Salinity	Shoots	Pigeon pea	Reduction in the DNA methylation in shoots	Awana et al. (2019)
17.	Salinity	Root	Chickpea	Accumulation of miRNAs at root apex	Khandal et al. (2017)
18.	Salinity	Glyma11g02400	Soybean	Demethylation	Song et al. (2012)
19.	Salinity	Glyma16g27950, Glyma20g30840	Soybean	Hypomethylation	Song et al. (2012)
20.	Salinity	DNMT	Chickpea	Increased expression of DNMT (CaDNMT2)	Garg et al. (2014)
21.	Seed development	Cotyledon (*WOX*, *CUC*, *CLAVATA*, *PIN1* genes)	Soybean	DNA methylation	An et al. (2017), Lin et al. (2017)
22.	Flower development	Expressed in shoot apical meristem (SAM)	Soybean	124 histone modifiers	Liew et al. (2013)
23.	Flowering time regulation	Flower development	Peanuts	Methylation of the *FWA* promoter region	Chan et al. (2006)
24.	Flower and pod initiation	Flower development	Peanuts	Differential expression of various methylation genes (DRM2, MET1 DNA methylases, *DMS3*, *DRD1*, *MORC1*, and *IDN2*)	Wang et al. (2018)

In soybean, Raju et al. (2018) have proposed a breeding strategy using the *MSH1* gene system for the improvement of yield and stability by inducing epigenetic variations. The soybean memory lines (wild type and *msh1* acquired) were crossed to develop the epi-lines having wide variations for various yield-related traits. The identified epi-types showed low epi-type × environment (e × E) interactions and thus more stability under varied environments expressing different abiotic stresses (Varotto et al. 2020). The novel epigenetic variations induced by the *MSH1* suppression were found to be inherited for at least three generations and can be used for enhancement and stabilization of the overall yield of soybean crops. In addition, several metabolic pathway genes regulating improved adaptation and plasticity (across generations) of the plant are also identified (Fujimoto et al. 2012; Raju et al. 2018; Robertson and Wolf 2012).

In association with classical genetic approaches, the novel sequencing technologies have helped in understanding the epigenetic process at the whole-genome level. Epigenome profiling and epigenome editing will help in the creation of novel epiallelic variants through DNA methylation and chromatin modifications (Springer and Schmitz 2017). Breeders are now preferring to use the mapping of epigenetic marks at genome-wide level (epigenomics), and also identification of epigenetic targets to modify the plants' epigenomic variability to make them more resilient and climate smart (Lane et al. 2014). There is a need to do large-scale cross-species generation and comparison of epigenetic data in legumes, especially in response to abiotic stresses (Lane et al. 2014). Epigenetic modifications can be attempted either globally or at a specific locus using emerging techniques like CRISPR/Cas9 and dCas (Hilton et al. 2015; Moradpour and Abdulah 2020). The knowledge about the activation and repression of specific chromatin regions (using DNA-binding domains like Zn fingers, TALEs, dCas9) under specific abiotic stress can be used for the gene-specific activation or repression as per the need for the imposition of abiotic stress tolerance in crop plants (Bilichak and Kovalchuk 2016). In addition, a sound prediction model about the impact of epigenetic variations on the plant's overall performance is needed (Colicchio et al. 2015; Yaodong Hu et al. 2015).

4.5 Conclusions and Future Prospects

In legumes, the epigenetic studies are still in infancy and are mainly targeting the identification of key epigenetic factors in the plant's developmental and stress-related processes. A major reason for this could be the poor annotation of most of the legume genomes, which are incidentally full of numerous high-copy-number genes having overlapping or distinct functions (Windels et al. 2021). However, in times to come, we expect tremendous growth in legume epigenetic studies for various traits including abiotic stress tolerance. Stably inherited natural or induced epigenetic variations can be used to create climate-smart crops (Vriet et al. 2015). Otherwise, most stress-induced epigenetic modifications show reversion once there is no stress. Still, some of the modifications do show stable inheritance as epigenetic-mediated stress memory does result in long-term adaptations (Sudan et al. 2018). However, detailed studies are needed to find the factors regulating the epiallele stability in crop plants for their further use in a breeding program (Hofmeister et al. 2017). There is a need to develop various mathematical models for the identification of heritable epigenetic phenotypes, for the enhanced efficiency of the breeding program (Tal et al. 2010). Besides, epi-genotyping procedures can be developed for the identification of newly formed epialleles and their inheritance pattern (Hofmeister et al. 2017). Also, more precise epi-mutagenesis and targeted epigenome editing are needed for targeted epigenome editing (Johnson et al. 2014; Springer and Schmitz 2017).

The regions associated with the transposable elements are more prone to methylation under abiotic stress situations. Thus, these regions should be targeted to understand the trend of epigenetic changes at the whole-genome level through cytosine methylation studies (Bruce et al. 2007). Differential DNA methylation has been recorded in different tissues of the soybean (Song et al. 2013), but it is unclear whether the differences were spontaneous or developmentally controlled by differentially methylated regions (DMRs) (Salgotra and Gupta 2019). We expect an increase in the functional studies of various key epigenetic factors that can be enhanced by the recent developments in the CRISPER technologies via the generation of several epigenetic mutants at least in major legume crops. Thus, a better understanding generated about the epigenetic mechanism along with the identification of epialleles will potentially boost the plant's ability to cope with various abiotic stresses.

There is a need to modify the active DNA demethylation through CRISPER/Cas9 technology to the genes involved in the demethylation pathway. In the future, we need very precise control on the DNA methylation and demethylation of specific genes as epigenome engineering, for targeted abiotic stress tolerance breeding in the legumes (Springer and Schmitz 2017; Stricker et al. 2017). In soybean, most of the DNA methyltransferase genes were found to be expressed at low levels in seed and seem to contribute to the silencing of certain mC genes in the seed tissues. There is a need to do deep analysis about the mC pattern in different tissues under various abiotic stresses to gain an insight into the role of gene methylation, resulting in novel epigenetic gene regulation (Garg et al. 2014). The details of abiotic stress management strategies in legumes using epigenetic approaches are presented in Fig. 4.3.

Although several legume crops (viz., mung bean, lentil, peanut, chickpea, cowpea, pea, Medicago, pigeon pea, lotus, soybean, beans, etc.) have been sequenced, epigenetic studies concerning abiotic stress tolerance are limited to a few species like soybean, chickpea, pigeon pea, cowpea, and beans. There is an urgent need to study more legumes for abiotic stress tolerance using epigenetic approaches. For this, a joint research platform may be developed by various national and international organizations like the International Crops Research Institute for the Semi-Arid Tropics (ICRISAT, India), the International Center for Agriculture Research in the Dry Areas (ICARDA, Lebanon), the International Center for Tropical Agriculture which is an international research and development organization (CIAT, Colombia), and Indian Agricultural Research Institute (IARI, India), working for the

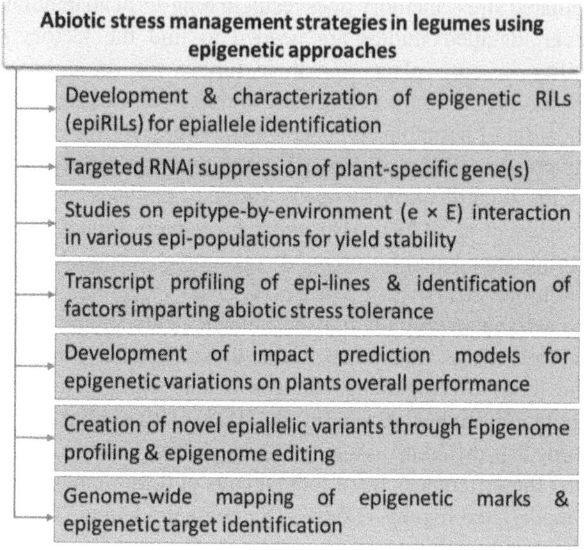

Fig. 4.3 Comprehensive abiotic stress management strategies in legumes using epigenetic approaches

improvement of various legume crops for targeted improvement using epigenetic approaches. In the initial stage, crops like mung bean, lentil, peanut, and pea can be targeted, and at a later stage depending on the availability of whole-genome information, more legumes can be added.

References

Abid G, Mingeot D, Muhovski Y, Mergeai G, Aouida M, Abdelkarim S, Aroua I, El Ayed M, M'hamdi M, Sassi K (2017) Analysis of DNA methylation patterns associated with drought stress response in faba bean (*Vicia faba* L.) using methylation-sensitive amplification polymorphism (MSAP). Environ Exp Bot 142:34–44

Ahmad S, Nawade B, Sangh C, Mishra GP, Bosamia TC, Kumar N, Dobaria JR, Gajera HP (2020) Identification of novel QTLs for late leaf spot resistance and validation of a major rust QTL in peanut (*Arachis hypogaea* L.). 3 Biotech 10(10):1–13

Al-Lawati A, Al-Bahry S, Victor R, Al-Lawati AH, Yaish MW (2016) Salt stress alters DNA methylation levels in alfalfa (*Medicago* spp.). Genet Mol Res 15(1):15018299

An YC, Goettel W, Han Q, Bartels A, Liu Z, Xiao W (2017) Dynamic changes of genome-wide DNA methylation during soybean seed development. Sci Rep 7(1):1–14

Arya H, Singh MB, Bhalla PL (2017) Molecular genomic investigations on floral induction using soybean as a model system. In: International symposium on flowering, fruit set and alternate bearing, vol 1229, pp 47–56

Awana M, Yadav K, Rani K, Gaikwad K, Praveen S, Kumar S, Singh A (2019) Insights into salt stress-induced biochemical, molecular and epigenetic regulation of spatial responses in Pigeonpea (*Cajanus cajan* L.). J Plant Growth Regul 38(4):1545–1561

Barrera-Figueroa BE, Gao L, Diop NN, Wu Z, Ehlers JD, Roberts PA, Close TJ, Zhu J-K, Liu R (2011) Identification and comparative analysis of drought-associated microRNAs in two cowpea genotypes. BMC Plant Biol 11(1):1–11

Barrera-Figueroa BE, Gao L, Wu Z, Zhou X, Zhu J, Jin H, Liu R, Zhu J-K (2012) High throughput sequencing reveals novel and abiotic stress-regulated microRNAs in the inflorescences of rice. BMC Plant Biol 12(1):1–11

Berger SL (2007) The complex language of chromatin regulation during transcription. Nature 447(7143):407–412

Bhalani H, Thankappan R, Mishra GP, Sarkar T, Bosamia TC, Dobaria JR (2019) Regulation of antioxidant mechanisms by AtDREB1A improves soil-moisture deficit stress tolerance in transgenic peanut (*Arachis hypogaea* L.). PLoS One 14(5):e0216706

Bhardwaj J, Mahajan M, Yadav SK (2013) Comparative analysis of DNA methylation polymorphism in drought sensitive (HPKC2) and tolerant (HPK4) genotypes of horse gram (*Macrotyloma uniflorum*). Biochem Genet 51(7):493–502

Bilichak A, Kovalchuk I (2016) Transgenerational response to stress in plants and its application for breeding. J Exp Bot 67(7):2081–2092

Bosamia TC, Mishra GP, Thankappan R, Dobaria JR (2015) Novel and stress relevant EST derived SSR markers developed and validated in peanut. PLoS One 10(6):e0129127

Bosamia TC, Dodia SM, Mishra GP, Ahmad S, Joshi B, Thirumalaisamy PP, Kumar N, Rathnakumar AL, Sangh C, Kumar A (2020) Unraveling the mechanisms of resistance to Sclerotium rolfsii in peanut (*Arachis hypogaea* L.) using comparative RNA-Seq analysis of resistant and susceptible genotypes. PLoS One 15(8):e0236823

Boyko A, Blevins T, Yao Y, Golubov A, Bilichak A, Ilnytskyy Y, Hollander J, Meins F Jr, Kovalchuk I (2010) Transgenerational adaptation of Arabidopsis to stress requires DNA methylation and the function of Dicer-like proteins. PLoS One 5(3):e9514

Bruce TJA, Matthes MC, Napier JA, Pickett JA (2007) Stressful "memories" of plants: evidence and possible mechanisms. Plant Sci 173(6):603–608

Budak H, Hussain B, Khan Z, Ozturk NZ, Ullah N (2015) From genetics to functional genomics: improvement in drought signaling and tolerance in wheat. Front Plant Sci 6:1012

Cañas LA, Beltrán JP (2018) Model legumes: functional genomics tools in Medicago truncatula. In: Functional genomics in *Medicago Truncatula*. Humana, New York, pp 11–37

Chan SW-L, Henderson IR, Jacobsen SE (2005) Gardening the genome: DNA methylation in *Arabidopsis thaliana*. Nat Rev Genet 6(5):351–360

Chan SW-L, Zhang X, Bernatavichute YV, Jacobsen SE (2006) Two-step recruitment of RNA-directed DNA methylation to tandem repeats. PLoS Biol 4(11):e363

Chinnusamy V, Zhu J-K (2009) Epigenetic regulation of stress responses in plants. Curr Opin Plant Biol 12(2):133–139

Colicchio JM, Miura F, Kelly JK, Ito T, Hileman LC (2015) DNA methylation and gene expression in *Mimulus guttatus*. BMC Genomics 16(1):1–15

Dasgupta U, Mishra GP, Dikshit HK, Mishra DC, Bosamia T, Roy A, Bhati J, Aski M, Kumar RR, Singh AK (2021) Comparative RNA-Seq analysis unfolds a complex regulatory network imparting yellow mosaic disease resistance in mungbean [*Vigna radiata* (L.) R. Wilczek]. PLoS One 16(1):e0244593

De la Rosa C, Covarrubias AA, Reyes JL (2019) A dicistronic precursor encoding miR398 and the legume-specific miR2119 coregulates CSD1 and ADH1 mRNAs in response to water deficit. Plant Cell Environ 42(1):133–144

Deal RB, Henikoff S (2011) Histone variants and modifications in plant gene regulation. Curr Opin Plant Biol 14(2):116–122

Deans C, Maggert KA (2015) What do you mean, "epigenetic"? Genetics 199(4):887–896

Do Kim K, El Baidouri M, Abernathy B, Iwata-Otsubo A, Chavarro C, Gonzales M, Libault M, Grimwood J, Jackson SA (2015) A comparative epigenomic analysis of polyploidy-derived genes in soybean and common bean. Plant Physiol 168(4):1433–1447

Folsom JJ, Begcy K, Hao X, Wang D, Walia H (2014) Rice fertilization-Independent Endosperm1 regulates seed size under heat stress by controlling early endosperm development. Plant Physiol 165(1):238–248
Fujimoto R, Sasaki T, Ishikawa R, Osabe K, Kawanabe T, Dennis ES (2012) Molecular mechanisms of epigenetic variation in plants. Int J Mol Sci 13(8):9900–9922
Garg R, Kumari R, Tiwari S, Goyal S (2014) Genomic survey, gene expression analysis and structural modeling suggest diverse roles of DNA methyltransferases in legumes. PLoS One 9(2):e88947
Gibney ER, Nolan CM (2010) Epigenetics and gene expression. Heredity 105(1):4–13
Greaves IK, Groszmann M, Wang A, Peacock WJ, Dennis ES (2014) Inheritance of trans chromosomal methylation patterns from Arabidopsis F1 hybrids. Proc Natl Acad Sci 111(5):2017–2022
Gutzat R, Scheid OM (2012) Epigenetic responses to stress: triple defense? Curr Opin Plant Biol 15(5):568–573
Hajyzadeh M, Turktas M, Khawar KM, Unver T (2015) miR408 overexpression causes increased drought tolerance in chickpea. Gene 555(2):186–193. https://doi.org/10.1016/j.gene.2014.11.002
Hauben M, Haesendonckx B, Standaert E, Van Der Kelen K, Azmi A, Akpo H, Van Breusegem F, Guisez Y, Bots M, Lambert B (2009) Energy use efficiency is characterized by an epigenetic component that can be directed through artificial selection to increase yield. Proc Natl Acad Sci 106(47):20109–20114
Hilton IB, D'ippolito AM, Vockley CM, Thakore PI, Crawford GE, Reddy TE, Gersbach CA (2015) Epigenome editing by a CRISPR-Cas9-based acetyltransferase activates genes from promoters and enhancers. Nat Biotechnol 33(5):510–517
Hofmeister BT, Lee K, Rohr NA, Hall DW, Schmitz RJ (2017) Stable inheritance of DNA methylation allows creation of epigenotype maps and the study of epiallele inheritance patterns in the absence of genetic variation. Genome Biol 18(1):1–16
Hossain MS, Kawakatsu T, Do Kim K, Zhang N, Nguyen CT, Khan SM, Batek JM, Joshi T, Schmutz J, Grimwood J (2017) Divergent cytosine DNA methylation patterns in single-cell, soybean root hairs. New Phytol 214(2):808–819
Hsieh T-F, Ibarra CA, Silva P, Zemach A, Eshed-Williams L, Fischer RL, Zilberman D (2009) Genome-wide demethylation of Arabidopsis endosperm. Science 324(5933):1451–1454
Hu Y, Zhang LU, He S, Huang MIN, Tan J, Zhao LIN, Yan S, Li HUI, Zhou KUN, Liang Y (2012) Cold stress selectively unsilences tandem repeats in heterochromatin associated with accumulation of H3K9ac. Plant Cell Environ 35(12):2130–2142
Hu Y, Morota G, Rosa GJM, Gianola D (2015) Prediction of plant height in *Arabidopsis thaliana* using DNA methylation data. Genetics 201(2):779–793
Ivashuta S, Uchiyama K, Gau M, Shimamoto Y (2002) Linear amplification coupled with controlled extension as a means of probe amplification in a cDNA array and gene expression analysis during cold acclimation in alfalfa (*Medicago sativa* L.). J Exp Bot 53(367):351–359
Johnson LM, Bostick M, Zhang X, Kraft E, Henderson I, Callis J, Jacobsen SE (2007) The SRA methyl-cytosine-binding domain links DNA and histone methylation. Curr Biol 17(4):379–384
Johnson LM, Du J, Hale CJ, Bischof S, Feng S, Chodavarapu RK, Zhong X, Marson G, Pellegrini M, Segal DJ (2014) SRA- and SET-domain-containing proteins link RNA polymerase V occupancy to DNA methylation. Nature 507(7490):124–128
Kang YJ, Bae A, Shim S, Lee T, Lee J, Satyawan D, Kim MY, Lee S-H (2017) Genome-wide DNA methylation profile in mungbean. Sci Rep 7(1):1–8
Kapazoglou A, Drosou V, Argiriou A, Tsaftaris AS (2013) The study of a barley epigenetic regulator, HvDME, in seed development and under drought. BMC Plant Biol 13(1):1–16
Khandal H, Parween S, Roy R, Meena MK, Chattopadhyay D (2017) MicroRNA profiling provides insights into post-transcriptional regulation of gene expression in chickpea root apex under salinity and water deficiency. Sci Rep 7(1):1–14
Kouzarides T (2007) Chromatin modifications and their function. Cell 128(4):693–705

Kulcheski FR, de Oliveira LFV, Molina LG, Almerão MP, Rodrigues FA, Marcolino J, Barbosa JF, Stolf-Moreira R, Nepomuceno AL, Marcelino-Guimarães FC (2011) Identification of novel soybean microRNAs involved in abiotic and biotic stresses. BMC Genomics 12(1):1–17
Kurdyukov S, Mathesius U, Nolan KE, Sheahan MB, Goffard N, Carroll BJ, Rose RJ (2014) The 2HA line of Medicago truncatula has characteristics of an epigenetic mutant that is weakly ethylene insensitive. BMC Plant Biol 14(1):1–16
La H, Ding B, Mishra GP, Zhou B, Yang H, del Rosario Bellizzi M, Chen S, Meyers BC, Peng Z, Zhu J-K (2011) A 5-methylcytosine DNA glycosylase/lyase demethylates the retrotransposon Tos17 and promotes its transposition in rice. Proc Natl Acad Sci 108(37):15498–15503
Labra M, Ghiani A, Citterio S, Sgorbati S, Sala F, Vannini C, Ruffini-Castiglione M, Bracale M (2002) Analysis of cytosine methylation pattern in response to water deficit in pea root tips. Plant Biol 4(06):694–699
Lämke J, Bäurle I (2017) Epigenetic and chromatin-based mechanisms in environmental stress adaptation and stress memory in plants. Genome Biol 18(1):1–11
Lämke J, Brzezinka K, Altmann S, Bäurle I (2016) A hit-and-run heat shock factor governs sustained histone methylation and transcriptional stress memory. EMBO J 35(2):162–175
Lane AK, Niederhuth CE, Ji L, Schmitz RJ (2014) pENCODE: a plant encyclopedia of DNA elements. Annu Rev Genet 48:49–70
Lavin M, Herendeen PS, Wojciechowski MF (2005) Evolutionary rates analysis of Leguminosae implicates a rapid diversification of lineages during the tertiary. Syst Biol 54(4):575–594
Law JA, Jacobsen SE (2010) Establishing, maintaining and modifying DNA methylation patterns in plants and animals. Nat Rev Genet 11(3):204–220
Lee RC, Feinbaum RL, Ambros V (1993) The C. elegans heterochronic gene lin-4 encodes small RNAs with antisense complementarity to lin-14. Cell 75(5):843–854
Lei M, Zhang H, Julian R, Tang K, Xie S, Zhu J-K (2015) Regulatory link between DNA methylation and active demethylation in Arabidopsis. Proc Natl Acad Sci 112(11):3553–3557
Li W-X, Oono Y, Zhu J, He X-J, Wu J-M, Iida K, Lu X-Y, Cui X, Jin H, Zhu J-K (2008) The Arabidopsis NFYA5 transcription factor is regulated transcriptionally and posttranscriptionally to promote drought resistance. Plant Cell 20(8):2238–2251
Li W, Wang P, Li Y, Zhang K, Ding F, Nie T, Yang X, Lv Q, Zhao L (2015) Identification of microRNAs in response to different day lengths in soybean using high-throughput sequencing and qRT-PCR. PLoS One 10(7):e0132621
Liang X, Hou X, Li J, Han Y, Zhang Y, Feng N, Du J, Zhang W, Zheng D, Fang S (2019) High-resolution DNA methylome reveals that demethylation enhances adaptability to continuous cropping comprehensive stress in soybean. BMC Plant Biol 19(1):1–17
Liew LC, Singh MB, Bhalla PL (2013) An RNA-seq transcriptome analysis of histone modifiers and RNA silencing genes in soybean during floral initiation process. PLoS One 8(10):e77502
Lin J-Y, Le BH, Chen M, Henry KF, Hur J, Hsieh T-F, Chen P-Y, Pelletier JM, Pellegrini M, Fischer RL (2017) Similarity between soybean and Arabidopsis seed methylomes and loss of non-CG methylation does not affect seed development. Proc Natl Acad Sci 114(45):E9730–E9739
Liu C, Lu F, Cui X, Cao X (2010) Histone methylation in higher plants. Annu Rev Plant Biol 61: 395–420
Lobaton JD, Miller T, Gil J, Ariza D, de la Hoz JF, Soler A, Beebe S, Duitama J, Gepts P, Raatz B (2018) Resequencing of common bean identifies regions of inter–gene pool introgression and provides comprehensive resources for molecular breeding. Plant Genome 11(2):170068
Long SP, Ort DR (2010) More than taking the heat: crops and global change. Curr Opin Plant Biol 13(3):240–247
Malabarba J, Windels D, Xu W, Verdier J (2021) Regulation of DNA (de)methylation positively impacts seed germination during seed development under heat stress. Genes 12(3):457
Mantri N, Basker N, Ford R, Pang E, Pardeshi V (2013) The role of micro-ribonucleic acids in legumes with a focus on abiotic stress response. Plant Genome 6(3):plantgenome2013-05

Matzke MA, Mosher RA (2014) RNA-directed DNA methylation: an epigenetic pathway of increasing complexity. Nat Rev Genet 15(6):394–408

Maxwell EK, Ryan JF, Schnitzler CE, Browne WE, Baxevanis AD (2012) MicroRNAs and essential components of the microRNA processing machinery are not encoded in the genome of the ctenophore *Mnemiopsis leidyi*. BMC Genomics 13(1):1–11

Mayer BF, Ali-Benali MA, Demone J, Bertrand A, Charron J (2015) Cold acclimation induces distinctive changes in the chromatin state and transcript levels of COR genes in *Cannabis sativa* varieties with contrasting cold acclimation capacities. Physiol Plant 155(3):281–295

McClung CR, Davis SJ (2010) Ambient thermometers in plants: from physiological outputs towards mechanisms of thermal sensing. Curr Biol 20(24):R1086–R1092

Meister G, Tuschl T (2004) Mechanisms of gene silencing by double-stranded RNA. Nature 431(7006):343–349

Meyer P (2015) Epigenetic variation and environmental change. J Exp Bot 66(12):3541–3548

Migicovsky Z, Yao Y, Kovalchuk I (2014) Transgenerational phenotypic and epigenetic changes in response to heat stress in *Arabidopsis thaliana*. Plant Signal Behav 9(2):e27971

Mishra GP, Radhakrishnan T, Kumar A, Thirumalaisamy PP, Kumar N, Bosamia TC, Nawade B, Dobaria JR (2015) Advancements in molecular marker development and their applications in the management of biotic stresses in peanuts. Crop Prot 77:74–86

Mishra GP, Dikshit HK, Kumari J, Tripathi K, Devi J, Aski M, Mehra R, Sarker A, Kumar S (2020) Identification and characterization of novel penta-podded genotypes in the cultivated lentil. Crop Sci 60(4):1974–1985

Mishra GP, Dikshit HK, Tontang MT, Stobdan T, Sangwan S, Aski MS, Singh A, Kumar RR, Tripathi K, Kumar S (2021) Diversity in phytochemical composition, antioxidant capacities, and nutrient contents among mungbean and lentil microgreens when grown at plain-altitude region (Delhi) and high-altitude region (Leh-Ladakh), India. Front Plant Sci 12:1485

Mishra GP, Dikshit HK, Kukreja B, Aski M, Yadava DK, Sarker A, Kumar S (2022) Historical overview of biofortification in crop plants and its implications. In: Biofortification of staple crops. Springer, New York, pp 31–61

Mochida K, Yoshida T, Sakurai T, Yamaguchi-Shinozaki K, Shinozaki K, Tran L-SP (2010) LegumeTFDB: an integrative database of *Glycine max*, *Lotus japonicus* and *Medicago truncatula* transcription factors. Bioinformatics 26(2):290–291

Moradpour M, Abdulah SNA (2020) CRISPR/dC as9 platforms in plants: strategies and applications beyond genome editing. Plant Biotechnol J 18(1):32–44

Mosher RA, Schwach F, Studholme D, Baulcombe DC (2008) PolIVb influences RNA-directed DNA methylation independently of its role in siRNA biogenesis. Proc Natl Acad Sci 105(8): 3145–3150

Nawade B, Mishra GP, Radhakrishnan T, Dodia SM, Ahmad S, Kumar A, Kumar A, Kundu R (2018) High oleic peanut breeding: achievements, perspectives, and prospects. Trends Food Sci Technol 78:107–119

Naydenov M, Baev V, Apostolova E, Gospodinova N, Sablok G, Gozmanova M, Yahubyan G (2015) High-temperature effect on genes engaged in DNA methylation and affected by DNA methylation in Arabidopsis. Plant Physiol Biochem 87:102–108

Neri F, Rapelli S, Krepelova A, Incarnato D, Parlato C, Basile G, Maldotti M, Anselmi F, Oliviero S (2017) Intragenic DNA methylation prevents spurious transcription initiation. Nature 543(7643):72–77

Ni Z, Hu Z, Jiang Q, Zhang H (2013) GmNFYA3, a target gene of miR169, is a positive regulator of plant tolerance to drought stress. Plant Mol Biol 82(1):113–129

Niederhuth CE, Schmitz RJ (2014) Covering your bases: inheritance of DNA methylation in plant genomes. Mol Plant 7(3):472–480

Oakey H, Cullis B, Thompson R, Comadran J, Halpin C, Waugh R (2016) Genomic selection in multi-environment crop trials. G3 6(5):1313–1326

Pandey G, Sharma N, Pankaj Sahu P, Prasad M (2016) Chromatin-based epigenetic regulation of plant abiotic stress response. Curr Genomics 17(6):490–498

Parrilla-Doblas JT, Roldán-Arjona T, Ariza RR, Córdoba-Cañero D (2019) Active DNA demethylation in plants. Int J Mol Sci 20(19):4683
Patel KG, Mandaliya VB, Mishra GP, Dobaria JR, Thankappan R (2016) Transgenic peanut overexpressing mtlD gene confers enhanced salinity stress tolerance via mannitol accumulation and differential antioxidative responses. Acta Physiol Plant 38(7):1–14
Patel KG, Thankappan R, Mishra GP, Mandaliya VB, Kumar A, Dobaria JR (2017) Transgenic peanut (*Arachis hypogaea* L.) overexpressing mtlD gene showed improved photosynthetic, physio-biochemical, and yield-parameters under soil-moisture deficit Stress in lysimeter system. Front Plant Sci 8:1881
Pauli H, Gottfried M, Dullinger S, Abdaladze O, Akhalkatsi M, Alonso JLB, Coldea G, Dick J, Erschbamer B, Calzado RF (2012) Recent plant diversity changes on Europe's mountain summits. Science 336(6079):353–355
Pecinka A, Chevalier C, Colas I, Kalantidis K, Varotto S, Krugman T, Michailidis C, Vallés M-P, Muñoz A, Pradillo M (2020) Chromatin dynamics during interphase and cell division: similarities and differences between model and crop plants. J Exp Bot 71(17):5205–5222
Penterman J, Zilberman D, Huh JH, Ballinger T, Henikoff S, Fischer RL (2007) DNA demethylation in the Arabidopsis genome. Proc Natl Acad Sci 104(16):6752–6757
Peñuelas J, Rutishauser T, Filella I (2009) Phenology feedbacks on climate change. Science 324(5929):887–888
Peterson CL, Laniel M-A (2004) Histones and histone modifications. Curr Biol 14(14):R546–R551
Popova OV, Dinh HQ, Aufsatz W, Jonak C (2013) The RdDM pathway is required for basal heat tolerance in Arabidopsis. Mol Plant 6(2):396–410
Rajkumar MS, Gupta K, Khemka NK, Garg R, Jain M (2020) DNA methylation reprogramming during seed development and its functional relevance in seed size/weight determination in chickpea. Commun Biol 3(1):1–13
Raju SKK, Shao M, Sanchez R, Xu Y, Sandhu A, Graef G, Mackenzie S (2018) An epigenetic breeding system in soybean for increased yield and stability. Plant Biotechnol J 16(11): 1836–1847
Rakei A, Maali-Amiri R, Zeinali H, Ranjbar M (2016) DNA methylation and physio-biochemical analysis of chickpea in response to cold stress. Protoplasma 253(1):61–76
Ramesh SV, Govindasamy V, Rajesh MK, Sabana AA, Praveen S (2019) Stress-responsive miRNAome of *Glycine max* (L.) Merrill: molecular insights and way forward. Planta 249(5): 1267–1284
Ramu VS, Paramanantham A, Ramegowda V, Mohan-Raju B, Udayakumar M, Senthil-Kumar M (2016) Transcriptome analysis of sunflower genotypes with contrasting oxidative stress tolerance reveals individual- and combined-biotic and abiotic stress tolerance mechanisms. PLoS One 11(6):e0157522
Reddy VRP, Das S, Dikshit HK, Mishra GP, Aski M, Meena SK, Singh A, Pandey R, Singh MP, Tripathi K (2020) Genome-wide association analysis for phosphorus use efficiency traits in mungbean (*Vigna radiata* L. Wilczek) using genotyping by sequencing approach. Front Plant Sci 11:1546
Robertson AL, Wolf DE (2012) The role of epigenetics in plant adaptation. Trends Evol Biol 4(1): e4
Salgotra RK, Gupta M (2019) Exploring the role of epigenetics in cereal and leguminous crops exposed to abiotic stress. In: Epigenetics in plants of agronomic importance: fundamentals and applications. Springer, New York, pp 149–170
Saraswat S, Yadav AK, Sirohi P, Singh NK (2017) Role of epigenetics in crop improvement: water and heat stress. J Plant Biol 60(3):231–240
Sarkar T, Thankappan R, Kumar A, Mishra GP, Dobaria JR (2014) Heterologous expression of the AtDREB1A gene in transgenic peanut-conferred tolerance to drought and salinity stresses. PLoS One 9(12):e110507

Sarkar T, Thankappan R, Kumar A, Mishra GP, Dobaria JR (2016) Stress inducible expression of AtDREB1A transcription factor in transgenic peanut (*Arachis hypogaea* L.) conferred tolerance to soil-moisture deficit stress. Front Plant Sci 7:935

Schmitz RJ, He Y, Valdés-López O, Khan SM, Joshi T, Urich MA, Nery JR, Diers B, Xu D, Stacey G (2013) Epigenome-wide inheritance of cytosine methylation variants in a recombinant inbred population. Genome Res 23(10):1663–1674

Sen S, Chakraborty J, Ghosh P, Basu D, Das S (2017) Chickpea WRKY70 regulates the expression of a homeodomain-leucine zipper (HD-Zip) I transcription factor CaHDZ12, which confers abiotic stress tolerance in transgenic tobacco and chickpea. Plant Cell Physiol 58(11): 1934–1952

Shanker A, Venkateswarlu B (2011) Abiotic stress in plants: mechanisms and adaptations. BoD–Books on Demand

Shui X-R, Chen Z-W, Li J-X (2013) MicroRNA prediction and its function in regulating drought-related genes in cowpea. Plant Sci 210:25–35

Sigman MJ, Slotkin RK (2016) The first rule of plant transposable element silencing: location, location, location. Plant Cell 28(2):304–313

Slotkin RK, Vaughn M, Borges F, Tanurdžić M, Becker JD, Feijó JA, Martienssen RA (2009) Epigenetic reprogramming and small RNA silencing of transposable elements in pollen. Cell 136(3):461–472

Song Y, Ji D, Li S, Wang P, Li Q, Xiang F (2012) The dynamic changes of DNA methylation and histone modifications of salt responsive transcription factor genes in soybean. PLoS One 7(7): e41274

Song Q-X, Lu X, Li Q-T, Chen H, Hu X-Y, Ma B, Zhang W-K, Chen S-Y, Zhang J-S (2013) Genome-wide analysis of DNA methylation in soybean. Mol Plant 6(6):1961–1974

Sosa-Valencia G, Romero-Pérez PS, Palomar VM, Covarrubias AA, Reyes JL (2017a) Insights into the function of the phasiRNA-triggering miR1514 in response to stress in legumes. Plant Signal Behav 12(3):e1284724

Sosa-Valencia G, Palomar M, Covarrubias AA, Reyes JL (2017b) The legume miR1514a modulates a NAC transcription factor transcript to trigger phasiRNA formation in response to drought. J Exp Bot 68(8):2013–2026

Springer NM, Schmitz RJ (2017) Exploiting induced and natural epigenetic variation for crop improvement. Nat Rev Genet 18(9):563–575

Stępiński D (2012) Levels of DNA methylation and histone methylation and acetylation change in root tip cells of soybean seedlings grown at different temperatures. Plant Physiol Biochem 61:9–17

Stricker SH, Köferle A, Beck S (2017) From profiles to function in epigenomics. Nat Rev Genet 18(1):51–66

Sudan J, Raina M, Singh R (2018) Plant epigenetic mechanisms: role in abiotic stress and their generational heritability. 3 Biotech 8(3):1–12

Suter L, Widmer A (2013) Phenotypic effects of salt and heat stress over three generations in *Arabidopsis thaliana*. PLoS One 8(11):e80819

Tal O, Kisdi E, Jablonka E (2010) Epigenetic contribution to covariance between relatives. Genetics 184(4):1037–1050

Tang K, Lang Z, Zhang H, Zhu J-K (2016) The DNA demethylase ROS1 targets genomic regions with distinct chromatin modifications. Nat Plants 2(11):1–10

Timko MP, Rushton PJ, Laudeman TW, Bokowiec MT, Chipumuro E, Cheung F, Town CD, Chen X (2008) Sequencing and analysis of the gene-rich space of cowpea. BMC Genomics 9(1):1–20

Trindade I, Capitão C, Dalmay T, Fevereiro MP, dos Santos DM (2010) miR398 and miR408 are up-regulated in response to water deficit in Medicago truncatula. Planta 231(3):705–716

Varotto S, Tani E, Abraham E, Krugman T, Kapazoglou A, Melzer R, Radanović A, Miladinović D (2020) Epigenetics: possible applications in climate-smart crop breeding. J Exp Bot 71(17): 5223–5236

Varshney RK, Song C, Saxena RK, Azam S, Yu S, Sharpe AG, Cannon S, Baek J, Rosen BD, Tar'an B (2013) Draft genome sequence of chickpea (*Cicer arietinum*) provides a resource for trait improvement. Nat Biotechnol 31(3):240–246

Venterea R (2014) Climate change 2014: mitigation of climate change. J Environ Qual 38:837

Vlasova A, Capella-Gutiérrez S, Rendón-Anaya M, Hernández-Oñate M, Minoche AE, Erb I, Câmara F, Prieto-Barja P, Corvelo A, Sanseverino W (2016) Genome and transcriptome analysis of the Mesoamerican common bean and the role of gene duplications in establishing tissue and temporal specialization of genes. Genome Biol 17(1):1–18

Vriet C, Hennig L, Laloi C (2015) Stress-induced chromatin changes in plants: of memories, metabolites and crop improvement. Cell Mol Life Sci 72(7):1261–1273

Wahid A, Gelani S, Ashraf M, Foolad MR (2007) Heat tolerance in plants: an overview. Environ Exp Bot 61(3):199–223

Walther G-R, Post E, Convey P, Menzel A, Parmesan C, Beebee TJC, Fromentin J-M, Hoegh-Guldberg O, Bairlein F (2002) Ecological responses to recent climate change. Nature 416(6879):389–395

Wang P, Shi S, Ma J, Song H, Zhang Y, Gao C, Zhao C, Zhao S, Hou L, Lopez-Baltazar J (2018) Global Methylome and gene expression analysis during early Peanut pod development. BMC Plant Biol 18(1):1–13

Wassenegger M, Heimes S, Riedel L, Sänger HL (1994) RNA-directed de novo methylation of genomic sequences in plants. Cell 76(3):567–576

Wibowo A, Becker C, Marconi G, Durr J, Price J, Hagmann J, Papareddy R, Putra H, Kageyama J, Becker J (2016) Hyperosmotic stress memory in Arabidopsis is mediated by distinct epigenetically labile sites in the genome and is restricted in the male germline by DNA glycosylase activity. Elife 5:e13546

Wierzbicki AT, Ream TS, Haag JR, Pikaard CS (2009) RNA polymerase V transcription guides ARGONAUTE4 to chromatin. Nat Genet 41(5):630–634

Windels D, Dang TT, Chen Z, Verdier J (2021) Snapshot of epigenetic regulation in legumes. Legume Sci 3(3):e60

Wong CE, Singh MB, Bhalla PL (2013) The dynamics of soybean leaf and shoot apical meristem transcriptome undergoing floral initiation process. PLoS One 8(6):e65319

Wu T, Pi E-X, Tsai S-N, Lam H-M, Sun S-M, Kwan YW, Ngai S-M (2011) GmPHD5 acts as an important regulator for crosstalk between histone H3K4 di-methylation and H3K14 acetylation in response to salinity stress in soybean. BMC Plant Biol 11(1):1–13

Xu C, Tian J, Mo B (2013) siRNA-mediated DNA methylation and H3K9 dimethylation in plants. Protein Cell 4(9):656–663

Yaish MW, Peng M, Rothstein SJ (2014) Global DNA methylation analysis using methyl-sensitive amplification polymorphism (MSAP). In: Arabidopsis protocols. Springer, New York, pp 285–298

Yaish MW, Al-Lawati A, Al-Harrasi I, Patankar HV (2018) Genome-wide DNA Methylation analysis in response to salinity in the model plant caliph medic (*Medicago truncatula*). BMC Genomics 19(1):1–17

Yan Y, Zhang Y, Yang K, Sun Z, Fu Y, Chen X, Fang R (2011) Small RNAs from MITE-derived stem-loop precursors regulate abscisic acid signaling and abiotic stress responses in rice. Plant J 65(5):820–828

Yang X, Kundariya H, Xu Y-Z, Sandhu A, Yu J, Hutton SF, Zhang M, Mackenzie SA (2015) MutS HOMOLOG1-derived epigenetic breeding potential in tomato. Plant Physiol 168(1):222–232

Zeller G, Henz SR, Widmer CK, Sachsenberg T, Rätsch G, Weigel D, Laubinger S (2009) Stress-induced changes in the *Arabidopsis thaliana* transcriptome analyzed using whole-genome tiling arrays. Plant J 58(6):1068–1082

Zemach A, Kim MY, Hsieh P-H, Coleman-Derr D, Eshed-Williams L, Thao K, Harmer SL, Zilberman D (2013) The Arabidopsis nucleosome remodeler DDM1 allows DNA methyltransferases to access H1-containing heterochromatin. Cell 153(1):193–205

Zhang H, Zhu J-K (2011) RNA-directed DNA methylation. Curr Opin Plant Biol 14(2):142–147

Zhang H, Zhu J-K (2012) Active DNA demethylation in plants and animals. Cold Spring Harb Symp Quant Biol 77:161–173

Zhang X, Yazaki J, Sundaresan A, Cokus S, Chan SW-L, Chen H, Henderson IR, Shinn P, Pellegrini M, Jacobsen SE (2006) Genome-wide high-resolution mapping and functional analysis of DNA methylation in Arabidopsis. Cell 126(6):1189–1201

Zhang H, Lang Z, Zhu J-K (2018a) Dynamics and function of DNA methylation in plants. Nat Rev Mol Cell Biol 19(8):489–506

Zhang B, Su L, Hu B, Li L (2018b) Expression of AhDREB1, an AP2/ERF transcription factor gene from peanut, is affected by histone acetylation and increases abscisic acid sensitivity and tolerance to osmotic stress in Arabidopsis. Int J Mol Sci 19(5):1441

Zhao T, Zhan Z, Jiang D (2019) Histone modifications and their regulatory roles in plant development and environmental memory. J Genet Genomics 46(10):467–476

Zhu J-K (2009) Active DNA demethylation mediated by DNA glycosylases. Annu Rev Genet 43: 143–166

Zilberman D, Coleman-Derr D, Ballinger T, Henikoff S (2008) Histone H2A.Z and DNA methylation are mutually antagonistic chromatin marks. Nature 456(7218):125–129

Zong W, Zhong X, You J, Xiong L (2013) Genome-wide profiling of histone H3K4-tri-methylation and gene expression in rice under drought stress. Plant Mol Biol 81(1):175–188

Morphophysiological and Molecular Diversity in Mung Bean (*Vigna radiata* L.)

5

Rakesh Pathak, Pooja Panchariya, Manoj Choudhary, Kantilal Solanki, Reena Rani, R. K. Kakani, and Rajwant K. Kalia

Abstract

Mung bean [*Vigna radiata* (L.) Wilczek] is one of the upsurging, highly economical, nutritive Asiatic leguminous crops. The crop is getting higher attention in terms of the consumption and production worldwide being an important source of amino acids, proteins, dietary fibre and unsaturated fatty acids. It possesses folate and iron in significant amount along with several phytochemicals. The short life cycle and nitrogen-fixing ability make it more suitable for sowing along with other crops. In spite of several advantages, it has got less attention in terms of development of morphophysiological and molecularly diverse varieties. Mung bean has a small genome, and fortunately it has been sequenced; therefore, it may be utilized as an exemplary plant to understand other legumes. Development of wild mung bean pool from diverse origins and environmental conditions would help to conserve the genetic wealth of the crop. Higher yields, shorter maturity period, higher harvest index, photoperiod insensitivity, resistance to major insect pests/diseases, compact canopy and synchronous maturity are some of the important objectives for crop improvement in mung bean. This chapter reviews the morphophysiological and molecular diversity of mung bean and also gives an insight about mutagenesis, plant protection and abiotic stresses associated with the crop.

Keywords

Mung bean · Diversity · Biometric · Physiological · Molecular level

R. Pathak (✉) · P. Panchariya · M. Choudhary · K. Solanki · R. Rani · R. K. Kakani · R. K. Kalia
ICAR-Central Arid Zone Research Institute, Jodhpur, Rajasthan, India

P. Muthu Arjuna Samy et al. (eds.), *Legumes: Physiology and Molecular Biology of Abiotic Stress Tolerance*, https://doi.org/10.1007/978-981-19-5817-5_5

5.1 Introduction

Mung bean [*Vigna radiata* (L.) Wilczek], also known as green gram, golden gram, green bean or mash bean, is an important fast-growing, highly economical, nutritive, multipurpose leguminous crop cultivated in tropical and subtropical regions of Asia (Tah 2006; Yang et al. 2008). It is a self-pollinating species belonging to the genus *Vigna* of Fabaceae family. The crop is mainly grown in frost-free regions from Asia to Africa, South America and Australia (Nair et al. 2012). India is one of the largest producers of mung bean and shares about 50% of global annual production (Nair et al. 2012).

Being a leguminous crop, mung bean is an important source of amino acids, proteins, dietary fibre and unsaturated fatty acids (Hou et al. 2019). It is easily digestible, produces low flatulence as compared to other legumes and contains higher folate and iron (Keatinge et al. 2011). The crop makes the soil fertile and improves its texture (Graham and Vance 2003). Similarly, it has also been observed that the cereals intercropped with mung bean have a lesser incidence of pest infestation and have a higher yield due to the availability of nitrogen fertilizer (Yaqub et al. 2010).

Owing to higher vitamin, calcium, iron and phosphorus content as compared to other leguminous crops, mung bean is a preferred nutritive food. The presence of amino acids, proteins, polyphenols and oligosaccharides in the crop has been exploited for antioxidant, antitumor, anti-inflammatory and antimicrobial activities (Anjum et al. 2011; Randhir et al. 2004). Mung bean has also been reported to contain several phytochemicals, viz., steroids, triterpenoids, glycosides, flavonoids, alkaloids, polyphenols, tannins, saponins, daidzin, daidzein, ononin, formononetin, isoformononetin, quercetin, kaempferol, myricetin, rhamnetin, etc. (Priya et al. 2012; Ramesh et al. 2011; Tang et al. 2014).

Mung bean has also been reported to contain a good amount of antifungal proteins (Solanki et al. 2018) that can be used against human and plant pathogens. Mung bean seeds possess alkaloids, coumarin and phytosterol that support the physiological metabolism in human beings. The seeds are also free from anti-nutritional factors, viz., trypsin inhibitors, phytohemagglutinins and tannins (Xin et al. 2003).

Mung bean has been used as a model crop for physiological studies (Musgrave et al. 1988) and for understanding the beginning and expansion of adventitious roots (Norcini et al. 1985; Tripepi et al. 1983). The rooting bioassay of this plant has also been used to assess the root-promoting potential of growth regulators (Kling et al. 1988). Mung bean is used globally for human consumption, cattle feed and medicinal purposes (Jo et al. 2006). Its sprouts and splits are very nutritious, and as a component of soups, noodles, cake or ice cream fillings, it is commonly used in human foods. Its haulm, green and dry fodder are used as nutritious animal feed (Garg et al. 2004). Studies have revealed its importance in the treatment of hepatitis, gastritis, etc., and it has antihypertensive, antidiabetic and anticancer properties (Kumar and Singhal 2009). Keeping in view the importance of the crop, the consumption of mung bean has increased considerably along with its production

(Shanmugasundaram et al. 2009). Therefore, mung bean is considered among cash crops and has attracted the interest of researchers.

5.2 Origin

Mung bean has diploid ($2n = 2x = 22$) chromosome numbers. Vavilov (1951) proposed Central Asian regions as the basic genetic centre of mung bean and India as the centre of its domestication (Singh et al. 1970; Smartt 1985). The diversity data and archaeological confirmations also suggested India to be the origin place of mung bean (Fuller and Harvey 2006; Jain and Mehra 1980), although the wild relatives of mung bean have been reported from the subtropical and tropical provinces of northern and eastern Australia (Lawn and Cottrell 1988). Studies carried out based on protein and enzyme variability suggest that modern mung bean has several series of domestication (Lambrides and Godwin 2007; Viña and Tomooka 1994).

5.3 Genetic Resources

Availability of germplasms having superior alleles and wide genetic diversity is one of the prerequisites for a sustainable breeding programme. Therefore, numerous organizations have collected mung bean germplasm to sustain the genetic resources. To facilitate the effective utilization and easier access to genetic resources, germplasms have been conserved in China, India, Korea and the USA. Asian Vegetable Research and Development Center has established a core collection of about 1700 mung bean accessions. These accessions have been morphologically and molecularly characterized (Shanmugasundaram et al. 2009). Germplasms having variable characteristics are the most important resource for crop improvement and play an important role in widening the genetic background of cultivars.

5.4 Cultivation

Mung bean is a short-day crop and is generally grown during the rainy seasons. It takes about 90–120 days to mature. It is the third most important leguminous crop after chickpea and pigeon pea cultivated in India (Ahmad and Belwal 2019). Mung bean is globally cultivated on nearly seven million hectares and is mostly limited to Asian countries (Nair et al. 2019). The total production of the crop in India from 2018 to 2019 was 2455.37 thousand tonnes with an average productivity of 516 kg per hectare (Anonymous 2020), suggesting that India is one of the largest producers of mung bean.

The production and partitioning of dry matter potential in mung bean are an outcome of several growth stages of the plant. The changes in the growth stage mainly depend upon the temperature and photoperiod. Manipulation in the process of the growth stage in context to the environmental conditions may lead to grain yield improvement. The time taken for mung bean crop to mature is an important yield factor. The duration may change with the environmental conditions, sowing time and cropping season. It helps to determine the suitability of crops under various cropping systems. Mung bean is sensitive to photoperiod, and flowering in the crop is influenced by the duration of light (Aggarwal and Poehlman 1977). It has been reported that short days lead to early flowering, while long days result in delayed flowering (Aggarwal and Poehlman 1977). A higher yield can be realized from the crops grown under proper drainage conditions in sandy loam soil, while higher humidity and excessive rainfall may lead to several diseases and lower yields in mung bean (Oelke et al. 1990). The determinants influencing the crop duration in mung bean have been discussed by several workers (Robertson et al. 2002; Summerfield and Lawn 1987).

Mung bean has broad and trifoliate leaves that overlap horizontally bounding the light into the canopy. It has been noticed that the mung bean plants having narrow leaves capture maximum light and give comparatively higher yields (Lee et al. 2004). Mung bean has epigeal germination, and cotyledons have to arise from the soil for the growth of the seedling. However, low-moisture conditions and crusting of soil under higher temperatures may limit this type of germination (Cook et al. 1995), resulting in poor germination and simultaneously poor establishment (Harris et al. 2005). Seedling vigour may be important under such conditions, but no relation could be noticed between seedling vigour and crop yield in mung bean (TeKrony and Egli 1991). A plant stand of about 30 plants under each square meter is considered significant to provide higher yields in mung bean (Rachaputi et al. 2015).

The flowering and pod maturity in mung bean do not take place evenly, and differences between these two incidences are higher (Tah and Saxena 2009), leading to non-synchronous maturity and yield losses (Alam Mondal et al. 2011). Early and uniform maturity of a crop has a positive effect on the grain yield; however, this important characteristic is not known in the case of mung bean (Chen et al. 2008). High-yielding, uniform-maturity and disease-resistant varieties are of choice for the successful cultivation of mung bean (Tomooka et al. 2005), while low-yielding potential, poor harvest index and vulnerability to diseases and biotic and abiotic stresses (Srinives et al. 2007) are some of the major challenges in its cultivation. Wild species of mung bean may serve as a better genetic material as the cultivated germplasm may have lost many alleles during the process of domestication and/or breeding programmes (Hyten et al. 2006). Therefore, beneficial alleles from uncultivated species have been accustomed to the crop improvement in mung bean (Nair et al. 2012).

5.5 Genetic Variability

Self-pollinated crops generally have composite floral structures and low natural variability. Therefore, the selection of such plants for crop improvement becomes difficult; nevertheless, estimation of the phenotypic coefficient of variation (PCV), genotypic coefficient of variation (GCV), heritability estimates and genetic advance

(GA) provides immense opportunity to choose better genotypes. Estimation of these variabilities reveals the influence of several gene effects operating towards total variability for the desired traits. Several studies reported the importance of GCV, PCV, heritability and GA in the improvement of traits in different crops (Denton and Nwangburuka 2011; Johnson et al. 1955; Kim et al. 2015a). Evaluation of phenotypic or genotypic variability offers better insight into the utilization of available germplasm resources (Bisht et al. 1998; Schafleitner et al. 2015). Wide variation in morphological traits, viz., number of pods per plant, yield per plant, 100-seed weight, fruit-setting capacity, flowering period, maturity, number of pod-bearing peduncles, plant height, primary branches, length of branches, nodule and leaf pattern, has been observed in mung bean (Bisht et al. 1998).

The study on the inheritance of narrow trifoliate leaves in mung bean revealed the inheritance of larger leaflets over smaller leaflets (Dwivedi and Singh 1985). Lobed leaf shape was found dominant over the entire leaf shape, while pentafoliate leaf was reported to be an inherited characteristic in mung bean (Chhabra 1990). The inheritance of dullness and shininess of leaf surface suggest that it is governed by a digenic interaction (Bhadra et al. 1991) with the dominance of dullness over shininess. Inheritance of plant and flower bud colour in mung bean suggested that the dark purple colour of the plant was dominant over the green plant colour; similarly, purple flower buds showed dominance over the green flower buds (Khattak et al. 2000). It was found that black and green seed colour is governed by similar genes; however, black seed is dominant over green seeds (Chen et al. 2001). The occurrence of anthocyanin is a dominant character so is the black-colour seed coat over the green colour (Chen et al. 2001). The study also revealed that the genes responsible for purple petiole and black seed colour have higher lineages. Single recessive gene was observed to control mung bean yellow mosaic virus resistance in the crop with susceptible behaviour being dominant over the resistant behaviour (Win et al. 2021).

PCV and GCV along with heritability estimates provide an insight into the improvement of requisite characters (Burton and de Devane 1953). Mung bean has been reported with higher PCV and GCV for seed yield and pod numbers (Makeen et al. 2007), plant height, pod numbers and grain yield, while it was low with respect to days to 50% flowering (Anand and Anandhi 2016). Primary branches, pod numbers, seed yield and clusters showed higher GCV and PCV in mung bean (Asari et al. 2019). Higher PCV and GCV were reported for 100-seed weight, flowering period, seed length and seed breadth (Tripathi et al. 2020). The number of pods, seed yield and number of clusters have been recorded with high PCV and GCV in mung bean (Salman et al. 2021), suggesting the presence of higher variabilities for these traits, and therefore, there are more opportunities for further improvement using several genetic influences.

Heritability is the amount of phenotypic variance among different genotypes due to the effect of inherited genes. The estimation of heritability is done to find the similarity between the genotypes (Falconer and Mackay 2005). It also explores the association between phenotypic and genotypic variance (Lourenço et al. 2017). Heritability in combination with genetic advance gives better insight into the desired genotype (Nwangburuka and Denton 2012). The traits presenting higher heritability along with higher genetic advance in mung bean may be enhanced by the selection method (Degefa et al. 2014) because these characteristics are under the influence of additive gene action. The influence of both additive and non-additive gene effects has been reported for several traits in mung bean (Khattak et al. 2002). Days to first pod maturity (Khattak et al. 2001) and seed yield (Sharma 1999) exhibited higher heritability. It has also been reported that the additive gene effect governs the seed yield in mung bean (Joseph and Santhoshkumar 2000).

The number of pods, plant height and test weight had a high value of heritability coupled with a higher genetic advance in mung bean, suggesting the influence of additive gene effect in their manifestation (Makeen et al. 2007). Higher variability was recorded during the assessment of genetic diversity among yield-attributing traits comprised of 9 qualitative and 21 quantitative characters among 340 cultivated mung bean collections (Yimram et al. 2009). Several yield-attributing traits showed higher genetic variability and heritability (Yimram et al. 2009). The number of seeds, seed yield and biomass yield exhibited higher heritability coupled with higher genetic advance, suggesting their importance in the selection of mung bean for better yield potentials (Degefa et al. 2014).

Genetic architecture of synchronous pod maturation and yield-related traits in mung bean were studied, and domination of additive and environmental components for days to flowering, pod maturation, synchrony in pod maturation and yield-related characters were recorded (Iqbal et al. 2014). The study suggested that inter-crossing of F2-generation plants having earliness and synchronized pod maturation along with high-yielding potential and their subsequent selection may be useful for manipulation of complex inherited characters in the development of mung bean lines for plant improvement (Iqbal et al. 2014). Seed yield, plant height and number of pods exhibited high values for heritability (Anand and Anandhi 2016).

High heritability coupled with higher genetic advance was reported for plant height, number of primary branches, number of clusters, number of pods and seed yield, signifying the dominance of additive gene action (Asari et al. 2019). Higher heritability was reported for seed dimension-related traits, days to 80% maturity, 100-seed weight, days to 50% flowering, pod length and days to initial maturity, suggesting that these traits are appropriate for mung bean breeding (Tripathi et al. 2020). Pod numbers, seed yield, clusters, number of branches, seeds and height had high heritability coupled with high genetic advance, suggesting the influence of additive genes in the inheritance of these morphological characters (Salman et al. 2021).

5.6 Mutation

Mutation is an unexpected genetic modification caused by variation in the gene sequences, leading to alteration in several plant characteristics including height, branches, flowers, pods, etc. It may occur naturally or may be induced artificially. The natural mutation is sudden, and its frequency is very low; therefore, it cannot be considered realistic. Hence, artificial methods of mutation were discovered to create variability in the crops. The introduction of mutations has played an important role in the field of genetic studies and plant breeding (Raina et al. 2016). The mutation is considered a promising tool for evolution, and induced mutagenesis is an ideal methodology for the creation of required genetic variability in crops (Auti 2012; Dubinin 1962). It may be induced using physical and chemical mutagens either individually or in combination. Various physical and chemical mutagens have been recognized in various crops (Pathak 2015; Shah et al. 2008). X-rays and gamma rays are generally applied as physical mutagens, while ethyl methane sulphonate (EMS), diethyl sulphonate (DES), sodium azide (SA), methyl methane sulphonate (MMS), nitrosoguanidine (NG), nitroso-methyl urea (NMU), etc. are the chemical mutagens used for creating variability. The genetic material generated through mutagenesis and the mutants with better desired characteristics may be included in the breeding programmes. Several attempts have been undertaken to improve the genetic variability in mung bean using different mutation techniques. The variability lost during the adaptation or evolution of a crop can be refurbished or renewed with the help of induced mutations. Selection of morphologically varied mutants, viz., plant type, chlorophyll, leaf, flower and seed-type mutants, has enhanced genetic variability and showed higher level of resistance towards abiotic and biotic stresses (Mounika 2020).

5.6.1 Mutations Induced Through Physical Factors

Physical mutagenesis is an effective method for creating variability for crop improvement in self-pollinated crops including mung bean (Sarkar and Kundagrami 2018; Shah et al. 2008). Irradiation with ionizing or non-ionizing rays is used to induce physical mutation. It was started with X-rays, but at the later stage, gamma rays got more popular (Auerbach and Robson 1946) due to better effects over plant growth and development by stimulating cytological, genetical, biochemical, physiological as well as morphological variabilities (Gunckel and Sparrow 1967). The influence of gamma irradiation on morphological and cytological changes in mung bean was recorded, wherein decreased seed germination, seedling survivability and growth rate were observed with increased doses of gamma rays (Subramanian 1980). Dosage of 10–30 kR gamma rays was reported to be appropriate to obtain earliness, synchrony in the maturity and resistance towards yellow mosaic disease in mung bean (Singh and Chaturvedi 1982). Substantial variability for the number of clusters was recorded with 10, 30 and 40 kR gamma radiation in different mung bean genotypes (Tah 2006), and a 16–20% increase was observed over the control. Mung bean varieties treated with 10–40 Gy gamma rays resulted in mutants having synchronous maturity (Tah and Saxena 2009).

Gamma rays were applied to create synchrony in the pod maturity, and the obtained mutants exhibited synchronous pod maturity along with variegated leaves (Sangsiri et al. 2007). The shallow rooting system of high-yielding and MYMV-resistant mung bean variety (Samrat) was improved using 450 Gy gamma rays, and a long-root mutant possessing a root length of 71 cm was identified in the M_2

generation (Dhole and Reddy 2010). The mutant showed better performance in terms of water uptake as compared to 'Samrat' and survived better under drought conditions. Gamma rays (300, 400 and 500 Gy) and EMS (10, 20 and 30 mM) were applied to screen the yellow vein mosaic virus disease-resistant mutants in mung bean, and several disease-resistant mutants were identified in M3 generation (Vairam et al. 2016). Gamma radiation was applied to advance the genetic constitution of mung bean, and 20 mutants from M5 progeny having an early maturing period and high yield potential were identified (Sarkar and Kundagrami 2018). Four doses of gamma rays (100, 200, 300 and 400 Gy) were applied to improve genetic variation in mung bean varieties, and mutants showing higher harvest index were isolated in M7 generation from 200 and 400 Gy dosages of gamma rays (Dewanjee and Sarkar 2018). The mutants having potential characteristics may be released as a variety, or the potential character may be transformed in other varieties to get better yields in mung bean (Pratap et al. 2020).

5.6.2 Mutations Induced Through Chemical Factors

Mutations carried out by irradiation of ionizing rays may lead to chromosomal aberrations; therefore, chemical mutagens were taken as a substitute to create variabilities. Chemical mutagens have become more popular as no specific equipment is involved during their applications, and it is comparatively easy to induce. Compared to physical mutagens, it induced point mutations causing single base pair changes (Sikora et al. 2011). Two important groups of chemical mutagens, viz., alkylating agents and base analogues, are usually applied for creating mutations. However, out of these chemical mutagens, alkylating agents such as EMS and NMU are generally used to induce mutation in crops. Various chemical mutagens, viz., ethyl methane sulphonate (EMS), sodium azide (SA) and hydrazine hydrate (HZ), have been used in mung bean (Auti and Apparao 2009; Khan and Goyal 2009; Wani 2006). Variation in seed size of mung bean was observed when it was treated with EMS and nitroso-methyl carbamide (Singh and Chaturvedi 1982). Higher seed yield, fertile branches and pods were reported in mung bean mutant lines acquired after the application of EMS and HZ (Wani 2006). EMS induces mutations more efficiently in mung bean as compared to gamma rays (Singh and Rao 2007). The crop duration in the M2 generation of mung bean was reduced with the help of SA mutagen (Lavanya et al. 2011).

5.6.3 Mutations Induced Through Physical and Chemical Factors

Physical and chemical mutagens individually have several advantages and induce random changes in the genome. However, the genetic variability induced by the combination of physical and chemical mutagens is comparatively more efficient, and the possibility of obtaining the required characteristics is significantly higher (Raina et al. 2017). A combination of lower doses of physical and chemical mutagen is more acceptable for artificial mutation (Medina et al. 2004). Both effectiveness and efficacy are important parameters for mutagens. The effectiveness and efficacy give information regarding the rate of point mutations concerning dosage and other biological effects, respectively, induced by the mutagen (Konzak et al. 1965). It relies upon the genotype and the mutagen. Varied effectiveness and efficiency of mutagens have been reported in several crops including mung bean (Wani et al. 2017). EMS and gamma rays were applied to create variability in mung bean and subsequently for the development of novel cultivars having higher yields and resistance towards insect pests (Khan and Goyal 2009; Wani 2006). Mung bean seeds treated with different concentrations of SA and EMS and different doses of gamma radiation were grown to study mutagenesis in mung bean (Auti and Apparao 2009), and several viable morphological and physiological mutants were obtained.

Seeds of mung bean were treated with gamma rays (10–60 KR) and EMS (0.1–0.4%) alone and in various combinations, and several chlorophyll and morphological mutants were identified in the M2 generation (Kumar et al. 2009). Chlorophyll-deficient mutants are considered genetic markers and are used to study the photosynthesis process (Rungnoi et al. 2010). Maximum mutations were recorded with EMS followed by gamma rays and their combinations. Higher numbers of albina-, chlorina- and viridis-type chlorophyll mutants were observed with the treatment of EMS, MMS and SA in mung bean (Khan and Siddiqui 1993). Similarly, albina, xantha, viridish, sectorial and chlorina mutants have also been recognized by Singh and Rao (2007) in mung bean. Chlorophyll mutation in mung bean has also been observed with gamma radiations and EMS alone and in its combinations (Kumar et al. 2009), wherein maximum frequency was recorded with EMS followed by gamma rays and their combinations. A higher number of chlorophyll mutation was observed when 300 Gy gamma rays were used in combination with 10 mM EMS in M2 generation (Vairam et al. 2016). Bifoliate, tetrafoliate and pentafoliate leaves have been reported in mung bean with the treatment of EMS (Auti and Apparao 2009). Mutation in flower colour has also been reported by various workers. Comb-like flowers having pollen sterility have been reported in mung bean upon mutation (Sangsiri et al. 2005). Variations in seed shape, seed size and seed colour were observed in mung bean mutants developed through treatment with gamma rays, EMS and SA (Auti and Apparao 2009).

5.7 Genotype × Environment Interaction and Stability

Improvement in the quality and quantity of crops coupled with enhanced stability over the varied environmental conditions is the most important requirement in the breeding programme. The best varieties always have higher yields along with better stability (Eberhart and Russell 1966). Genotype × environment (G × E) interaction suggests the variable responses of a trait of genotypes evaluated under different environments. It also reveals the comparative suitability of a genotype within a particular environmental situation (Allard 1960). The genotype may acquire stability alone or may be due to the buffering effect of the population; however, the yield is validated due to the effect of G × E interactions (Allard and Bradshaw 1964). Nevertheless, the comparison of varieties in a chain of environments provides relatively different positions resulting in difficulties to identify superior varieties (Eberhart and Russell 1966). The comparative performance of genotypes differs from one environment to another, and it can be articulated as a linear function of an environmental variable (Pathak 2015; Tan et al. 1979). Therefore, to assess the stability of a variety for the desired trait, an understanding of G × E interactions is essential. Stable varieties have great significance in several crops including mung bean for cultivation in variable environmental conditions (Verma et al. 2008). Variable performance of a variety towards different environmental conditions compels to search novel breeding materials under multi-environmental trials for years to evaluate their stability for desired traits (Fehr 1987; Kang 1993). A decrease in the interactions between genotype and environment is necessary to find a stable genotype that has less interrelation with the environment wherein it is cultivated. Significance of genotypes upon environment and adaptation of varieties towards yield and yield-attributing traits with respect to stability has been thoroughly underlined by several workers in mung bean (Abbas et al. 2008; Dwivedi 2006; Mahalingam et al. 2018). While highlighting the importance, it was suggested that the environment and G × E interactions must be considered during the designing and selection of materials for breeding in mung bean (Singh et al. 2009).

Stable varieties of mung bean have been identified over the years under varied environmental conditions by several researchers (Abbas et al. 2008; Baraki et al. 2020; Raturi et al. 2012b), and the prominence of some genotypes over the environment was also observed (Mahalingam et al. 2018). The environment imposes a higher impact on several characteristics of mung bean including flowering time, pod formation as well as yields. Kamannavar and Vijaykumar (2011) assessed G × E interactions in mung bean cultivars grown in different agro-climatic zones and reported that genotype, environment and G × E interaction were significant for all the characters signifying the existence of variabilities for genotype and environment along with non-linear influence of genotypes over the environment. However, the partitioning of interaction into linear and non-linear components suggests the involvement of both predictable and unpredictable sources of variables. Non-significant G × E interaction was recorded for 100-seed weight, suggesting the variable response of genotypes towards variable environmental conditions (Revanappa and Kajjidoni 2004). On the basis of stability analysis and their influences, Henry and Mathur (2007) categorized the genotypes for favourable, adverse and variable environmental conditions.

Raturi et al. (2012a, b) reported significant G × E interactions for 1000-seed weight, days to 50% flowering, number of seeds per pod and number of primary branches revealing varied responses of genotypes to varied environments. Significant G × E interactions have been recorded for seed yield among genotypes of mung bean grown under varied environmental conditions (Baraki et al. 2020). A crossover G × E interaction is usually observed if genotypes are evaluated under multi-location trials. Studies suggest that the variation in the seed yield of mung bean due to G × E interactions is inherited, and the genotypes perform differently to the varied environmental situations of the site of sowing (Baraki et al. 2020; Waniale et al. 2014). Therefore, mung bean genotypes may essentially be tested at multi-locations.

5.8 Correlation and Path Analysis

The morphophysiological characteristics of a genotype depend on several factors, and therefore, several aspects are taken into account during the selection of a genotype including the fact related to the association of characters and the influence of direct and indirect effects of each trait. Correlation provides the information with respect to the association between the traits, but it does not reveal the cause and/or consequence of association (Roy 2000), while the path coefficient analysis gives a better insight into the influence of one trait on another during identification of a predictor variable (Akanda and Mundt 1996). Thus, path analysis informs about the cause and reveals the comparative influence of the traits, while correlation analysis just provides reciprocal relation of traits (Dewey and Lu 1959).

The findings on correlation coefficient in mung bean recommend that a plant with more number of branches, clusters, pods and higher number of seeds in a pod is anticipated to provide higher seed yields. Thus, an increase in the number of branches and pods may be culminated into higher seed yield as branches bear pods and pods bear seeds. The association between seed yields was significantly positive with the number of branches, number of pods and total biomass in mung bean (Nawab et al. 2001), indicating the influence of these traits on the seed yield. The number of pods and plant height had a significantly positive association with seed yield (Makeen et al. 2007; Upadhaya et al. 1980); similarly, these traits along with test weight showed a maximum direct effect on the seed yield (Makeen et al. 2007). A significant positive association was observed between seed yield and days to 50% flowering, primary branches, secondary branches, clusters, pods, pod length, seeds, pod mass, pod wall mass, seed mass, shelling percentage, seed and harvest index (Singh and Kumar 2014), suggesting that these traits may be useful for selecting genotypes for yield improvement in mung bean. Seed yield had highly significant and positive correlations with pods, clusters and seed numbers (Singh and Kumar 2014), whereas days to maturity had a negative association with seed yield. The study also showed that seed yield had no significant association with protein content.

Number of clusters and number of pods showed a significantly positive association with seed yield, suggesting that these are the most important components for crop improvement in mung bean (Anand and Anandhi 2016; Asari et al. 2019). Similarly, the study also revealed a positive and direct impact of days to 50% flowering, test weight, number of clusters, number of pods and number of primary branches on seed yield (Asari et al. 2019), suggesting that emphasis may be given on these traits during the crop improvement in mung bean. Seed weight was reported to be negatively associated with seed roundedness, days to first flowering, days to 50% flowering, flowering period and days to maturity (Tripathi et al. 2020), while pod length showed a positive correlation with seed weight, seed area and seed dimensions.

5.9 Genetic Divergence

Quantification of divergence within the characters required to be improved gives the understanding to find suitable parents for breeding programmes (Mahalanobis 1936). It was suggested that the measurement of the metric distance between population centroids may help in the consideration of high-yielding parents having wider genetic divergence that are found beneficial in the development of high-yielding hybrids (Murty and Arunachalam 1966). The analysis also measures the magnitude of divergence and simultaneously provides an understanding of the evolutionary patterns in terms of the comparative influence of various traits on the entire divergence functioning at intra- and inter-cluster levels. Genetic divergence studies help in the identification of suitable parents for hybridization during crop improvement (Mohammadi and Prasanna 2003) as the involvement of genetically different parents brings gene constellation in the progressive generations.

Several studies have been carried out to find the nature and extent of genetic divergence in mung bean using Mahalanobis D^2 statistics (Goyal et al. 2021; Rahim et al. 2010; Ramana and Singh 1987; Ramanujam et al. 1974; Sen and De 2017), and it was concluded that the genotypes grouped in different clusters with higher statistical distances may be utilized in the hybridization programmes for crop improvement in mung bean. The comparative influence of each character on the total genetic divergence, the clusters having the highest statistical distance and the collection of at least one genotype from such clusters are some of the most significant points for the identification of parents using D^2 statistics. It has been observed that there is no relation between geographic and genetic diversity in mung bean (Naidu and Satyanarayana 1991; Raje and Rao 2000; Tripathi et al. 2020).

5.10 Plant Protection

Mung bean is susceptible to several viral, bacterial and fungal diseases leading to major economic losses to the crop (Mbeyagala et al. 2017; Pandey et al. 2018; Singh et al. 2000). Cercospora leaf spot, powdery mildew, anthracnose, dry root rot, web blight, fusarium wilt and Alternaria leaf spot are major fungal diseases (Pandey et al. 2018); halo blight, bacterial leaf spot and tan spot are the important bacterial diseases; while mung bean yellow mosaic disease (MYMD) is a major viral disease (Nair et al. 2017) found in mung bean. Maximum yield losses in mung bean have been reported due to MYMD (Karthikeyan et al. 2014) followed by several fungal diseases (Bhat et al. 2014; Maheshwari and Krishna 2013; Shukla et al. 2014). Effect of several bactericides and fungicides in the seed treatment and foliar spray along with the influence of good agronomic practices have been reported to combat these infections (Pandey et al. 2018). The use of disease-resistant varieties and the employment of integrated disease management are the best cost-effective ways to control the incidence of diseases in mung bean.

5.10.1 Viral Diseases

Mung bean yellow mosaic virus (MYMV) is a major threat to mung bean cultivation. The reference genome of this virus is available (Morinaga et al. 1993). The virus is comprised of two DNAs of about 2.7 kb. There are several views concerning genetic resistance associated with MYMV. It was suggested that it is controlled by a solo recessive gene (Reddy 2009), a dominant gene (Sandhu et al. 1985), while others reported that it is controlled by two recessive genes and a complementary recessive gene (Ammavasai et al. 2004; Dhole and Reddy 2012; Pal et al. 1991). The infected plant shows yellow-coloured spots on the young leaves that become yellow mosaic shape in the later stage, and simultaneously drooping of leaves takes place after the entire yellowing and drying of the leaves. Presently, fully resistant varieties to MYMV are unavailable. However, resistant varieties exhibit high variability and depend on climatic conditions (Nair et al. 2017) as the virus is transmitted through whitefly. The occurrence, distribution and transmission of this vector are well known that may help to cope with the spread of the virus. The variation in pathogen because of several other factors makes its control more cumbersome (Alam et al. 2014).

5.10.2 Fungal Diseases

Cercospora leaf spot (CLS) disease caused by fungus *Cercospora canescens* is one of the important foliar diseases in mung bean. The disease may reduce the yield up to 40%. There is chaos on the genetic basis of CLS-resistant gene, whether it is monogenic or multigenic. It has been reported that CLS resistance is governed by a single dominant gene (Lee 1980); besides this, studies also suggest the presence of quantitative genetic control (Chankaew et al. 2011) and a single recessive gene influence (Mishra et al. 1988) in respect to CLS resistance in mung bean. Variability among *C. canescens* strains is a major problem in crop breeding as it varies in the same region and within the same host including mung bean. Variable mycelial characteristics have also been reported with CLS (Joshi et al. 2006).

5.10.3 Bacterial Diseases

Blight caused by *Xanthomonas axonopodis* is a distressing bacterial disease in mung bean. Seeds are the primary source of bacteria, and therefore proper treatment of seeds before sowing is the best practice to control the disease (Baker and Smith 1966). A bacterial disease showing symptoms of marginal and veinal necrosis of leaves caused by *Curtobacterium flaccumfaciens* subsp. *flaccumfaciens* has been reported (Wood and Easdown 1990). The pathogen does not cause any wilting. The disease can be generally seen in rainfed crops suffering with water stress. Another bacterial disease showing the symptoms of necrotic spots on the leaves and collapsing of the upper part of the stem was observed in mung bean, and it was reported that the disease is caused by *Pseudomonas syringae* pv. *syringae* (George and Tripepi 1990). Besides this, irregular necrotic spots encircled with slender chlorotic and water-soaked radiance are seen on the leaves of mung bean that may result in blight. The disease is caused by *X. axonopodis* pv. *phaseoli* and may lead to severe loss to the crop (Osdaghi 2014). Necrotic spots surrounded with yellow halo caused by *P. syringae* pv. *phaseolicola* have been observed in China (Sun et al. 2017). A foliar disease caused by *P. syringae* pv. *tabaci* showing resemblance to wildfire has also been reported in mung bean (Sun et al. 2017). The disease initially appears in the form of small rounded light green patches that becomes brown from the centre during later stage due to necrosis of parenchymatic tissues. The necrosis proceeds quickly, and the brown spot encircled with watery lesion increases in length and width. The severity of infection may lead to deformation and drooping of leaves.

5.10.4 Nematodes

Nematodes have destructive effect on agriculture. Several nematodes, viz., *Rotylenchulus reniformis*, *Meloidogyne incognita*, *Bitylenchus vulgaris*, *Basirolaimus indicus*, *B. seinhorsti*, *Helicotylenchus indicus*, *H. retusus*, *Tylenchorhynchus mashhoodi* and *Tylenchus* sp., have been reported to infest mung bean (Ali 1995). *Heterodera vigni* is also known to infect mung bean crops, resulting in higher yield loss and dry matter content. Population-monitoring system (Saxena and Reddy 1987) and oil extracted from herbs (Sangwan et al. 1990; Siddiqui and Mahmood 1996) are considered better approaches to getting rid of nematodes in mung bean.

5.10.5 Insect Pests

Several insect pests are known to infest mung bean from its sowing to storage and lead to severe yield losses. Some of the insect pests found on mung bean are stem fly, thrips, aphids, whitefly, pod borer complex, pod bugs and bruchids (Swaminathan et al. 2012). They may directly attack the crop or work as vectors of diseases. Bean fly (*Ophiomyia phaseoli*) is the important pest found on mung bean. Besides *O. phaseoli*, other species of bean flies such as *Melanagromyza sojae* and *O. centrosematis* also infest mung bean crops (Talekar 1990). The flies attack the crop within a week after the germination, and under severe conditions, it may lead to complete loss of the crop (Chiang and Talekar 1980). Whitefly (*Bemisia tabaci*) is another pest that affects the crop directly and indirectly. It feeds on phloem and excretes honeydew on the plant that becomes black sooty moulds; besides this, it is the well-known vector of MYMV. Thrips also infest the crop at different stages. Several thrips, i.e., seedling thrips (*Thrips palmi* and *Thrips tabaci*) and flowering thrips (*Caliothrips indicus* or *Megalurothrips* spp.), are found on the crop. Spotted pod borer (*Maruca vitrata*) is also an important pest found on mung bean crops grown in tropical and subtropical regions. The larvae of this pod borer attack the flower, stem, peduncle and pod of mung bean (Sharma 1999). Azuki bean weevil (*Callosobruchus chinensis*) and cowpea weevil (*Callosobruchus maculatus*) are some of the most serious pests of mung bean in the field, while bruchids are the serious pests found in storage conditions (Somta et al. 2007; Tomooka et al. 1992).

5.11 Physiology and Abiotic Stresses

Abiotic stresses have an adverse effect on plant growth and productivity, leading to major economic losses (Ye et al. 2017). These stresses may include several atmospheric issues along with drought, flooding, radiation, salinity, temperature, etc. The effect of climatic aberrations over the periods also reduced crop yields (Boyer et al. 2013; Rosenzweig et al. 2014). Mung bean is highly sensitive to salinity, drought and fluctuating temperatures during the flowering and pod formation stages, leading to severe yield losses. Understanding of physiological limits influencing the seed yield in mung bean is critical, and it should be properly identified before devising solutions.

5.11.1 Water Stress and Drought

Mung bean is generally grown under limited soil moisture conditions and does not require any additional input. Nevertheless, its growth is highly influenced by the availability of moisture in the field. However, it is highly susceptible to waterlogging conditions (Singh and Singh 2011). It was observed that water stress during the flowering stage resulted in 50–60% yield reduction (El Nakhlawy et al. 2018) in mung bean, and the study also revealed that seed formation was the most sensitive stage to water stress. Further, studies also suggest that the extreme drought conditions may lead to a reduction of plant biomass, pod numbers and consequently great toll on seed yield (Kumar and Sharma 2009). A decline in the pace of pod initiation, its development (Begg 1980) and flower shedding (Moradi et al. 2009) are the significant impacts of water stress during the reproductive growth of the crop. Drought condition during the reproductive stage has a negative effect on flowering and simultaneously leads to a reduction in the yield (Raza et al. 2012).

Drought conditions during flowering and podding stages may lead to 31–57% and 26% yield reduction, respectively (Nadeem et al. 2019). Drought condition leads to the production of destructive superoxide molecules that damages cells, and this oxidative stress depends mainly upon the level of ascorbic acid and glutathione pools (Anjum et al. 2015). Heat and cold stress are highly dangerous to different growth stages and may result in higher yield losses. The optimum temperature for plant growth is 28–30 °C. Higher temperatures (>45 °C) during the flowering stage may lead to flower shedding. Several developmental stages of mung bean including germination, seed emergence, vegetative phases, flowering stage and pod/seed setting stage are highly sensitive to temperature extremity (HanumanthaRao et al. 2016). Crops grown during February or March months face major problems of water stress due to insufficient or no rainfall; hence, sowing of short-duration varieties may be preferred to avoid the stress (Pratap et al. 2013). Mung bean varieties/lines having tolerance against several abiotic stresses, viz., drought, heat and salt, have been identified over the period (Bindumadhava et al. 2018; Dutta et al. 2016; Dutta and Bera 2008; Manasa et al. 2017; Sharma et al. 2016).

5.11.2 Salt Stress

Salt stress adversely affects seed germination, biomass and shoot and root growth along with several yield-attributing traits (Ahmed 2009; Promila and Kumar 2000). Lesser seed germination was observed in mung bean with the increasing salinity levels (Kandil et al. 2012; Maliwal and Paliwal 1982). It may be due to the fact that salinity evades water uptake or causes toxic effects, resulting in a reduction of seed germination (Murillo-Amador et al. 2002). Salt stress is usually exhibited as a general stunning of plant growth. Symptoms of salt injury such as chlorosis and necrosis have also been reported in mung bean due to increased levels of salinity (Reddy 1982; Wahid et al. 2004). Significant variability was observed for growth, yield, yield components and chemical composition in mung bean seeds under different salinity levels (Mohamed and El-Kramany 2005). Mung bean plants have been reported to have higher proline content in the root and shoot due to increased salinity and or salinity stress (Misra and Gupta 2006). It was suggested that salt stress may affect the filling of seeds in the pods of mung bean, leading to a reduction in the number of seeds in the pods and simultaneously a reduction in the yield potential (Ahmed 2009). Yield variability in mung bean upon salt stress has also been noticed by various workers (Hossain et al. 2008; Jahan et al. 2020). It has been observed that salt-stressed plants of mung bean had a higher concentration of sodium and chloride ions in their leaves, roots and shoots and a lower concentration of potassium and calcium ions as compared to the non-stressed plants (Mohammed 2007). Owing to this condition, the electrolyte leakage in mung bean was comparatively higher (Alharby et al. 2019). A decrease in seed germination, plant height, shoot and root length, dry matter, biomass, and root, stem and leaf weights has been reported in mung bean due to an increase in salt stress (Mohamed and El-Kramany 2005; Mohammed 2007). It has been observed that 50 mM NaCl significantly affected the yield of mung bean (Saha et al. 2010). Accumulation of a higher quantity of salt leads to a reduction of the osmotic ability of soil sap, resulting in water stress in plants and consequently nutritive deficiency and oxidative stresses (Tavakkoli et al. 2011) along with reduction in photosynthesis rate. This may also stimulate physiological and metabolic pathways (Misra and Dwivedi 2004) of the cells. Reduction in root length due to salt stress impedes the uptake and supply of nutrients. Number of nodules also reduced with the increase in salinity; however, their size increased due to salinity (Naher and Alam 2010). Pre-treatment of mung bean with sub-lethal dosage of sodium chloride may help in adaptation of the crop to the lethal levels of salinity (Saha et al. 2010). There is comparatively little work available on the development of salt-tolerant varieties of mung bean. Decline in relative water content, cellular dehydration and osmotic stress have been observed in mung bean due to salt stress (Singh et al. 2021). The biometric, morphophysiological, biochemical and biophysical characters in mung bean were highly affected due to salt stress (Kumar et al. 2012). It suggests that salt stress imposes water insufficiency in plants and may cause physiological drought. It has been reported that salinity tolerance depends on the genotype and different growth stages; hence, salt tolerance at seedling stage may not suggest that it may show tolerance at maturity stage (Sehrawat et al. 2013). Salinity has different responses in the plant, which can be manifested at tissue, canopy, physiological or molecular level (HanumanthaRao et al. 2016).

5.11.3 Other Abiotic Stresses

Rising application of synthetic fertilizers and higher human interference along with the mixing of contaminated industrial effluents have deteriorated the cultivated land, and indirectly the crops are grown on it. The water or air pollutants are significant threats to crop cultivation as they have a higher concentration of heavy metals (Lagerverff and Specht 1970). The metal accretion in the soil is increasing continuously due to uncontrolled usage of fertilizers, pesticides, industrial waste and sewage (Harland et al. 2000). Soil pollution due to heavy metals is very hazardous because heavy metals cannot be despoiled naturally and may remain in the ecosystem for a longer time and simultaneously in the food chain (Igwe et al. 2005). Lethal impacts of heavy metals have been observed on the soil microflora (Pawlowska and Charvat 2004) along with amendment of the variability, quantity and entire activity of the microbial communities (Smejkalova et al. 2003). Besides heavy metal contamination, air pollution has higher concentration of sulphur dioxide, nitrogen dioxide and ozone, which also have deleterious effects on biomass, seed quality and yield potential of crops including mung bean (Agrawal et al. 2003, 2006). The toxic effect of heavy metals on mung bean seed germination was studied, and delayed germination was observed with a higher concentration of lead (Ashraf and Ali 2007). The study also suggests that silver was more toxic followed by lead and zinc. A decrease in the biomass and quality of seeds was reported due to air pollutants such as sulphur dioxide, nitrogen dioxide and ozone in mung bean (Agrawal et al. 2006). Heavy metal nickel adversely influences the photosynthetic pigments and yield in mung bean (Ahmad et al. 2007). It also supports the deposition of sodium, potassium and calcium ions.

An increased level of proline in the plant is suggestive of abiotic stress. The level of proline was tested in mung bean under cadmium, cobalt, lead and zinc stress (Saradhi 1991), and cadmium was found as the most poisonous metal triggering proline production. Cadmium increases glutathione reductase activity (Gill and Tuteja 2010), inhibits photosynthetic activity (Wahid et al. 2008) and affects the activity and structure of chloroplast (Wahid et al. 2007) in mung bean. Cadmium and lead induce changes in growth, biochemical attributes and mineral accumulation (Ashraf et al. 2016), while mercury induces changes in germination and biochemical attributes (Saminathan 2013) in mung bean, suggesting that heavy metal-contaminated soil exhibits negative impacts on the development, production and protein content in the crop. Sharma et al. (2021) observed that cadmium had amended several morphological and biochemical characteristics of mung bean. It also affected the chlorophyll, carbohydrate, protein, polyphenol and antioxidant profile of the crop.

5.12 Tissue Culture and Genetic Transformation

The development of plants through the tissue culture technique permits the transfer of genes into plant cells (Chandra and Pental 2003). The transgenic exploration in mung bean is sluggish owing to its recalcitrant behaviour towards tissue culture and lower frequency of regeneration after transformation (Eapen 2008; Varshney et al. 2015). However, regeneration protocols for mung bean have been developed through embryogenesis (Sivakumar et al. 2010), organogenesis (Himabindu et al. 2014) and axillary bud proliferation using cotyledonary node explants (Sagare and Mohanty 2015; Yadav et al. 2010). Successful transformation in mung bean has also been reported in which transgenes were effectively inherited and conceded to the following generations (Baloda and Madanpotra 2017).

Genetic transformation in mung bean was initially carried out in hypocotyls and primary leaves (Jaiwal et al. 2001), and a binary vector (selection marker: neomycin phosphotransferase and reporter gene: beta-glucuronidases) was successfully incorporated. Later, Saini et al. (2007) developed morphologically normal and fertile transgenic plants of mung bean comprising two transgenes, bialaphos resistance and alpha-amylase inhibitor, using cotyledonary node explants. A pathogenesis-related gene (*bjnpr1*) isolated from mustard was introduced into mung bean, and it was observed that the transgenic mung bean plants exhibited resistance against fungal diseases (Vijayan and Kirti 2012). Similarly, *annexin1bj* gene was successfully incorporated into mung bean, and the consequently developed transgenic plants revealed better tolerance against drought stress (Yadav et al. 2012). Transformation of mung bean plants for salt and drought tolerance was carried out by introducing a gene for an osmoprotectant glycine betaine (Saraswat et al. 2017), and transformation and expression of the transgene (*codA* gene) were realized. Modification in the DNA structure of food crops is usually unacceptable; therefore, genetically engineered food crops have always been viewed with a question mark despite several advantages.

5.13 Genetic Markers and Biotechnology

Several molecular markers, viz., restriction fragment length polymorphism (RFLP), random amplified polymorphic DNA (RAPD), simple sequence repeat (SSR) and single nucleotide polymorphism (SNP) markers, have been employed to study the genetic diversity in mung bean. The molecular markers have also been used for the construction of linkage maps focusing on yield, nutritional aspects and disease resistance.

Mung bean has a smaller (~600 Mb) genome and takes a lesser period to complete its life cycle; therefore, it is comparatively more suitable to apply other approaches for crop improvement. It was observed that maximum genes found in mung bean showed synteny with the genes found in soybean (Kang et al. 2014). *Vigna radiata* variety VC1973A was genetically sequenced, transcriptome sequences of 22 accessions were obtained (Kang et al. 2014) and relatedness of two homologous genomes of *V. reflex-pilosa* (a wild species) was outlined. The study enhanced the understanding of the evolution of *Vigna* species that may enable crop improvement in mung bean. The molecular markers facilitate the identification of loci linked to the desirable characteristics, and their tracking is more accurate and effective as compared to traditional breeding (Collard and Mackill 2008).

Bruchid resistance in mung bean was analysed using RFLP markers (Young et al. 1992), and 153 RFLP markers were categorized into 14 linkage groups having an average interval of 9.3 cM. Further, RFLP markers were used to prepare a linkage map of mung bean comprising 11 linkage groups, and an interspecific hybrid population between *V. radiata* ssp. *radiata* and *V. radiata* ssp. *sublobata* was obtained (Menancio-Hautea et al. 1992). Humphry et al. (2002) exploited RFLP markers to construct a genetic map using recombinant inbred populations of 80 mung bean accessions derived from a cultivated variety and *V. radiata* subsp. *sublobata*. The map included 13 linkage groups with an average distance of 3 cM, and a highly conserved marker order was reported between mung bean and *Lablab purpureus*. Transfer of bruchid beetle resistance allele (Somta et al. 2008; Tomooka et al. 1992) and yellow mosaic disease resistance allele (Basak et al. 2005; Gill et al. 1983) from wild mung bean is an example of marker-assisted breeding in mung bean.

RAPD markers were applied to assess the genetic diversity among uncultivated and cultivated *Vigna* species, namely *V. angularis*, *V. umbellata*, *V. radiata*, *V. aconitifolia* and *V. mungo* (Kaga et al. 1996). A genetic map was prepared using RFLP and RAPD markers using F2 populations obtained by crossing *V. radiata* ssp. *radiata* and *V. radiata* ssp. *sublobata*. Lambrides et al. (2000) grouped all the 67 accessions in 12 linkage groups having 691.7 cM intervals. Kaga and Ishimoto (1998) also used RFLP and RAPD markers to prepare a linkage map and identified the genes accountable for bruchid resistance. Genetic maps showing the information on several morphophysiological and agronomic traits of cultivated and wild accessions of mung bean have been constructed (Isemura et al. 2012; Wang et al. 2016) that will facilitate the understanding of important traits of interest in both cultivated and wild mung bean accessions. RAPD and inter-simple sequence repeat (ISSR) markers were used to assess genetic diversity in mung bean germplasm (Chattopadhyay et al. 2005), wherein ISSR markers were found to be competent as compared to RAPD markers. Yu et al. (1999) employed simple sequence repeat (SSR) to assess microsatellite efficacy as genetic markers in mung bean and 61 simple repetitive DNA sequences having 23 motifs were recognized as prospective microsatellites.

Mung bean gene pools comprising 415 cultivated, 189 wild and 11 intermediate accessions were assessed to study the presence of genetic diversity using 19 SSR markers (Sangiri et al. 2008), and wide polymorphism was recorded among wild and cultivated pools. The study suggested that Australia and New Guinea were the diversity core for wild mung bean. In view of the higher diversity in mung bean accessions from South Asia, it was suggested that the crop may have been domesticated in South Asia (Sangiri et al. 2008). SSR markers linked to *Cercospora* leaf spot (Yundaeng et al. 2021) and powdery mildew diseases (Chankaew et al. 2013; Kasettranan et al. 2010) have been identified, and quantitative trait loci (QTL) maps were prepared using these markers. A genetic linkage map was constructed, and a genetic analysis of domestication-related traits in mung bean was done using 430 SSR and EST-SSR markers (Isemura et al. 2012). The markers were grouped into 11 linkage groups with a total distance of 727.6 cM, and 105 QTLs including 38 domestication-related gene traits were distinguished. The study also revealed some useful QTLs for seed size, pod dehiscence and pod maturity in mung bean.

With the developments in next-generation sequencing, the attention of researchers has shifted to finding single nucleotide polymorphisms (SNPs). SNP markers are biallelic, codominant and universally distributed across the entire genome (Brumfield et al. 2003). Mung bean cultivars were sequenced to search for resistance to *Riptortus clavatus* and *Callosobruchus chinensis* (Moe et al. 2011), and 2098 SNPs were reported. Raturi et al. (2012a) characterized 44 genotypes of mung bean based on nuclear ribosomal DNA and RAPD polymorphism to assess the genetic diversity and relationships and reported 82% polymorphism with wide intraspecific variations. The study also revealed internal transcribed spacer (ITS) length variations, SNPs and insertions/deletions at the number of sites in nuclear rDNA region. Genome sequence of mung bean and its comprehensions into evolution within *Vigna* species were carried out (Kang et al. 2014), and genomic evidence of allopolyploid event was reported on the basis of de novo assembly of a tetraploid *Vigna* species (*V. reflexo-pilosa* var. *glabra*).

EST-based SSR markers have been exploited to study functional genomics in mung bean (Chavan and Gacche 2014; Chen et al. 2015; Moe et al. 2011). SSR motifs were recognized in 1848 EST sequences in mung bean, and it was observed that about 45% and 55% of these motifs were situated in coding and untranslated regions, respectively (Moe et al. 2011). Biotin-labelled oligo-probes and streptavidin-coated beads were applied to prepare an SSR-enriched library from mung bean genotypes, and 308,509 SSR motifs were identified (Wang et al. 2016). Illumina paired-end sequencing technology was used for transcriptome sequencing of mung bean genes, and identification of EST-SSR markers (Chen et al. 2015) and more than 103 million high-quality cDNA sequences was done.

Mung bean genome has been characterized using translational genomics to obtain genomic information from well-studied species (Isemura et al. 2012; Kim et al. 2014). The flowering gene in mung bean was recognized with the help of genome-wide evaluation between mung bean and *Arabidopsis*. It was observed that out of 207 genes that were related to flowering in *Arabidopsis*, 129 were homologous to mung bean genes (Kim et al. 2015b). In another study, it was also observed that these genes were near to the SSR markers on a genetic map (Isemura et al. 2012). Mung bean genome was also compared to the soybean genome, and it was noticed that five flowering-related genes in mung bean were homologous to soybean flowering genes (Kim et al. 2015b). The studies may lead to the functional characterization of genes of interest in mung bean. Application of several biotechnological tools may facilitate the introduction of beneficial genes in promising mung bean lines to increase genetic variability.

5.14 Conclusion and Prospects

Being an important leguminous crop owing to its high nutritional contents, several studies have been carried out in mung bean addressing yield-related traits including resistance to different diseases and domestication-related traits. The lack of genomic information has led to stagnation in mung bean breeding. However, after the publication of the reference genome sequence of mung bean in 2014, breeders have got a better opportunity to understand the genomic and genetic background of several agronomically important traits of the crop. Preparation of wild mung bean pool from diverse origins and environmental conditions is essentially required to conserve the genetic diversity of the crop. The yield of more than 20 quintals per hectare, maturity period between 60 and 75 days, higher harvest index, photoperiod insensitivity, resistance to major insect pests/diseases, compact canopy and synchronous maturity are some of the important objectives for crop improvement in mung bean. The inclusion of the ideotype approach may also be considered to attain sustainable yield.

References

Abbas G, Atta BM, Shah TM, Sadiq MS, Haq MA (2008) Stability analysis for seed yield in mungbean, *Vigna radiata* L. Wilczek. J Agric Res (Pakistan) 46(3):223–228

Aggarwal DV, Poehlman JM (1977) Effects of photoperiod and temperature on flowering in mungbean (*Vigna radiata* (L.) Wilczek). Euphytica 26(1):207–219

Agrawal M, Singh B, Rajput M, Marshall F, Bell JNB (2003) Effect of air pollution on peri-urban agriculture: a case study. Environ Pollut 126(3):323–329

Agrawal M, Singh B, Agrawal SB, Bell JNB, Marshall F (2006) The effect of air pollution on yield and quality of mung bean grown in peri-urban areas of Varanasi. Water Air Soil Pollut 169(1): 239–254

Ahmad S, Belwal V (2019) Study of correlation and path analysis for yield and yield attributing traits in mung bean [*Vigna radiata* (L.) Wilczek]. Int J Chem Stud 8(1):2140–2143

Ahmad MSA, Hussain M, Saddiq R, Alvi AK (2007) Mungbean: a nickel indicator, accumulator or excluder? Bull Environ Contam Toxicol 78(5):319–324

Ahmed S (2009) Effect of soil salinity on the yield and yield components of mungbean. Pak J Bot 41(1):263–268

Akanda SI, Mundt CC (1996) Path coefficient analysis of the effects of stripe rust and cultivar mixtures on yield and yield components of winter wheat. Theor Appl Genet 92(6):666–672

Alam Mondal MM, Ali Fakir MS, Juraimi AS, Hakim MA, Islam MM, Shamsuddoha ATM (2011) Effects of flowering behavior and pod maturity synchrony on yield of mungbean ['*Vigna radiata*' (L.) Wilczek]. Aust J Crop Sci 5(8):945–953

Alam AKM, Somta P, Srinives P (2014) Identification and confirmation of quantitative trait loci controlling resistance to mungbean yellow mosaic disease in mungbean [*Vigna radiata* (L.) Wilczek]. Mol Breed 34(3):1497–1506

Alharby HF, Al-Zahrani HS, Hakeem KR, Iqbal M (2019) Identification of physiological and biochemical markers for salt (NaCl) stress in the seedlings of mungbean [*Vigna radiata* (L.) Wilczek] genotypes. Saudi J Biol Sci 26(5):1053–1060

Ali SS (1995) Nematodes associated with mungbean in Gwalior district. Indian J Pulses Res 8:100–101

Allard RW (1960) Principles of plant breeding. Wiley, New York, p 485

Allard RW, Bradshaw AD (1964) Implications of genotype-environmental interactions in applied plant breeding 1. Crop Sci 4(5):503–508

Ammavasai S, Phogat DS, Solanki IS (2004) Inheritance of resistance to mungbean yellow mosaic virus (MYMV) in green gram [*Vigna radiata* (L.) Wilczek]. Am J Plant Sci 4(10):1924–1927

Anand G, Anandhi K (2016) Genetic variability, correlation and path analysis for yield and yield components in F6 families of Greengram (*Vigna radiata* (L.) Wilczek) under rainfed condition. Electron J Plant Breed 7(2):434–437

Anjum NA, Umar S, Iqbal M, Khan NA (2011) Cadmium causes oxidative stress in mung bean by affecting the antioxidant enzyme system and ascorbate-glutathione cycle metabolism. Russ J Plant Physiol 58(1):92–99

Anjum NA, Umar S, Aref IM, Iqbal M (2015) Managing the pools of cellular redox buffers and the control of oxidative stress during the ontogeny of drought-exposed mungbean (*Vigna radiata* L.)—role of sulfur nutrition. Front Environ Sci 2:66

Anonymous (2020). https://www.indiastat.com/table/agriculturedata/2/moong-greengram/19571/446269/data.aspx

Asari T, Patel BN, Patel R, Patil GB, Solanki C (2019) Genetic variability, correlation and path coefficient analysis of yield and yield contributing characters in mung bean [*Vigna radiata* (L.) Wilczek]. IJCS 7(4):383–387

Ashraf R, Ali TA (2007) Effect of heavy metals on soil microbial community and mung beans seed germination. Pak J Bot 39(2):629

Ashraf MY, Roohi M, Iqbal Z, Ashraf M, Öztürk M, Gücel S (2016) Cadmium (Cd) and lead (Pb) induced changes in growth, some biochemical attributes, and mineral accumulation in two cultivars of mung bean [*Vigna radiata* (L.) Wilczek]. Commun Soil Sci Plant Anal 47(4): 405–413

Auerbach C, Robson JM (1946) Chemical production of mutations. Nature 157(3984):302

Auti SG (2012) Lhb mutant–a novel mutant of mungbean (*Vigna radiata* (L.) Wilczek) induced by gamma radiation. Bioremed Biodivers Bioavail 6(Special Issue 1):87–93

Auti SG, Apparao BJ (2009) Induced mutagenesis in mungbean (*Vigna radiata* (L.) Wilczek). Induced plant mutations in the genomics era. Food and Agriculture Organization of the United Nation, Rome, Italy, pp 97–100

Baker KF, Smith SH (1966) Dynamics of seed transmission of plant pathogens. Annu Rev Phytopathol 4(1):311–332

Baloda A, Madanpotra S (2017) Transformation of mungbean plants for salt and drought tolerance by introducing a gene for an osmoprotectant glycine betaine. J Plant Stress Physiol 3:5–11

Baraki F, Gebregergis Z, Belay Y, Berhe M, Zibelo H (2020) Genotype × environment interaction and yield stability analysis of mung bean (*Vigna radiata* (L.) Wilczek) genotypes in Northern Ethiopia. Cogent Food Agric 6(1):1729581

Basak J, Kundagrami S, Ghose TK, Pal A (2005) Development of Yellow Mosaic Virus (YMV) resistance linked DNA marker in *Vigna mungo* from populations segregating for YMV-reaction. Mol Breed 14(4):375–383

Begg JE (1980) Morphological adaptations of leaves to water stress. In: Adaptation of plants to water and high temperature stress. Wiley, New York

Bhadra SK, Akhter MI, Quashem A (1991) Genetics of seed lustre and joint inheritance of seed-coat colour and seed lustre in mungbean (*Vigna radiata* (L.) Wilczek). Bangladesh J Bot (Bangladesh) 20(1):61–64

Bhat FA, Mohiddin FA, Bhat HA (2014) Reaction of greengram (*Vigna radiata*) to *Cercospora canescens* (ell.) and mart. Indian J Agric Res 48(2):140–144

Bindumadhava H, Sharma L, Nair RM, Nayyar H, Riley JJ, Easdown W (2018) High-temperature-tolerant mungbean (*Vigna radiata* L.) lines produce better yields when exposed to higher CO_2 levels. J Crop Improv 32(3):418–430

Bisht IS, Mahajan RK, Patel DP (1998) The use of characterisation data to establish the Indian mungbean core collection and assessment of genetic diversity. Genet Resour Crop Evol 45(2): 127–133

Boyer JS, Byrne P, Cassman KG, Cooper M, Delmer D, Greene T, Gruis F, Habben J, Hausmann N, Kenny N (2013) The US drought of 2012 in perspective: a call to action. Glob Food Sec 2(3):139–143

Brumfield RT, Beerli P, Nickerson DA, Edwards SV (2003) The utility of single nucleotide polymorphisms in inferences of population history. Trends Ecol Evol 18(5):249–256

Burton GW, de Devane EH (1953) Estimating heritability in tall fescue (*Festuca arundinacea*) from replicated clonal material 1. Agron J 45(10):478–481

Chandra A, Pental D (2003) Regeneration and genetic transformation of grain legumes: an overview. Curr Sci 84(3):381–387

Chankaew S, Somta P, Sorajjapinun W, Srinives P (2011) Quantitative trait loci mapping of Cercospora leaf spot resistance in mungbean, *Vigna radiata* (L.) Wilczek. Mol Breed 28(2): 255–264

Chankaew S, Somta P, Isemura T, Tomooka N, Kaga A, Vaughan DA, Srinives P (2013) Quantitative trait locus mapping reveals conservation of major and minor loci for powdery mildew resistance in four sources of resistance in mungbean [*Vigna radiata* (L.) Wilczek]. Mol Breed 32(1):121–130

Chattopadhyay K, Ali MN, Sarkar HK, Mandal N, Bhattacharyya S (2005) Diversity analysis by RAPD and ISSR markers among the selected mungbean [*Vigna radiata* (L.) Wilczek] genotypes. Indian J Genet Plant Breed 65(3):173–175

Chavan SP, Gacche R (2014) Identification and characterization of EST-SSRs in mungbean. WebmedCentral Biotechnol 5(3):WMC004598

Chen H, Liu X, Chen HM, Liu XH (2001) Inheritance of seed color and lustre in mungbean (*Vigna radiata*). Agric Sci Technol Hunan 2:8–12

Chen L-R, Markhart AH, Shanmugasundaram S, Lin T-Y (2008) Early developmental and stress responsive ESTs from mungbean, *Vigna radiata* (L.) Wilczek, seedlings. Plant Cell Rep 27(3): 535–552

Chen H, Wang L, Wang S, Liu C, Blair MW, Cheng X (2015) Transcriptome sequencing of mung bean (*Vigna radiate* L.) genes and the identification of EST-SSR markers. PLoS One 10(4): e0120273

Chhabra AK (1990) Inheritance of lobed and pentafoliate leaf in mungbean. Indian J Pulses Res 3(1):69–72

Chiang HS, Talekar NS (1980) Identification of sources of resistance to the beanfly and two other agromyzid flies in soybean and mungbean. J Econ Entomol 73(2):197–199

Collard BCY, Mackill DJ (2008) Marker-assisted selection: an approach for precision plant breeding in the twenty-first century. Philos Trans R Soc B Biol Sci 363(1491):557–572

Cook SMF, Gupta SC, Woodhead T, Larson WE (1995) Soil physical constraints to establishment of mungbeans (*Vigna radiata* L. Wilczek) in paddy rice (*Oryza sativa* L.) soils. Soil Tillage Res 33(1):47–64

Degefa I, Petros Y, Andargie M (2014) Genetic variability, heritability and genetic advance in Mungbean (*Vigna radiata* L. Wilczek) accessions. Plant Sci Today 1(2):94–98

Denton OA, Nwangburuka CC (2011) Heritability, genetic advance and character association in six yield related characters of Solanum anguivi. Asian J Agric Res 5(3):201–207

Dewanjee S, Sarkar KK (2018) Evaluation of performance of induced mutants in mungbean [*Vigna radiata* (L.) Wilczek]. Legum Res 41(2):213–217

Dewey DR, Lu K (1959) A correlation and path-coefficient analysis of components of crested wheatgrass seed production 1. Agron J 51(9):515–518

Dhole VJ, Reddy KS (2010) Gamma rays induced moisture stress tolerant long root mutant in mungbean (*Vigna radiata* L. Wilczek). Electron J Plant Breed 1(5):1299–1305

Dhole VJ, Reddy KS (2012) Genetic analysis of resistance to mungbean yellow mosaic virus in mungbean (*Vigna radiata*). Plant Breed 131(3):414–417

Dubinin NP (1962) Problems of radiation genetics (Vol. 5376). US Atomic Energy Commission, Division of Technical Information

Dutta P, Bera AK (2008) Screening of mungbean genotypes for drought tolerance. Legum Res 31(2):145–148

Dutta P, Bandopadhyay P, Bera AK (2016) Identification of leaf based physiological markers for drought susceptibility during early seedling development of mungbean. Am J Plant Sci 7(14): 1921

Dwivedi N (2006) Germplasm characterization in indigenous green gram [*Vigna radiata* (L.) Wilczek] from arid and semi-arid region, India. J Arid Legumes 3(1):55–59

Dwivedi S, Singh DP (1985) Inheritance of narrow trifoliate leaf in mungbean [*Vigna radiata* (L.) Wilczek]. SABRAO J 17:177–180

Eapen S (2008) Advances in development of transgenic pulse crops. Biotechnol Adv 26(2): 162–168

Eberhart SA, Russell WA (1966) Stability parameters for comparing varieties 1. Crop Sci 6(1): 36–40

El Nakhlawy FS, Ismail SM, Basahi JM (2018) Optimizing mungbean productivity and irrigation water use efficiency through the use of low water-consumption during plant growth stages. Legum Res 41(1)

Falconer D, Mackay T (2005) Introduction to quantitative genetics, 4th edn. Pearson Prentice Hall, Harlow

Fehr WR (1987) Vols. 1 and 2 Principles of cultivar development. Macmillan Publishing Co., New York

Fuller DQ, Harvey EL (2006) The archaeobotany of Indian pulses: identification, processing and evidence for cultivation. Environ Archaeol 11(2):219–246

Garg DD, Arya RS, Sharma T, Dhuria RK (2004) Effect of replacement of sewan straw (*Lasiurus scindicus*) by moong (*Phaseolus aureus*) chara on rumen and haemato-biochemical parameters in sheep. Vet Pract 5(1):70–73

George MW, Tripepi RR (1990) Bacterial leaf spot and stem collapse of mung bean caused by *Pseudomonas syringae* pv. *syringae*. Plant Dis 74(5):394

Gill SS, Tuteja N (2010) Reactive oxygen species and antioxidant machinery in abiotic stress tolerance in crop plants. Plant Physiol Biochem 48(12):909–930

Gill AS, Verma MM, Dhaliwal HS, Sandhu TS (1983) Interspecific transfer of resistance to mungbean yellow mosaic virus from *Vigna mungo* to *Vigna radiata*. Curr Sci 52(1):31–33

Goyal L, Intwala CG, Modha KG, Acharya VR (2021) Association and diversity analysis for yield attributing traits in advance generation of green gram (*Vigna radiata* (L.) Wilczek). IJCS 9(1): 1934–1939

Graham PH, Vance CP (2003) Legumes: importance and constraints to greater use. Plant Physiol 131(3):872–877

Gunckel JE, Sparrow AH (1967) Ionizing radiation: biochemical, physiological and morphological aspects of their effects on plants. Rutgers Univ, New Brunswick, NJ

HanumanthaRao B, Nair RM, Nayyar H (2016) Salinity and high temperature tolerance in mungbean [*Vigna radiata* (L.) Wilczek] from a physiological perspective. Front Plant Sci 7:957

Harland BJ, Taylor D, Wither A (2000) The distribution of mercury and other trace metals in the sediments of the Mersey Estuary over 25 years 1974–1998. Sci Total Environ 253(1–3):45–62

Harris D, Breese WA, Rao JK (2005) The improvement of crop yield in marginal environments using 'on-farm' seed priming: nodulation, nitrogen fixation, and disease resistance. Aust J Agric Res 56(11):1211–1218

Henry A, Mathur BK (2007) Genetic diversity and stability studies in promising mungbean selections. J Arid Legumes 4(2):130–133

Himabindu Y, Reddy MC, Chandrasekhar T (2014) In vitro regeneration of green gram (*Vigna radiata* (L.) Wilczek) cultivar Vamban-2 using cotyledonary nodes. *CIBTech*. J Biotechnol 3(4):11–15

Hossain MM, Miah MNA, Rahman MA, Islam MA, Islam MT (2008) Effect of salt stress on growth and yield attributes of mungbean. Bangladesh Res Pub J 1(4):324–336

Hou D, Yousaf L, Xue Y, Hu J, Wu J, Hu X, Feng N, Shen Q (2019) Mung bean (*Vigna radiata* L.): bioactive polyphenols, polysaccharides, peptides, and health benefits. Nutrients 11(6):1238

Humphry M, Konduri V, Lambrides C, Magner T, McIntyre C, Aitken E, Liu C (2002) Development of a mungbean (*Vigna radiata*) RFLP linkage map and its comparison with lablab (*Lablab purpureus*) reveals a high level of colinearity between the two genomes. Theor Appl Genet 105(1):160–166

Hyten DL, Song Q, Zhu Y, Choi I-Y, Nelson RL, Costa JM, Specht JE, Shoemaker RC, Cregan PB (2006) Impacts of genetic bottlenecks on soybean genome diversity. Proc Natl Acad Sci 103(45):16666–16671

Igwe JC, Nnororm IC, Gbaruko BC (2005) Kinetics of radionuclides and heavy metals behaviour in soils: implications for plant growth. Afr J Biotechnol 4(13)

Iqbal J, Ahsan M, Saleem M, Asghar ALI (2014) Genetic architecture of synchronous pods maturation and yield related traits in mungbean [*Vigna radiata* (L.) Wilczek]. J Appl Biol Sci 8(3):85–92

Isemura T, Kaga A, Tabata S, Somta P, Srinives P, Shimizu T, Jo U, Vaughan DA, Tomooka N (2012) Construction of a genetic linkage map and genetic analysis of domestication related traits in mungbean (*Vigna radiata*). PLoS One 7(8):e41304

Jahan I, Rahman MM, Tuzzohora MF, Hossain MA, Begum SN, Burritt DJ, Hossain MA (2020) Phenotyping of mungbean (*Vigna radiata* L.) genotypes against salt stress and assessment of variability for yield and yield attributing traits. J Plant Stress Physiol 6:7–17

Jain HK, Mehra KL (1980) Evolution, adaptation, relationships, and uses of the species of Vigna cultivated in India. In: Evolution, adaptation, relationships, and uses of the species of Vigna cultivated in India, pp 459–468

Jaiwal PK, Kumari R, Ignacimuthu S, Potrykus I, Sautter C (2001) Agrobacterium tumefaciens-mediated genetic transformation of mungbean (*Vigna radiata* L. Wilczek)—a recalcitrant grain legume. Plant Sci 161(2):239–247

Jo B-K, Ahn G-W, Jeong J-H, Hwang Y-I (2006) Clinical studies on the anti-irritation effects of mung bean (*Phaseolus aureus*) extract in cosmetics. SÖFW J 132(1–2)

Johnson HW, Robinson HF, Comstock RE (1955) Estimates of genetic and environmental variability in soybeans. Agron J 47(7):314–318

Joseph J, Santhoshkumar AV (2000) Genetic analysis of quantitative characters through bi-parental mating in green gram [*Vigna radiata* (L.) Wilczek]. Res Crops 1(2):221–225

Joshi A, Souframanien J, Chand R, Pawar SE (2006) Genetic diversity study of Cercospora canescens (Ellis & Martin) isolates, the pathogen of Cercospora leaf spot in legumes. Curr Sci 90:564–568

Kaga A, Ishimoto M (1998) Genetic localization of a bruchid resistance gene and its relationship to insecticidal cyclopeptide alkaloids, the vignatic acids, in mungbean (*Vigna radiata* L. Wilczek). Mol Gen Genet 258(4):378–384

Kaga A, Tomooka N, Egawa Y, Hosaka K, Kamijima O (1996) Species relationships in the subgenus Ceratotropis (genus Vigna) as revealed by RAPD analysis. Euphytica 88(1):17–24

Kamannavar PY, Vijaykumar AG (2011) Genotype x Environment interaction in Mungbean (*Vigna radiata* (L.) Wilczek) cultivars grown in different agro-climatic zones of Karnataka. Electron J Plant Breed 2(4):501–505

Kandil AA, Arafa AA, Sharief AE, Ramadan AN (2012) Genotypic differences between two mungbean varieties in response to salt stress at seedling stage. Int J Agric Sci 4(7):278

Kang MS (1993) Simultaneous selection for yield and stability in crop performance trials: consequences for growers. Agron J 85(3):754–757

Kang YJ, Kim SK, Kim MY, Lestari P, Kim KH, Ha B-K, Jun TH, Hwang WJ, Lee T, Lee J (2014) Genome sequence of mungbean and insights into evolution within Vigna species. Nat Commun 5(1):1–9

Karthikeyan A, Shobhana VG, Sudha M, Raveendran M, Senthil N, Pandiyan M, Nagarajan P (2014) Mungbean yellow mosaic virus (MYMV): a threat to green gram (*Vigna radiata*) production in Asia. Int J Pest Manag 60(4):314–324

Kasettranan W, Somta P, Srinives P (2010) Mapping of quantitative trait loci controlling powdery mildew resistance in mungbean (*Vigna radiata* (L.) Wilczek). J Crop Sci Biotechnol 13(3): 155–161

Keatinge JDH, Easdown WJ, Yang RY, Chadha ML, Shanmugasundaram S (2011) Overcoming chronic malnutrition in a future warming world: the key importance of mungbean and vegetable soybean. Euphytica 180(1):129–141

Khan S, Goyal S (2009) Improvement of mungbean varieties through induced mutations. African J Plant Sci 3(8):174–180

Khan S, Siddiqui BA (1993) Chlorophyll mutations in *Vigna radiata* (L.) Wilczek. II. mutagenic effectiveness and efficiency of chemical mutagens. Pak J Bot 2(2):161–166

Khan M, Nawab K, Khan A, Baloch MS (2001) Genetic variability and correlation studies in mungbean. J Biol Sci 1:117–119

Khattak GSS, Haq MA, Ashraf M, Awan MA (2000) Inheritance of plant and flower bud colour in mungbean (*Vigna radiata* (L.) Wilczek). Songkhlanakarin J Sci Technol 22(2):249–251

Khattak GSS, Haq MA, Ashraf M, Tahir GR, Marwat EUK (2001) Detection of epistasis, and estimation of additive and dominance components of genetic variation for synchrony in pod maturity in mungbean (*Vigna radiata* (L.) Wilczek). Field Crop Res 72(3):211–219

Khattak GS, Haq MA, Ashraf M, Khan AJ, Zamir R (2002) Genetic architecture of secondary yield components in mungbean (*Vigna radiata* (L.) Wilczek). Breed Sci 52(4):235–241

Kim SK, Lee T, Kang YJ, Hwang WJ, Kim KH, Moon J-K, Kim MY, Lee S-H (2014) Genome-wide comparative analysis of flowering genes between Arabidopsis and mungbean. Genes Genomics 36(6):799–808

Kim KD, El Baidouri M, Abernathy B, Iwata-Otsubo A, Chavarro C, Gonzales M, Libault M, Grimwood J, Jackson SA (2015a) A comparative epigenomic analysis of polyploidy-derived genes in soybean and common bean. Plant Physiol 168(4):1433–1447

Kim SK, Nair RM, Lee J, Lee S-H (2015b) Genomic resources in mungbean for future breeding programs. Front Plant Sci 6:626

Kling GJ, Meyer MM Jr, Seigler D (1988) Rooting cofactors in five Acer species. J Am Soc Hortic Sci 113(2):252–257

Konzak C, Nilan R, Wagner J (1965) Efficient chemical mutagenesis. Rad Bot 5:49–70

Kumar A, Sharma KD (2009) Physiological responses and dry matter partitioning of summer mungbean (*Vigna radiata* L.) genotypes subjected to drought conditions. J Agron Crop Sci 195(4):270–277

Kumar VL, Singhal A (2009) Germinating seeds of the mung bean, *Vigna radiata* (Fabaceae), as a model for the preliminary evaluation of cytotoxic effects of drugs. Biocell 33(1):19–24

Kumar A, Parmhansh P, Prasad R (2009) Induced chlorophyll and morphological mutations in mungbean (*Vigna radiata* L. Wilczek). Legum Res 32(1):41–45

Kumar SB, Prakash M, Sathiya N et al (2012) Breeding for salinity tolerance in mung bean. APCBEE Procedia 4:30–35

Lagerverff J, Specht A (1970) Contamination of roadside soil with Cd, Ni, Pb and Zn. Environ Sci Technol 4:583–586

Lambrides CJ, Godwin ID (2007) Mungbean. In: Pulses, sugar and tuber crops. Springer, New York, pp 69–90

Lambrides CJ, Lawn RJ, Godwin ID, Manners J, Imrie BC (2000) Two genetic linkage maps of mungbean using RFLP and RAPD markers. Aust J Agric Res 51(4):415–425

Lavanya GR, Yadav L, Babu GS, Paul PJ (2011) Sodium azide mutagenic effect on biological parameters and induced genetic variability in mungbean. J Food Legumes 24(1):46–49

Lawn RJ, Cottrell A (1988) Wild mungbean and its relatives in Australia. Biologist 35:267–273

Lee YB (1980) Inheritance study on resistance to Cercospora leaf spot in mungbean. Asian Vegetable Research and Development Center, Shanhua, Taiwan

Lee YS, Lee JY, Kim DK, Yoon CY, Bak GC, Park IJ, Bang GP, Moon JK, Oh YJ, Min KS (2004) A new high-yielding mungbean cultivar, "Samgang" with lobed leaflet. Korean J Breed 36:183–184

Lourenço VM, Rodrigues PC, Pires AM, Piepho H-P (2017) A robust DF-REML framework for variance components estimation in genetic studies. Bioinformatics 33(22):3584–3594

Mahalanobis PC (1936) On the generalized distance in statistics. National Institute of Science of India

Mahalingam A, Manivannan N, Lakshmi Narayanan S, Sowmya Sree S (2018) AMMI analysis of phenotypic stability in greengram [*Vigna radiata* (L.) Wilczek] genotypes over seasons. Crop Res 53:131–134

Maheshwari SK, Krishna H (2013) Field efficacy of fungicides and bio-agents against Alternaria leaf spot of Mung bean. Ann Plant Protect Sci 21(2):364–367

Makeen K, Abrahim G, Jan A, Singh AK (2007) Genetic variability and correlations studies on yield and its components in mungbean (*Vigna radiata* (L.) Wilczek). J Agron 6(1):216–218

Maliwal GL, Paliwal KV (1982) Salt tolerance of some mungbean (*Vigna radiata*), urdbean (*Vigna mungo*) and guar (*Cyamopsis tetragonoloba*) varieties at germination and early growth stages. Legum Res 5(1):23–30

Manasa R, Bindumadhava H, Nair RM, Prasad TG, Shankar AG (2017) Screening mungbean (*Vigna radiata* L.) lines for salinity tolerance using salinity induction response technique at seedling and physiological growth assay at whole plant level. Int J Plant Anim Environ Sci. https://doi.org/10.21276/ijpaes

Mbeyagala KE, Amayo R, Obuo JP, Pandey AK, War AR, Nair RM (2017) A manual for mungbean (greengram) production in Uganda. Natl Agric Res Org 32

Medina FIS, Amano E, Tano S (2004) FNCA mutation breeding manual. Forum for Nuclear Cooperation in Asia

Menancio-Hautea D, Kumar L, Danesh D, Young ND (1992) A genome map for mungbean [*Vigna radiata* (L.) Wilczek] based on DNA genetic markers ($2n = 2x = 22$). In: Genetic maps. Cold Spring Harbor Laboratory Press, Cold Spring Harbor, NY, pp 6–259

Mishra SP, Asthana AN, Yadav L (1988) Inheritance of Cercospora leaf spot resistance in mung bean, *Vigna radiata* (L.) Wilczek. Plant Breed 100(3):228–229

Misra N, Dwivedi UN (2004) Genotypic difference in salinity tolerance of green gram cultivars. Plant Sci 166(5):1135–1142

Misra N, Gupta AK (2006) Interactive effects of sodium and calcium on proline metabolism in salt tolerant green gram cultivar. Am J Plant Physiol 1(1):1–12

Moe KT, Chung J, Cho Y, Moon J, Ku J, Jung J, Lee J, Park Y (2011) Sequence information on simple sequence repeats and single nucleotide polymorphisms through transcriptome analysis of mungbean. J Integr Plant Biol 53(1):63–73

Mohamed MH, El-Kramany MF (2005) Salinity tolerance of some mungbean varieties. J Appl Sci Res 1:78–84
Mohammadi SA, Prasanna BM (2003) Analysis of genetic diversity in crop plants—salient statistical tools and considerations. Crop Sci 43(4):1235–1248
Mohammed A (2007) Physiological aspects of mungbean plant (*Vigna radiata* L. Wilczek) in response to salt stress and gibberellic acid treatment. Res J Agric Biol Sci 3:200–213
Moradi A, Ahmadi A, Hoseinzadeh A (2009) Agronomic and physiological interaction of Mung bean (Partov) to sever and light stress in different stages. J Agric Res 12:659–671
Morinaga T, Ikegami M, Miura K (1993) The nucleotide sequence and genome structure of mung bean yellow mosaic geminivirus. Microbiol Immunol 37(6):471–476
Mounika Y (2020) Effects of mutagens on various traits in mung bean (*Vigna radiata* L. Wilczek). Int J Curr Microbiol App Sci 11:394–402
Murillo-Amador B, López-Aguilar R, Kaya C, Larrinaga-Mayoral J, Flores-Hernández A (2002) Comparative effects of NaCl and polyethylene glycol on germination, emergence and seedling growth of cowpea. J Agron Crop Sci 188(4):235–247
Murty BR, Arunachalam V (1966) The nature of divergence in relation to breeding systems in some crop plants. Indian J Genet Plant Breed 26:188–198
Musgrave ME, Gerth WA, Scheld HW, Strain BR (1988) Growth and mitochondrial respiration of mungbeans (*Phaseolus aureus* Roxb.) germinated at low pressure. Plant Physiol 86(1):19–22
Nadeem M, Li J, Yahya M, Sher A, Ma C, Wang X, Qiu L (2019) Research progress and perspective on drought stress in legumes: a review. Int J Mol Sci 20(10):2541
Naher N, Alam AK (2010) Germination, growth and nodulation of mungbean (*Vigna radiata* L.) as affected by sodium chloride. Int J Sustain Crop Prod 5(2):8–11
Naidu NV, Satyanarayana A (1991) Studies on genetic divergence over environments in mungbean (*Vigna radiata* (L.) Wilczek). Indian J Genet Plant Breed 51(04):454–460
Nair RM, Schafleitner R, Kenyon L, Srinivasan R, Easdown W, Ebert AW, Hanson P (2012) Genetic improvement of mungbean. SABRAO J Breed Genet 44:177–190
Nair RM, Götz M, Winter S, Giri RR, Boddepalli VN, Sirari A, Bains TS, Taggar GK, Dikshit HK, Aski M (2017) Identification of mungbean lines with tolerance or resistance to yellow mosaic in fields in India where different begomovirus species and different Bemisia tabaci cryptic species predominate. Eur J Plant Pathol 149(2):349–365
Nair RM, Pandey AK, War AR, Hanumantharao B, Shwe T, Alam A, Pratap A, Malik SR, Karimi R, Mbeyagala EK (2019) Biotic and abiotic constraints in mungbean production—progress in genetic improvement. Front Plant Sci 10:1340
Norcini JG, Heuser CW, Hamilton RH (1985) Changes in free and conjugated indole-3-acetic acid during initiation and early development of adventitious roots in mung bean. J Am Soc Hortic Sci. https://doi.org/10.21273/JASHS.110.4.528
Nwangburuka CC, Denton OA (2012) Heritability, character association and genetic advance in six agronomic and yield related characters in leaf *Corchorus olitorius*. Int J Agric Res 7(7):367–375
Oelke EA, Oplinger ES, Putnam DH, Durgan BR, Doll JD, Undersander DJ (1990) Millets, alternative field crops manual. University of Wisconsin and University of Minnesota Cooperative Extension
Osdaghi E (2014) Occurrence of common bacterial blight on mungbean (*Vigna radiata*) in Iran caused by *Xanthomonas axonopodis* pv. *phaseoli*. New Dis Rep 30(9):588–2044
Pal SS, Dhaliwal HS, Bains SS (1991) Inheritance of resistance to yellow mosaic virus in some Vigna species. Plant Breed 106(2):168–171
Pandey AK, Burlakoti RR, Kenyon L, Nair RM (2018) Perspectives and challenges for sustainable management of fungal diseases of mungbean [*Vigna radiata* (L.) R. Wilczek var. radiata]: a review. Front Environ Sci 6:53
Pathak R (2015) Clusterbean: physiology, genetics and cultivation. Springer, New York
Pawlowska TE, Charvat I (2004) Heavy-metal stress and developmental patterns of arbuscular mycorrhizal fungi. Appl Environ Microbiol 70(11):6643–6649

Pratap A, Sen Gupta D, Singh BB, Kumar S (2013) Development of super early genotypes in greengram [*Vigna radiata* (L.) Wilczek]. Legum Res 36(2)

Pratap A, Douglas C, Prajapati U, Kumari G, War AR, Tomar R, Pandey AK, Dubey S (2020) Breeding progress and future challenges: biotic stresses. In: The mungbean genome. Springer, New York, pp 55–80

Priya AJP, SudhaLakshmi GY, Banu F, Gopalakrishnan S, Dhanalakshmi P, Sagadevan E, Manimaran A, Arumugam P (2012) Phytochemical screening and antibacterial activity of *Vigna radiata* L. against bacterial pathogens involved in food spoilage and food borne diseases. J Acad Ind Res 1(6):355–359

Promila K, Kumar S (2000) *Vigna radiata* seed germination under salinity. Biol Plant 43(3): 423–426

Rachaputi RCN, Chauhan Y, Douglas C, Martin W, Krosch S, Agius P, King K (2015) Physiological basis of yield variation in response to row spacing and plant density of mungbean grown in subtropical environments. Field Crop Res 183:14–22

Rahim MA, Mia AA, Mahmud F, Zeba N, Afrin KS (2010) Genetic variability, character association and genetic divergence in Mungbean ('*Vigna radiata*' L. Wilczek). Plant Omics 3(1):1–6

Raina A, Laskar RA, Khursheed S, Amin R, Tantray YR, Parveen K, Khan S (2016) Role of mutation breeding in crop improvement-past, present and future. Asian Res J Agric 2(2):1–13

Raina A, Laskar RA, Khursheed S, Khan S, Parveen K, Amin R (2017) Induce physical and chemical mutagenesis for improvement of yield attributing traits and their correlation analysis in chickpea. Int Lett Nat Sci 61:14–22

Raje RS, Rao SK (2000) Association analysis for yield and its components in mungbean (*Vigna radiata* L. Wilczek). Legum Res 23(1):42–48

Ramana MV, Singh DP (1987) Genetic divergence in mungbean. Genome 30:835–837

Ramanujam S, Tiwari AS, Mehra RB (1974) Genetic divergence and hybrid performance in mung bean. Theor Appl Genet 45(5):211–214

Ramesh CK, Rehman A, Prabhakar BT, Vijay Avin BR, Aditya Rao SJ (2011) Antioxidant potentials in sprouts vs. seeds of *Vigna radiata* and *Macrotyloma uniflorum*. J Appl Pharm Sci 1(07):99–103

Randhir R, Lin Y-T, Shetty K (2004) Stimulation of phenolics, antioxidant and antimicrobial activities in dark germinated mung bean sprouts in response to peptide and phytochemical elicitors. Process Biochem 39(5):637–646

Raturi A, Singh SK, Sharma V, Pathak R (2012a) Molecular characterization of *Vigna radiata* (L.) Wilczek genotypes based on nuclear ribosomal DNA and RAPD polymorphism. Mol Biol Rep 39(3):2455–2465

Raturi A, Singh SK, Sharma V, Pathak R (2012b) Stability and environmental indices analyses for yield attributing traits in Indian *Vigna radiata* genotypes under arid conditions. Asian J Agric Sci 4(2):126–133

Raza MH, Sadozai GU, Baloch MS, Khan EA, Din I, Wasim K (2012) Effect of irrigation levels on growth and yield of mungbean. Pak J Nutr 11(10):876

Reddy CV (1982) Note on the effect of saline water irrigation on green gram. Curr Agric 6:183–185

Reddy KS (2009) A new mutant for yellow mosaic virus resistance in Mungbean (*Vigna radiata* (L.) Wilczek) variety SML-668 by Recurrent Gamma-ray Irradiation. In: Induced plant mutation in the genomics era, pp 361–362

Revanappa S, Kajjidoni ST (2004) Genotype x environment interaction for seed yield and its components in advance breeding lines of blackgram (*Vigna mungo* L. Hepper). Legum Res 35: 1–2

Robertson MJ, Carberry PS, Huth NI, Turpin JE, Probert ME, Poulton PL, Bell M, Wright GC, Yeates SJ, Brinsmead RB (2002) Simulation of growth and development of diverse legume species in APSIM. Aust J Agric Res 53(4):429–446

Rosenzweig C, Elliott J, Deryng D, Ruane AC, Müller C, Arneth A, Boote KJ, Folberth C, Glotter M, Khabarov N (2014) Assessing agricultural risks of climate change in the 21st century in a global gridded crop model intercomparison. Proc Natl Acad Sci 111(9):3268–3273

Roy D (2000) Plant breeding: analysis and exploitation of variation. Alpha Science Int'l Ltd.

Rungnoi O, Chanprem S, Toojinda T, Godwin I, Lambrides C, Srinives P (2010) Characterization, inheritance, and molecular study of opaque leaf mutant in mungbean (*Vigna radiata* (L.) Wilczek). J Crop Sci Biotechnol 13(4):219–226

Sagare DB, Mohanty IC (2015) In vitro regeneration system in green gram (*Vigna radiata* L., Wilczek, cv. Sujata): a recalcitrant legume crop. Res J Agric Sci 6(1):64–67

Saha P, Chatterjee P, Biswas AK (2010) NaCl pretreatment alleviates salt stress by enhancement of antioxidant defense system and osmolyte accumulation in mungbean (*Vigna radiata* L. Wilczek). Indian J Exp Biol 48(6):593–600

Saini R, Singh RP, Jaiwal PK (2007) Agrobacterium tumefaciens mediated transfer of Phaseolus vulgaris α-amylase inhibitor-1 gene into mungbean *Vigna radiata* (L.) Wilczek using bar as selectable marker. Plant Cell Rep 26(2):187–198

Salman MAS, Anuradha C, Sridhar V, Babu ER, Pushpavalli S (2021) Genetic variability for yield and its related traits in green gram [*Vigna radiata* (L.) Wilczek]. Legum Res 1:5

Saminathan B (2013) Mercury induced changes in germination and biochemical analysis of green gram (*Vigna radiata* (L.) Wilczek). Int J Environ Biol 3(4):151–156

Sandhu TS, Brar JS, Sandhu SS, Verma MM (1985) Inheritance of resistance to mungbean yellow mosaic virus in green gram. J Res Punjab Agric Univ 22(4):607–611

Sangiri C, Kaga A, Tomooka N (2008) Genetic diversity of the mung bean (*Vigna radiata*, Leguminosae) genepool on the basis of microsatellite analysis. Aust J Bot 55:837–847

Sangsiri C, Sorajjapinun W, Srinives P (2005) Gamma radiation induced mutations in mungbean. Sci Asia 31:251–255

Sangsiri C, Sorajjapinun W, Srinives P (2007) Inheritance and ultrastructure of variegated leaf mutant in mungbean (*Vigna radiata* (L.) Wilczek). Thai J Agric Sci 40:159–166

Sangwan NK, Verma BS, Verma KK, Dhindsa KS (1990) Nematicidal activity of some essential plant oils. Pestic Sci 28(3):331–335

Saradhi PP (1991) Proline accumulation under heavy metal stress. J Plant Physiol 138(5):554–558

Saraswat S, Yadav AK, Sirohi P, Singh NK (2017) Role of epigenetics in crop improvement: water and heat stress. J Plant Biol 60(3):231–240

Sarkar M, Kundagrami S (2018) Selection of high yielding, extra short duration lines of mungbean derived through gamma radiation. Indian J Genet 78(2):233–241

Saxena R, Reddy DDR (1987) Crop losses in pigeonpea and mungbean by pigeonpea cyst nematode, *Heterodera cajani*. Indian J Nematol 17(1):91–94

Schafleitner R, Nair RM, Rathore A, Wang Y, Lin C, Chu S, Lin P, Chang J-C, Ebert AW (2015) The AVRDC–The World Vegetable Center mungbean (*Vigna radiata*) core and mini core collections. BMC Genomics 16(1):1–11

Sehrawat N, Jaiwal PK, Yadav M, Bhat KV, Sairam RK (2013) Salinity stress restraining mungbean (*Vigna radiata* (L.) Wilczek) production: gateway for genetic improvement. Int J Agric Crop Sci 6(9):505

Sen M, De DK (2017) Genetic divergence in mung bean. Legum Res 40(1):16–21

Shah TM, Mirza JI, Haq MA, Atta BM (2008) Induced genetic variability in chickpea (*Cicer arietinum* L.). II. Comparative mutagenic effectiveness and efficiency of physical and chemical mutagens. Pak J Bot 40(2):605–613

Shanmugasundaram S, Keatinge JDH, Hughes J (2009) Counting on beans: mungbean improvement in Asia. IFPRI

Sharma RN (1999) Heritability and character association in non segregating populations of Mungbean (*Vigna radiata* L. Wilczek). J Interacademicia 3(1):5–10

Sharma L, Priya M, Bindumadhava H, Nair RM, Nayyar H (2016) Influence of high temperature stress on growth, phenology and yield performance of mungbean [*Vigna radiata* (L.) Wilczek] under managed growth conditions. Sci Hortic 213:379–391

Sharma P, Gautam A, Kumar V, Khosla R, Guleria P (2021) Naringenin reduces Cd-induced toxicity in *Vigna radiata* (mungbean). Plant Stress 1:100005

Shukla V, Baghel S, Maravi K, Singh SK (2014) Yield loss assessment in mungbean [*Vigna radiata* (L.) Wilczek] caused by anthracnose [*Colletotrichum truncatum* (schw.) Andrus and Moore]. Bioscan 9(3 Suppl):1233–1235

Siddiqui ZA, Mahmood I (1996) Biological control of plant parasitic nematodes by fungi: a review. Bioresour Technol 58(3):229–239

Sikora P, Chawade A, Larsson M, Olsson J, Olsson O (2011) Mutagenesis as a tool in plant genetics, functional genomics, and breeding. Int J Plant Genomics 2011:314829

Singh V, Chaturvedi S (1982) Quantitative variation in mung bean induced by EMS, NMU, gamma rays and their combined treatments. J Cytol Genet 17:47–52

Singh J, Kumar P (2014) The interrelationship of various traits with seed yield in mungbean (*Vigna radiata* (L.) Wilczek). Electron J Plant Breed 5(4):869–874

Singh B, Rao GT (2007) Induced chlorophyll mutations in greengram (*Vigna radiata* (L) Wilczek). Legum Res 30(4):308–310

Singh A, Singh AK (2011) Fruits and vegetables with longer shelf-life for extended availability in the markets. In: Horti-Business. Delhi Agri-Horticultural Society, pp 131–135

Singh HB, Joshi BS, Thomas TA (1970) In: Kachroo P (ed) Pulse crops of India. ICAR, New Delhi

Singh BR, Chandra S, Ram S (2000) Evaluation of mungbean varieties against yellow mosaic virus. Ann Plant Prot Sci 8:270–271

Singh SK, Singh IP, Singh BB, Singh O (2009) Stability analysis in mungbean (*Vigna radiata* (L.) Wilczek). Legum Res 32(2):108–112

Singh CM, Singh P, Tiwari C, Purwar S, Kumar M, Pratap A, Singh S, Chugh V, Mishra AK (2021) Improving drought tolerance in mungbean (*Vigna radiata* L. Wilczek): morpho-physiological, biochemical and molecular perspectives. Agronomy 11(8):1534

Sivakumar P, Gnanam R, Ramakrishnan K, Manickam A (2010) Somatic embryogenesis and regeneration of *Vigna radiata*. Biol Plant 54(2):245–251

Smartt J (1985) Evolution of grain legumes. III. Pulses in the genus Vigna. Exp Agric 21(2):87–100

Smejkalova M, Mikanova O, Boruvka L (2003) Effects of heavy metal concentrations on biological activity of soil micro-organisms. Plant Soil Environ 49(7):321–326

Solanki DS, Kumar S, Parihar K, Sharma K, Gehlot P, Singh SK, Pathak R (2018) Purification and characterization of a novel thermostable antifungal protein with chitinase activity from mung bean. J Environ Biol 39(3):406–412

Somta P, Ammaranan C, Ooi PA-C, Srinives P (2007) Inheritance of seed resistance to bruchids in cultivated mungbean (*Vigna radiata*, L. Wilczek). Euphytica 155(1):47–55

Somta P, Musch W, Kongsamai B, Chanprame S, Nakasathien S, Toojinda T, Sorajjapinun W, Seehalak W, Tragoonrung S, Srinives P (2008) New microsatellite markers isolated from mungbean (*Vigna radiata* (L.) Wilczek). Mol Ecol Resour 8(5):1155–1157

Srinives P, Somta P, Somta C (2007) Genetics and breeding of resistance to bruchids (*Callosobruchus* spp.) in Vigna crops: a review. NU Int J Sci 4(1):1–17

Subramanian D (1980) Effect of gamma-radiation in Vigna [cowpeas, India]. Indian J Genet Plant Breed 40(1):187–194

Summerfield RJ, Lawn RJ (1987) Environmental modulation of flowering in mung bean (*Vigna radiata*): a reappraisal. Exp Agric 23(4):461–470

Sun S, Zhi Y, Zhu Z, Jin J, Duan C, Wu X, Xiaoming W (2017) An emerging disease caused by *Pseudomonas syringae* pv. *phaseolicola* threatens mung bean production in China. Plant Dis 101(1):95–102

Swaminathan R, Singh K, Nepalia V (2012) Insect-pests of green gram *Vigna radiata* (L.) Wilczek and their management. Agric Sci 10:197–222

Tah PR (2006) Studies on gamma ray induced mutations in mungbean [*Vigna radiata* (L.) Wilczek]. Asian J Plant Sci 5(1):61–70

Tah PR, Saxena S (2009) Induced synchrony in pod maturity in mungbean (*Vigna radiata*). Int J Agric Biol 11(3):321–324

Talekar NS (1990) Agromyzid flies of food legumes in the tropics. Wiley Eastern Ltd., New Delhi

Tan W, Tan G, Walton PD (1979) Regression analysis of genotype-environment interaction in smooth bromegrass 1. Crop Sci 19(3):393–396

Tang D, Dong Y, Ren H, Li L, He C (2014) A review of phytochemistry, metabolite changes, and medicinal uses of the common food mung bean and its sprouts (*Vigna radiata*). Chem Cent J 8(1):1–9

Tavakkoli E, Fatehi F, Coventry S, Rengasamy P, McDonald GK (2011) Additive effects of Na^+ and Cl^- ions on barley growth under salinity stress. J Exp Bot 62(6):2189–2203

TeKrony DM, Egli DB (1991) Relationship of seed vigor to crop yield: a review. Crop Sci 31(3): 816–822

Tomooka N, Lairungreang C, Nakeeraks P, Egawa Y, Thavarasook C (1992) Development of bruchid-resistant mungbean line using wild mungbean germplasm in Thailand. Plant Breed 109(1):60–66

Tomooka N, Vaughan DA, Kaga A (2005) Mungbean [*Vigna radiata* (L.) Wilczek]. In: Genetic resources, chromosome engineering, and crop improvement: grain legumes, vol 1, pp 325–345

Tripathi K, Meena SK, Panwar BS, Lal H, Rana JC, Singh K (2020) Understanding genetic variability in the mungbean (*Vigna radiata* L.) genepool. Ann Appl Biol 177(3):346–357

Tripepi RR, Heuser CW, Shannon JC (1983) Incorporation of tritiated thymidine and uridine into adventitious-root initial cells of *Vigna radiata*. J Am Soc Hortic Sci 108(3):469–474

Upadhaya LP, Singh RB, Agarwal RK (1980) Character associations in greengram populations of different maturity groups. Indian J Agric Sci 50(6):473–476

Vairam N, Lavanya SA, Muthamilan M, Vanniarajan C (2016) Screening of M3 mutants for yellow vein mosaic virus resistance in greengram [*Vigna radiata* (L.) Wilczek]. Int J Plant Sci 11(2): 265–269

Varshney RK, Kudapa H, Pazhamala L, Chitikineni A, Thudi M, Bohra A, Gaur PM, Janila P, Fikre A, Kimurto P (2015) Translational genomics in agriculture: some examples in grain legumes. Crit Rev Plant Sci 34(1–3):169–194

Vavilov NI (1951) The origin, variation, immunity and breeding of cultivated plants. Agron J 72(6): 102

Verma SK, Tuteja OP, Monga D (2008) Evaluation for genotype × environment interactions in relation to stable genetic male sterility-based Asiatic cotton (*Gossypium arboreum*) hybrids of north zone. Indian J Agric Sci 78(4):375–378

Vijayan S, Kirti PB (2012) Mungbean plants expressing BjNPR1 exhibit enhanced resistance against the seedling rot pathogen, Rhizoctonia solani. Transgenic Res 21(1):193–200

Viña AC, Tomooka N (1994) Genetic diversity in mungbean (*Vigna radiata* (L) Wilczek) based on two enzyme systems. Philippine J Crop Sci 19(1):1–9

Wahid A, Hameed M, Rasul E (2004) Salt-induced injury symptoms, changes in nutrient and pigment composition, and yield characteristics of mungbean. Int J Agric Biol 6:1143–1152

Wahid A, Gelani S, Ashraf M, Foolad MR (2007) Heat tolerance in plants: an overview. Environ Exp Bot 61(3):199–223

Wahid A, Ghani A, Javed F (2008) Effect of cadmium on photosynthesis, nutrition and growth of mungbean. Agron Sustain Dev 28(2):273–280

Wang L, Wu C, Zhong M, Zhao D, Mei L, Chen H, Wang S, Liu C, Cheng X (2016) Construction of an integrated map and location of a bruchid resistance gene in mung bean. Crop J 4(5): 360–366

Wani MR (2006) Estimates of genetic variability in mutated populations and the scope of selection for yield attributes in *Vigna radiata* (L.) Wilczek. Egypt J Biol 8

Wani MR, Dar AR, Tak A, Amin I, Shah NH, Rehman R, Baba MY, Raina A, Laskar R, Kozgar MI (2017) Chemo-induced pod and seed mutants in mungbean (*Vigna radiata* L. Wilczek). SAARC J Agric 15(2):57–67

Waniale A, Wanyera N, Talwana H (2014) Morphological and agronomic traits variations for mungbean variety selection and improvement in Uganda. Afr Crop Sci J 22(2):123–136

Win KS, Win S, Htun TM, Win NKK, Oo KS (2021) Inheritance and gene effect of resistance to mungbean yellow mosaic virus (MYMV) in Mungbean [*Vigna radiata* (L.) wilczek]. Legum Res 1:5

Wood BA, Easdown WJ (1990) A new bacterial disease of mung bean and cowpea for Australia. Australas Plant Pathol 19(1):16–21

Xin C, Sorajjapinun W, Reiwthongchum S, Srinives P (2003) Identification of parental mungbean lines for production of hybrid varieties. CMU J 2(2):97–105

Yadav SK, Krishna MG, Maheswari M, Vanaia M, Venkateswarlu B (2010) High frequency induction of multiple shoots and plant regeneration from cotyledonary nodal explant of mung bean [*Vigna radiata* (L) Wilczek]. J Plant Biochem Biotechnol 19(2):267–270

Yadav SK, Katikala S, Yellisetty V, Kannepalle A, Narayana JL, Maddi V, Mandapaka M, Shanker AK, Bandi V, Bharadwaja KP (2012) Optimization of Agrobacterium mediated genetic transformation of cotyledonary node explants of *Vigna radiata*. Springerplus 1(1):1–8

Yang JK, Yuan TY, Zhang WT, Zhou JC, Li YG (2008) Polyphasic characterization of mung bean (*Vigna radiata* L.) rhizobia from different geographical regions of China. Soil Biol Biochem 40(7):1681–1688

Yaqub M, Mahmood T, Akhtar M, Iqbal MM, Ali S (2010) Induction of mungbean [*Vigna radiata* (L.) Wilczek] as a grain legume in the annual rice-wheat double cropping system. Pak J Bot 42(5):3125–3135

Ye H, Liu S, Tang B, Chen J, Xie Z, Nolan TM, Jiang H, Guo H, Lin H-Y, Li L (2017) RD26 mediates crosstalk between drought and brassinosteroid signalling pathways. Nat Commun 8(1):1–13

Yimram T, Somta P, Srinives P (2009) Genetic variation in cultivated mungbean germplasm and its implication in breeding for high yield. Field Crop Res 112(2–3):260–266

Young ND, Kumar L, Menancio-Hautea D, Danesh D, Talekar NS, Shanmugasundarum S, Kim DH (1992) RFLP mapping of a major bruchid resistance gene in mungbean (*Vigna radiata*, L. Wilczek). Theor Appl Genet 84(7):839–844

Yu K, Park SJ, Poysa V (1999) Abundance and variation of microsatellite DNA sequences in beans (Phaseolus and Vigna). Genome 42(1):27–34

Yundaeng C, Somta P, Chen J, Yuan X, Chankaew S, Chen X (2021) Fine mapping of QTL conferring Cercospora leaf spot disease resistance in mungbean revealed TAF5 as candidate gene for the resistance. Theor Appl Genet 134(2):701–714

Molecular Characterization and Mapping of Stress Resistance Genes Using SNP Platform in Legumes

6

Kandiah Pakeerathan

Abstract

Legumes play a vital role in agriculture and food security. Biotic and abiotic stresses are major hurdles for legume production and lower the current productivity per unit area. There is an obligation to accelerate genetic improvement of most food legumes by introducing alleles conferring resistance to pests and pathogens, adaptation to abiotic stresses, and high yield potential. The tapping of potential resistance alleles present in the landraces and wild relatives and its exploitation in legume resistance breeding programs with the aid of next-generation molecular breeding approaches are the quickest ways to develop high-yielding elite legume varieties with long-lasting resistance. This chapter attempts to explore the advanced molecular approaches in germplasm characterization, marker-assisted genomic selection, molecular mapping of biotic stress resistance gene(s)/QTLs using single nucleotide polymorphism (SNP) markers, mining of SNPs using various next-generation sequencing (NGS) platforms, marker-assisted selection, and marker-assisted pyramiding of resistance genes in elite germplasm. This chapter also highlights major qualitative/quantitative resistant trait/loci and linked SNP markers, and recently published highly saturated SNP(s) linkage/consensus map information of 13 important food legumes. Genetic, genomic, and marker information elucidated in this chapter will be a guide to the researchers and students who are interested in advanced molecular plant breeding and to address the global challenge of ensuring food security in the face of scarce natural resources and unpredicted climate change-induced stress.

K. Pakeerathan (✉)
Department of Agricultural Biology, Faculty of Agriculture, University of Jaffna, Kilinochchi, Sri Lanka
e-mail: pakeerathank@univ.jfn.ac.lk

P. Muthu Arjuna Samy et al. (eds.), *Legumes: Physiology and Molecular Biology of Abiotic Stress Tolerance*, https://doi.org/10.1007/978-981-19-5817-5_6

Keywords

Biotic stress · Gene · Genome · Legumes · Molecular mapping · Molecular markers · QTLs · SNP · Stress resistance

6.1 Introduction to Legumes

There is an 80% probability that the world population will hit 9.6 billion in the year 2030, and 12.3 billion in 2100 (Gerland et al. 2014). Therefore, the continuously increasing population constantly demands to increase the current food production by 2% every year to double food production in 2050 (Janni et al. 2020). The land under cultivation is limited, and the only option is to increase the productivity of rice, wheat, and food legumes per unit land area through intensive cultivation of genetically improved high-yielding varieties.

Legumes belonging to the family Leguminosae/Fabaceae are important food crops all around the globe. Legumes containing over 18,000 species are divided into three subfamilies Mimosoideae, Caesalpinioideae, and Papilionoideae (Varshney et al. 2007, 2009). A variety of essential amino acid-rich domesticated food legumes have been cultivated, for centuries, to satisfy the 33–35% of human's dietary protein requirement (Sharma et al. 2013; Van Kessel and Hartley 2000; Vance et al. 2000). Due to the richest nutritional value, legumes contribute 27% of the world's primary crop production with a cultivation extent of more than 15% of the total arable land (<180 million hectares) (source: FAO Database). Therefore, legumes are cultivated as the third-rank crop next to cereals and oilseeds and play a vital role in the sustainability of the environment, agriculture, and animal production, and human health and food security (Kudapa et al. 2013; Vance et al. 2000; Varshney et al. 2007).

Among the food legumes, soybean (*Glycine max*), common bean (*Phaseolus vulgaris*), pigeon pea (*Cajanus cajan*), peanut or groundnut (*Arachis hypogaea*), cowpea (*Vigna unguiculata*), chickpea (*Cicer arietinum*), pea (*Pisum sativum*), mung bean (*Vigna radiata*), lentil (*Lens culinaris*), faba bean (*Vicia faba*), and lupin (*Lupinus luteus*) constitute important components of the human diet in developing countries (FAO 2010). Despite having an important role in food security, the majority of these legume crops demonstrate low productivity due to biotic and abiotic stresses (Dwivedi et al. 2017). For example, drought is an important abiotic stress constraint, and major biotic stresses include cyst nematode in soybean, anthracnose, angular leaf spot, bean rust, bacterial blight in common bean, Ascochyta blight, and Fusarium wilt in chickpea (Fritsche-Neto et al. 2019; Garg et al. 2018). Thus, it is necessary to enhance our genetic knowledge of specific aspects of defense/stress responses of germplasm to improve crop productivity. Towards this aim, emerging genomics technology can be applied to identify candidate genes or key loci controlling stress tolerance or resistance (Kankanala et al. 2019). Subsequently, these genes can be used in genetic modification or molecular breeding programs to develop improved varieties with enhanced resistance/tolerance to stress (Kudapa et al. 2013).

The narrow genetic base of cultivars coupled with low utilization of genetic resources is the major factor limiting grain legume production and productivity globally (Sharma et al. 2013; Upadhyaya et al. 2010; Varshney et al. 2009). It is therefore important to identify genes in food legumes conferring resistance to biotic stresses and tolerance to abiotic stresses that can be used both to understand molecular mechanisms of plant response to the environment and to exploit new and diverse sources for the genetic enhancement of grain legumes (Pandey et al. 2012, 2016; Varshney et al. 2007). Wild progenitors with enhanced levels of resistance/tolerance to multiple stresses provide important sources of genetic diversity for crop improvement (Varshney et al. 2007, 2019). However, their exploitation for cultivar improvement is limited by cross-incompatibility barriers and linkage drags. Pre-breeding provides a unique opportunity through the introgression of desirable genes from wild germplasm into genetic backgrounds readily used by the breeders with minimum linkage drag (Upadhyaya et al. 2010; Varshney et al. 2017). To overcome these bottlenecks, pre-breeding activities using promising landraces, wild relatives, and popular cultivars have been initiated at the International Crops Research Institute for the Semi-Arid Tropics (ICRISAT) to develop new gene pools in food legumes with a high frequency of useful genes, wider adaptability, and a broad genetic base (Sharma et al. 2013). The availability of genomic and molecular marker information will greatly assist in reducing linkage drags and increasing the efficiency of introgression (Sharma et al. 2013). Recent advances in genomics offer a range of approaches such as the sequencing and resequencing of genomes and transcriptomes, gene expression microarray as well as RNA-seq-based gene expression profiling, and map-based cloning for the identification and isolation of biotic and abiotic stress-responsive genes in several crop legumes (Kudapa et al. 2013; Pandey et al. 2016; Varshney et al. 2020). These candidate biotic stress-associated genes should provide insights into the molecular mechanisms of gene expression when host-pathogen interact, exact gene location, and tightly linked SNP markers to develop resistance gene(s)-introgressed legume varieties to reduce pesticide use and increase productivity.

6.1.1 Stress Resistance in Legumes

Biotic and abiotic stresses are major hurdles of crop legume cultivation. Plant tolerance and plant resistance are ways that plants deal with stressors in their environment. Resistance and tolerance are plants' best defense mechanisms. At the most basic level, the difference between tolerance and resistance is related to how the plant defends itself. Tolerance means plant's strategies that help it to survive despite dangers within their local environment. Contrastingly, plant resistance starts at the environmental or genetic level (Agrios 2005).

6.1.2 Tolerance

Tolerance is a plant's ability to grow and produce an acceptable yield despite a pest attack. Tolerance is typically attributed to plant vigor, regrowth of damaged tissue, and a plant's ability to produce additional stems/branches—factors that enable a plant to avoid, tolerate, or recover from damage from inclement weather, pests, or herbivores, under conditions that would typically cause a greater amount of injury to other plants of the same species (Acquaah 2007).

6.1.3 Resistance

It means that a plant is immunized from a particular stressor—typically, a biotrophic pathogen infection. The host (i.e., the plant) has a resistance gene that prevents the proliferation of the pathogen, while a pathogen typically contains an avirulence gene that triggers plant immunity. Two main types of resistance exist: ecological resistance/pseudo-resistance [host evasion, induced resistance, escape] and genetic resistance: (a) resistance based on the number of genes [monogenic, oligogenic, polygenic] and (b) resistance based on the biotype reaction [vertical resistance and horizontal resistance] (Agrios 2005).

6.2 Breeding Strategies for Characterization of Stress Resistance Genes

6.2.1 Germplasm Characterization

Breeding for crop improvement in resistance/tolerance involves the transfer of genes from one genetic background to another, or combining genes from different sources with the hope that the new cultivar will combine the best of both parents while being distinct from both. When a plant breeder has decided on the combination of traits that are to be incorporated in a new cultivar to be developed, the next crucial step is to find donor sources consisting of an appropriate gene(s) for desired characters (Varshney et al. 2009). In the early conventional crop domestication and breeding, breeders targeted to improve the only yield component of the cultivars; therefore, early improved cultivar's genetic diversity is narrowed for the specific traits governing biotic and abiotic stresses (Pandey et al. 2012).

Since the sixteenth century, individual or group botanists have collected 7.4 million germplasms with diversified characters that are being conserved and managed in more than 1750 national and international "gene banks" (germplasm repositories/germplasm banks) (FAO 2010). Among the collections, 13% are in the 11 of the CGIAR centers' germplasm collections such as Biodiversity International, CIAT, CIMMYT, CIP, ICARDA, the World Agroforestry Center (formerly ICRAF), ICRISAT, IITA, ILRI, INIBAP, IRRI, and AfricaRice (formerly WARDA) and are managed on behalf of the world community (FAO 2010). Among the total accessions collected, 17% are food legumes (Table 6.1). ICRISAT maintains the world's largest collections of chickpea (20%) and groundnut (12%), and ICARDA houses the world's largest collections of lentil (19%), faba bean (21%), and vetches (16%) (Pandey et al. 2012; Upadhyaya et al. 2010). CIAT is responsible for the world's largest collections of beans (14%). The nature of the accessions (for example whether they comprise advanced cultivars, breeding lines, landraces, wild relatives, etc.) is known for about half of the material conserved ex situ, and of these, about 17% are advanced cultivars, 22% breeding lines, 44% landraces, and 17% wild or weedy species (FAO 2010).

The tapping of those conserved ancestors, wild types, and tribes consisting of rich sources of diverse alleles may be vital for future crop improvement. From conserved germplasms, <2% has been characterized, and few uses have led to major crop improvements due to the limited or no availability of potential trait or other genetic information (Upadhyaya et al. 2010; Varshney et al. 2020). Since the economic value of a cultivar is determined by its phenotypic characteristics, to identify the desired gene of interest present in those germplasms, mass-level in situ and ex situ screening is expected, and good knowledge of the genetic constitution of the sources showing clear phenotypic differences is further taken into consideration in genetic evaluation/genotyping to facilitate crop breeding (Leng et al. 2017; Saeed and Darvishzadeh 2017). In the OMICs era, with the availability of closely linked marker(s), there are possibilities to characterize the desired allele or locus without mass-level in situ or ex situ screening. The available genomic information and genome-wide association studies facilitate to precisely dissect the genetic architecture of plant traits, and mining and genome-wide search of SNP markers, large-scale mapping of agronomically important quantitative trait loci, gene cloning and characterization, mining of elite alleles/haplotypes, and exploitation of natural variations (Leng et al. 2017).

In the absence of the desired trait in diversified germplasm, mutation is an indispensable option to create new stress resistance allele(s) in the germplasms. Most mutations are recessive, so the selection is made in M2, and for polygenic traits, the selection is done in M3. Beneficial mutations occur at very low frequencies (Tran et al. 2020). Natural mutations are randomly induced and are recurrent. Mutations have mostly pleiotropic effects (Johannes and Schmitz 2019). Mutations are also being created artificially using both radiation and chemical mutagens. Mutation through radiation can be categorized into, 1. Particulate radiation: Germplasm are mutated using alpha rays, beta rays, and fast and thermal neutrons; 2. Non-particulate radiation: X-rays, gamma rays, and Non-ionizing ultraviolet (UV) rays are used to mutate the germplasm. In the chemical method of mutation, a) Alkylating agents: sulfur and nitrogen mustards, epoxides, ethyl methane sulphonate (EMS), methyl methane sulfonate (MMS), etc.; b) Acridine dyes: acriflavine, proflavine, Acridine orange, Acridine yellow, ethidium bromide, Base analogs: 5-bromouracil, 5-chlorouracil; c) Others chemical agents: nitrous acid, hydroxylamine, sodium azide are used as chemical mutagens (Acquaah 2007).

Table 6.1 Summary of germplasm and genomic information of 13 important food legumes

Crop	Number of accessions available	Ploidy level and genome size	Genome-wide SNPs, indel markers, and structural variants (SVs)	Number of germplasm sources screened	Reference(s)
Soybean (*Glycine max*)	229,944	$2n = 40$, 1.1–1.15 Gb	205,614 SNPs	Re-sequenced a total of 17 wild and 14 cultivated soybean genomes	Lam et al. (2010)
			3.87 million high-quality single nucleotide polymorphisms (SNPs)	High-depth resequencing of 10 cultivated and 6 wild accessions	Chung et al. (2014)
			5,102,244 SNPs and 707,969 indels	25 new and 30 published whole-genome re-sequencing accessions	Li et al. (2013)
			5,835,185 SNPs and 1,329,844 indels	28 Brazilian cultivars	Maldonado dos Santos et al. (2019), Zhou et al. (2015)
			9,790,744 single nucleotide polymorphisms (SNPs) and 876,799 indels	302 re-sequenced wild, landrace, and improved soybean accessions	
			180,961 SNPs containing Axiom® Soya SoyaSNP array	High-depth genome re-sequencing of 16 soybean accessions and low-depth genome re-sequencing of 31 soybean accessions, 47 accessions	Lee et al. (2015)
			10 million SNPs, including 35% not previously reported	106 accessions representing wild, landraces, and elite lines	Valliyodan et al. (2016)
			SoySNP50K array	6 cultivated and 2 wild accessions	Song et al. (2013)
			SoySNP50K array	18,480 domesticated soybean and 1168 wild soybean accessions	Song et al. (2015a)
			SoySNP6K BeadChip array containing 5376 SNPs	92 RILs involving soybean cultivars "Maryland 96-5722" and "Spencer"	Akond et al. (2013)

			SoySNP1.5K chip array GoldenGate assay	Selected from 2435 random SNPs evenly covering the genome from the Soybean SNP database	Shen et al. (2012)
			BARCSoySNP6K	Clustering of the 96 wild, 96 landrace, and 96 elite cultivars	Song et al. (2020)
Common bean (*Phaseolus vulgaris*)	261,963	$2n = 22$, 32,970 Mbp	6286 DArT Seq high-density SNPs	188 accessions, including landraces and cultivars from Andean and Mesoamerican gene pools	Valdisser et al. (2017)
			44,875 SNPs, 3633 indels tagging 11,027 genes	18 genotypes of wild and domesticated accessions	Ariani et al. (2016)
			768-SNP Illumina GoldenGate assay	Six common bean and two tepary bean accessions	Gujaria-Verma et al. (2016)
			BARCBean6K_3 BeadChip containing 6000 SNPs	Genotyping 365 dry bean and 134 snap bean accessions	Song et al. (2015b)
			84,416 SNPs	363 common bean accessions form USDA core collection	Wen et al. (2019)
Pigeon pea (*Cajanus cajan*)	40,820	$2n = 2 \times = 22$, 833.07 Mbp	4,686,422 SNPs and 779,254 indels	20 accessions belonging to primary and secondary gene pools	Kumar et al. (2016)
			Whole-genome re-sequence 15.1 million SNPs, 0.9 million small insertions, and 1.2 million small deletions (indels of 1–5 bp in length)	292 accessions from the reference set, including 117 breeding lines, 166 landraces, 2 others, and 7 accessions from three wild species	Sinha et al. (2020)
			62 K SNP chip array "CcSNPnks" for Affymetrix GeneTitan® platform	Re-sequencing of 45 diverse genotypes	Sinha et al. (2020)
Groundnut (*Arachis hypogaea*)	128,435	$2n = 2x = 20$, 2890 Mbp (cultivated)	Affymetrix 60 K SNP array ±	20 cultivated accessions	Clevenger et al. (2017)

(continued)

Table 6.1 (continued)

Crop	Number of accessions available	Ploidy level and genome size	Genome-wide SNPs, indel markers, and structural variants (SVs)	Number of germplasm sources screened	Reference(s)
		$2n = 4x = 40$, 1260 Mbp (diploid genome)	High-density SNP array "Axiom_Arachis" with 58 K SNPs	RNA-sequencing of 41 groundnut accessions and wild diploid ancestors	Pandey et al. (2017a)
Cowpea (*Vigna unguiculata*)	65,323	$2n = 2x = 22$, 620 Mb	1048 SNPs	768 accessions	Xiong et al. (2016)
			Illumina Cowpea iSelect Consortium Array with 51,128 SNPs	36 diverse cowpea accessions, 5 biparental RIL populations	Muñoz-Amatriaín et al. (2017)
			20,000 DArTseq SNP	94 cowpea genotypes	Nkhoma et al. (2020)
Chickpea (*Cicer arietinum*)	98,313	$2n = 16$, 740 Mbp	2,058,566 SNPs and 292,588 indels	35 accessions representing 16 mapping populations	Thudi et al. (2016)
			CicArVarDB containing 1,965,803 SNPs and indels	90 accessions	Doddamani et al. (2015)
			82,489 SNPs	93 wild and cultivated accessions	Bajaj et al. (2015)
			827,411 SNPs	69 chickpea genotypes (48 Australian varieties, 16 advanced breeding lines, 4 landraces, and 1 accession)	Li et al. (2017)
			ApeKI GBS library	29 desi and 28 Kabuli chickpea accessions	Pavan et al. (2017)
			Kabuli reference genome (16,376)- and de novo (8029)-based GBS (genotyping-by-sequencing) assays	92 *desi* and *Kabuli* chickpea accessions	Tyagi et al. (2016)

			"Axiom® CicerSNP Array"	Diverse set of 429 chickpea accessions from other studies were used (unpublished)	Roorkiwal et al. (2018)
Pea (*Pisum sativum*)	94,001	$2n = 2x = 14$, 4.45 Gbp	131,850 SNPs	4 accessions	Boutet et al. (2016)
			GenoPea 13.2 K SNP Array	12 RIL population	Tayeh et al. (2015)
Mung bean (*Vigna radiate*)	6700	$2n = 2x = 22$, 579 Mb	236,998 SNPs as well as 8896 indels	2 accessions	Jiao et al. (2016)
			1993 SNPs	22 accessions of 18 *Vigna* species and protein sets of *Glycine max*	Kang et al. (2014)
			5288 DArT seg polymorphic SNP markers	Minicore collection of 297 accessions	Breria et al. (2020)
Black gram (*Vigna mungo*)	534		3675 SNP markers using high-density map	RIL population	Somta et al. (2019)
Faba bean (*Vicia faba*)	43,695	$2n = 12$, 13 Gb	1728 SNP markers	3 mapping population	Khazaei et al. (2021), Carrillo-Perdomo et al. (2020)
			687 SNP	6 mapping populations	Webb et al. (2016)
Lentil (*Lens culinaris*)	58,405	$2n = 2x = 14$, 4 Gbp	Consensus map using 9793 DArT SNP markers	3 RIL populations	Ates et al. (2018)
			RNA-seq transcriptome identified 6306 quality polymorphic markers (SNPs and short indels)	RIL population	Polanco et al. (2019)
			188-plex genotyping-by-sequencing yielded 6693 SNPs	184 *Lens culinaris* accessions	Pavan et al. (2019)
			RNA-seq identified 130,073 SNPs	6 diverse lentil accessions	Wang et al. (2020)

(continued)

Table 6.1 (continued)

Crop	Number of accessions available	Ploidy level and genome size	Genome-wide SNPs, indel markers, and structural variants (SVs)	Number of germplasm sources screened	Reference(s)
Lupin (*Lupinus luteus*)	38,050	$2n = 52$	High-density consensus linkage map based on 4062 markers: new, transcriptome-anchored markers	RIL population	Książkiewicz et al. (2017)
Lathyrus (*Lathyrus sativus*)	26,066	$2n = 14$, 8.2 Gbp	RNA-sequencing-derived 126 E-SSRs, 5 ITAPs, and 196 SNPs	RIL population	Santos et al. (2018)

6.3 Genetic Analysis and Selection Methods for Stress Resistance in Legumes

Natural variation present in plants is a precious and sustainable resource of the phenotypic and genetic diversity within plant species that provides beneficial traits for plant breeding (Alqudah et al. 2020). Legume breeding for pest and disease resistance is a continuous process because the quick evolution of new virulent pathotypes can overwhelm the characterized resistance if selection pressure is high (Bansal et al. 2008). To overcome the risk, various long-term conventional breeding strategies are being applied to breed cultivars with multiple biotic resistance (Bakhtiar et al. 2014; Sharma and Sharma 2014). But in conventional plant breeding, genetic variation is usually identified phenotypically (Xu 2010). Therefore, the postulation of resistance genes through phenotypic evaluation is a method to screen the unknown resistance (Collard et al. 2005; Liu et al. 2013a).

6.3.1 Screening Methods

In the conventional breeding, screening of phenotypic variation(s) achieved through various selection methods may be **pedigree**: for selection for resistance to biotic stresses; **bulk method:** used for the development of high-yielding and short-duration varieties; **modified bulk method:** for selection of traits such as abiotic stresses, seed size, earliness, and plant type, and **single seed descent method:** for selection of traits such as biotic and abiotic disease resistance (Bansal et al. 2015). The weakness of the conventional method of characterization is it being time consuming and laborious (Singh et al. 2017). Moreover, primitive breeding approaches for stress resistance in legumes are rigorous and wrathless under field condition, and uttermost care has to be taken to minimize the deleterious effect; for example, selection of resistance against one pathogen may lead to the susceptibility to another (Janni et al. 2020; Johnsson et al. 2019; Varshney et al. 2020).

6.3.2 Marker-Assisted Genomic Selection

Genetics advancement after the discovery of DNA as a hereditary material and its sequence on chromosomes opened a new era in the field of molecular genetics. Thereafter, breeders started to chase molecular breeding to characterize the stress resistance cultivars quickly through various forms of marker-assisted selection (Bohra et al. 2014). Marker-assisted genomic selection (MAGS) is a very important pre-molecular breeding step, but its success is dependent on how much genomic information is available related to that germplasm (Pavan et al. 2017). In the pre-genomics era, with the poor genetic knowledge, the mass-level field screening method was the only technique to select quality germplasm by assessing through obvious phenotypic variation which leads to genetic drift, and the primary gene pool has been narrowed (Pandey et al. 2016). Recent advances in molecular genetics generated vast genomic information (Table 6.1) with the help of bioinformatics tools ends up with a rapid characterization of single or multiple QTL, mapping of desired gene/QTLs, cloning (Setia et al. 2008).

The success of the marker-assisted genomic selection (MAGS) depends on when the marker should be closely linked with the target gene and express the high level of genetic polymorphism, co-dominance (differentiate the heterozygous and homozygous), clear distinct allele features, even distribution on the entire genome (high genome coverage), neutral selection without pleiotropism, easy detection, low cost for marker development and genotyping, and high duplicability (Eagles et al. 2001; Kumar et al. 2019). Different forms of marker-assisted genomic selection methods are being used in nonconventional plant breeding. Each method has its own advantages in which purpose crop is planned for breeding. Marker-assisted genomic selection has reduced the unwanted mass-level field or nursery or greenhouse screening time and cost. Next-generation plant breeders need to utilize this novel MAGS technique to develop stress-resistant high-yielding elite legume varieties/cultivars quickly (Kumar et al. 2019).

6.3.3 Gene Postulation

Gene postulation is a classical method of detecting the presence of a particular qualitative gene(s) in crop cultivar with the aid of NILs (Admassu et al. 2012). It is a fast and simple method of all gene analysis (Li et al. 2011). Great knowledge in the identification of previously characterized resistance gene(s), which are conferring resistance against different pathotypes, is playing an important role in the accuracy of gene postulation (Singh et al. 2001) as well as gene pyramiding (Mebrate et al. 2008). The principle behind gene postulation is the gene-for-gene interaction between the host and the pathogen genotypes to determine the probability of the presence of the resistance gene (Kolmer 1996). A well-characterized collection of pathotypes with diversified avirulence gene combinations is used to postulate resistance genes in the host (Qamar et al. 2008) on the basis of phenotypic expression as infection types (Soliman et al. 2012; Vanzetti et al. 2011). Most probably, a single-gene cultivar or near-isogenic lines (NIL) carrying a known gene is used as a comparison with cultivars consisting of unknown single or polygenic resistance (Mebrate et al. 2008). The success of gene postulation depends on the availability of diversified pathotypes and hosts (Li et al. 2011). Multi-pathotype testing is highly recommended to postulate the all-stage resistance (ASR) in greenhouses. The postulation of adult plant resistance is difficult using this method. To determine the level of resistance as disease index or disease severity, pathogen/host-specific indexing methods are used. In the absence of well-characterized pathotypes in most legumes, the current application of this technique is very minimal, but well developed in wheat and barley (Randhawa et al. 2016).

6.3.4 Genetic Analysis

Genetic analysis is commonly practiced to determine the number of gene(s)/QTLs segregated in a cultivar or germplasm. For conducting genetic analysis, resistance parent is crossed with susceptible parent and F_1 plants are selfed to get F_2 population or backcrossed with a susceptible parent to produce BCF1. The number of segregating resistance genes can then be determined by phenotyping the F_3 or BCF2 families with specific pathotypes in seedlings and also evaluating the segregating families for adult-plant resistance in field tests using a representative mixture of rust pathotypes (Kolmer 1996). If a population is segregating for more than one gene, then isolation and characterization of single-gene F_3 families are the most important steps. The major advantages of using BCF_2 populations compared to F_3 families are that smaller population sizes are required, and resistance genes can be isolated within families that are segregating in single-gene ratios (1:2:1). In these families, plants with the lowest infection type can be progeny tested to obtain lines that are homozygous for resistance. Homozygous lines can be tested with a collection of isolates to determine if the resistance is a previously identified gene or an uncharacterized resistance gene. An additional advantage of the backcross method is that segregating resistances can be evaluated in the background with 75% of the susceptible recurrent parent. This can be very helpful in evaluating adult plants in field tests from crosses in which the two parents vary for maturity or vernalization response (Kolmer 1996).

6.4 Population Development

6.4.1 Development of Mapping Population

In legumes, the conventional method of gene transfers or gene combinations is by crossing or sexual hybridization. This procedure causes genes from the two parents to be assembled into a new genetic matrix. It follows that if parents are not genetically compatible, gene transfer by sexual means cannot occur at all or, at best, may be fraught with complications (Acquaah 2007). The product of hybridization is called a hybrid. Sexual hybridization can occur naturally through agents of pollination, but artificial sexual hybridization is the most common conventional method of generating a segregating population for selection in the breeding of flowering species. Hybridization involves **single cross**: used to transfer resistance against biotic and abiotic stresses; **three-way crosses**: the progenies of three-way crosses are more variable with a wide genetic base than single crosses; and **multiple crosses**: the cultivars developed from multiple crosses are expected to have wider adaptation for a range of environments (Acquaah 2007; Collard et al. 2005).

For the phenotyping and construction of a genetic map, the development of a mapping population is essential. Mapping population means, for self-pollinated crops, the segregating population developed from the crossing of two homozygous contrasting parents showing polymorphism for the trait of interest (Collard et al.

2005; Reynolds 2001). These genotypes, although should be having sufficient polymorphism, should not be distantly related because it causes sterility of progenies and segregation distortion (Kumar et al. 2010). The number of lines used for mapping will vary from 50 to 250, and it depends on whether we are going to use it for preliminary mapping or high-resolution mapping and the number of targeted traits segregated in the mapping population (Collard et al. 2005; Mohan et al. 1997; Xu 2010).

Two different types of mapping population are commonly used for oligo-gene/ QTL mapping, such as **doubled haploid**: generating plants by chromosome doubling (Kumar et al. 2010), and **recombinant inbred lines (RIL)**: germplasms (parents) showing contrast phenotypic variation for a particular trait are crossed to get F_1 plants. The F_1 plants are harvested individually to get F_2 seeds. Seeds from a single F_2 plant are space planted in the field, and seeds from every individual line are harvested (F_3 population) separately (Fig. 6.1). Single seeds from individual F_3 family are planted, the single spike is harvested to get F_4 generation, the process is repeated for further two to three generations, and seeds from individual plants are harvested as bulk in F_7 generation [recombinant inbred lines (RILs)] (inbreeding from individual F_2 plants until F_6–F_8 are obtained) (Collard et al. 2005; Kumar et al. 2019).

MutMap population: Conventional gene mapping is a rigorous process and requires large mapping population and a lot of molecular markers spanned across the entire chromosome or linkage group. Generating population by crossing parent carrying the desired allele with the same parent carrying muted desired allele is called "MutMap population" (Tran et al. 2020; Tribhuvan et al. 2018). This is made possible by generating a backcross population of the mutant genotype with the parent (wild type), thereby removing the false SNPs and retaining only the SNPs linked to the mutant phenotype. With the emergence of re-sequencing techniques, quick mapping of genes has become possible with reduced time and cost by using approaches like SHOREmap, NGM, and MutMap methodologies. Among these, MutMap is widely used because it is more focused on causal SNPs (Tran et al. 2020; Tribhuvan et al. 2018; Yuan et al. 2017). Improved and specialized methods of MutMap like MutMap-Gap and QTL-Seq have also emerged to expand the horizon of application of the MutMap approach. All these methods are akin to bulked segregant analysis popularly employed for mapping simply inherited traits. These methods escape the requirement of genotyping all the individuals of the mapping population and generation of high-density linkage maps for mapping of the gene for the trait of interest (Tribhuvan et al. 2018; Yuan et al. 2017). However, in most situations, the F_1 is selfed (to give an F_2) to generate recombinants (as a result of recombination of the parental genomes) or a segregating population, in which selection is practiced.

Legume genomics is advancing quickly, but, due to the large genome size of many legume species, accurate positioning of QTL governing resistance to pest and diseases is still difficult in the biparental mapping population. To overcome this problem, association studies using multiparent advanced generation intercross populations (MAGIC) and nested association mapping population (NAMP) are

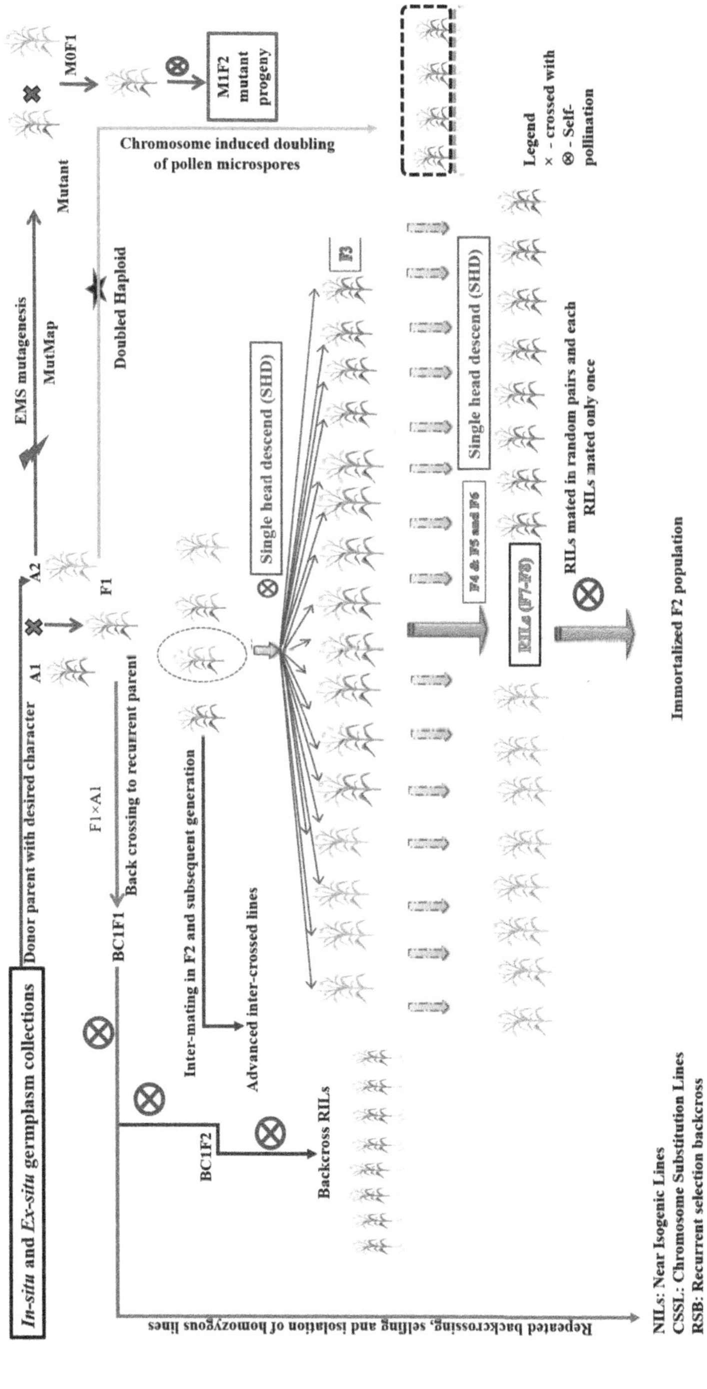

Fig. 6.1 Population development pipeline for mapping gene(s)/QTLs in food legumes

currently being popular to identify the QTL and marker of a complex trait (s) (Varshney et al. 2006).

6.5 Molecular Breeding of Legumes in Genomics Era

Integration of molecular technology is indispensable for quick identification of allele adaptation to various stresses, and its successful exploitation in elite germplasm is essential to ensure food security for the rapidly increasing world population (Jacob et al. 2016). The success of any molecular breeding depends on the selection of appropriate genetic material and the application of suitable molecular tools. The fast development of NGS technologies has facilitated swift sequencing and re-sequencing of several hundred potential lines, development of haplotype map (HapMaps), high-density SNP-based genetic maps, a range of marker genotyping platforms, and identification of markers associated with a variety of agronomic traits in these legume crops. Due to the need-based accurate application of these OMIC tools in legume resistance breeding programs, many improved varieties have been released throughout the globe over the past decade through marker-assisted selection, marker-assisted backcrossing, marker-assisted pyramiding, and gene-editing approaches (Varshney et al. 2020). Molecular breeding further emphasizes independent or a combination of parental selection, enhancing genetic diversity in breeding programs, forward breeding for early generation, and genomic selection using a sequence-based breeding approach (Varshney et al. 2020).

Moreover, next-generation powerful statistical genetic methods and crop breeding technology, genomic selection, transcriptome mapping (expressed sequence tags—ESTs, serial analysis of gene expression—SAGE, massively parallel signature sequencing—MPSS, microarray), genomics (whole-genome sequencing, NGS, and genotyping-by-sequencing), and allele mining approaches have been proposed to identify gene/s, transcription factors (TF), microRNA (miRNA), and quantitative trait loci (QTLs) responsible for stress resistance (Varshney et al. 2020). Genome-wide association study (GWAS) is one of these useful methods, and it is successfully used to identify candidate genes for many important traits in many crops as it tests the association between the marker type (e.g., SNP) and the phenotype of a target trait (Alqudah et al. 2020).

6.5.1 Molecular Markers for Selection of Stress-Resistant Genes

Molecular markers (markers are characters) used in legume breeding programs can be classified into morphological, biochemical, and DNA-based markers (Collard et al. 2005; Eagles et al. 2001). Morphological markers or classical markers are used to postulate the presence of the gene by phenotypic characterization or visual observation (Collard et al. 2005; Xu 2010). But its application in legume's resistance breeding program is limited. Biochemical markers are actually proteins (isozymes). These isozymes are structural variants of enzymes and can be used as markers in gene mapping. But their application is also limited to legume breeding. DNA markers are very prominent compared to other marker types because of their abundance and high polymorphism. DNA markers are selectively neutral, located in noncoding regions of the DNA, and not affected by the environment (Collard et al. 2005; Eagles et al. 2001). A variety of molecular markers are being used in molecular breeding, but depending on the detection and throughput, molecular markers can be classified as low-throughput hybridization-based markers like RFLP; medium-throughput PCR-based markers such as RAPD, AFLP, and SSR; and high-throughput sequence-based markers like SNPs (Davey et al. 2011; Mammadov et al. 2012). DNA molecular markers, especially simple sequence repeats (SSRs) and single nucleotide polymorphisms (SNPs), are used widely for the construction of linkage maps, mapping of quantitative trait loci (QTL), map-based gene cloning, marker-assisted selection, exploration of population diversity, etc. in all major crops.

6.6 High-Throughput Technology and SNP Discovery

6.6.1 Sequencing for SNP discovery

SNP available in the organism is discovered through sequencing and comparing of genomic DNA or cDNA (complementary or copy DNA) or in silico alignment of sequenced data from two or more individuals of a species. The determination of the base sequence of a DNA fragment is called sequencing. Methods used for the sequencing of DNA can be broadly classified into first-generation sequencing and NGS (Singh and Singh 2015).

6.6.2 First-Generation DNA Sequencing

In this method, chemical or *E. coli* DNA polymerase *I* is used to modify the bases at the breakpoints of the DNA fragment. This method of DNA sequencing was also called Sanger–Coulson method. This method is useful for sequencing 15–200 nucleotides but is more laborious and needs the preparation of template, enzyme, and gel electrophoresis. Therefore, the application of this technique was not suitable for sequencing an organism with a higher ploidy level like faba bean (Sanger et al. 1977).

6.6.3 Next-Generation Sequencing (NGS)

NGS opened a pathway for discovery, sequencing, and genotyping of thousands to hundred thousands of markers through parallelized library preparation of genomic DNA without using restriction enzymes (Davey et al. 2011). The development of high-throughput genotyping platforms for the screening of millions of SNPs was difficult and lengthy and involved high costs for crops with a large genome (Gb) (Sonah et al. 2013). Therefore, the application of NGS had limitations for species with large complex genomes such as barley and wheat. To overcome this problem, several sequencing techniques were emerged using NGS as a base platform by combining restriction enzymes as a versatile tool such as reduced representation libraries (RRLs), complexity reduction of polymorphic sequences (CRoPS), restriction site-associated DNA sequencing (RAD-seq), sequence-based polymorphic marker technology (SBP), low-coverage multiplexed shotgun genotyping (MSG), and genotyping-by-sequencing (GBS) (Davey et al. 2011; Yang et al. 2012). Among them, GBS is now being widely used in legume research as a molecular tool. Initially, GBS has been developed as a tool for association studies and genomic-assisted breeding in a range of species including those with complex genomes. GBS uses restriction enzymes for targeted complexity reduction followed by multiplex sequencing to produce high-quality polymorphism data at a relatively low per-sample cost of the desired population (Sonah et al. 2013). Continuous optimization has led to innovative third- and fourth-generation platforms such as single-molecule real-time (SMRT) sequencing by PacBio, nanopore sequencing, etc. (Meera Krishna et al. 2019). As a consequence, there has been a sharp increase in the number of genomes being published and other genome-based studies since 2012. Many of these platforms, e.g., microarray-based GS, involve the partial representation of the genome, and these can be utilized even in the absence of prior knowledge on WGS.

6.6.4 SNP Genotyping and Validation

Many NGS platforms, for example, NGS-derived transcriptome sequences, have an option for parallel sequencing of many germplasms from different populations, and through which millions of genome-wide SNPs are being discovered in many crops including food legumes (Sim et al. 2012). At the same time, advancement in modern chemistries developed diversified typical genotyping platforms for SNP validation such as Illumina's BeadArray technology-based Golden Gate (GG) and Infinium assays, Life Technologies' TaqMan assay coupled with OpenArray platform (TaqMan OpenArray Genotyping system, Product bulletin), and KBiosciences' Competitive Allele Specific PCR (KASPar) combined with the SNP line platform (SNP Line XL; http://www.kbioscience.co.uk). The choice of chemistry and genotyping platform varies with the length of SNP context sequence, the overall number of SNPs to genotype, and the number of SNPs that need validation, but most of these chemistries still remain cost effective (Mammadov et al. 2012). By using one or more of the NG platforms, recently developed SNP bead chip arrays with genome-wide validated SNPs for 12 food legumes are summarized in Table 6.1.

6.7 Molecular Mapping of Stress Resistance Gene(s)/QTL(s) Using SNP Markers

6.7.1 Genetic Maps of Legumes

When two parents with distinct alleles at many loci are crossed, variation can be created by crossover events during meiosis. These novel variations yield descendants with unique phenotypes different from their parents. However, when two loci are located closely on the same chromosome, the probability of crossover events between them falls, and the recombinant genotype becomes relatively rare. The crossover rate increases in proportion to the distance between genes, so crossover rate data allow the estimation of loci distances on the chromosome. Plant biologists use three distinctive types of "maps" such as cytological or cytogenetic maps, linkage/genetic maps, and physical maps for OMIC studies (Rana et al. 2019). In conventional genetics, chromosomes were identified cytogenetically only in the availability of deletion stocks. Once the chromosome was identified, the location of a particular gene was confirmed and mapped using the nullisomic or haploid method. Chromosome deletion, translocation, trisomic, monosomic, and nullisomic lines serve as valuable tools for cytogenetic mapping (Endo and Gill 1996). In a physical map, the genes/molecular markers are depicted in the same order as they occur in the chromosomes, but the distances between adjacent genes/markers are depicted in terms of base pairs. Physical and genetic maps are a collection of genetic markers and gene loci. The distance between locus is based on the genetic linkage information in genetic maps, while physical maps use actual physical distances usually measured in the number of base pairs (Singh and Singh 2015). While genetic maps often offer insights into the nature of different regions of the chromosome, the physical map could be a more "accurate" representation of the genome. A genetic map constructed by this way shows the relative locations of morphological or molecular markers in a particular chromosome (Collard et al. 2005). On such maps, one map unit is defined as having a crossover rate of 1% and is called a centimorgan (cM). If a genetic map is available, the genotype/phenotype correspondence of individuals in a segregating population can be calculated by comparing the phenotype and the marker genotype (Singh and Singh 2015).

The construction of consensus/genetic map(s) of plants was highly dependent on a variety of abovementioned DNA-based molecular markers (Song et al. 2005). In the twenty-first century, with the discovery of SNP as minimum single base pair variation in all organisms through the human genome project, genome-wide SNP discovery and mining were started using NGS technology (Elshire et al. 2011). Revolutionized genomic and transcriptomic approaches further boomed with automated NGS platforms and bioinformatics tools (Varshney et al. 2011). Several modified NGS methods, such as reduced representation sequencing using reduced representation libraries (RRLs) or complexity reduction of polymorphic sequences (CRoPS) and restriction site-associated DNA sequencing (RAD-seq), for genome-wide SNP marker development and genotyping use restriction enzyme digestion of target genomes to reduce the complexity of the target. Identified SNP markers through modified NGS methods are more cheaper, abundant, amenable, and reliable and have reduced the complexity in genotyping whether draft-sequenced genome is available or not (Davey et al. 2011). Now, whole/partial genome of many legume crops, such as soybean (*Glycine max*), peanut or groundnut (*Arachis hypogaea*), chickpea (*Cicer arietinum*), cowpea (*Vigna unguiculata*), common bean (*Phaseolus vulgaris*), pigeon pea (*Cajanus cajan*), pea (*Pisum sativum*), lentil (*Lens culinaris*), faba bean (*Vicia faba*), mung bean (*Vigna radiata*), and lupin (*Lupinus luteus*), have been sequenced and SNP-rich consensus maps are available for public use (Alqudah et al. 2020; Davey et al. 2011). The genetics, genomic, transcriptomic marker, and map sequence data can be accessed through one of the web portals summarized in Table 6.2.

6.8 Mapping a Gene or QTL

To use marker-assisted selection, marker-assisted backcrossing/introgression, and marker-assisted pyramiding in conventional breeding programs, markers tightly linked to the gene(s) controlling target trait must be identified first. Therefore, different molecular mapping approaches and efforts have been used to find out the genetic distance between molecular markers and genes controlling qualitative and quantitative traits of interest (Kumar et al. 2011; Rafalski 2010).

6.8.1 Oligo-Gene Mapping (Single-Gene Mapping)

The one or few genes present in plants showing larger phenotypic effects are called "qualitative traits," and its phenotypic expression is relatively little affected by the environment. The purpose of mapping an oligo-gene(s) with molecular markers is to identify closely linked marker(s) to the oligo-gene(s) for marker-assisted selection for the concerned trait (Collard et al. 2005; Collard and Mackill 2008).

In the last century, chromosome location was determined by studying the progenies of crosses developed by crossing test cultivars with monosomic series (Endo and Gill 1996). But the recent advances in NGS technology, transcriptome sequencing, and whole-genome sequencing/re-sequencing of many food legume crops have been partially or fully completed. This NGS technology has opened up to construct chromosome-wise highly saturated SNP maps. With the aid of this genomic information, the chromosomal location of an unknown gene or QTL can be detected by genome-by-sequencing/genotyping-by-sequencing rapidly. Moreover, for the oligo-gene mapping, to reduce the genotyping work, and to facilitate identification of markers that are most likely to be closely linked to the targeted locus/gene (s) governing the target trait, bulk segregant analysis (BSA) (Michelmore et al. 1991), selective DNA pooling (Darvasi and Soller 1994; Lee et al. 2014), bulked segregant RNA-seq, chromosome-targeted selective genotyping, MutMap, etc. are currently being used by researchers. In all methods, contrasting phenotypes carrying

Table 6.2 Genomic, marker, map, sequence, and bioinformatics databases for grain legumes

Database	Description	URL
GenBank	General public sequence repository	http://www.ncbi.nlm.nih.gov/genbank/
EMBL	General public sequence repository	http://www.ebi.ac.uk/embl/
DDBJ	General public sequence repository	http://www.ddbj.nig.ac.jp
UniProt	Protein sequences and functional information	http://www.uniprot.org/
NCBI	Biomedical and genomic information	http://www.ncbi.nlm.nih.gov/
Gene Index Project	Transcriptome repository	http://compbio.dfci.harvard.edu/tgi/
GOLD	Repository of genome databases	http://genomesonline.org/cgi-bin/GOLD/bin/gold.cgi
Phytozome	Genomic plant database	http://www.phytozome.net/
PlantGDB	Genomic plant database	http://www.plantgdb.org
CropNet	Genomic plant database	http://ukcrop.net/
Pulse Crop Database	Pulse	https://www.pulsedb.org/
Phytozome 10.2	*Glycine max*	http://phytozome.jgi.doe.gov/
LIS—Legume Information System	*Cajanus cajan*	http://legumeinfo.org/gbrowse cajca1.0 http://cicar.comparative-legumes.org/ http://plantgenomics.snu.ac.kr/ https://genebank.ciat.cgiar.org/genebank/
International Initiative for Pigeonpea Genomics (IIPG)	Pigeon pea (*Cajanus cajan*)	http://www.icrisat.org/gt-bt/IIPG/home.html
CicArVarDB	Chickpea SNP-indel database	http://cicarvardb.icrisat.org/
ACPFG Bioinformatics	SNP discovery	http://autosnpdb.appliedbioinformatics.com.au/
Crop Genomics Lab	Mung bean	http://plantgenomics.snu.ac.kr/mediawiki-1.21.3/index.php/Main_Page

bulks or lines are screened with previously mapped markers in the consensus or highly saturated linkage maps (Singh 2015).

6.8.1.1 Bulked Segregant Analysis (BSA)

It is a simple technique to screen the polymorphic markers for the two parental lines carrying contrasting phenotypes for a particular locus of interest (Michelmore et al. 1991). To carry out the BSA, DNA of homozygous resistant (HR) and homozygous susceptible (HS) lines of segregating population derived from the single cross is pooled separately. Moreover, artificial F_1 bulk is also prepared by mixing of DNA from randomly selected lines of the same cross. These three bulks (HR, HS, and F_1) are genotyped with a large number of markers using high-throughput NGS platform. Generated data will be analyzed using GenomeStudio. The sequence of SNPs which showed strong linkage will be converted to develop PCR-based markers, e.g., Kompetitive allele-specific PCR (KASP) primers. In legumes, different high-throughput platforms mentioned in Table 6.1 are used to achieve this task, for example, BARCBean6K_3 BeadChip containing 6000 SNPs (Song et al. 2015b) and Diversity Arrays Technology (DArTSeq) seg high-density SNP for common bean BSA (Valdisser et al. 2017). By using the BSA approach, countable number resistant gene(s) and closely linked SNP markers for the resistant traits have been identified (Vuong et al. 2016).

6.8.1.2 Selective Genotyping

Selective genotyping is defined as genotyping of the selected individuals carrying contrasting phenotypes from the segregating population to identify the genomic location of the trait of interest and markers (Darvasi and Soller 1994). This technique selects genotypes to act as true representatives of the population and remarkably decreases the number of individuals to be genotyped, cost, and time. Selective genotyping is suitable for mapping a quantitative trait or qualitative trait (Bariana et al. 2016; Lee et al. 2014). By using the selective genotyping approach, a considerable number of resistant gene(s), closely linked SNP markers for the resistant traits (Table 6.2), and chromosomal location of the crops have been identified as well as mapped for example I gene resistance to bean common mosaic virus resistance in common bean (Bello et al. 2014), soybean seed protein and oil QTLs (Phansak et al. 2016), and soybean rust (Vuong et al. 2016).

6.8.1.3 Bulked Segregant RNA-Seq (BSR-Seq)

Bulked segregant RNA-Seq (BSR-Seq) is also the quickest method to map a gene or QTL showing larger phenotypic variation. This is a modified method of BSA, and instead of DA, high-quality RNA is isolated from bulks showing contrasting phenotype from the F_2 population and sequenced by RNA-seq technology (Varshney et al. 2019). RNA sequence data provides information about the approximate number of copies of each RNA sequence present in the sample assuming that the numbers of reads of various sequences reflect their relative concentrations in the sample. The RNA sequence data are used to mine or discover a large number of polymorphic SNP markers present in that progeny. By using this technique, several resistance genes in legumes have been located on the chromosome as well as mapped on a particular chromosome such as rust and late leaf spot resistance in the groundnut (Hyten et al. 2009; Pandey et al. 2017a).

6.8.1.4 Single-Gene Mapping Procedure

Linked markers identified by BSA or selective genotyping are converted into PCR-based markers checked on a few homogenous and heterogeneous lines of that particular mapping population to check the parental polymorphism (Fig. 6.2). Polymorphic markers are genotyped in the mapping population (Bansal et al. 2014, 2015; Randhawa et al. 2014, 2015). Then the marker and trait association is analyzed with the LOD value of 3 through sophisticated software such as MapManager QTX (Manly et al. 2001) and MapDisto (Lorieux 2012) to calculate the recombination fraction using the Kosambi mapping function (Kosambi 1944). The genetic linkage map is drawn with the help of MapChart software (Voorrips 2002) to show the graphical representation of map order, and marker distance is defined in centimorgan (cM). If the markers are not closed, new markers will be designed using different marker designing software (e.g., Primer 3).

6.9 QTL Mapping

Quantitative trait locus (QTL) means the region within the genome where the gene responsible for the particular quantitative trait is located. Individual QTL may be called as "minor" ($R2 < 10\%$) or "major" ($R2 > 10\%$) based on the proportion of phenotypic variation explained as "R2" value, and for instance, major QTL may be called as QTL if the phenotypic variation is constant over the different environment (Collard et al. 2005). In another way, QTL can be classified into "suggestive," "significant," and "highly significant" to ensure the "true hints of linkage" and to eliminate a "flood of false-positive claims" (Center 1995). Detection of QTL was impossible by conventional phenotypic evaluation and was possible after the discovery of genetic markers. QTL analysis means examining the association between the genotype (markers) and the phenotype. Therefore, accurate phenotypic evaluation is very important (Collard et al. 2005). The principle behind the QTL analysis is that during the chromosome crossover, the targeted trait and the closely linked marker(s) are co-segregated together into the progeny, thus allowing analysis in the progeny. Three methods are widely used to detect the QTLs, such as single marker scan, simple interval mapping, and composite interval mapping. Composite interval mapping is becoming popular because it allows the analysis of linked QTLs as well as additional markers in the linear statistical system. Nowadays, statistical packages QTL Cartographer (Wang 2007), MapManager QTX (Manly et al. 2001), and MapChart (Voorrips 2002) are publically available to perform the QTL analysis and allow to discover more QTLs very easily in a short period.

6.9.1 Mapping a QTL(s): Procedure

Marker data generated from genotyping-by-sequencing or genome-by-sequencing (using one of the abovementioned NG platforms) will be cleaned by removing markers that show the same calls for the parents (monomorphic markers), more than 10% missing data, and segregation distortion (Fig. 6.2). Markers that are not assigned to any chromosome will be distributed to the best matching chromosome by using the option "distribute" in MAP MANAGER Version QTXb20 (Manly et al.

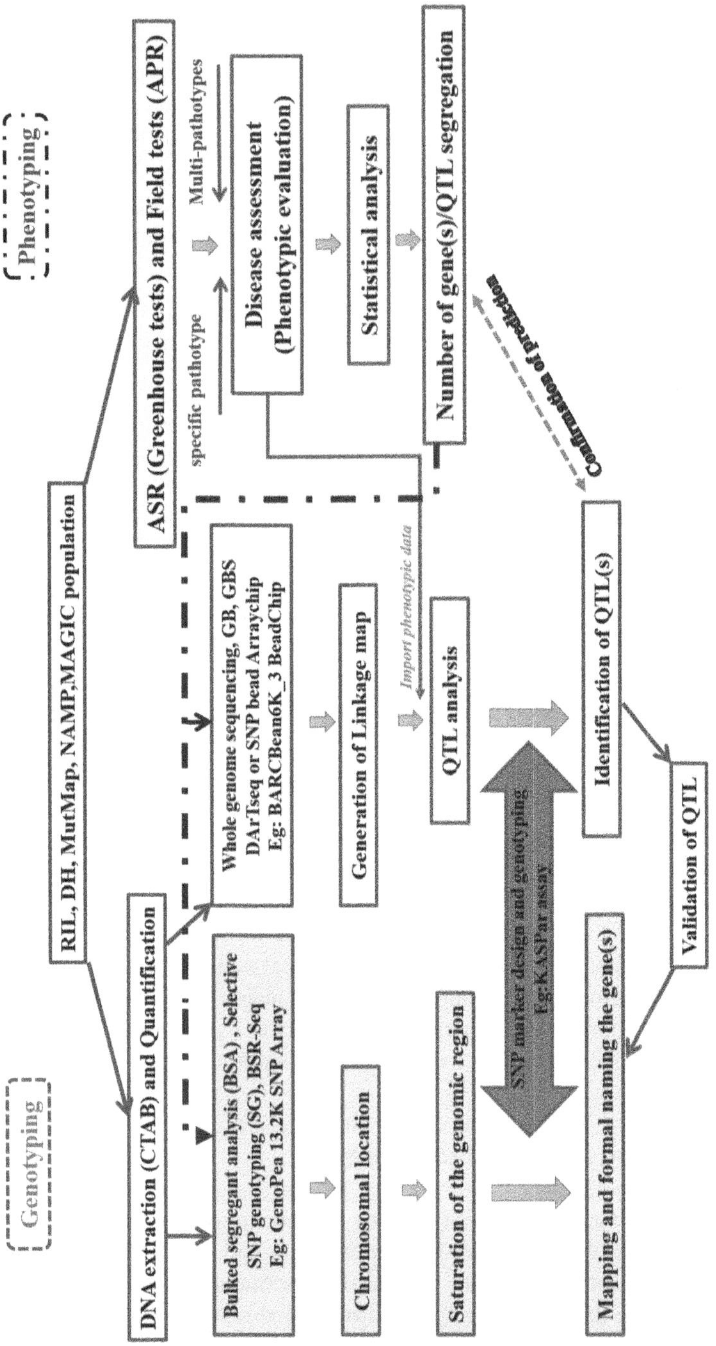

Fig. 6.2 Phenotyping and genotyping pipeline to map gene(s) or QTL(s)

2001). Thereafter, redundant markers will be taken out by using the command "hide redundant loci." A genetic linkage map will be constructed with the help of MAP MANAGER Version QTXb20 (Manly et al. 2001) or MapDisto (Lorieux 2012) using Kosambi mapping function to convert recombination fraction into centimorgan (cM) (Kosambi 1944). QTL Cartographer 2.5 (any QTL mapping software) can be used to analyze marker-trait associations using a composite interval mapping function. To declare the significant QTL, the LOD threshold value is calculated by 1000 permutations at a $P = 0.05$, and walking speed is set at 1 cM. The proportion of phenotypic variance explained (R2) by the QTL is used to determine the effectiveness of QTL. Thereafter, QTL figures will be drawn by using any map drawing software.

Legumes encounter multiple stresses including wilt, Ascochyta blight (lentil, chickpea), rust (pea and lentil), powdery mildew (pea), terminal drought, heat and salinity (chickpea, lentil, pigeon pea), and waterlogging (pigeon pea) during their life cycle (Kumar et al. 2011). There are lots of QTL discovery and mapping research in legumes that have been published over the decades using RFLP, AFLP, RAPD, SCAR, and SSR (Kumar et al. 2011; Varshney et al. 2013). After the discovery of SNP as a single base pair variant, SNPs have become the dominant marker for detailed characterization of many QTLs associated with the biotic resistance in many crops including food legumes with the aid of GBS (Dwivedi et al. 2017). Characterized QTL(s) governing biotic stresses using the NGS platform and peaked/linked/flanking SNP(s) are summarized in Table 6.2, which will be helpful for further mapping studies and formal naming.

6.10 Marker-Assisted Backcrossing and Gene Pyramiding

Once these are identified, the next approach is to transfer them into elite cultivars. In modern plant breeding, a molecular technique that uses molecular markers to track genes from germplasm or to select trait of interest indirectly is called marker-assisted selection (MAS) (Goutam et al. 2015). Markers that are linked with genes are used as flags to help breeders select the best gene combinations, and breeders are now using these markers to pyramid genes into the new varieties to provide more durable resistance (Kumar et al. 2019). With the use of markers, varieties are selected faster without infecting lines and without the confound influence of the environment (Tyagi et al. 2014). The transgenic approach is feasible to engineer traits that are controlled by one or a few major genes and QTLs not easily amenable through transformation. For this purpose, we can use high-throughput technology, marker-assisted selection (MAS), marker-assisted backcrossing (MABC), and gene pyramiding to elucidate thousands of genes or even entire genomes (Rana et al. 2019; Taran et al. 2013; Varshney et al. 2020).

6.10.1 Marker-Assisted Backcrossing

Marker-assisted backcrossing aims transfer of targeted desired one or two traits without disturbing the remaining all other native traits of target cultivar (Kumar et al. 2010). During the past two decades, in nonconventional wheat breeding, usage of molecular markers as the most effective tools proved to be easy, quick, and important to avoid unnecessary delays and helped to identify, isolate, stack, and map several genes simultaneously (Ali et al. 2010; Asad et al. 2012; Liu et al. 2013b; Reynolds 2001; Reynolds et al. 2012). With the availability of diagnostic polymorphic SNP marker(s), marker-assisted backcrossing is suitable to track the major gene conferring resistance, and for minor or complex traits that are generally controlled by QTL with major phenotypic effects. Marker-assisted recurrent selection (MARS) is a highly preferred method to deploy well-characterized resistance genes in elite legume cultivars (Todorovska et al. 2009). There are several success histories of MARS introgression in food legumes, and for example, Varshney et al. (2014) successfully backcrossed Fusarium wilt race *1* and Ascochyta blight resistance genes in C 214 (an elite cultivar of chickpea). Several SNPs summarized in Table 6.3 are already being used for marker-assisted selection and marker-assisted backcrossing.

6.10.2 Gene Pyramiding

However, genetic improvement for single biotic stress using single-gene-based resistance does not result in permanent gains in productivity because of the emergence of increasingly more virulent races/biotypes in nature (Ali et al. 2003). However, genetic improvement for single biotic stress using single-gene-based resistance does not result in permanent gains in productivity because of the emergence of increasingly more virulent races/biotypes in nature (Kumar et al. 2011). Genetic improvement for a single biotic or abiotic stress using single-gene-based resistance does not result in permanent gains in productivity because of the time-to-time emergence of highly virulent races/biotypes in nature (Ali et al. 2003). Once resistance gene/QTL is characterized or postulated with closely linked SNP markers, breeders prefer gene pyramiding as the best strategy to stack multiple resistance genes in elite germplasm to extend the durability of the characterized resistance gene (s) (Bansal et al. 2011; Simons et al. 2011).

Gene pyramiding means the stacking or encompassing of more than one resistance gene/QTL characterized or mapped in different parents into a cultivar to express the polygenic/multigene resistance (Collard and Mackill 2008; Joshi and Nayak 2010). Pyramiding of resistance genes through traditional phenotypic based technology is difficult when different resistance genes produce similar infection types (Khan et al. 2005; Suresh and Malathi 2013). However, marker-assisted selection facilitates the identification of the successfully pyramided genes into the target cultivars with multiple target genes (Bin et al. 2012; Joshi and Nayak 2010; Simons et al. 2011). Moreover, success in pyramiding of genes that are resistant to

Table 6.3 SNP marker(s) associated with resistance QTL(s)/gene for ten important food legumes

Crop	Resistance Trait	QTL/Genes	SNP Marker/Marker ID/Position	Chromosome	References
Soybean	*Sclerotinia* stem rot	*Glyma.01 g048000*	*5594765*	01	Boudhrioua et al. (2020)
		QTL	*13,651,235*	15	Bastien et al. (2014)
		QTL	*S18_14,327,556*	18	Wei et al. (2017)
	Cyst nematode	*Rhg1*	*GSM 381, GSM383*	18	Tran et al. (2019)
		Rhg4	*GSM 191*	08	Tran et al. (2019)
	Southern stem canker	QTL	*GBSRdm370*	14	Maldonado dos Santos et al. (2019)
		QTL	*44,735,630*	03	Iquira et al. (2015)
	Rust	*Rpp3*	*BARC-051071-10973, BARC-023203-03824, BARC-061709-17355, BARC-024739-05617*	06	Vuong et al. (2016)
		Rpp4	*BARC-016867-02359_BARC048761-10703, Satt288_BARC-024489-04936*	18	Vuong et al. (2016)
	Meloidogyne incognita	*genes/QTL*	*Gm10_831916* *Gm10_981062* *Gm10_1536434*	10	Dubiela et al. (2019)
	Bacterial leaf pustule	*rxp*	*SNUSNP17_12*	17	Kim et al. (2010)
	Phytophthora root rot	*RpsZS18*	*Indelwz1-7*	02	Zhong et al. (2018)
	Mung bean yellow mosaic India virus	*QTL*	*SNP 18–1861613*	18	Yadav et al. (2015)
Common bean	Anthracnose	QTL	*BARCPV_1.0_Chr02_23542475_A_G*	02	Fritsche-Neto et al. (2019)
	Angular leaf spot resistance	QTL	*BARCPV_1.0_Chr10_20935383_C_T*	10	Fritsche-Neto et al. (2019)
	Anthracnose and angular leaf spot	QTL	*SNP scaffold00021_89379*	07	Perseguini et al. (2016)
	Bean fly	QTL	*SNP (SNP372_8196616)*	01	Ojwang et al. (2019)
Chickpea	*Ascochyta* blight	*AB-Q-SR-4-1*	*ICCM0068 and CaM1158*	04	Garg et al. (2018)

(continued)

Table 6.3 (continued)

Crop	Resistance Trait	QTL/Genes	SNP Marker/Marker ID/Position	Chromosome	References
		AB-Q-SR-4-1 and AB-Q-APR-4-1	*CKAM0847 CKAM0964*	04	Garg et al. (2018)
		QTL	*Ca4: 15,920,939*	04	Li et al. (2017)
		qAB2.1	*Ca2-ABA-RCav1sc520.1p50440*	02	Deokar et al. (2019)
		qAB2.2	*Cav1sc246.1p121732–Cav1sc689.1p195825*	02	Deokar et al. (2019)
		qAB2.3	*Ca2-GDSL2Ca2-PEI*	02	Deokar et al. (2019)
		qAB3.1	*SCA3_15444471SCA3_21346384*	03	Deokar et al. (2019)
		qAB4.1	*SCA4_6022188SCA4_8584363*	04	Deokar et al. (2019)
		qAB4.2	*SCA4_21540251Ca4-ER2*	04	Deokar et al. (2019)
		qAB5.1	*SCA5_20906121 Ca5-BTB*	05	Deokar et al. (2019)
		qAB6.1	*SCA6_54348174-Cav1sc30.1p60427*	06	Deokar et al. (2019)
	Rust	*Ruv1*	*2_01772 and 2_03292*	09	Wu et al. (2018)
	Fusarium wilt	*Foc5*	*23349051, 23487393, 24129368, 24536633, 24828612, 23213288*	02	Caballo et al. (2019)
	Fusarium wilt *race 1*	*FW-Q-APR-2-1*	*TR19 and H2B061*	02	Garg et al. (2018)
Pea	*Sclerotinia sclerotiorum*	QTL	*Chr5LG3_569106574, Chr5LG3_561561067*	05	Ashtari Mahini et al. (2020)
	Pea weevil	*BpSI.III*	*3537674 and 3542227*	04	Aznar-Fernández et al. (2020)
Mung bean	Mung bean yellow mosaic virus (MYMV)	*qMYMV4-1*	*Vr04:14504302 and 15788321*	04	Mathivathana et al. (2019)
	Bruchids	QTL	*3:10,830,930 and 5:5,730,691*	05	Schafleitner et al. (2016)
		qSD05	*VigSNP_05_18 and VigSNP_05_19*	05	Mariyammal et al. (2019)
		qAE08	*VigSNP_08_45 and VigSNP_08_46*	08	Mariyammal et al. (2019)

Black gram	Bruchids	*qVmunBr6.1 and qVmunBr6.2 a*	*14881, 9514 and 15884 flanking*	06	Somta et al. (2019)
Faba bean	Rust	*Uvf-2*	*KASP_Vf_0703*	03	Ijaz (2018)
		Uvf-3	*KASP_Ac × F165*	05	Ijaz (2018)
Lentil	*Ascochyta* blight	*AB_IH1-*	*PBA_LC_0629 and SNP_20005010*	02	Bhadauria et al. (2017), Sudheesh et al. (2016)
		AB_N2	*Vf_0843, Vf_0673*	06	Sudheesh et al. (2016)
		AB_IH2.1/ AB_IH2.2-	*SNP_20000505 and SNP_20000553*	03	Bhadauria et al. (2017), Sudheesh et al. (2016)
		AB_N1 1	*Vf_1122, Vf_0382*	01	Sudheesh et al. (2016)
		qANTH0-3	*Contig119649p37483*	03	Bhadauria et al. (2017)
		qANTH0-5.1 and qANTH0-5.2	*Contig23353p125770 Contig27270p11193*	05	Bhadauria et al. (2017)
		qSB-2.1	*Contig271180p29128*	02	Bhadauria et al. (2017)
		qSB-2.2	*Contig313227p47568*	02	Bhadauria et al. (2017)
		qSB-3	*Contig406212p17766*	03	Bhadauria et al. (2017)
Blue lupin	*Colletotrichum gloeosporioides*	*Lanr1*	*AnSeq3 and AnSeq4*		Yang et al. (2012)
Pigeon pea	Sterility mosaic disease	*qSMD11.1*	*S11_30004779–S11_36027138*	11	Saxena et al. (2017)
Peanut	Rust	QTL	*GMRQ517, GMRQ786, and GMRQ843*	03	Pandey et al. (2017b)
	Late leaf spot	QTL	*GMLQ975*	03	Pandey et al. (2017b)
	Bacteria wilt	*qBWRB02-1*	*AhEXZ253127–AhEXZ369227*	02	Luo et al. (2020)
		qBWRB01-1	*AhEXZ281542–AhMXZ205274*	01	Luo et al. (2020)
		qBWRB01-2	*AhEXZ117519–AhMXZ288911*	01	Luo et al. (2020)
	Late leaf spot	*qLLS2015-B05-2 qLLS2015-B05-1*	*8860908–8866178 9173668–9177669*	05	Han et al. (2018)

(continued)

Table 6.3 (continued)

Crop	Resistance Trait	QTL/Genes	SNP Marker/Marker ID/Position	Chromosome	References
	Early leaf spot	*ELS2017-A03, and qELS2016-B04*	*133776796–133780539 132549922–132552596*	03	Han et al. (2018)
	Web blotch	*qWBRA03*	*Chr3_bin366–Chr3_bin367*	03	Liu et al. (2020)
		qWBRA04	*Chr4_bin756–Chr4_bin757*	04	Liu et al. (2020)
		qWBRA05	*Chr5_bin760–Chr5_bin761*	05	Liu et al. (2020)
		qWBRA13	*Chr13_bin2385–Chr13_bin2386*	13	Liu et al. (2020)
		qWBRA14	*Chr14_bin2746–Chr14_bin2747*	14	Liu et al. (2020)
		qWBRA16	*Chr16_bin3084–Chr16_bin3085*	16	Liu et al. (2020)
		qWBRA17	*Chr17_bin3253–Chr17_bin3254*	17	Liu et al. (2020)
		qWBRA19	*Chr19_bin3690–Chr19_bin3691*	19	Liu et al. (2020)
	Rust	QTL	*Ah-280*	08	Leal-Bertioli et al. (2015)

more than one stress depends on the number of genes to be transferred, the close linkage between transferred gene(s) and markers, the number of genetic sources used in breeding, and the nature of the germplasm (Joshi and Nayak 2010). So far, resistance genes have been successfully pyramided into elite food legume varieties to provide durable resistance, for example pyramiding respective *Rsv* genes from different loci (*Rsv1*, *Rsv3*, and *Rsv4*) through marker-assisted selection (MAS) (Saghai Maroof et al. 2008).

References

Acquaah G (2007) Principles of plant genetics and breeding. First Published. Blackwell Publishing Ltd., Australia

Admassu B, Friedt W, Ordon F (2012) Stem rust seedling resistance genes in Ethiopian wheat cultivars and breeding lines. Afr Crop Sci J 20(3):149–162

Agrios GN (2005) Plant pathology. Elsevier, Amsterdam

Akond M, Liu S, Schoener L, Anderson JA, Kantartzi SK, Meksem K, Song Q, Wang D, Wen Z, Lightfoot DA (2013) A SNP-based genetic linkage map of soybean using the SoySNP6K Illumina Infinium BeadChip genotyping array. Plant Genet Genomics Biotechnol 1(3):80–89

Ali M, Kumar S, Singh N (2003) Chickpea research in India, 1st edn. Indian Institute of Pulses Research

Ali M, Ji WQ, Hu YG, Baloch GM, Zhong H, Wang CY (2010) Molecular implications from SSR markers for stripe rust (*Puccinia striiformis* f. sp. *tritici*) resistance gene in bread wheat line N95175. Pak J Bot 42(1):383–390

Alqudah AM, Sallam A, Baenziger PS, Börner A (2020) GWAS: fast-forwarding gene identification and characterization in temperate cereals: lessons from barley–a review. J Adv Res 22:119–135

Ariani A, Berny Mier y Teran JC, Gepts P (2016) Genome-wide identification of SNPs and copy number variation in common bean (*Phaseolus vulgaris* L.) using genotyping-by-sequencing (GBS). Mol Breed 36(7):1–11

Asad MA, Xia X, Wang C, He Z (2012) Molecular mapping of stripe rust resistance gene YrSN104 in Chinese wheat line Shaannong 104. Hereditas 149(4):146–152

Ashtari Mahini R, Kumar A, Elias EM, Fiedler JD, Porter LD, McPhee KE (2020) Analysis and identification of QTL for resistance to *Sclerotinia sclerotiorum* in pea (*Pisum sativum* L.). Front Genet 11:1436

Ates D, Aldemir S, Alsaleh A, Erdogmus S, Nemli S, Kahriman A, Ozkan H, Vandenberg A, Tanyolac B (2018) A consensus linkage map of lentil based on DArT markers from three RIL mapping populations. PLoS One 13(1):e0191375

Aznar-Fernández T, Barilli E, Cobos MJ, Kilian A, Carling J, Rubiales D (2020) Identification of quantitative trait loci (QTL) controlling resistance to pea weevil (*Bruchus pisorum*) in a high-density integrated DArTseq SNP-based genetic map of pea. Sci Rep 10(1):1–12

Bajaj D, Das S, Badoni S, Kumar V, Singh M, Bansal KC, Tyagi AK, Parida SK (2015) Genome-wide high-throughput SNP discovery and genotyping for understanding natural (functional) allelic diversity and domestication patterns in wild chickpea. Sci Rep 5(1):1–17

Bakhtiar F, Afshari F, Najafian G, Mohammadi M (2014) Backcross-breeding and doubled-haploid facilitated introgression of stripe rust resistance in bread wheat. Arch Phytopathol Plant Protect 47(14):1675–1685

Bansal UK, Bossolini E, Miah H, Keller B, Park RF, Bariana HS (2008) Genetic mapping of seedling and adult plant stem rust resistance in two European winter wheat cultivars. Euphytica 164(3):821–828

Bansal UK, Forrest KL, Hayden MJ, Miah H, Singh D, Bariana HS (2011) Characterisation of a new stripe rust resistance gene Yr47 and its genetic association with the leaf rust resistance gene Lr52. Theor Appl Genet 122(8):1461–1466

Bansal U, Bariana H, Wong D, Randhawa M, Wicker T, Hayden M, Keller B (2014) Molecular mapping of an adult plant stem rust resistance gene Sr56 in winter wheat cultivar Arina. Theor Appl Genet 127(6):1441–1448

Bansal UK, Muhammad S, Forrest KL, Hayden MJ, Bariana HS (2015) Mapping of a new stem rust resistance gene Sr49 in chromosome 5B of wheat. Theor Appl Genet 128(10):2113–2119

Bariana H, Forrest K, Qureshi N, Miah H, Hayden M, Bansal U (2016) Adult plant stripe rust resistance gene Yr71 maps close to Lr24 in chromosome 3D of common wheat. Mol Breed 36(7):1–10

Bastien M, Sonah H, Belzile F (2014) Genome wide association mapping of *Sclerotinia sclerotiorum* resistance in soybean with a genotyping-by-sequencing approach. Plant Genome 7(1):plantgenome2013-10

Bello MH, Moghaddam SM, Massoudi M, McClean PE, Cregan PB, Miklas PN (2014) Application of in silico bulked segregant analysis for rapid development of markers linked to Bean common mosaic virus resistance in common bean. BMC Genomics 15(1):1–13

Bhadauria V, Ramsay L, Bett KE, Banniza S (2017) QTL mapping reveals genetic determinants of fungal disease resistance in the wild lentil species Lens ervoides. Sci Rep 7(1):1–9

Bin B, Yan R, Xia X, Du J, Gang Z, Ling W, Zhu H, He Z, Wang C (2012) Mapping of quantitative trait loci for adult plant resistance to stripe rust in German wheat cultivar Ibis. J Integr Agric 11(4):528–536

Bohra A, Pandey MK, Jha UC, Singh B, Singh IP, Datta D, Chaturvedi SK, Nadarajan N, Varshney RK (2014) Genomics-assisted breeding in four major pulse crops of developing countries: present status and prospects. Theor Appl Genet 127(6):1263–1291

Boudhrioua C, Bastien M, Torkamaneh D, Belzile F (2020) Genome-wide association mapping of *Sclerotinia sclerotiorum* resistance in soybean using whole-genome resequencing data. BMC Plant Biol 20(1):1–9

Boutet G, Carvalho SA, Falque M, Peterlongo P, Lhuillier E, Bouchez O, Lavaud C, Pilet-Nayel M-L, Rivière N, Baranger A (2016) SNP discovery and genetic mapping using genotyping by sequencing of whole genome genomic DNA from a pea RIL population. BMC Genomics 17(1): 1–14

Breria CM, Hsieh C-H, Yen T-B, Yen J-Y, Noble TJ, Schafleitner R (2020) A SNP-based genome-wide association study to mine genetic loci associated to salinity tolerance in mungbean (*Vigna radiata* L.). Genes 11(7):759

Caballo C, Madrid E, Gil J, Chen W, Rubio J, Millan T (2019) Saturation of genomic region implicated in resistance to *Fusarium oxysporum* f. sp. *ciceris* race 5 in chickpea. Mol Breed 39 (2):1–11

Carrillo-Perdomo E, Vidal A, Kreplak J, Duborjal H, Leveugle M, Duarte J, Desmetz C, Deulvot C, Raffiot B, Marget P (2020) Development of new genetic resources for faba bean (*Vicia faba* L.) breeding through the discovery of gene-based SNP markers and the construction of a high-density consensus map. Sci Rep 10(1):1–14

Center C (1995) Genetic dissection of complex traits: guidelines for interpreting and reporting linkage results. Nat Genet 11(3):241–247

Chung W-H, Jeong N, Kim J, Lee WK, Lee Y-G, Lee S-H, Yoon W, Kim J-H, Choi I-Y, Choi H-K (2014) Population structure and domestication revealed by high-depth resequencing of Korean cultivated and wild soybean genomes. DNA Res 21(2):153–167

Clevenger J, Chu Y, Chavarro C, Agarwal G, Bertioli DJ, Leal-Bertioli SCM, Pandey MK, Vaughn J, Abernathy B, Barkley NA (2017) Genome-wide SNP genotyping resolves signatures of selection and tetrasomic recombination in peanut. Mol Plant 10(2):309–322

Collard BCY, Mackill DJ (2008) Marker-assisted selection: an approach for precision plant breeding in the twenty-first century. Philos Trans R Soc B Biol Sci 363(1491):557–572

Collard BCY, Jahufer MZZ, Brouwer JB, Pang ECK (2005) An introduction to markers, quantitative trait loci (QTL) mapping and marker-assisted selection for crop improvement: the basic concepts. Euphytica 142(1):169–196

Darvasi A, Soller M (1994) Selective DNA pooling for determination of linkage between a molecular marker and a quantitative trait locus. Genetics 138(4):1365–1373

Davey JW, Hohenlohe PA, Etter PD, Boone JQ, Catchen JM, Blaxter ML (2011) Genome-wide genetic marker discovery and genotyping using next-generation sequencing. Nat Rev Genet 12(7):499–510

Deokar A, Sagi M, Tar'an B (2019) Genome-wide SNP discovery for development of high-density genetic map and QTL mapping of Ascochyta blight resistance in chickpea (*Cicer arietinum* L.). Theor Appl Genet 132(6):1861–1872

Doddamani D, Khan AW, Katta MAVS, Agarwal G, Thudi M, Ruperao P, Edwards D, Varshney RK (2015) CicArVarDB: SNP and InDel database for advancing genetics research and breeding applications in chickpea. Database 2015:bav078

Dubiela C, Montecelli TDN, Lazzari F, de Souto ER, Schuster I (2019) Development and validation of SNP assays for the selection of resistance to Meloidogyne incognita in soybean. Crop Breed Appl Biotechnol 19:102–109

Dwivedi SL, Scheben A, Edwards D, Spillane C, Ortiz R (2017) Assessing and exploiting functional diversity in germplasm pools to enhance abiotic stress adaptation and yield in cereals and food legumes. Front Plant Sci 8:1461

Eagles HA, Bariana HS, Ogbonnaya FC, Rebetzke GJ, Hollamby GJ, Henry RJ, Henschke PH, Carter M (2001) Implementation of markers in Australian wheat breeding. Aust J Agric Res 52(12):1349–1356

Elshire RJ, Glaubitz JC, Sun Q, Poland JA, Kawamoto K, Buckler ES, Mitchell SE (2011) A robust, simple genotyping-by-sequencing (GBS) approach for high diversity species. PLoS One 6(5): e19379

Endo TR, Gill BS (1996) The deletion stocks of common wheat. J Hered 87(4):295–307

FAO (2010) The Second Report on the State of the World's Plant Genetic Resources for Food and Agriculture. Commission on Genetic Resources for Food and Agriculture (CGRFA). Food and Agriculture Organization of the United Nations (FAO), Rome

Fritsche-Neto R, de Souza TLPO, Pereira HS, de Faria LC, Melo LC, Novaes E, Brum IJB, Jannink J-L (2019) Association mapping in common bean revealed regions associated with anthracnose and angular leaf spot resistance. Sci Agric 76:321–327

Garg T, Mallikarjuna BP, Thudi M, Samineni S, Singh S, Sandhu JS, Kaur L, Singh I, Sirari A, Basandrai AK (2018) Identification of QTLs for resistance to Fusarium wilt and Ascochyta blight in a recombinant inbred population of chickpea (*Cicer arietinum* L.). Euphytica 214(3):1–11

Gerland P, Raftery AE, Ševčíková H, Li N, Gu D, Spoorenberg T, Alkema L, Fosdick BK, Chunn J, Lalic N (2014) World population stabilization unlikely this century. Science 346(6206): 234–237

Goutam U, Kukreja S, Yadav R, Salaria N, Thakur K, Goyal AK (2015) Recent trends and perspectives of molecular markers against fungal diseases in wheat. Front Microbiol 6:861

Gujaria-Verma N, Ramsay L, Sharpe AG, Sanderson L-A, Debouck DG, Tar'an B, Bett KE (2016) Gene-based SNP discovery in tepary bean (*Phaseolus acutifolius*) and common bean (*P. vulgaris*) for diversity analysis and comparative mapping. BMC Genomics 17(1):1–16

Han S, Yuan M, Clevenger JP, Li C, Hagan A, Zhang X, Chen C, He G (2018) A SNP-based linkage map revealed QTLs for resistance to early and late leaf spot diseases in peanut (*Arachis hypogaea* L.). Front Plant Sci 9:1012

Hyten DL, Smith JR, Frederick RD, Tucker ML, Song Q, Cregan PB (2009) Bulked segregant analysis using the GoldenGate assay to locate the Rpp3 locus that confers resistance to soybean rust in soybean. Crop Sci 49(1):265–271

Ijaz U (2018) Molecular mapping and microscopic analysis of faba bean-Uromyces viciae-fabae host-pathogen interaction

Iquira E, Humira S, François B (2015) Association mapping of QTLs for sclerotinia stem rot resistance in a collection of soybean plant introductions using a genotyping by sequencing (GBS) approach. BMC Plant Biol 15(1):1–12

Jacob C, Carrasco B, Schwember AR (2016) Advances in breeding and biotechnology of legume crops. Plant Cell Tissue Organ Cult 127(3):561–584

Janni M, Gullì M, Maestri E, Marmiroli M, Valliyodan B, Nguyen HT, Marmiroli N (2020) Molecular and genetic bases of heat stress responses in crop plants and breeding for increased resilience and productivity. J Exp Bot 71(13):3780–3802

Jiao K, Li X, Guo W, Yuan X, Cui X, Chen X (2016) Genome re-sequencing of two accessions and fine mapping the locus of lobed leaflet margins in mungbean. Mol Breed 36(9):1–12

Johannes F, Schmitz RJ (2019) Spontaneous epimutations in plants. New Phytol 221(3):1253–1259

Johnsson M, Gaynor RC, Jenko J, Gorjanc G, De Koning D-J, Hickey JM (2019) Removal of alleles by genome editing (RAGE) against deleterious load. Genet Sel Evol 51(1):1–18

Joshi RK, Nayak S (2010) Gene pyramiding-a broad spectrum technique for developing durable stress resistance in crops. Biotechnol Mol Biol Rev 5(3):51–60

Kang YJ, Kim SK, Kim MY, Lestari P, Kim KH, Ha B-K, Jun TH, Hwang WJ, Lee T, Lee J (2014) Genome sequence of mungbean and insights into evolution within Vigna species. Nat Commun 5(1):1–9

Kankanala P, Nandety RS, Mysore KS (2019) Genomics of plant disease resistance in legumes. Front Plant Sci 10:1345

Khan RR, Bariana HS, Dholakia BB, Naik SV, Lagu MD, Rathjen AJ, Bhavani S, Gupta VS (2005) Molecular mapping of stem and leaf rust resistance in wheat. Theor Appl Genet 111(5):846–850

Khazaei H, O'Sullivan DM, Stoddard FL, Adhikari KN, Paull JG, Schulman AH, Andersen SU, Vandenberg A (2021) Recent advances in faba bean genetic and genomic tools for crop improvement. Legume Sci 3(3):e75

Kim DH, Kim KH, Van K, Kim MY, Lee S-H (2010) Fine mapping of a resistance gene to bacterial leaf pustule in soybean. Theor Appl Genet 120(7):1443–1450

Kolmer JA (1996) Genetics of resistance to wheat leaf rust. Annu Rev Phytopathol 34(1):435–455

Kosambi DD (1944) The estimation of map distances from recombination values. Ann Eugenics 12: 172–175

Książkiewicz M, Nazzicari N, Yang H, Nelson MN, Renshaw D, Rychel S, Ferrari B, Carelli M, Tomaszewska M, Stawiński S (2017) A high-density consensus linkage map of white lupin highlights synteny with narrow-leafed lupin and provides markers tagging key agronomic traits. Sci Rep 7(1):1–15

Kudapa H, Ramalingam A, Nayakoti S, Chen X, Zhuang W-J, Liang X, Kahl G, Edwards D, Varshney RK (2013) Functional genomics to study stress responses in crop legumes: progress and prospects. Funct Plant Biol 40(12):1221–1233

Kumar J, Pratap A, Solanki RK (2010) Basics of molecular genetic mapping and QTL analysis in plants. Adv Biotechnol:35–52

Kumar J, Choudhary AK, Solanki RK, Pratap A (2011) Towards marker-assisted selection in pulses: a review. Plant Breed 130(3):297–313

Kumar V, Khan AW, Saxena RK, Garg V, Varshney RK (2016) First-generation HapMap in Cajanus spp. reveals untapped variations in parental lines of mapping populations. Plant Biotechnol J 14(8):1673–1681

Kumar J, Kumar S, Gupta DS, Dubey S, Gupta S, Gupta P (2019) Molecular marker assisted gene pyramiding. In: Singh M (ed) Lentils. Elsevier, Amsterdam, pp 125–139

Lam H-M, Xu X, Liu X, Chen W, Yang G, Wong F-L, Li M-W, He W, Qin N, Wang B (2010) Resequencing of 31 wild and cultivated soybean genomes identifies patterns of genetic diversity and selection. Nat Genet 42(12):1053–1059

Leal-Bertioli SCM, Cavalcante U, Gouvea EG, Ballén-Taborda C, Shirasawa K, Guimarães PM, Jackson SA, Bertioli DJ, Moretzsohn MC (2015) Identification of QTLs for rust resistance in the peanut wild species Arachis magna and the development of KASP markers for marker-assisted selection. G3 5(7):1403–1413

Lee H-I, Ho H-A, Kao C-H (2014) A new simple method for improving QTL mapping under selective genotyping. Genetics 198(4):1685–1698

Lee Y, Jeong N, Kim JH, Lee K, Kim KH, Pirani A, Ha B, Kang S, Park B, Moon J (2015) Development, validation and genetic analysis of a large soybean SNP genotyping array. Plant J 81(4):625–636

Leng P, Lübberstedt T, Xu M (2017) Genomics-assisted breeding–a revolutionary strategy for crop improvement. J Integr Agric 16(12):2674–2685

Li M, Jing F, Cao S, Lin R, Cheng G, Yu Y, Chen W, Xu S (2011) Postulation of seedlings resistance genes to yellow rust in commercial wheat cultivars from Yunnan province in China. Agric Sci China 10(11):1723–1731

Li Y-h, Zhao S, Ma J, Li D, Yan L, Li J, Qi X, Guo X, Zhang L, He W (2013) Molecular footprints of domestication and improvement in soybean revealed by whole genome re-sequencing. BMC Genomics 14(1):1–12

Li Y, Ruperao P, Batley J, Edwards D, Davidson J, Hobson K, Sutton T (2017) Genome analysis identified novel candidate genes for Ascochyta blight resistance in chickpea using whole genome re-sequencing data. Front Plant Sci 8:359

Liu M, Zhang C, Yuan C, Zhang L, Huang L, Wu J, Wang J, Zheng Y, Zhang H, Liu D (2013a) Stripe rust resistance in germplasm. Crop Sci 53(5):2014–2020

Liu Z, Bowden RL, Bai G (2013b) Molecular markers for leaf rust resistance gene Lr42 in wheat. Crop Sci 53(4):1566–1570

Liu H, Sun Z, Zhang X, Qin L, Qi F, Wang Z, Du P, Xu J, Zhang Z, Han S (2020) QTL mapping of web blotch resistance in peanut by high-throughput genome-wide sequencing. BMC Plant Biol 20(1):1–11

Lorieux M (2012) MapDisto: fast and efficient computation of genetic linkage maps. Mol Breed 30(2):1231–1235

Luo H, Pandey MK, Zhi Y, Zhang H, Xu S, Guo J, Wu B, Chen H, Ren X, Zhou X (2020) Discovery of two novel and adjacent QTLs on chromosome B02 controlling resistance against bacterial wilt in peanut variety Zhonghua 6. Theor Appl Genet 133(4):1133–1148

Maldonado dos Santos JV, Ferreira EGC, Passianotto ALL, Brumer BB, Dos Santos AB, Soares RM, Torkamaneh D, Arias CAA, Belzile F, Abdelnoor RV (2019) Association mapping of a locus that confers southern stem canker resistance in soybean and SNP marker development. BMC Genomics 20(1):1–13

Mammadov J, Aggarwal R, Buyyarapu R, Kumpatla S (2012) SNP markers and their impact on plant breeding. Int J Plant Genomics 2012:728398

Manly KF, Cudmore RH Jr, Meer JM (2001) Map Manager QTX, cross-platform software for genetic mapping. Mamm Genome 12(12):930–932

Mariyammal I, Seram D, Samyuktha SM, Karthikeyan A, Dhasarathan M, Murukarthick J, Kennedy JS, Malarvizhi D, Yang T-J, Pandiyan M (2019) QTL mapping in *Vigna radiata* × *Vigna umbellata* population uncovers major genomic regions associated with bruchid resistance. Mol Breed 39(7):1–13

Mathivathana MK, Murukarthick J, Karthikeyan A, Jang W, Dhasarathan M, Jagadeeshselvam N, Sudha M, Vanniarajan C, Karthikeyan G, Yang T-J (2019) Detection of QTLs associated with mungbean yellow mosaic virus (MYMV) resistance using the interspecific cross of *Vigna radiata* × *Vigna umbellata*. J Appl Genet 60(3):255–268

Mebrate SA, Dehne H, Pillen K, Oerke E (2008) Postulation of seedling leaf rust resistance genes in selected Ethiopian and German bread wheat cultivars. Crop Sci 48(2):507–516

Meera Krishna B, Khan MA, Khan ST (2019) Next-generation sequencing (NGS) platforms: an exciting era of genome sequence analysis. In: Microbial genomics in sustainable agroecosystems. Springer, New York, pp 89–109

Michelmore RW, Paran I, Kesseli R (1991) Identification of markers linked to disease-resistance genes by bulked segregant analysis: a rapid method to detect markers in specific genomic regions by using segregating populations. Proc Natl Acad Sci 88(21):9828–9832

Mohan M, Nair S, Bhagwat A, Krishna TG, Yano M, Bhatia CR, Sasaki T (1997) Genome mapping, molecular markers and marker-assisted selection in crop plants. Mol Breed 3(2): 87–103

Muñoz-Amatriaín M, Mirebrahim H, Xu P, Wanamaker SI, Luo M, Alhakami H, Alpert M, Atokple I, Batieno BJ, Boukar O (2017) Genome resources for climate-resilient cowpea, an essential crop for food security. Plant J 89(5):1042–1054

Nkhoma N, Shimelis H, Laing MD, Shayanowako A, Mathew I (2020) Assessing the genetic diversity of cowpea [*Vigna unguiculata* (L.) Walp.] germplasm collections using phenotypic traits and SNP markers. BMC Genet 21(1):1–16

Ojwang PPO, Eldridge T, Corredor-Moreno P, Njung'e V (2019) Genome-wide association study of resistance to bean fly and population structure of market classes of common bean. BioRxiv 2019:633545

Pandey MK, Monyo E, Ozias-Akins P, Liang X, Guimarães P, Nigam SN, Upadhyaya HD, Janila P, Zhang X, Guo B (2012) Advances in Arachis genomics for peanut improvement. Biotechnol Adv 30(3):639–651

Pandey MK, Roorkiwal M, Singh VK, Ramalingam A, Kudapa H, Thudi M, Chitikineni A, Rathore A, Varshney RK (2016) Emerging genomic tools for legume breeding: current status and future prospects. Front Plant Sci 7:455

Pandey MK, Agarwal G, Kale SM, Clevenger J, Nayak SN, Sriswathi M, Chitikineni A, Chavarro C, Chen X, Upadhyaya HD (2017a) Development and evaluation of a high density genotyping 'Axiom_Arachis' array with 58 K SNPs for accelerating genetics and breeding in groundnut. Sci Rep 7(1):1–10

Pandey MK, Khan AW, Singh VK, Vishwakarma MK, Shasidhar Y, Kumar V, Garg V, Bhat RS, Chitikineni A, Janila P (2017b) QTL-seq approach identified genomic regions and diagnostic markers for rust and late leaf spot resistance in groundnut (*Arachis hypogaea* L.). Plant Biotechnol J 15(8):927–941

Pavan S, Lotti C, Marcotrigiano AR, Mazzeo R, Bardaro N, Bracuto V, Ricciardi F, Taranto F, D'Agostino N, Schiavulli A (2017) A distinct genetic cluster in cultivated chickpea as revealed by genome-wide marker discovery and genotyping. Plant Genome 10(2):plantgenome2016-11

Pavan S, Bardaro N, Fanelli V, Marcotrigiano AR, Mangini G, Taranto F, Catalano D, Montemurro C, De Giovanni C, Lotti C (2019) Genotyping by sequencing of cultivated lentil (*Lens culinaris* Medik.) highlights population structure in the Mediterranean gene pool associated with geographic patterns and phenotypic variables. Front Genet 10:872

Perseguini JMKC, Oblessuc PR, Rosa JRBF, Gomes KA, Chiorato AF, Carbonell SAM, Garcia AAF, Vianello RP, Benchimol-Reis LL (2016) Genome-wide association studies of anthracnose and angular leaf spot resistance in common bean (*Phaseolus vulgaris* L.). PLoS One 11(3): e0150506

Phansak P, Soonsuwon W, Hyten DL, Song Q, Cregan PB, Graef GL, Specht JE (2016) Multi-population selective genotyping to identify soybean [*Glycine max* (L.) Merr.] seed protein and oil QTLs. G3 6(6):1635–1648

Polanco C, Sáenz de Miera LE, González AI, García P, Fratini R, Vaquero F, Vences FJ, Pérez de la Vega M (2019) Construction of a high-density interspecific (*Lens culinaris* × *L. odemensis*) genetic map based on functional markers for mapping morphological and agronomical traits, and QTLs affecting resistance to Ascochyta in lentil. PLoS One 14(3):e0214409

Qamar M, Ahmad SD, Shah AH, Wellings CR, Batool F (2008) Postulation of stripe rust resistant genes in some Australian bread wheat cultivars and their response to temperature. Pak J Bot 40(6):2573–2585

Rafalski JA (2010) Association genetics in crop improvement. Curr Opin Plant Biol 13(2):174–180

Rana M, Sood A, Hussain W, Kaldate R, Sharma TR, Gill RK, Kumar S, Singh S (2019) Gene pyramiding and multiple character breeding. In: Lentils. Elsevier, Amsterdam, pp 83–124

Randhawa M, Bansal U, Valárik M, Klocová B, Doležel J, Bariana H (2014) Molecular mapping of stripe rust resistance gene Yr51 in chromosome 4AL of wheat. Theor Appl Genet 127(2): 317–324

Randhawa MS, Bariana HS, Mago R, Bansal UK (2015) Mapping of a new stripe rust resistance locus Yr57 on chromosome 3BS of wheat. Mol Breed 35(2):1–8
Randhawa M, Bansal U, Lillemo M, Miah H, Bariana H (2016) Postulation of rust resistance genes in Nordic spring wheat genotypes and identification of widely effective sources of resistance against the Australian rust flora. J Appl Genet 57(4):453–465
Reynolds MP (2001) Application of physiology in wheat breeding. CIMMYT
Reynolds MP, Pask AJD, Mullan DM (2012) Physiological breeding I: interdisciplinary approaches to improve crop adaptation. CIMMYT
Roorkiwal M, Jain A, Kale SM, Doddamani D, Chitikineni A, Thudi M, Varshney RK (2018) Development and evaluation of high-density Axiom® Cicer SNP Array for high-resolution genetic mapping and breeding applications in chickpea. Plant Biotechnol J 16(4):890–901
Saeed A, Darvishzadeh R (2017) Association analysis of biotic and abiotic stresses resistance in chickpea (*Cicer* spp.) using AFLP markers. Biotechnol Biotechnol Equip 31(4):698–708
Saghai Maroof MA, Jeong SC, Gunduz I, Tucker DM, Buss GR, Tolin SA (2008) Pyramiding of soybean mosaic virus resistance genes by marker-assisted selection. Crop Sci 48(2):517–526
Sanger F, Nicklen S, Coulson AR (1977) DNA sequencing with chain-terminating inhibitors. Proc Natl Acad Sci 74(12):5463–5467
Santos C, Almeida NF, Alves ML, Horres R, Krezdorn N, Leitão ST, Aznar-Fernández T, Rotter B, Winter P, Rubiales D (2018) First genetic linkage map of Lathyrus cicera based on RNA sequencing-derived markers: key tool for genetic mapping of disease resistance. Hortic Res 5:45
Saxena RK, Kale SM, Kumar V, Parupali S, Joshi S, Singh V, Garg V, Das RR, Sharma M, Yamini KN (2017) Genotyping-by-sequencing of three mapping populations for identification of candidate genomic regions for resistance to sterility mosaic disease in pigeonpea. Sci Rep 7(1):1–10
Schafleitner R, Huang S, Chu S, Yen J, Lin C, Yan M, Krishnan B, Liu M, Lo H, Chen C, Chen L, Wu D, Bui T, Ramasamy S, Tung C, Nair R (2016) Identification of single nucleotide polymorphism markers associated with resistance to bruchids (*Callosobruchus* spp.) in wild mungbean (*Vigna radiata* var. *sublobata*) and cultivated *V. radiata* through genotyping by sequencing and quantitative trait locus a. BMC Plant Biol 16(1):159
Setia RC, Setia N, Harash N (2008) The 'omics' technologies and crop improvement. In: Crop improvement: strategies and applications. International Publishing House Pvt. Ltd, New Delhi, pp 1–17
Sharma P, Sharma RB (2014) Molecular characterization of wheat germplasm for leaf and stem rust resistance genes using linked SSR markers. Can J Plant Breed 2:15–27
Sharma S, Upadhyaya HD, Varshney RK, Gowda CLL (2013) Pre-breeding for diversification of primary gene pool and genetic enhancement of grain legumes. Front Plant Sci 4:309
Shen H, He H, Li J, Chen W, Wang X, Guo L, Peng Z, He G, Zhong S, Qi Y (2012) Genome-wide analysis of DNA methylation and gene expression changes in two Arabidopsis ecotypes and their reciprocal hybrids. Plant Cell 24(3):875–892
Sim S-C, Durstewitz G, Plieske J, Wieseke R, Ganal MW, Van Deynze A, Hamilton JP, Buell CR, Causse M, Wijeratne S (2012) Development of a large SNP genotyping array and generation of high-density genetic maps in tomato. PLoS One 7(7):e40563
Simons K, Abate Z, Chao S, Zhang W, Rouse M, Jin Y, Elias E, Dubcovsky J (2011) Genetic mapping of stem rust resistance gene Sr13 in tetraploid wheat (*Triticum turgidum* ssp. *durum* L.). Theor Appl Genet 122(3):649–658
Singh R (2015) Medicinal plants: a review. J Plant Sci 3(1):50–55
Singh BD, Singh AK (2015) Marker-assisted plant breeding: principles and practices. Springer, New York
Singh D, Park RF, McIntosh RA (2001) Postulation of leaf (brown) rust resistance genes in 70 wheat cultivars grown in the United Kingdom. Euphytica 120(2):205–218
Singh VK, Khan AW, Saxena RK, Sinha P, Kale SM, Parupalli S, Kumar V, Chitikineni A, Vechalapu S, Sameer Kumar CV (2017) Indel-seq: a fast-forward genetics approach for

identification of trait-associated putative candidate genomic regions and its application in pigeonpea (*Cajanus cajan*). Plant Biotechnol J 15(7):906–914

Sinha P, Singh VK, Saxena RK, Khan AW, Abbai R, Chitikineni A, Desai A, Molla J, Upadhyaya HD, Kumar A (2020) Superior haplotypes for haplotype-based breeding for drought tolerance in pigeonpea (*Cajanus cajan* L.). Plant Biotechnol J 18(12):2482–2490

Soliman NEK, Abdelbacki AMM, Najeeb MAA, Omara RI (2012) Geographical distribution of physiologic races of Puccinia triticina and postulation of resistance genes in new wheat cultivars in Egypt. Int J Phytopathol 1(1):73–80

Somta P, Chen J, Yundaeng C, Yuan X, Yimram T, Tomooka N, Chen X (2019) Development of an SNP-based high-density linkage map and QTL analysis for bruchid (*Callosobruchus maculatus* F.) resistance in black gram (*Vigna mungo* (L.) Hepper). Sci Rep 9(1):1–9

Sonah H, Bastien M, Iquira E, Tardivel A, Legare G, Boyle B, Belzile F (2013) An improved genotyping by sequencing (GBS) approach offering increased versatility and efficiency of SNP discovery and genotyping. PLoS One 8:e54603

Song QJ, Shi JR, Singh S, Fickus EW, Costa JM, Lewis J, Gill BS, Ward R, Cregan PB (2005) Development and mapping of microsatellite (SSR) markers in wheat. Theor Appl Genet 110(3): 550–560

Song Q, Hyten DL, Jia G, Quigley CV, Fickus EW, Nelson RL, Cregan PB (2013) Development and evaluation of SoySNP50K, a high-density genotyping array for soybean. PLoS One 8(1): e54985

Song Q, Hyten DL, Jia G, Quigley CV, Fickus EW, Nelson RL, Cregan PB (2015a) Fingerprinting soybean germplasm and its utility in genomic research. G3 5(10):1999–2006

Song Q, Jia G, Hyten DL, Jenkins J, Hwang E-Y, Schroeder SG, Osorno JM, Schmutz J, Jackson SA, McClean PE (2015b) SNP assay development for linkage map construction, anchoring whole-genome sequence, and other genetic and genomic applications in common bean. G3 5(11):2285–2290

Song Q, Yan L, Quigley C, Fickus E, Wei H, Chen L, Dong F, Araya S, Liu J, Hyten D (2020) Soybean BARCSoySNP6K: an assay for soybean genetics and breeding research. Plant J 104(3):800–811

Sudheesh S, Rodda MS, Davidson J, Javid M, Stephens A, Slater AT, Cogan NOI, Forster JW, Kaur S (2016) SNP-based linkage mapping for validation of QTLs for resistance to Ascochyta blight in lentil. Front Plant Sci 7:1604

Suresh S, Malathi D (2013) Gene pyramiding for biotic stress tolerance in crop plants. Wkly Sci Res J 1:1–14

Taran B, Warkentin TD, Vandenberg A (2013) Fast track genetic improvement of Ascochyta blight resistance and double podding in chickpea by marker-assisted backcrossing. Theor Appl Genet 126(6):1639–1647

Tayeh N, Aluome C, Falque M, Jacquin F, Klein A, Chauveau A, Bérard A, Houtin H, Rond C, Kreplak J (2015) Development of two major resources for pea genomics: the GenoPea 13.2 K SNP Array and a high-density, high-resolution consensus genetic map. Plant J 84(6):1257–1273

Thudi M, Khan AW, Kumar V, Gaur PM, Katta K, Garg V, Roorkiwal M, Samineni S, Varshney RK (2016) Whole genome resequencing reveals genome-wide variations among parental lines of 16 mapping populations in chickpea (*Cicer arietinum* L.). BMC Plant Biol 16(1):53–64

Todorovska E, Christov N, Slavov S, Christova P, Vassilev D (2009) Biotic stress resistance in wheat—breeding and genomic selection implications. Biotechnol Biotechnol Equip 23(4): 1417–1426

Tran DT, Steketee CJ, Boehm JD Jr, Noe J, Li Z (2019) Genome-wide association analysis pinpoints additional major genomic regions conferring resistance to soybean cyst nematode (*Heterodera glycines Ichinohe*). Front Plant Sci 10:401

Tran QH, Bui NH, Kappel C, Dau NTN, Nguyen LT, Tran TT, Khanh TD, Trung KH, Lenhard M, Vi SL (2020) Mapping-by-sequencing via MutMap identifies a mutation in ZmCLE7 underlying fasciation in a newly developed EMS mutant population in an elite tropical maize inbred. Genes 11(3):281

Tribhuvan KU, Kumar K, Sevanthi AM, Gaikwad K (2018) MutMap: a versatile tool for identification of mutant loci and mapping of genes. Indian J Plant Physiol 23(4):612–621

Tyagi S, Mir RR, Kaur H, Chhuneja P, Ramesh B, Balyan HS, Gupta PK (2014) Marker-assisted pyramiding of eight QTLs/genes for seven different traits in common wheat (*Triticum aestivum* L.). Mol Breed 34(1):167–175

Tyagi AK, Bajaj D, Sharma S, Upadhyaya HD, Parida SK, Kumar V, Gowda CLL, Das S (2016) Genetic dissection of seed-iron and zinc concentrations in chickpea. Sci Rep 6:24050

Upadhyaya HD, Yadav D, Dronavalli N, Gowda CLL, Singh S (2010) Mini core germplasm collections for infusing genetic diversity in plant breeding programs. Electron J Plant Breed 1(4):1294–1309

Valdisser PAMR, Pereira WJ, Almeida Filho JE, Müller BSF, Coelho GRC, de Menezes IPP, Vianna JPG, Zucchi MI, Lanna AC, Coelho ASG (2017) In-depth genome characterization of a Brazilian common bean core collection using DArTseq high-density SNP genotyping. BMC Genomics 18(1):1–19

Valliyodan B, Qiu D, Patil G, Zeng P, Huang J, Dai L, Chen C, Li Y, Joshi T, Song L (2016) Landscape of genomic diversity and trait discovery in soybean. Sci Rep 6(1):1–10

Van Kessel C, Hartley C (2000) Agricultural management of grain legumes: has it led to an increase in nitrogen fixation? Field Crop Res 65(2–3):165–181

Vance CP, Graham PH, Allan DL (2000) Biological nitrogen fixation: phosphorus-a critical future need? In: Nitrogen fixation: from molecules to crop productivity. Springer, New York, pp 509–514

Vanzetti LS, Campos P, Demichelis M, Lombardo LA, Aurelia PR, Vaschetto LM, Bainotti CT, Helguera M (2011) Identification of leaf rust resistance genes in selected Argentinean bread wheat cultivars by gene postulation and molecular markers. Electron J Biotechnol 14(3):9

Varshney RK, Hoisington DA, Tyagi AK (2006) Advances in cereal genomics and applications in crop breeding. Trends Biotechnol 24(11):490–499

Varshney RK, Hoisington DA, Upadhyaya HD, Gaur PM, Nigam SN, Saxena K, Vadez V, Sethy NK, Bhatia S, Aruna R (2007) Molecular genetics and breeding of grain legume crops for the semi-arid tropics. In: Genomics-assisted crop improvement. Springer, New York, pp 207–241

Varshney RK, Close TJ, Singh NK, Hoisington DA, Cook DR (2009) Orphan legume crops enter the genomics era! Curr Opin Plant Biol 12(2):202–210

Varshney RK, Bansal KC, Aggarwal PK, Datta SK, Craufurd PQ (2011) Agricultural biotechnology for crop improvement in a variable climate: hope or hype? Trends Plant Sci 16(7):363–371

Varshney RK, Mohan SM, Gaur PM, Gangarao N, Pandey MK, Bohra A, Sawargaonkar SL, Chitikineni A, Kimurto PK, Janila P (2013) Achievements and prospects of genomics-assisted breeding in three legume crops of the semi-arid tropics. Biotechnol Adv 31(8):1120–1134

Varshney RK, Mohan SM, Gaur PM, Chamarthi SK, Singh VK, Srinivasan S, Swapna N, Sharma M, Pande S, Singh S, Kaur L (2014) Marker-assisted backcrossing to introgress resistance to Fusarium Wilt Race 1 and Ascochyta Blight in C 214, an Elite Cultivar of Chickpea. Plant Genome 7(1):plantgenome2013.2010.0035. https://doi.org/10.3835/plantgenome2013.10.0035

Varshney RK, Saxena RK, Upadhyaya HD, Khan AW, Yu Y, Kim C, Rathore A, Kim D, Kim J, An S (2017) Whole-genome resequencing of 292 pigeonpea accessions identifies genomic regions associated with domestication and agronomic traits. Nat Genet 49(7):1082–1088

Varshney RK, Pandey MK, Bohra A, Singh VK, Thudi M, Saxena RK (2019) Toward the sequence-based breeding in legumes in the post-genome sequencing era. Theor Appl Genet 132(3):797–816

Varshney RK, Sinha P, Singh VK, Kumar A, Zhang Q, Bennetzen JL (2020) 5Gs for crop genetic improvement. Curr Opin Plant Biol 56:190–196

Voorrips R (2002) MapChart: software for the graphical presentation of linkage maps and QTLs. J Hered 93(1):77–78

Vuong TD, Walker DR, Nguyen BT, Nguyen TT, Dinh HX, Hyten DL, Cregan PB, Sleper DA, Lee JD, Shannon JG (2016) Molecular characterization of resistance to soybean rust (*Phakopsora pachyrhizi* Syd. & Syd.) in soybean cultivar DT 2000 (PI 635999). PLoS One 11(12):e0164493

Wang S (2007) Windows QTL cartographer 2.5. http://statgen.ncsu.edu/qtlcart/WQTLCart.html

Wang D, Yang T, Liu R, Li N, Wang X, Sarker A, Zhang X, Li R, Pu Y, Li G (2020) RNA-Seq analysis and development of SSR and KASP markers in lentil (*Lens culinaris Medikus* subsp. *culinaris*). Crop J 8(6):953–965

Webb A, Cottage A, Wood T, Khamassi K, Hobbs D, Gostkiewicz K, White M, Khazaei H, Ali M, Street D (2016) A SNP-based consensus genetic map for synteny-based trait targeting in faba bean (*Vicia faba* L.). Plant Biotechnol J 14(1):177–185

Wei W, Mesquita ACO, Figueiró AA, Wu X, Manjunatha S, Wickland DP, Hudson ME, Juliatti FC, Clough SJ (2017) Genomewide association mapping of resistance to a Brazilian isolate of *Sclerotinia sclerotiorum* in soybean genotypes mostly from Brazil. BMC Genomics 18(1):1–16

Wen L, Chang H-X, Brown PJ, Domier LL, Hartman GL (2019) Genome-wide association and genomic prediction identifies soybean cyst nematode resistance in common bean including a syntenic region to soybean Rhg1 locus. Hortic Res 6:9

Wu X, Wang B, Wu X, Lu Z, Li G, Xu P (2018) SNP marker-based genetic mapping of rust resistance gene in the vegetable cowpea landrace ZN016. Legume Res 41(2):222–225

Xiong H, Shi A, Mou B, Qin J, Motes D, Lu W, Ma J, Weng Y, Yang W, Wu D (2016) Genetic diversity and population structure of cowpea (*Vigna unguiculata* L. Walp). PLoS One 11(8): e0160941

Xu Y (2010) Molecular plant breeding. CABI

Yadav CB, Bhareti P, Muthamilarasan M, Mukherjee M, Khan Y, Rathi P, Prasad M (2015) Genome-wide SNP identification and characterization in two soybean cultivars with contrasting mungbean yellow mosaic India virus disease resistance traits. PLoS One 10(4):e0123897

Yang H, Tao Y, Zheng Z, Li C, Sweetingham MW, Howieson JG (2012) Application of next-generation sequencing for rapid marker development in molecular plant breeding: a case study on anthracnose disease resistance in *Lupinus angustifolius* L. BMC Genomics 13(1):1–12

Yuan JH, Li JH, Yuan JJ, Jia KL, Li SF, Deng CL, Gao WJ (2017) The application of MutMap in forward genetic studies based on whole-genome sequencing. Hereditas 39(12):1168–1177

Zhong C, Sun S, Yao L, Ding J, Duan C, Zhu Z (2018) Fine mapping and identification of a novel Phytophthora root rot resistance locus RpsZS18 on chromosome 2 in soybean. Front Plant Sci 9: 44

Zhou Z, Jiang Y, Wang Z, Gou Z, Lyu J, Li W, Yu Y, Shu L, Zhao Y, Ma Y (2015) Resequencing 302 wild and cultivated accessions identifies genes related to domestication and improvement in soybean. Nat Biotechnol 33(4):408–414

Genomics of Abiotic Stress in Rice bean (*Vigna umbellata*) 7

A. Karthikeyan, V. G. Renganathan, M. Pandiyan, and N. Senthil

Abstract

Rice bean (*Vigna umbellata*) also known as red bean and mambi bean is a multipurpose crop cultivated mainly in South and Southeast Asian countries. It possesses several nutritional and climate-resilient attributes; however, the potential value of the crop is taken too lightly, and rice bean has remained an underutilized crop. Rice bean is widely grown in marginal lands under water-deficit conditions and environmental stresses. However, the impact of abiotic stresses in rice bean is still unknown. So far, few studies have investigated the rice bean responses to abiotic stresses (i.e., drought, salinity, and heavy metal). Though they are preliminary studies, no comprehensive studies were performed to understand the responses in terms of genetics, physiology, and gene regulatory networks on rice bean under abiotic stress. Recent advances in genomics enable

A. Karthikeyan (✉)
Department of Biotechnology, Centre of Excellence in Innovation, Agricultural College and Research Institute, Tamil Nadu Agricultural University, Madurai, Tamil Nadu, India

Subtropical Horticulture Research Institute, Jeju National University, Jeju, South Korea

V. G. Renganathan
Department of Biotechnology, Centre of Excellence in Innovation, Agricultural College and Research Institute, Tamil Nadu Agricultural University, Madurai, Tamil Nadu, India

M. Pandiyan
Agricultural College and Research Institute, Tamil Nadu Agricultural University, Vazhavachanur, Tamil Nadu, India

N. Senthil
Department of Biotechnology, Centre of Excellence in Innovation, Agricultural College and Research Institute, Tamil Nadu Agricultural University, Madurai, Tamil Nadu, India

Department of Plant Molecular Biology and Bioinformatics, Center for Plant Molecular Biology and Biotechnology, Tamil Nadu Agricultural University, Coimbatore, Tamil Nadu, India

P. Muthu Arjuna Samy et al. (eds.), *Legumes: Physiology and Molecular Biology of Abiotic Stress Tolerance*, https://doi.org/10.1007/978-981-19-5817-5_7

us to improve the understanding of stress-related genes and their associated molecular regulatory networks in plants. However, genomic studies for abiotic stress in rice bean are limited or almost none when compared with other *Vigna* species. In this chapter, we summarize the studies conducted to understand the rice bean responses to abiotic stresses and detail the research progress in the area of rice bean genomics and the status as well as opportunities for genomic research for abiotic stress in rice bean.

Keywords

Abiotic stress · Genetics · Genomics · Physiology · Rice bean and *Vigna* spp.

7.1 Introduction

Legumes are the third largest family of flowering plants and possess an important nutritional source for both humans and animals. Despite their positive impact on world food and nutrition security, they have also played a dynamic part in world agriculture because of their unique capacity for symbiotic nitrogen fixation. But, so far, ten legume species alone have been domesticated and considered as a human diet. The *Vigna* genus is a major and large set of legumes comprising more than 200 cultivated and wild species, and mungbean (*V. radiata*), black gram (*V. mungo*), cowpea (*V. unguiculata*), adzuki bean (*V. angularis*), and rice bean (*V. umbellata*) are the most important species that contribute substantially to global production and nutritional security (Katoch 2020). Rice bean is also known as red bean or mambi bean and is a diploid ($2n = 22$) crop and mainly grown in South and Southeast Asian countries including India, Nepal, Bhutan, Myanmar, China, Vietnam, Thailand, and Indonesia. It is one of the eight *Vigna* species domesticated in Asia and has been found to have a very close relationship to the adzuki bean compared to the other *Vigna* species, possessing a similar evolutionary pattern (Kaga et al. 1996; Tomooka 2002; Tomooka et al. 2006). The nutritional value of rice bean is reasonably similar to other legumes. It is a potential source of proteins, vitamins, dietary fiber, minerals, and nutrients (Pattanayak et al. 2019). Rice bean is also recognized to be a climate-resilient crop because of its good adaptability to diverse environmental conditions and less susceptibility to environmental stresses compared to other *Vigna* species. Despite all the benefits, rice bean has remained a less utilized crop. Plants are often exposed to many abiotic stresses (i.e., drought, heat, salinity, cold, flooding, and heavy metal); thus, many researchers have studied the reduction in growth and productivity due to abiotic stresses. But the progress in understanding and improving the plant tolerance to environmental stresses has been slow due to the complexity of the trait (Araújo et al. 2015).

Rice bean is widely grown in marginal and unused land, water-deficit areas, and exhausted soils (Dhillon and Tanwar 2018). It has been designated as a future crop expected for domestication through farmers in marginal lands. So far, no studies have elaborately discussed the yield losses and the impact of abiotic stress in rice bean. Thus, it is essential to understand the responses of rice bean plants to abiotic stresses. It not only does help to improve the stress tolerance in rice bean but is also useful to build tolerance in its closely related species. Few studies have investigated the effect of drought, salinity, and heavy metal stresses on rice bean responses (Atta et al. 2020, 2021; Fan et al. 2014, 2015; Sritongtae et al. 2017; Wanek and Richter 1997). Comprehensive studies conducted to understand the responses in terms of genetics, physiology, and gene regulatory networks on rice bean under abiotic stresses are very limited. Genomics facilitates the study of the genetic base of traits and evaluates the performance of plants based on sequence information. Recent advances in genomics have provided a simple and cost-effective tool and help to improve our understanding of stress-related genes and their associated molecular regulatory network in various crop plants. Similarly, genomic tools facilitate revealing the major stress-inducible genes and transcription factors associated with abiotic stress tolerance in legume crops like mung bean, cowpea, and adzuki bean (Breria et al. 2020; Kumar et al. 2020; Ravelombola et al. 2021; Zhu et al. 2020; Zuo et al. 2017). With this backdrop, this chapter aims to discuss the efforts made to know the rice bean responses in terms of physiology and genetics to abiotic stresses and detail the research developments in the field of rice bean genomics, as well as the status and opportunities in genomic research for abiotic stress in rice bean.

7.2 Genetic Resources of Rice bean

The rice bean germplasm accessions have been conserved in more than 24 countries, with the largest contribution from India and Nepal. The National Bureau of Plant Genetic Resources (NBPGR); the Indian Institute for Pulses Research (IIPR), India; Nepal Agricultural Research Council (NARC), Nepal; the Institute of Crop Genetic Resources (ICGR), China; and the World Vegetable Center (AVRDC), Taiwan, are maintaining over 4000 accessions from various parts of the world. With the series of conservation and evaluation of rice bean programs, the NBPGR, India, conserves 1993 indigenous collections (ICs) and 96 exotic collections (ECs). Out of these, 23 accessions were released as improved cultivars across the country (Pattanayak et al. 2019). Globally, about 4892 rice bean accessions are conserved in various gene banks (Table 7.1). However, most of the accessions had no passport data due to the lack of systematic phenotypic and genotypic characterization across different gene banks. According to Pattanayak et al. (2019), comparative report on rice bean characterization and evaluation across the world revealed accessions of Nepal, which showed higher number of pods per plant and high yield per plant, while Indian accessions were maximum in the number of branches per plant, pod length, and seeds per pod. Apart from this, several F_2 and recombinant inbred line (RIL) populations have been developed from the cross made between rice bean and its close relatives (Isemura et al. 2010; Kaga et al. 2000; Mathivathana et al. 2019; Somta et al. 2006). However, no registration data are available for the rice bean breeding population in the public domain. So far, large-scale germplasm screening against abiotic stresses has not been conducted. To accomplish major gaps in genetic

Table 7.1 Major organizations across the globe conserving the rice bean (till February 2021)

S. No.	Organization	Number of accessions
1	National Bureau of Plant Genetic Resources, India	2071
2	Institute of Crop Genetic Resources, China	1363
3	National Agriculture and Food Research Organization, Japan	399
4	World Vegetable Center, Taiwan	351
5	Nepal Agricultural Research Council, Nepal	300
6	National Plant Genetic Resource Laboratory, Philippines	161
7	Agricultural Research Service-Germplasm Resources Information Network, USA	147

Source: Pattanayak et al. (2019)

enhancement of rice bean, the Department of Biotechnology, India, has a multi-institutional project with the National Institute of Plant Genome Research (NIPGR), ICAR-National Bureau of Plant Genetic Resource (NBPGR), and ICMR-National Institute of Nutrition (NIN) entitled "Integrated genomic strategy for accelerating domestication of Rice bean" under the mission program "Genetic Enhancement of Minor Pulses" (ricebeanportal.com). This multi-institutional project is being implemented to decipher the domestication pattern as well as the development of molecular markers, and gene regulatory mechanisms controlling desirable morpho-agronomic and nutritional quality traits via next-generation sequencing (NGS) tools that are crucial for genetic enhancement, popularization, commercialization, and domestication of rice bean.

7.3 Physiology and Genetics of Abiotic Stress

Abiotic stresses (drought, heat, salinity, flooding, cold, and metal) are generally interlinked with each other and exist in the form of osmotic stress, malfunction of ion distribution, and plant cell homeostasis. For instance, drought can cause damage at various growth stages of the plants, with diverse effects on plant function, and therefore needs different tolerance mechanisms (Langridge et al. 2006). Many abiotic stresses including heat, salinity, and high level of toxic solutes and nutrient deficiency frequently happen during drought stress, and they are also varied based on the location and time (Fleury et al. 2010; Mittler 2006; Salekdeh et al. 2009). It was seen that plants exhibit a variety of mechanisms and combinations of mechanisms to tolerate each of these stresses. Plant physiology and genetic studies have improved our knowledge of plant responses to abiotic stresses and the foundation of major changes in tolerance levels.

To date, very few studies have been conducted to know the physiological response of rice bean plants to drought, salinity, cold, and metal stresses using a limited number of germplasms. Many researchers stated that understanding the physiological responses of plants to abiotic stress is important to reduce the harmful effect of the abiotic stress and increase the yield. So far, different physiological responses to abiotic stress have been identified in plants. Also, numerous key physiological traits alleviate the effect of drought stress on legume crops (Valliyodan and Nguyen 2008). Genes regulating these physiological changes are vital for plant breeders as they are valuable sources to genetically enhance the abiotic stress tolerance by breeding program. However, in the rice bean, attempts to dissect the physiological response of the various abiotic stress responses are still in the initial phases. For instance, Nandeshwar et al. (2017) standardized salinity concentration at 120 mM NaCl for effective salt tolerance screening. Based on that, the rice bean cultivars KRB-77, KRB-10, and KRB-273 were reported to have the best performance against saline stress. Similarly, drought can affect plants, primarily the germination, creating altered metabolic functions followed by reduction of photosynthetic pigments, carbohydrate biosynthesis, nutrient metabolism, and growth promoters. Atta et al. (2021) recorded that the speed of germination was highly reduced at 18% PEG and 200 mM NaCl concentration over that of control in rice bean cultivars. During drought, saline, and metal stress, the seed protein content was invariably found higher even at medium to higher doses (Atta et al. 2021). However, when compared to saline, drought produces a more drastic effect on biochemical and physiological parameters of germinating seed. Similarly, for metal stress, at the same equimolar concentrations, the highest intensity of copper stress produces more detrimental effects on water uptake % in germination than lead stress.

Aluminum (Al) toxicity is a widespread problem in crop plants growing in acidic soils (Kochian 1995). Nitric oxide (NO), on the other hand, is a signaling molecule modulating the physiological and biochemical function in turn in the adaptation of plant's heavy metal toxicity. The defense response in plants is caused by the simultaneous and balanced synthesis of NO and ROS, according to experimental findings (Bright et al. 2006; Zaninotto et al. 2006). However, in rice bean, aluminum exposure of rice bean seedling to 25 μm for 24 h caused a 35% inhibition of root elongation compared to control. In general, at certain concentrations, NO triggers the signal transduction pathways of defense mechanisms to Al toxicity in crop plants (Wang and Yang 2005), while in rice bean, the exogenous application of NO plays a destructive role in mediating aluminum-induced toxicity (Zhou et al. 2012). In another study, Atta et al. (2020) justified the varied detrimental effect of drought and salinity stress on the growth and physiology of rice bean seedlings. During drought, the membrane structure of the cell wall is disorganized due to the decreased content of starch, total soluble sugar, and phenol content of leaf; similarly, saline stress decreases the photosynthetic pigment, relative water content (RWC), and leaf protein. Chhetri and Lama (2013) identified the black cultivar of *V. umbellata* to be cold tolerant, while the grey cultivar exhibited high salt tolerance. The reason for the response to the stress was justified as the accumulation of proline and malondialdehyde (MDA) that gradually increased during saline in the grey cultivar, and it was double the time at low temperature (chilling) in the black cultivar of rice bean. During drought stress, the accumulation of low-molecular-mass compounds such as ononitol or pinitol has been reported in several legume species (Keller and Ludlow 1993). These low molecular compounds increased during abiotic stress due to diminished catabolism in crop plants (Wanek and Richter 1997). At the same time, several legume species show reduced myoinositol and its substrate ononitol in various plant parts (Keller and Ludlow 1993). In contrast, in *V. umbellata*, the ononitol content was found to have significantly enhanced level in both leaves and stems during the drought stress, and they were metabolically very stable and did not decrease within 4 days of rewatering (Wanek and Richter 1997). Similarly, heavy metal, at limited concentrations, acts as a cofactor for various metabolites involved in growth and development (Chhetri et al. 2004). However, at higher concentrations, it alters metabolic functions and has many deleterious effects. For instance, increased lead concentrations cause reduced DNA, RNA, and protein in rice embryos (Hilmy et al. 1985). Membrane integrity was highly affected with an increased dose of lead and cadmium metals, which causes decreased peroxidase activity that controls/regulates the membrane permeability (Chhetri et al. 2004). The study also reported that there is considerable variation in tolerance against metal toxicity in rice bean cultivars, which needs to be validated across the location.

Genetic analysis is a better option to explore the abiotic stress tolerance mechanism in the plant, and the information obtained from the genetic analysis could be useful for marker-assisted breeding (MAB) or genetic modification. For genetic studies, discovering the naturally occurring variation of abiotic stress tolerance in cultivars, landraces, and wild relatives of crops, and analyzing the traits, which are involved in tolerance, is important. Also, potential genetic materials and consistent and precise phenotyping methods to measure the traits are essential (Berger et al. 2010; Roy et al. 2011). Some components of abiotic stress are difficult to measure, whereas some traits are easy. Recently, advances in phenomics improved the measurement of traits and also assist the genetic analysis of abiotic stress tolerance in different crops. But in the case of rice bean, no genetic studies were performed to elucidate the abiotic stress tolerance mechanism. The studies conducted in many crops revealed that the tolerance to abiotic stresses such as drought, heat, salinity, flooding, cold, and metal is typically a complex quantitative trait that is influenced by many genetic and environmental interactions (Witcombe et al. 2008). Any kind of selection for these stresses is needed to evaluate more than one time and in more than one environment in the specific area because abiotic stress (i.e., drought) tolerance has low heritability. To solve the low heritability of abiotic stress tolerance, plant breeders have integrated genomic tools into their programs and attempted to improve abiotic stress tolerance. Combining information from physiology, genetics, and genomics could be useful to discover the abiotic stress-tolerant genotypes possessing the maximum number of genes governing abiotic stress tolerance.

7.4 Genomic Resources in Rice bean

7.4.1 Genome Sequences

The whole-genome sequences (WGS) offer a genetic blueprint of the species and assist the development of genomic tools that would accelerate the progress of better cultivars by the use of MAB. Recent advances in genomics helped to obtain the

WGS for the important *Vigna* species including mung bean (Kang et al. 2014), cowpea (Lonardi et al. 2019), and adzuki bean (Kang et al. 2015; Yang et al. 2015). This genomic breakthrough has facilitated the knowledge of their genome structure and also aided to develop an impressive number of genomic tools for *Vigna* species that offer better chances to identify the genes and their relationship with the development of precise phenotypes. The rice bean genome is expected to publish soon, and the unpublished genome has the size of 414 Mb and consists of 31,276 highly confidential genes from 15,521 scaffolds. It was close to adzuki bean, mung bean, and cowpea, though adzuki bean is the closest (Kaul et al. 2019).

7.4.2 Molecular Markers and Transcriptomes

Molecular markers such as RFLP, RAPD, ISSR, SSR, SRAP, and SNP are indispensable genomic tools to study the genetic variation and identify the genomic regions/genes associated with a specific trait (Karthikeyan et al. 2014). Among the various marker systems, SSRs and SNPs are the ideal markers of choice for a variety of applications, particularly in marker-assisted breeding (MAB) (Ashokkumar et al. 2020). Unfortunately, the number of markers from rice bean on a public platform is very limited. Comparative genomic studies enormously assisted rice bean improvement. Primarily, random markers and SSR markers from closely related species including adzuki bean, mung bean, and cowpea were successfully utilized in rice bean genetic studies (Bajracharya et al. 2008; Isemura et al. 2010; Muthusamy et al. 2008; Somta et al. 2006; Tian et al. 2013). However, SSR markers from azuki bean and mung bean showed a better transferability compared to cowpea. The reason for this fact is that cowpea is a member of the subgenus *Vigna*, whereas adzuki bean and others including mung bean and rice bean are members of Ceratotropis, a subgenera of genus *Vigna*, also known as Asian *Vigna*. Genetic diversity among the 10 rice bean landraces was studied with the support of 74 RAPD and 37 ISSR markers, and the diversity between the landraces was explored as well as it was confirmed that RAPD and ISSR marker systems were suitable to analyze the genetic variation in rice bean (Muthusamy et al. 2008).

Similarly, 112 rice bean germplasms originated from India and Nepal were evaluated using 35 adzuki bean markers (Bajracharya et al. 2008). Later, Tian and coworkers detailed the first large-scale genetic variation analysis of cultivated and wild rice bean accessions that originated from 16 Asian countries using adzuki bean SSR markers. In this study, 472 rice bean accessions were screened using 13 SSR markers. The results show that the gene diversity in cultivated accessions was around 83% of the wild counterparts (Tian et al. 2013). Also, the rice bean accessions from South Asia have a high level of outcrossing. In another study, out of 2540 mung bean-derived SSRs screened for polymorphism in rice bean accessions, 787 were amplified successfully and 47 produced the polymorphism. Further, these 47 SSR markers were evaluated in a rice bean core collection composed of 230 rice bean genotypes originally maintained at the Chinese Academy of Agricultural Science, Beijing, China. The population structure study divided the 230 rice bean genotypes into 6 major clusters; the grouping of clusters was fairly following the geographical regions across the 12 states where rice bean is obtained (Wang et al. 2015).

The same research group developed the rice bean SSR-enriched library and then sequenced it. Further, annotation of flanking regions of SSR-containing sequences over chromosomes of mung bean and common bean revealed about 261,458 SSRs distributed in 433,562 reads. Of these, dinucleotide repeats (48.8%) were dominant and found to be in a compound form, followed by trinucleotides. Among the obtained flanking sequences, 2928 sequences were corresponding to gene models from the *Arabidopsis thaliana* protein database and were also consistent with 608 nonredundant gene ontology terms, biological processes (64.6%), molecular functions (24.2%), and cellular components (11.2%). Further, homologue analysis of rice bean SSR flanking sequences revealed that 1595 and 5000 sequences were similar to mung bean and common bean, respectively. Eventually, an SSR validation study revealed that 58 of 220 SSR primers will be able to be used in rice bean and 53 were transmitted to mung bean. However, 11 exhibited polymorphism in 32 rice bean genotypes (Wang et al. 2016).

Chen and coworkers used the Illumina sequencing platform, generated about 26 million high-quality rice bean cDNA sequences, successfully arranged them into 71,929 unigenes, as well as identified 3011 genic SSRs. Among the unigenes, 38,840 were annotated to the previously reported proteins in the NCBI database. Moreover, 30,170, 25,451, and 21,982 were placed into gene ontology, Swiss-Prot, and KOG database categories. Also, 9301 were mapped onto 118 pathways via the KEGG pathway database. Among the 3011 genic SSRs, 300 SSRs were randomly chosen and validated in 32 rice bean accessions, which revealed 23 informative markers (Chen et al. 2016). In another study, a microsatellite library was constructed using A 5′-anchored PCR, where 28 SSR markers were developed to study the genetic variation of 65 rice bean accessions originated from Northeast India. The results showed 179 alleles with an average of 6.39 alleles per locus, and the population structure study revealed three different clusters. Also, the high genetic variation among the accessions, with a high outcrossing rate, was obtained (Iangrai et al. 2017).

7.4.3 Genetic Linkage Maps

Several genetic linkage maps were constructed in rice bean using RFLP, RAPD, AFLP, SSR, SRAP, and SNP markers. The first interspecific linkage map was constructed using an F_2 population developed from a cross made between *V. angularis* and *V. umbellata* and 1 phenotypic marker, 114 RFLP, and 74 RAPD markers. This map size was 1702 cM and contained 14 LGs, each LG has four markers, and the regular interval among the markers was 9.7 cM (Kaga et al. 2000). Later, the intraspecific linkage map between wild and cultivated rice bean was developed using 223 SSR and 103 AFLP markers. The map covers a total distance of 796.1 cM in the rice bean genome with an average marker interval of 2.5 cM. Also, a high level of colinearity was seen in the marker position among the rice bean and azuki bean linkage maps (Isemura et al. 2010). An interspecific F_2 population comprising 74 plants derived from *V. umbellata* and *V. nakashimae* was used to construct the linkage map. The linkage map covering 11 linkage groups was established using 74 RFLPs and 101 SSRs and spanned for a total length of 652 cM (Somta et al. 2006). Venkataramana et al. (2016) developed the linkage map between rice bean genotypes using SSR and SRAP markers covering ten linkage groups with a genetic distance of 872.1 cM. The mean interval size between the markers was 32.05 cM. Recently, SNP-based genetic linkage map was developed in interspecific recombinant inbred lines (RIL) of *Vigna radiata* and *V. umbellata*. The map composed 538 SNP markers, consisted of 11 linkage groups, and covered for 1291.7 cM with a mean marker interval of 2.40 cM (Mathivathana et al. 2019). These maps were successfully used to map the QTLs and discover the candidate genes related to bruchids and mung bean yellow mosaic virus resistance and domestication-related traits.

7.5 Status and Opportunities of Genomic Research for Abiotic Stress in Rice bean

In the past 10 years, a slow progress of genomic research was made to elucidate the abiotic stress tolerance mechanism in rice bean, except the studies conducted to understand the metal stress tolerance. Subtractive hybridization technology was combined with reverse northern blot and qRT-PCR analyses to identify the differentially expressed genes (DEGs) in the rice bean root apex in response to low (5 M) and high (25 M) aluminum (Al) concentrations at the early stages of Al stress. The results showed that *VuMATE1* 393 genes exhibited an early response to Al stress. These genes were grouped into different physiological and molecular categories. The upregulated genes were mainly associated with metabolism and energy, signal transduction and transcription, and transport, while downregulated genes were associated with protein translation, processing, and degradation. Comparative transcriptional profiling analysis showed possible genes related to citrate secretion and detailed many novel features of the molecular functions describing Al toxicity and tolerance (Fan et al. 2014). In another study, the involvement of *VuSTOP1* in regulating the expression of *VuMATE1* was studied in Al^{3+} and H^+ stress. The experimental data revealed that *VuSTOP1* has an important role in the tolerance to H^+, whereas it plays a minor role in the tolerance to Al^{3+}. The differential transcriptional regulation of *VuSTOP1* and *VuMATE1* discloses a complex regulatory system guiding the expression of *VuMATE1* (Fan et al. 2015). Moreover, Al stress caused an increase of ABA in the root apex of rice bean that likely controlled the tolerance to Al. At the same time, it was not related to the reported tolerance mechanisms. Genome-wide transcriptome analysis showed that almost one-third of the responsive genes were common among the ABA treatments and Al stress. Additional studies revealed that the transcription factor, ABI5, plays a major role in Al tolerance. This is the first report that detailed the transcriptomic profiling of ABA-arbitrated Al tolerance mechanisms (Fan et al. 2019). Apart from these three studies conducted

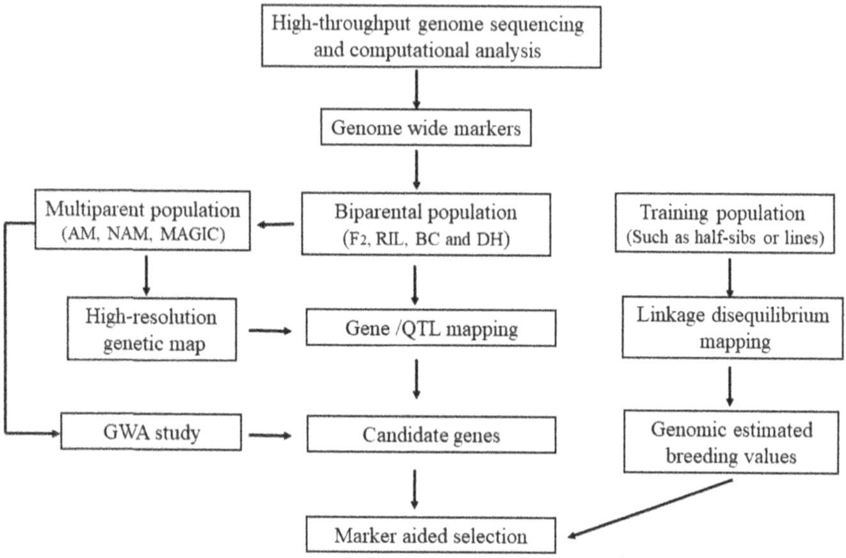

Fig. 7.1 Schematic representation of genomic-based molecular breeding strategies appropriate for rice bean improvement

for Al tolerance in rice bean, no major study was performed in the view of genomic application to improve the understanding of abiotic stress tolerance in rice bean.

Genomics improves the precision of breeding programs and is useful to discover novel genes related to abiotic stress tolerance. Different molecular breeding approaches are established based on the development in population genotyping (Fig. 7.1). Many researchers used advanced genotyping platforms in soybean, common bean, mung bean, and cowpea to discover the QTL/gene(s) related to drought, salinity, and flooding tolerance (Ali et al. 2020; Breria et al. 2020; Diaz et al. 2020; Ravelombola et al. 2021; Ren et al. 2020; Wu et al. 2019). However, these approaches have not yet been utilized in rice bean. Thus, applying genomic tools in breeding could assist the breeders to discover the genes that are involved in various abiotic stress tolerance in rice bean. Plant responses to abiotic stresses are multifaceted and linked with many biological and physiological functions that consist of the up- and downregulation of genes. Comparative transcriptome analysis, by RNA sequencing, offers comprehensive data to know the changing aspects of interaction among the plant and stress. Recently, transcriptome comparisons among the tolerant and susceptible genotypes to drought and cold stress in mung bean, cowpea, and adzuki bean (Kumar et al. 2020; Zhu et al. 2020; Zuo et al. 2017) provided better information on the mechanism and genes related to tolerance to stress. So far, the transcriptome response of rice bean to abiotic stresses (excluding Al stress) still awaits transcriptome analysis. Thus, performing such kind of study assists to identify the DEGs controlling the rice bean response to drought or salinity and is also useful for thoroughly understanding the mechanisms and genes related to tolerance. This knowledge is valuable to rice bean breeders to know the complex interaction among rice bean and abiotic stresses.

7.6 Future Perspectives

Rice bean has remained a less utilized crop despite its nutritional and agronomic benefits and could not gain much attention from scientists and farmers. When we compare the progress toward developing genomic resources in rice bean with other *Vigna* species, it is very slow. Though several efforts have been made in genome research, it is still far behind the other *Vigna* and legume species. The first version of the rice bean genome sequence will be published soon, and it is going to be a game-changer to rice bean molecular breeding programs. The rice bean genome sequence will facilitate the development of genome-wide markers, cost-effective genotyping platforms, construction of high-density maps for fine mapping, and discovery of candidate genes. Also, the genome sequence prerequisites to be accompanied with transcriptome analysis to know each gene position in the genome, gene functions, and expression patterns in various tissues, growth stages, and stress factors. Abiotic stress poses complex inheritance that often makes it difficult to identify the genes with different biological functions. Thus, core and mini-core collections and biparental and multiparent populations have to be developed to assess the genetic basis of abiotic stress tolerance in rice bean as well as to discover the genomic regions linked to stress tolerance by the linkage-based QTL mapping, genome-wide association study (GWAS), and genomic selection (GS), as well as for the detection of candidate genes. Rice bean is a source of genes accountable for stress tolerance. Thus, discovering the candidate genes and unveiling the molecular mechanism of plant responses to stress in inherently stress-tolerant crops such as rice bean will help develop highly stress-tolerant cultivars as well as are useful to build tolerance in other *Vigna* species.

Acknowledgments The first author acknowledges the support of the Science and Engineering Research Board (SERB), Department of Science and Technology (DST), Government of India (GOI), for the DST-SERB NPDF fellowship program (PDF/2016/003676).

References

Ali MJ, Xing G, He J, Zhao T, Gai J (2020) Detecting the QTL-allele system controlling seed-flooding tolerance in a nested association mapping population of soybean. Crop J 8:781–792. https://www.sciencedirect.com/science/article/pii/S2214514120301082

Araújo SS, Beebe S, Crespi M, Delbreil B, González EM, Gruber V et al (2015) Abiotic stress responses in legumes: strategies used to cope with environmental challenges. Crit Rev Plant Sci 34:237–280. https://doi.org/10.1080/07352689.2014.898450

Ashokkumar K, Govindaraj M, Karthikeyan A, Shobhana VG, Warkentin TD (2020) Genomics-integrated breeding for carotenoids and folates in staple cereal grains to reduce malnutrition. Front Genet 11:414. https://www.frontiersin.org/article/10.3389/fgene.2020.004141

Atta K, Chettri P, Pal AK (2020) Physiological and biochemical changes under salinity and drought stress in ricebean [*Vigna umbellata* (Thunb.) Ohwi and Ohashi] seedlings. Int J Environ Clim Chang:58–64

Atta K, Pal AK, Jana K (2021) Effects of salinity, drought and heavy metal stress during seed germination stage in ricebean [*Vigna umbellata* (Thunb.) Ohwi and Ohashi]. Plant Physiol Rep. https://doi.org/10.1007/s40502-020-00542-41

Bajracharya J, Singh S, Dangol B et al (2008) Food Security through Ricebean Research in India and Nepal (FOSRIN) Report 2. Identification of polymorphic markers in ricebean (*Vigna umbellata*). http://www.ricebean.org/deliverables/deliverable%207%20-%20molecular%20diversity.pdf

Berger B, Parent B, Tester M (2010) High-throughput shoot imaging to study drought responses. J Exp Bot 61:3519–3528. https://doi.org/10.1093/jxb/erq201

Breria C, Hsieh C-H, Yen T-B, Yen J-Y, Noble T, Schafleitner R (2020) A SNP-based genome-wide association study to mine genetic loci associated to salinity tolerance in mungbean (*Vigna radiata* L.). Genes (Basel) 11:759

Bright J, Desikan R, Hancock JT, Weir IS, Neill SJ (2006) ABA-induced NO generation and stomatal closure in Arabidopsis are dependent on H_2O_2 synthesis. Plant J 45:113–122. https://doi.org/10.1111/j.1365-313X.2005.02615.x

Chen H, Liu L, Wang L, Wang S, Cheng X (2016) VrDREB2A, a DREB-binding transcription factor from Vigna radiata, increased drought and high-salt tolerance in transgenic Arabidopsis thaliana. J Plant Res 129:263–273

Chhetri DR, Lama S (2013) Chilling and salinity induced metabolic responses in different cultivars of *Vigna umbellata* T. J Environ Social Sci Res 2:1–8

Chhetri DR, Modak S et al (2004) Physiological and biochemical responses of two rice-bean (*Vigna umbellata* T.) cultivars to heavy metal stress. Environ Ecol 22:27–33

Dhillon PK, Tanwar B (2018) Rice bean: a healthy and cost-effective alternative for crop and food diversity. Food Secur 10:525–535. https://doi.org/10.1007/s12571-018-0803-6

Diaz S, Ariza-Suarez D, Izquierdo P, Lobaton JD, de la Hoz JF, Acevedo F et al (2020) Genetic mapping for agronomic traits in a MAGIC population of common bean (*Phaseolus vulgaris* L.) under drought conditions. BMC Genomics 21:799. https://doi.org/10.1186/s12864-020-07213-616

Fan W, Lou HQ, Gong YL, Liu MY, Wang ZQ, Yang JL et al (2014) Identification of early Al-responsive genes in rice bean (*Vigna umbellata*) roots provides new clues to molecular mechanisms of Al toxicity and tolerance. Plant Cell Environ 37:1586–1597. https://doi.org/10.1111/pce.12258

Fan W, Lou HQ, Gong YL, Liu MY, Cao MJ, Liu Y et al (2015) Characterization of an inducible C_2H_2-type zinc finger transcription factor *VuSTOP1* in rice bean (*Vigna umbellata*) reveals differential regulation between low pH and aluminum tolerance mechanisms. New Phytol 208: 456–468. https://doi.org/10.1111/nph.13456

Fan W, Xu JM, Wu P, Yang ZX, Lou HQ, Chen WW et al (2019) Alleviation by abscisic acid of Al toxicity in rice bean is not associated with citrate efflux but depends on ABI5-mediated signal transduction pathways. J Integr Plant Biol 61:140–154. https://doi.org/10.1111/jipb.12695

Fleury D, Jefferies S, Kuchel H, Langridge P (2010) Genetic and genomic tools to improve drought tolerance in wheat. J Exp Bot 61:3211–3222

Hilmy AM, Shabana MB, Daabees AY (1985) Effects of cadmium toxicity upon the in vivo and in vitro activity of proteins and five enzymes in blood serum and tissue homogenates of *Mugil cephalus*. Comp Biochem Physiol Part C Comp Pharmacol 81:145–153. https://www.sciencedirect.com/science/article/pii/0742841385901069

Iangrai B, Pattanayak A, Khongwir DEA, Pale G, Gatphoh EM, Das A et al (2017) Development and characterization of a new set of genomic microsatellite markers in rice bean (*Vigna umbellata* (Thunb.) Ohwi and Ohashi) and their utilization in genetic diversity analysis of collections from North East India. PLoS One 12:1–18. https://doi.org/10.1371/journal.pone.0179801

Isemura T, Kaga A, Tomooka N, Shimizu T, Vaughan D (2010) The genetics of domestication of rice bean, *Vigna umbellata*. Ann Bot 106:927–944

Kaga A, Ohnishi M, Ishii T, Kamijima O (1996) A genetic linkage map of azuki bean constructed with molecular and morphological markers using an interspecific population (*Vigna angularis* X *V. nakashimae*). Theor Appl Genet 93:658–663. https://doi.org/10.1007/BF00224059. PMID: 24162391

Kaga A, Ishii T, Tsukimoto K, Tokoro E, Kamijima O (2000) Comparative molecular mapping in Ceratotropis species using an interspecific cross between azuki bean (*Vigna angularis*) and rice bean (*V. umbellata*). Theor Appl Genet 100:207–213. https://doi.org/10.1007/s001220050028

Kang YJ, Kim S, Kim M, Lestari P, Kim KH, Ha B-K et al (2014) Genome sequence of mungbean and insights into evolution within *Vigna* species. Nat Commun 5. https://doi.org/10.1038/ncomms6443

Kang YJ, Satyawan D, Shim S, Lee T, Lee J, Hwang W et al (2015) Draft genome sequence of adzuki bean, *Vigna angularis*. Sci Rep 5:8069. https://doi.org/10.1038/srep08069

Karthikeyan A, Shobhana VG, Sudha M, Raveendran M, Senthil N, Pandiyan M et al (2014) Mungbean yellow mosaic virus (MYMV): a threat to green gram (*Vigna radiata*) production in Asia. Int J Pest Manag 60:314–324. https://doi.org/10.1080/09670874.2014.982230

Katoch R (2020) Rice bean: a potential underutilized legume. In: Ricebean. Springer, New York, pp 55–79

Kaul T, Eswaran M, Thangaraj A, Meyyazhagan A, Nehra M, Raman NM et al (2019) Rice Bean (*Vigna umbellata*) draft genome sequence: unravelling the late flowering and unpalatability related genomic resources for efficient domestication of this underutilized crop. bioRxiv 2019:816595

Keller F, Ludlow MM (1993) Carbohydrate metabolism in drought-stressed leaves of pigeonpea (*Cajanus cajan*). J Exp Bot 44:1351–1359. https://doi.org/10.1093/jxb/44.8.1351

Kochian LV (1995) Cellular mechanisms of aluminum toxicity and resistance in plants. Annu Rev Plant Physiol Plant Mol Biol 46:237–260. https://doi.org/10.1146/annurev.pp.46.060195.001321

Kumar S, Ayachit G, Sahoo L (2020) Screening of mungbean for drought tolerance and transcriptome profiling between drought-tolerant and susceptible genotype in response to drought stress. Plant Physiol Biochem 157:229–238

Langridge P, Paltridge N, Fincher G (2006) Functional genomics of abiotic stress tolerance in cereals. Brief Funct Genomic Proteomic 4:343–354

Lonardi S, Muñoz-Amatriaín M, Liang Q, Shu S, Wanamaker SI, Lo S et al (2019) The genome of cowpea (*Vigna unguiculata* [L.] Walp.). Plant J 98:767–782. https://doi.org/10.1111/tpj.14349

Mathivathana MK, Jayakodi M, Adhimoolam K, Jang W, Dhasarathan M, Jagadeeshselvam N et al (2019) Detection of QTLs associated with mungbean yellow mosaic virus (MYMV) resistance using the interspecific cross of *Vigna radiata* × *Vigna umbellata*. J Appl Genet 60:255–268

Mittler R (2006) Abiotic stress, the field environment and stress combination. Trends Plant Sci 11: 15–19

Muthusamy S, Kanagarajan S, Ponnusamy S (2008) Efficiency of RAPD and ISSR markers system in accessing genetic variation of rice bean (*Vigna umbellata*) landraces. Electron J Biotechnol 11:32–41

Nandeshwar BC, Karande PT, Bhoite AG, Shinde SR, Shinde SB (2017) Effects of NaCl concentrations on germination and growth attributes at early seedling stages in ricebean {*Vigna umbellata* (Ihunb.) Ohwi and Ohashi} accessions. Int J Trop Agric 35:553–565

Pattanayak A, Roy S, Sood S, Iangrai B, Banerjee A, Gupta S et al (2019) Rice bean: a lesser known pulse with well-recognized potential. Planta 250:873–890. https://doi.org/10.1007/s00425-019-03196-1

Ravelombola W, Shi A, Huynh BL (2021) Loci discovery, network-guided approach, and genomic prediction for drought tolerance index in a multi-parent advanced generation intercross (MAGIC) cowpea population. Hortic Res 8:24

Ren H, Han J, Wang X, Zhang B, Yu L, Gao H et al (2020) QTL mapping of drought tolerance traits in soybean with SLAF sequencing. Crop J 8:977–989. https://www.sciencedirect.com/science/article/pii/S2214514120300714

Roy SJ, Tucker EJ, Tester M (2011) Genetic analysis of abiotic stress tolerance in crops. Curr Opin Plant Biol 14:232–239. https://www.sciencedirect.com/science/article/pii/S1369526611000148

Salekdeh GH, Reynolds M, Bennett J, Boyer J (2009) Conceptual framework for drought phenotyping during molecular breeding. Trends Plant Sci 14:488–496

Somta P, Kaga A, Tomooka N, Kashiwaba K, Isemura T, Chaitieng B et al (2006) Development of an interspecific Vigna linkage map between *Vigna umbellata* (Thunb.) Ohwi & Ohashi and *V. nakashimae* (Ohwi) Ohwi & Ohashi and its use in analysis of bruchid resistance and comparative genomics. Plant Breed 125:77–84

Sritongtae B, Sangsukiam T, Morgan MRA, Duangmal K (2017) Effect of acid pretreatment and the germination period on the composition and antioxidant activity of rice bean (*Vigna umbellata*). Food Chem 227:280–288. https://doi.org/10.1016/j.foodchem.2017.01.103

Tian J, Isemura T, Kaga A, Vaughan DA, Tomooka N, McIntyre CL (2013) Genetic diversity of the rice bean (*Vigna umbellata*) genepool as assessed by SSR markers. Genome 56:717–727

Tomooka N (2002) The Asian Vigna: genus *Vigna subgenus Ceratotropis* genetic resources. Springer Science & Business Media, New York. https://doi.org/10.1007/978-94-010-0314-8

Tomooka N, Kaga A, Vaughan DA (2006) The Asian Vigna (*Vigna subgenus Ceratotropis*) biodiversity and evolution. Plant Genome Biodivers Evol 1(Part C):87–126

Valliyodan B, Nguyen HT (2008) Genomics of abiotic stress in soybean. In: Genetics and genomics of soybean. Springer, New York, pp 343–372

Venkataramana PB, Gowda R, Somta P, Ramesh S, Mohan Rao A, Bhanuprakash K et al (2016) Mapping QTL for bruchid resistance in rice bean (*Vigna umbellata*). Euphytica 207:135–147

Wanek W, Richter A (1997) Biosynthesis and accumulation of D-ononitol in *Vigna umbellata* in response to drought stress. Physiol Plant 101:416–424

Wang Y-S, Yang Z-M (2005) Nitric oxide reduces aluminum toxicity by preventing oxidative stress in the roots of *Cassia tora* L. Plant Cell Physiol 46:1915–1923. https://doi.org/10.1093/pcp/pci202

Wang LX, Chen HL, Bai P, Wu JX, Wang SH, Blair MW et al (2015) The transferability and polymorphism of mung bean SSR markers in rice bean germplasm. Mol Breed 35:77

Wang L, Kim KD, Gao D, Chen H, Wang S, Lee SH et al (2016) Analysis of simple sequence repeats in rice bean (*Vigna umbellata*) using an SSR-enriched library. Crop J 4:40–47. https://doi.org/10.1016/j.cj.2015.09.004

Witcombe J, Hollington P, Howarth C, Reader S, Steele K (2008) Breeding for abiotic stresses for sustainable agriculture. Philos Trans R Soc Lond Ser B Biol Sci 363:703–716

Wu C, Mozzoni L, Moseley D, Hummer W, Ye H, Chen P et al (2019) Genome-wide association mapping of flooding tolerance in soybean. Mol Breed 40:1–14

Yang K, Tian Z, Chen C, Luo L, Zhao B, Wang Z et al (2015) Genome sequencing of adzuki bean (*Vigna angularis*) provides insight into high starch and low fat accumulation and domestication. Proc Natl Acad Sci U S A 112:13213–13218

Zaninotto F, La Camera S, Polverari A, Delledonne M (2006) Cross talk between reactive nitrogen and oxygen species during the hypersensitive disease resistance response. Plant Physiol 141: 379–383. http://www.plantphysiol.org/content/141/2/379.abstract22

Zhou Y, Xu XY, Chen LQ, Yang JL, Zheng SJ (2012) Nitric oxide exacerbates Al-induced inhibition of root elongation in rice bean by affecting cell wall and plasma membrane properties. Phytochemistry 76:46–51. https://doi.org/10.1016/j.phytochem.2011.12.004

Zhu Z, Chen H, Xie K, Liu C, Li L, Liu L et al (2020) Characterization of drought-responsive transcriptome during seed germination in adzuki bean (*Vigna angularis* L.) by PacBio SMRT and Illumina Sequencing. Front Genet 11:996

Zuo J, Wang Q, Han C, Ju Z, Cao D, Zhu B et al (2017) SRNAome and degradome sequencing analysis reveals specific regulation of sRNA in response to chilling injury in tomato fruit. Physiol Plant 160:142–154. https://doi.org/10.1111/ppl.1250964.Photosynthetica_phs-200001-0020.pdf

Genetics and Genomics of Drought and Heat Tolerance in Cowpea, Mung Bean and Black Gram

8

Dhanasekar Punniyamoorthy and Souframanien Jegadeesan

Abstract

Drought and heat stresses are the most important abiotic stresses that are the major stumbling blocks in the scientific endeavour to develop climate-smart legumes. These unpredictable abiotic stresses influence almost all facets of the plant right from germination to maturity and at extremes could be lethal. In India, with a predominant vegetarian population, pulses or grain legumes are the prime and affordable sources of dietary proteins. The protein-rich pulses play a significant role in alleviating protein malnutrition. There is an urgent need to sustain and accelerate the productivity of these pulses that are highly vulnerable to abiotic stresses, owing to their cultivation mostly under rainfed conditions with limited natural resources. The complex genetic architecture of the tolerance traits in conjunction with the ensuing climate change and narrow genetic base of the legumes poses a challenge to the breeding community in the redressal mechanism for ensuring food and nutritional security. Cowpea, mung bean and black gram are significant short-duration hardy pulses in India that are laden with the potential to perform better under challenging environmental vagaries in comparison to others. The limited success of conventional breeding in addressing the drought and heat stress tolerance could be negated by revisiting and adopting suitable holistic strategies to understand and breed for these complex traits. The prospect of developing resilience against drought and heat stress in these crops in the backdrop of the genetic and genomic resources currently available is discussed. With the advent of whole-genome sequencing, advanced phenotyping platforms and reducing cost of next-generation sequencing, it could be expected

D. Punniyamoorthy (✉) · S. Jegadeesan
Nuclear Agriculture and Biotechnology Division, Bhabha Atomic Research Centre, Mumbai, Maharashtra, India
e-mail: sekar@barc.gov.in

P. Muthu Arjuna Samy et al. (eds.), *Legumes: Physiology and Molecular Biology of Abiotic Stress Tolerance*, https://doi.org/10.1007/978-981-19-5817-5_8

that success strides witnessed in cereal crops like rice and wheat could be replicated in these orphans yet paramount crops as well.

Keywords

Cowpea · Mung bean · Black gram · Drought tolerance · Heat tolerance · Genetics · Genomics · Water stress · High-temperature stress

8.1 Introduction

Plants are unveiled to a variety of ecological stresses, which influence almost all facets of plants right from the time of germination to physiological maturity. Depending on their biological nature, stresses could be classified as biotic (pathogen, pests, weeds, etc.) or abiotic (excess or deficit moisture, high or low temperature, salinity, deficiency or toxicity of minerals, soil pH, etc.). These stresses interfere with the complete expression of the plant's genetic potential (Atkinson and Urwin 2012). The plant's sensitivity and response to these environmental stresses are highly complex and vary with the phenological stage of the plant, their genetic potential, and the extent and severity of the stresses (Zhu 2002). The erraticism of these stresses is further aggravated by the present-day climate change phenomenon. The farm breeding systems are necessitated to be more prolific, ensuring viable agricultural outputs counteracting climatic vagaries and safeguarding the food and nutritional security of the burgeoning population across the globe that is likely to cross over nine billion by the middle of this century (Santos et al. 2020).

Food legumes or pulses are indispensable in safeguarding food security as they serve as an inexpensive source of dietary proteins, essential vitamins and minerals (Bohra and Singh 2015). They form a unique and essential component of the diet by complementing staple cereals with proteins (two to three times that of wheat and rice). With their innate ability to grow on a range of edaphic and environmental conditions, pulses contribute vastly to the sustainability of the farming systems. Besides being an intrinsic unit of assorted cropping systems, they enrich soil productiveness through biological N fixation and help liberate soil-adsorbed phosphorus (Souframanien et al. 2020). In the global context, India preponderates with its highest production (25%), consumption (27%) and import (14%) of pulses. Despite the concerted research efforts of the plant breeders and the policies of the Government of India, 2.5 million tonnes of pulses were imported during 2018–2019 (DAC & FW 2020), ensuing a misfortune of 1.3 billion USD forex to the exchequer. Grain legumes are highly vulnerable to abiotic stresses, owing to their cultivation mostly under rainfed conditions with limited natural resources (Ahmad et al. 2005). Among the abiotic stresses, dearth of water and high temperature assume prime importance in the sustained endeavour to develop climate-smart legume crops. There has been meagre prosperity in the efforts to mitigate the repercussions of abiotic stresses in food legumes (Deshmukh et al. 2014). This slow progress could be attributed to the intricate genetic constitution of abiotic stress tolerance governed by several minor genes/quantitative loci coupled with environmental interference (Fleury et al. 2010). Further, breeding for abiotic stress tolerance is extremely complex as the crop plants are confronted simultaneously by several abiotic stresses.

Drought, a challenging constraint to legume productivity, is characterized by a lack of precipitation for a protracted period, resulting in water scarcity, which destructively affects the physiology and biochemistry of metabolic processes that eventually hamper crop yield. Extreme drought conditions could also ultimately lead to total crop failure. Another abiotic stress, as challenging as drought, is the high temperature. As per a prediction, 0.2 °C rise in the atmospheric temperature is expected each decade that would cause a 1.8–4.0 °C upsurge in the current temperature levels by the beginning of the next century (IPCC 2007). Heat stress triggered by high perpetual temperatures might hamper crop phenology, can diminish the yields and could prove mortal beyond certain levels (Janni et al. 2020). Lack of adequate water and high temperature are the principal unpredictable abiotic stresses which adversely impact global food production. According to an estimate, there could be a 50% yield decrement owing to these stresses, primarily in the arid and semi-arid zones (Nam et al. 2001). For sustained legume production, it is imperative to induce genetic tolerance against water stress and elevated temperature in legumes. Drought and/or heat tolerance are complex traits that endure the plants to survive, develop and substantially yield under water deficit (Singh and Matsui 2002) and/or high temperatures. Plants deploy one or more of the following three mechanisms to cope with drought and high-temperature stresses: escape (eluding the effects of stress by altering the crop ontogeny), avoidance (any plant machinery shunning the effects of stress) and tolerance (endurance under stress conditions) (Mitra 2001; Osmond et al. 1987).

Among the short-duration pulses in India, cowpea (*Vigna unguiculata* (L.) Walp.), mung bean (*Vigna radiata* (L.) Wilczek) and black gram (*Vigna mungo* (L.) Hepper) are pivotal legume crops, producing relatively higher yields than other cultivated legume species under stress conditions. Within these three crops, cowpea is the most drought and heat stress tolerant followed by mung bean and black gram (Singh et al. 2018). Nevertheless, the productivities of these crops are also highly disrupted under stress conditions as evident from the ramifications of drought stress on yield reduction in cowpea (Ahmed et al. 2010), mung bean and black gram (Kulkarni et al. 2016) that varied up to 66%, 71% and 74%, respectively. Therefore, it becomes imperative to address these issues and induce drought and temperature stress resilience into these crops before it escalates into an imperious situation warranted by climate change.

The limited success of the conventional breeding approaches in tackling the multigenic drought and heat stress tolerance traits advocates a cautious reconsideration in the game plan to comprehend and develop genotypes for these complex traits. A multidisciplinary, integrated approach involving sundry methodologies encompassing breeders, physiologists, and bioinformaticians would be beneficial considering the intricate interactions between various stresses and plant phenological traits, and amalgamating the genetic and genomic tools to effectively induce water and heat stress tolerance in food legumes. In this chapter, we have comprehensively reviewed the genetics and genomics of drought and heat stress tolerance in cowpea, mung bean and black gram. The prevalence of whole-genome sequence reckoners in these pulse crops (Jegadeesan et al. 2021; Kang et al. 2014; Lonardi et al. 2019) has extended great avenues to the pulse breeders for extensive exploration of stress tolerance traits.

8.2 Independent and Collective Effects of Drought and Heat Stress

Drought adversely affects crop production worldwide. Water stress prevails when low humidity levels in the soil and atmosphere coupled with high environmental temperature result in a disparity between evapotranspiration loss and water absorption from soil (Lipiec et al. 2013). The intricacy of drought stress alters the physiology, morphology, biochemistry and molecular biology of the plant system (Salehi-Lisar and Bakhshayeshan-Agdam 2016). Drought causes numerous catastrophic aftermaths by perturbing different plant metabolisms such as carbon fixation, cellular turgor potential, reactive oxygen species, leaf gas exchange, leaf morphogenesis, enzyme activity and electro-neutrality of ions and also negatively influences the characteristics and quantum of plant growth and yield. Water stress has also been reported to negatively impact seed traits such as germinability, viability, vigour, seed count, produce and other qualities. In grain legumes, drought is also well known to intervene in the biological nitrogen fixation process through its influence on the host-rhizobium symbiotic relationship and root nodulation as well. The response reaction of plants to water stress depends not only on the extent and duration of water dearth, but also on the nature of the species involved, its age and its ontogenic stage of drought exposure (Rao et al. 2006). The yield penalty varied from 34% to 68% in cowpea (Farooq et al. 2017) and 71–74% in mung bean and black gram (Kulkarni et al. 2016), as conditioned by the developmental timing of the water stress. Among the different stages of plant growth, the phase involving reproduction happens to be highly vulnerable to water stress. Jha et al. (2020a) may be referred for backreferences on the various effects of drought on plants. Mung bean is highly susceptible to water deficiency during the pod-filling phase. Yield reduction consequent to water stress was in the range of 10–33% in the course of the vegetative phase, while it varied by 5–27% amid flowering and 53–75% during the initial pod-filling stage in comparison to the unstressed plants. Mung bean compensated for a reduced number of pods during drought stress by redirecting carbon assimilates to the remnant pods, increasing individual grain size (Wenham et al. 2020).

Heat stress causes permanent harm to plant growth and development when the soil and air temperature rises for a minimum amount of time beyond a threshold level (Lamaoui et al. 2018). The threshold temperatures for heat stress in cowpea, mung bean and black gram are given in Table 8.1. Heat stress adversely affects a majority of the key plant growth parameters, including electron transport activity, photosystems I and II, respiration, chloroplast thylakoid membranes and biological

Table 8.1 Critical temperature range of cowpea, mung bean and black gram for high-temperature stress (updated from Sita et al. 2017)

Legume crop	Critical temperature (°C)	Reference
Cowpea	18–28	Laing et al. (1984), Craufurd et al. (1997)
Mung bean	28–35	Kumar et al. (2011)
Black gram	25–35	Shirsath and Bhosale Agro India Ltd. (2017)
	27–35	Divyaprasanth et al. (2020)

nitrogen fixation. The adversity of high-temperature stress in plants is more pronounced during the phase of reproduction in comparison to the pre-reproductive phase (Hall 1992). Male reproductive structures as against the female reproductive structures are more vulnerable to high-temperature stress. Also, pre-fertilization stages are highly susceptible to elevated temperatures than post-fertilization stages. Jha et al. (2017) may be referred for backreferences on the above ramifications of heat stress on plants. The whole reproductive course from gametogenesis to syngamy, embryogenesis and seed development are highly sensitive to heat stress. Disruption of microspore formation due to damaged tapetal layer and disparity of nutrients in the developing pollen leads to sterile reproductive organs. Impairment of fertilization due to reduced viable pollens, less receptive stigmas and poor pollen tube growth consequently lead to lower seed set, enhanced ovule abortion and shrunken seeds. Heat stress ultimately results in declined photosynthetic rates and diminished supply of photosynthates to developing seeds, causing severe yield losses (Sita et al. 2017 and references therein). In cowpea, it has been reported that for every incremental rise in night temperature beyond 16 °C, the number of pods and grain production were offset by 4–14% (Hall 2004). The grain production in cowpea was significantly affected when heat waves coincided with anthesis and pod-setting phases (Ntare 1991). Commonly, reduced pod bearing, abysmal harvest index, high flower abscission, pollen sterility, anther indehiscence, browning of seed coat and reduced biological nitrogen fixation beyond 40 °C have been described in cowpea subjected to heat stress (Jha et al. 2017 and references therein).

Also, under natural conditions, the legumes are mostly exposed to heat and water stresses concurrently during seed filling. Though these two stresses have overlapping effects, they act differently on various physiological processes. For example, in the investigation by Sehgal et al. (2017, 2019), the following was observed. RuBisCo activity and stomatal conductance were elevated under high temperatures but were depressed under water-stress conditions. Hydrolysis of sucrose, though increased independently under high-temperature and drought stresses, was found to be subdued under the combination of these stresses. Heat stress had contradicting effects on the starch levels in the leaves and seeds, wherein there was an increase in the former and a reduction in the latter. However, there was a drastic decline of the starch in seeds under water stress alone or combinedly with high temperature. The reduction in seed weights was more pronounced under water-stress conditions than under high temperatures. Therefore, it would be pertinent to decipher candidate genes discretely for high-temperature and water stresses in legumes.

8.3 Genetic Variability for Heat and Drought Tolerance

Tapping the genetic variability available in any crop germplasm could be ideally suitable for enhancing tolerance with yield stability to different abiotic stresses, inclusive of drought and heat. In this context, the landraces could come handy as they are a potential reservoir of rare alleles (Lopes et al. 2015). Though substantial progress has been accomplished in developing heat stress-tolerant cereal crops through the exploitation of natural genetic variation, it has remained comparatively untapped in grain legumes (Thudi et al. 2014). Therefore, there is a dire necessity to incorporate the stress tolerance-imparting genes/QTL(s) in these stress-sensitive legume cultivars with high yield potential.

The genus *Vigna* with significant diversity for drought tolerance might be exploited not only for genetic improvement of pulse crops against water stress but also to comprehend their mechanism of action (Iseki et al. 2018). Mai-Kodomi et al. (1999b) discerned type I and type II water stress tolerance in cowpea, wherein the former delayed senescence in both the true and primary leaves while in the latter the true leaves were relatively more tolerant to wilting in comparison to the cotyledonary leaves. IITA reported significant diversity among the 1200 germplasm lines concerning yield penalty in response to imposed water stress. They identified a total of 190 lines that showed enhanced levels of drought tolerance, which was further narrowed down to the best 10 for use in breeding (Boukar et al. 2019). Similarly, various researchers identified different cowpea genotypes for drought tolerance: 'C47' (Iran), 'C56' and 'C11' (Portugal) (Santos et al. 2020); 'PI293469', 'PI349674' and 'PI293568' (Ravelombola et al. 2018); '17-61', '17-86', 'Early Scarlet' and 'AR Black eye #1' (Cui et al. 2020); and 'C11', 'C18', 'C44', 'C46', 'C47', 'C50' and 'C54' (Carvalho et al. 2019b). The stress tolerance index was found to be the most superior benchmark for assessing genotypic variability for response to drought tolerance through biplot analysis (Batieno et al. 2016). In mung bean, the cultivars 'NM-2006' and 'NM-8005' were identified to be drought tolerant as assessed by seed germination parameters in conjunction with antioxidative potential and nutrient uptake of seedlings under water stress (Ali et al. 2018), and the variety 'Pratap' was found promising against drought by employing biochemical traits (Baroowa and Gogoi 2014). Based on high seed yield and physiological water-stress tolerance traits, the mung bean genotypes, *Vigna sublobata*, 'MCV-1', 'PLM-32', 'LGG-407', 'LGG-450', 'TM-96-2' and 'Sattya' genotypes (Bangar et al. 2019), and the Egyptian genotypes 'L4', 'L18', 'L19' and 'L21' (El-Nabarawy et al. 2016) were identified to possess drought tolerance. In black gram, the genotypes 'RU8-705', 'PALAVAYAL-LOCAL', 'T 9', 'PHM 8', 'ADT 3', 'CBG-09-06', 'VANNIYUR-LOCAL', 'CBG-09-13' (Prakash et al. 2018), 'Uttara', 'NP 16', 'PU 99', 'UH85-4' and 'No. 13/11' (Kumar et al. 2019) were identified as drought tolerant under rainfed conditions. The black gram genotype 'CBG-09-06' also performed well under moisture-stress conditions based on the dynamics of root and gas-exchange parameters and seems promising against drought stress. Morphophysiological and biochemical parameters imparted drought tolerance in the black gram genotypes 'PGRU95016', 'COBG05', 'IPU99209', 'IPU941', 'IPU243' (Gurumurthy et al. 2019) and 'T 9' (Baroowa and Gogoi 2014).

Concerning heat stress, significant genetic variation was unravelled in cowpea at pod-filling and anthesis stages under varied photoperiodic conditions. The genotype 'Prima' with a higher pod set was delineated as heat stress tolerant under hot and short days. The superior yields exhibited by two photoperiod-sensitive genotypes ('B 89-200' and 'TN 88-63') in response to hot short-day conditions rendered them valuable for developing elite cowpea genotypes with high yield potential (Ehlers and Hall 1998). Similarly, the possession of high-temperature tolerance at the reproductive phase enabled the genotype 'California Blackeye No. 27' ('CB27') to produce superior yields. Yield reduction was less pronounced in 'TVu 4552' and 'Prima' when exposed to high nocturnal temperatures at the flowering stage than in 'CB5'. A total of 268 cowpea accessions from the USA, India and Nigeria were categorized into eight discrete groups primarily based on their differential responses to high temperatures during flowering and pod-setting phases. Such classification may aid breeders in choosing appropriate genotypes for introgressive breeding aimed at incorporating heat tolerance in cowpea. Sunayana and Yadav (2016) identified mung bean genotypes 'MH 805', 'MH 736', 'MH 421', 'IPM 02-3', 'MH 721', 'MH 810', 'IPM 409-4', 'Ganga 8', 'IPM 03-3' and 'IPM 06-5' to possess drought- and heat-stress tolerance that could be used in breeding programmes. The black gram varieties 'J.L', 'PDU-1' (Dash and Shree 2013), 'VBG-07-001' and 'VBG-06-010' (Partheeban et al. 2017) were found to perform best in high-temperature regimes. Gupta et al. (2021) studied a panel of 97 diverse black gram genotypes for yield under stress and non-stress conditions in the field and identified 8 highly tolerant lines ('UPU 85-86', 'IPU 94-2', 'IPU 98/36', 'NO-5731', 'PGRU 95014', 'PGRU 95016', 'PLU 1', 'BGP 247'). Some of the important genotypes identified for drought and heat tolerance in cowpea, mung bean and black gram are enlisted in Table 8.2.

8.4 Genetics of Heat and Drought Tolerance

A priori knowledge on the genetics governing the inheritance of desirable traits is obligatory for any breeding programme. Drought tolerance in cowpea has largely been deciphered to be a complex, multigenic, quantitative trait and is one of the most difficult traits to study and characterize (Carvalho et al. 2017; Ravelombola et al. 2021). It is highly influenced by G × E interactions. The classical genetic studies like diallel or generation mean that analyses were largely based on yield or yield-attributing traits under irrigated or water-stressed conditions that have provided us with a fair idea of the genetic control of water-stress tolerance in cowpea. The wooden box technique (Mai-Kodomi et al. 1999b) for screening drought tolerance has been widely used for studying genetics in cowpea. Mai-Kodomi et al. (1999a) in their study found that the 'type I' and 'type 2' drought-tolerant reactions were governed by monogenic dominant genes, *Rds1* and *Rds2*, and that *Rds1* was dominant over *Rds2*. Olubunmi (2015) reported that grain yield and auxiliary

Table 8.2 Important drought- and heat-tolerant genotypes identified in cowpea, mung bean and black gram (modified from Jha et al. 2017, 2020a)

Crop	Tolerant resources	Basis of drought tolerance	Institute/source	Reference
Cowpea	Ein El Gazal	Early flowering	Institut Senegalais de Recherches Agricoles and University of California	Hall and Patel (1985)
	Mouride	Early flowering		Cisse et al. (1995)
	Melakh	Early flowering		Cisse et al. (1997)
	California Blackeye 5	High seed yield, high biological yield, early maturity	Delmarva Region of the USA	Dadson et al. (2005)
	Texas Cream 8, Elite Mississippi Silver			
	Gorom Local, Mouride TN88-63	Higher net photosynthesis		Hamidou et al. (2007)
	BRS-Paraguacu	Chlorophyll content, LAI	Embrapa Meio-Norte's germplasm bank	Bastos et al. (2011)
	Pingo-de-ouro-1-2 Pingo-de-ouro-2	Low reduction in pods/ plant and grain yield		
Mung bean	NM-2006, 8005	Higher activity of SOD and POD	The University of Agriculture, Faisalabad	Ali et al. (2018)
Black gram	CBG-09-13	Better root dynamics and gas exchange	Annamalai University, India	Prakash et al. (2018)
	VBN-4, K1	Increased synthesis of ABA, proline and lipid peroxidase	Tamil Nadu Agricultural University	Sai and Chidambaranathan (2019)
	UPU 85-86	Fast quenching of Fm, high antioxidant activity, high membrane stability, high ETR, leaf NBI	ICAR—Indian Institute of Pulses Research	Gupta et al. (2021)

Crop	Genotype name	Basis of heat stress tolerance	Yield (Y)/ survival (S) trait under HS	Reference
Cowpea	TN88-63	Pod set	Y	Ntare (1991)
Cowpea	CB27	Reproductive stage	Y	Ismail and Hall (1998), Ehlers et al. (2000)
Cowpea	B89-200 and TN88-63	High yield under heat stress	Y	Ehlers and Hall (1998)

(continued)

Table 8.2 (continued)

Crop	Genotype name	Basis of heat stress tolerance	Yield (Y)/ survival (S) trait under HS	Reference
Cowpea	Tvu 4552 and Prima	Seed yield	Y	Nielsen and Hall (1985)
Cowpea	–	Delayed leaf senescence	Y	Ismail et al. (2000)
Cowpea	Tvu 4552 and Prima Lower	Flower abscission	Y	Nielsen and Hall (1985)
Cowpea	IT93K-452-1, IT98K-1111-1, IT93K-693-2, IT97K-472-12, IT97K-472-25, IT97K-819-43, IT97K-499-38	Yield-related traits	Y	Timko and Singh (2008)
Mung bean	Binamoog-1	Antioxidant defence and methylglyoxal (MG) detoxification	S	Nahar et al. (2015)
Black gram	VBG-07-001, VBG-06-010	Cellular response post-TIR	Y	Partheeban et al. (2017)

water stress-adaptive characters were under the governance of additive and non-additive gene effects in cowpea. Though both the types of gene actions were observed, dominance and/or dominance × dominance effects played a predominant role in genetically controlling the traits related to water-stress tolerance (Olajide and Ilori 2018). The majority of the studies indicate the quantitative nature of drought tolerance traits in cowpea (Boukar et al. 2016; Muchero et al. 2009, 2013) and mung bean (Liu et al. 2017). However, the development of water stress-tolerant legume varieties is hindered by the passive selection response to various drought tolerance traits.

The poor understanding of the genetic mechanisms underlying high-temperature tolerance in grain legumes could be attributed to limited genetic inheritance studies and also to the complex nature of the trait. Genetic analyses to discern the genetics governing the high-temperature tolerance in grain legumes have been carried out on the basis of both classical and quantitative genetics. Initially, genetic inheritance of important agronomic traits contributing directly or indirectly to yield performance under heat stress and administered largely by a single dominant (Marfo and Hall 1992) or recessive gene (Hall 1993) has been worked out in cowpea. Browning of the seed coat (Patel and Hall 1988) and the abscission rate of reproductive organs (Rainey and Griffiths 2005) resulting from heat stress were reported to be under monogenic control in cowpea. Later, Marfo and Hall (1992) proclaimed two dominant genes in cowpea to primarily control most of the heritable tolerance to high temperature at the pod-filling stage. However, their findings also hinted towards QTLs controlling high-temperature tolerance, which was also reiterated by Lucas et al. (2013).

8.5 Breeding Strategies for Improving Drought and Heat Tolerance

Breeding for water-stress tolerance would be a tenable approach to subside the hazards of crop loss by enhancing the crop's ability to extricate water from the deeper layers of soil by altering the root morphology, by reducing the crop water requirements (improved water-use efficiency) or by improving the crop's endurance to withstand longer water-stress periods, consequently leading to improved yields under dryland conditions (Sofi et al. 2019). Breeding programmes involving biparental crosses have limited scope for drought tolerance improvement in legumes due to the narrow genetic base, and therefore, intercrosses involving multiparental advanced generations (MAGIC) should be exploited for introgression of drought tolerance and other desirable agronomic traits (Ravelombola et al. 2021). Classical breeding approaches involving intergeneric and interspecific crosses and induced mutations for isolating novel drought- and heat-tolerant traits could be capitalized on for stress tolerance improvement (Briglia et al. 2019). Iseki et al. (2018) in their studies highlighted that tolerance of domesticated species in genus *Vigna* could be enhanced through pre-breeding efforts. Physiological trait breeding has also been observed to bestow crops with better performance potential under water stress (Jha et al. 2020a). Heat-stress tolerance traits could be improved through different breeding strategies by exploiting the existing genetic variability in crop germplasm. The development of high temperature-tolerant cultivars could be accelerated by taking advantage of novel breeding techniques, such as developing multiparent advanced generation intercross populations, marker-assisted selections, accelerated breeding techniques and CRISPR/Cas9-based genome editing systems (Jha et al. 2020b). Inclusion of candidate genes transcribing heat-shock proteins in cultivar development programmes could be a judicious strategy in the redressal mechanism of high-temperature stress (Feder and Hofmann 1999). In cowpea, profuse flowering and copious pod bearing under high nocturnal temperatures and long photoperiods are used as selection criteria in breeding for high-temperature tolerance (Marfo and Hall 1992). Such breeding efforts culminated in the development of high temperature-tolerant high-yielding cowpea variety 'California Blackeye 27' ('CB27') (Janni et al. 2020). The mung bean selections with quick germination and rapid growth were observed to combat terminal heat stress probably due to improved population stand, earliness and increased yield (Hanumantharao et al. 2016). Selections with reduced leaf electrolyte leakage under high temperature were also associated with high pod set (Hall 2004). The yield-contributing characters such as number of pods, number of seeds and seed size in terms of test weight are suggested as three key traits, which could be beneficial in screening and breeding genotypes for high-temperature tolerance along with better seed yield in black gram (Anitha et al. 2016).

8.6 Screening of Target Traits for Drought- and Heat-Stress Tolerance

The intricacies involved in drought and high-temperature tolerance forbid the efficient screening of these traits in plants. It would be appropriate to have a gamut of selection indices so that several traits contributing to stress tolerance could be effectively screened and introgressed into the elite genetic background with good agronomical value. An array of parameters imparting water-stress tolerance in cowpea have been employed for screening cowpea genotypes (Carvalho et al. 2019a; Iseki et al. 2018; Matsui and Singh 2003; Muchero et al. 2008). Shoot biomass (Iseki et al. 2018), deep root systems (Matsui and Singh 2003) and increased root mass have been used widely (Santos et al. 2020). Slabbert et al. (2004) established other protocols for screening cowpea for water-stress tolerance like proline and ABA accruals, tetrazolium assays, membrane stability based on electrolyte leakage, relative water content (RWC), water potential and area of leaves, chlorophyll and carotenoid contents, chlorophyll fluorescence, enzymatic assays for studying antioxidative responses (SOD, glutathione reductase, ascorbate peroxidase) and wooden boxes for evaluating drought tolerance at early vegetative stage. The screening for the content of osmoprotectants (e.g. proline, trehalose, fructans, mannitol, glycine betaine) has also been used for screening water-stress tolerance (Carvalho et al. 2017). The agronomical traits such as early maturity and 'stay green' have also been widely used (Fatokun et al. 2012). The drought avoidance mechanisms such as stomatal closure, paraheliotropic movement of leaves and enhanced water conductivity of roots (Agbicodo et al. 2009) could also be exploited. Cowpea has been found to modify its metabolic activities under water-stress conditions for accommodating the demanding tolerance functions through the interactive shikimate and arginine/proline pathways, leading to manipulations in the levels of metabolites like proline, galactinol and quercetin 3-*O*-6′′-malonylglucoside (Goufo et al. 2017) with drought response manifestations. Ravelombola et al. (2018) and Verbree et al. (2015) have claimed water-stress tolerance to be correlated with stem diameter in cowpea seedlings. The depression of canopy temperature from the atmospheric temperature (CTD) expedites the screening of crop response to heat and water stresses (Sofi et al. 2019). Positive CTD resulting from cooler canopies has been correlated with high yield in various crops (Fischer et al. 1998; Singh and Kanemasu 1983). In mung bean, polyethylene glycol (PEG) has also been used to simulate water-stress conditions for laboratory screening (Islam et al. 2019). In black gram, several traits, viz. photosynthetic efficiency, conductance of stomata, rate of transpirational water loss, contents of photosynthetic pigments, prolines and activities of peroxidases, have been reported to be useful for screening water-stress tolerance (Gurumurthy et al. 2019).

Thiaw and Hall (2004) were of the view that a blend of traits rather than targeting a single trait proved beneficial in developing high temperature-tolerant cowpea genotypes. They observed that the traits such as abundant flowering and copious podding in combination led to the identification of summer-suitable cowpea genotypes with better performance potential under long photoperiods and hot conditions. Similarly, the selection of genotypes with high membrane stability and low electrolyte leakage proved beneficial for developing heat stress-tolerant winter-suitable cowpeas (Jha et al. 2017). The studies associated with stomatal performance and metabolite contents such as that of prolines and anthocyanins could efficiently discriminate the genotypes for high-temperature tolerance (Carvalho et al. 2019a, b). A number of traits affecting the physiology of plants such as rate of photosynthesis, germinability of pollen grains, cell membrane integrity, water potential in leaves and relative water content (RWC) have been explored extensively in discriminating high temperature-tolerant lines from the sensitive ones (Janni et al. 2020; Kumar et al. 2020a; Siddiqui et al. 2015). In mung bean, the genotypes with the genetic potential of retaining maximum flowers and sustaining higher productive pods during extremes of temperature (>40 °C) were found to be relatively more heat stress tolerant, and thus, these traits are valuable selection indices for screening (Khattak et al. 2006; Singh and Singh 2011). Alternatively, the preponderance of fertile pollen and sucrose synthase (*SuSy*) enzymatic activity have also been used for screening mung bean heat-tolerant lines at high temperature (40 °C). The identified heat-tolerant line also showed substantial variations for photoperiod-temperature photosynthetic response, Fv/Fm chlorophyll fluorescence parameter reflecting the maximum quantum efficiency of photosystem II, and photosynthetic electron transport rate (ETR) (Basu et al. 2019). As far as possible, an array of traits need to be screened for ascertaining the drought and heat tolerance of potential putative genotypes. Partheeban et al. (2017) standardized temperature induction response technique (TIR) in black gram (induction at 36–46 °C for 3 h and lethal temperature at 52 °C for 3 h) and studied the cellular response as a rapid and reliable technique for thermotolerance. Gupta et al. (2021) compared a set of highly heat-sensitive black gram genotypes with that of highly tolerant genotypes with respect to physiological and biochemical traits and found significant genotypic variability for leaf nitrogen balance index (NBI), chlorophyll (SPAD), epidermal flavonols and anthocyanin contents under 42/25 °C max/min temperature. The heat-tolerant lines also exhibited high membrane stability index, high electron transport rate, fast quenching of Fm following fluorescence kinetics and high antioxidation activity resulting in scavenging of ROS. The susceptible lines also displayed reduced quantum yield of PSII leading to reduced photosynthetic efficiency.

8.7 Genomics for Improving Drought and Heat Tolerance

Genomics pertains to the study of the complete genome of an organism; it involves DNA sequencing methods in conjunction with bioinformatics for sequencing, assembling and analysing their structural, functional and evolutionary aspects and opens avenues for mapping and editing of genomes. These genomic tools enable the generation of extensive and exhaustive data sets related to the differential gene expression patterns and proteomic and metabolomic differences resulting from exposure to water and high-temperature stresses. The various genomic tools available for genetic improvement of drought- and heat-stress tolerance in pulses are discussed below.

8.7.1 Quantitative Trait Locus (QTL) Mapping

Drought tolerance, being an intricate quantitative trait, is under the confluence of many genes and gene families and presents a highly unfeasible circumstance for simultaneous selection. Such quantitative traits could be discerned through QTL analysis. QTL analysis has been in vogue since 1997 to associate regions of chromosomes responsible for water-stress mitigation (Teulat et al. 1997). These QTLs are useful in mapping genes and finding applications in marker-assisted breeding programmes. Muchero et al. (2009) divulged the mapping of ten QTLs associated with water-stress tolerance in the cowpea seedlings of a RIL population under greenhouse conditions. These ten QTLs were confirmed through field experiments and accounted for 4.7–24.2% of phenotypic variance. Four QTLs pertaining to per-peduncle pod number under high-temperature conditions were also recognized and exploited in marker-assisted breeding (Lucas et al. 2013; Pottorff et al. 2014). Cloning QTL (first attempted by Salvi and Tuberosa 2005), though being technology, resource and time intensive, offers great scope in developing elite productive cultivars through marker-assisted selections. The hotspots within QTLs may harbour candidate genes and demonstrate differential gene expression. Similarly, the transcriptome atlases developed in various legume crops like common bean (PvGEA), cowpea (VuGEA), groundnut (AhGEA) and soybean could assist in deeper understanding of unannotated gene functions related to water stress (see Jha et al. 2020a for cross reference). Five QTLs concerning stay-green trait under water-stress conditions have been reported in cowpea (Muchero et al. 2013). Six stable QTLs associated with different water-stress tolerance traits were discerned and localized onto the SSR-based novel genetic linkage map developed in mung bean (Liu et al. 2017). Unfortunately, limited QTLs associated with high-temperature tolerance have been established in pulses. Five QTLs (*Cht-1*, *Cht-2*, *Cht-3*, *Cht-4* and *Cht-5*) conferring 11.5–18.1% of the total phenotypic variation for high-temperature tolerance have been identified in cowpea (Lucas et al. 2013). Significantly, these QTLs contained the candidate heat tolerance-imparting genes, viz. DNA J heat-shock proteins, heat-shock proteins (HSP) and heat-shock transcription factors (HSTF). Pottorff et al. (2014) discovered one major QTL *Hbs-1* explaining 77.3% of the total phenotypic variance and two minor QTLs *Hbs-2* and *Hbs-3* contributing 12.3% and 6.8% of the total variance, respectively, that were associated with high temperature-induced browning of seed testa in cowpea. Interestingly, the QTL *Hbs-1* was found to have syntenic correspondence with ethylene-forming enzyme (EFE) coding regions of other legumes, and SNP markers associated with *Hbs-1* gene were identified for use in MAS (Pottorff et al. 2014). In forthcoming years, such QTLs shall be promising in the efforts towards the development of drought- and high temperature-tolerant genotypes.

8.7.2 Association Studies

Genome-wide association studies (GWAS) are powerful tools for precisely detecting chromosomal regions tightly linked with the trait of interest by correlating the phenotypic and genotypic data generated on a relatively large set of natural variations existing in the germplasm (Brachi et al. 2011). The preponderance of molecular markers in pulses including cowpea (Muñoz-Amatriaín et al. 2017; Xu et al. 2017) has enabled in-depth study of water-stress tolerance. GWAS involving 383 genotypes and a biparental population-based QTL analysis independently confirmed the involvement of three prime QTLs (*Dro-1*, *Dro-3* and *Dro-7* for delayed senescence, biomass and grain yield) attributing stay-green trait for water-stress tolerance in cowpea (Muchero et al. 2013). Ravelombola et al. (2021) utilized GWAS on a MAGIC population comprising 305 F_8 RILs for evaluating drought tolerance index based on plant phenology, maturity, days to flowering, seed test weight and seed yield and identified many SNPs linked with water stress-associated plant growth habit (14), maturity (18), days to flower (5), seed test weight (5) and seed yield (35). The outcome of this study would aid in the comprehension of genes governing water stress tolerance and would be of help in genetic improvement of cowpea through marker-aided and genomic selections. GWAS was employed in mung bean, and eight *SnRK2* genes (sucrose non-fermenting-1-related protein kinase 2 family) were predicted. It was found that most of these *VrSnRK2* genes were upregulated post-induction of drought, suggestive of their role in water-stress mitigation response. The gene *SnRK2.6c* exhibited the highest differential gene expression (12-fold) under water-stress conditions, indicative of its crucial role in moderating the effects of drought stress (Fatima et al. 2020). Noble et al. (2017) have developed the mung bean nested association mapping (NAM) population that could prove to be a valuable resource for studying complex traits such as water and high-temperature stresses. As part of NAM, 560 mung bean, black gram and wild accessions (*Vigna sublobata* var. *sublobata*) have been genotyped. In addition, the mung bean diversity panel developed by Queensland University would be a treasured resource for accelerating genetic improvement of mung bean by providing deeper insights into the genetic architecture of traits of agronomic importance (Noble et al. 2018).

8.7.3 Comparative Genomics

The growing abundance of whole-genome reference sequences in various food crops including pulses has empowered legume researchers to gain deeper knowledge on the various traits of interest (Bohra and Singh 2015; Varshney et al. 2019). Concurrently, the reference genome sequences could act as a launch pad for performing comparative genomic studies for elucidating chromosomal regions imparting tolerance against water and high-temperature stresses in legumes. For example, the analyses of whole genomes have led to the recognition of 111 and 109 drought-responsive genes in pigeon pea (Varshney et al. 2012) and soybean (Schmutz et al.

2010), respectively. To this end, more recent whole-genome reference sequencing efforts in grain legumes have expounded unique genomic identities regulating important functional traits including water-stress tolerance in food legumes (referenced in Jha et al. 2020a, b). With the next-generation sequencing facilities and increasing availability of hybrid assemblers, relatively better quality genome assemblies are being made in those crops wherein quality reference genomes are unavailable. In legume crops like black gram, with the available information on genetic variation in conjunction with these assemblies, rapid strides in the development of promising varieties through marker-aided selections could be achieved. The comparative genomics in these orphan legume crops from the structural, functional and evolutionary perspectives should be possible in near future with the increasing availability of these valuable genomic resources (Pootakham et al. 2021).

8.7.4 Candidate Genes

Candidate genes with a potential role in alleviating water stress have been identified in several legumes including *CPRD8, CPRD12*, *CPRD14* and *CPRD22* genes in cowpea. *SKP1* (S-phase kinase-associated protein 1) gene is involved in proteolysis and has a determinative role in stress tolerance. The *Vigna radiata*-specific *SKP1* (*VrSKP1*) ORF comprising 550 bp and corresponding to 114 amino acids was isolated and cloned for the first time from water stress tolerant mung bean variety 'Pratap' and was found to be significantly upregulated, thereby validating their role as a candidate gene for drought tolerance (Bharadwaj et al. 2019). In water stress-tolerant cowpea, a unique drought-responsive element-binding protein 2-type transcription factor *VuDREB2A* mediates DRE-dependent expression of stress-responsive genes and confers enhanced drought resistance (Sadhukhan et al. 2014). Carvalho et al. (2017) have reviewed in detail the various candidate genes that have been identified in cowpea by various researchers. Glutathione reductase (GR) is involved in water-stress response in cowpea as was evident from the dual-targeted (*dtGR*) and cytosolic (*cGR*) glutathione reductase leaf-expressed genes. Among the plant explicit families of transcription factors (TF), APETALA2/ethylene-responsive element factor-binding proteins (*AP2*/*ERF*) play a determinative role in water-stress tolerance. In silico analysis performed by Labbo et al. (2018) revealed 71 AP2/ERF TFs in the *Vigna radiata* genome. Constituents of *DREB* subfamily are known to perform crucial functions in water-stress tolerance. The differential expression of *VrDREB* genes under water-stress conditions was studied, and five candidate genes (*VrDREB5*, *VrDREB12*, *VrDREB13*, *VrDREB22*, *VrDREB30*) were identified that were upregulated under water stress. The drought stress-induced genes like *VrP5CS*, *VrRAB18*, *VrDHN3*, *VrDREB* and *VrNCED* were highly expressed under combined phosphorus- and water-stress conditions, while *VrDHN3* and *VrNCED* were specific to drought (Meena et al. 2021). The list of various candidate genes identified in cowpea and mung bean is given in Table 8.3. Given the tremendous developments in the 'omics' field, it is expected that the identification of prime candidate genes having manifestations in the intricate abiotic stresses should be

Table 8.3 Candidate genes associated with drought and high-temperature tolerance identified in cowpea and mung bean (modified and updated from Carvalho et al. 2017)

Crop	Gene designation	Gene function	Author
Cowpea	*CPRD8, CPRD14, CPRD22, CPRD12*	Response to dehydration stress	Iuchi et al. (1996a)
	CPRD46	Neoxanthin cleavage enzyme involved in ABA biosynthesis	Iuchi et al. (1996b)
	VuNCED1	9-*Cis*-epoxycarotenoid dioxygenase involved in a key step of ABA biosynthesis	Ajayi et al. (2021), Iuchi et al. (2000)
	VuABA1	Zeaxanthin epoxidase involved in early step of ABA biosynthesis	Iuchi et al. (2000)
	VuPLD1	Putative phospholipase D, a major lipid-degrading enzyme in plant	El Maarouf et al. (1999)
	VuPAP-α, VuPAP-β	Putative phosphatidate phosphatase, important for the enzymatic cascade leading to membrane lipid degradation under environmental stresses or senescence	Marcel et al. (2000)
	VuPAT1	Galactolipid acyl hydrolase involved in membrane degradation induced by drought stress	Matos et al. (2001)
	VuC1	Protein inhibitor of cysteine proteinase belonging to the papain family	Diop et al. (2004)
	dtGR	Dual-targeted glutathione reductase, a key enzyme involved in detoxification of AOS	Contour-Ansel et al. (2006)
	cGR	Cytosolic glutathione reductase, a key enzyme involved in detoxification of AOS	Contour-Ansel et al. (2006)
	VucAPX	Cytosolic ascorbate peroxidase, a key enzyme involved in detoxification of AOS	D'Arcy-Lameta et al. (2006)
	VupAPX	Peroxisomal ascorbate peroxidase, a key enzyme involved in detoxification of AOS	D'Arcy-Lameta et al. (2006)
	VusAPX	Stromatic ascorbate peroxidase, a key enzyme involved in detoxification of AOS	D'Arcy-Lameta et al. (2006)
	VutAPX	Thylakoidal ascorbate peroxidase, a key enzyme involved in detoxification of AOS	D'Arcy-Lameta et al. (2006)
	GST	Glutathione-*S*-transferase, a well-recognized stress-related gene	Gazendam and Oelofse (2007)
	PR-1	Pathogenesis-related protein 1, a well-recognized stress-related gene	Gazendam and Oelofse (2007)
	VuNSR4	Digalactosyldiacylglycerol synthase 1	da Silva et al. (2012)
	VuNSR10	Kinase protein calcium dependent	da Silva et al. (2012)
	VuNSR44	CPRD12 protein	da Silva et al. (2012)
	VuNSR47	CPRD8 protein 'old yellow' enzyme	da Silva et al. (2012)

(continued)

Table 8.3 (continued)

Crop	Gene designation	Gene function	Author
	VuNSR49	CPRD65 protein	da Silva et al. (2012)
	VuDREB2A	DRE-dependent expression of stress-responsive genes	Sadhukhan et al. (2014)
	VuHsp17.7	sHSP family class I protein	Carvalho et al. (2019b)
	Hbs-1	Ethylene-forming enzymes EFE	Pottorff et al. (2014)
	Hbs-3	ACC synthase 1	Pottorff et al. (2014)
Mung bean	*VrbZIP*	Drought-responsive gene	Wang et al. (2018)
	codA	Improve abiotic stress tolerance	Baloda et al. (2017)
	VrWRKY	Enhance abiotic stress tolerance	Srivastava et al. (2018)
	VrSKP1	Ubiquitin-proteasome system component	Bharadwaj et al. (2019)

feasible in field crops. The whole-genome sequence of black gram could be used to identify candidate genes for water- and high temperature-stress tolerance using genome-wide association studies (Pootakham et al. 2021; Souframanien et al. 2020).

8.7.5 Genes for Heat-Shock Proteins

Under conditions of high temperature, plants produce unique types of molecular chaperon proteins commonly referred to as heat-shock proteins (HSPs). These 10–200 kDa HSPs prevent the functional proteins from getting aggregated and denatured, ensuring the effectiveness of various biological membranes and metabolic processes including photosynthesis, assimilate apportioning, water and nutrient balance in plants having tolerance to elevated temperatures. Therefore, the inclusion of genes underlying these HSPs in procreating cultivars with tolerance against high temperatures could be an important accommodative tactic for heat-stress redressal in plants (reviewed in Kumar et al. 2020a). Heat-shock transcription factors (Hsfs) are essential signal-transducing elements that mediate gene expression in retortion to various abiotic stresses. With the growing emergence of genomic resources now permitting functional analysis of genes, Li et al. (2019) dissected the mung bean *Hsfs* through genome-wide association and differential expression analyses. They studied the evolutionary and conserved domains of 24 *VrHsf* genes and categorized them into three sets (A, B and C). The promoters of these highly conserved *VrHsf* motifs are known to house *cis*-elements against multiple stresses. The *VrHsf* genes expressed differentially under varying stresses, suggestive of their plausible roles in stress alleviation. Thus, *Hsfs* motifs are considered to be vital in plants as regulatory elements transducing signals and facilitating the expression of different genes involved in tolerance against numerous abiotic stresses such as low temperature, water, salt and high temperature (reviewed in Li et al. 2019).

8.7.6 Genomic-Assisted Breeding

The increasing adoption of marker technologies in breeding programmes such as marker-assisted backcrossing (MABC) and marker-assisted recurrent selection (MARS) has resulted in the rapid advancement of water-stress and high-temperature tolerance-contributing traits in various crop species. The MAS/MABC is effective in transferring small number of major QTLs in grain legumes that have profound phenotypic influence (Varshney et al. 2019). However, MABC could be quite challenging in the improvement of intricate traits like yield under drought stress that are under the control of numerous minor QTLs having hitherto little phenotypic influence. To address this situation, researchers are progressively inclined towards genomic selections (GS), which could be performed due to the ease of access to millions of single nucleotide polymorphisms (SNPs) across the genome as a result of de-escalating cost of sequencing. Promising genotypes could be identified from the breeding population as GS permits swift, precise and effectual selection. The application of GS models in food legumes has divulged enhanced prediction efficiency for intricate traits (reviewed in Jha et al. 2020a, b). MABC has been used to introgress water stress-tolerant *Striga* and *Meloidogyne* nematode resistance QTLs into an extensively cultivated cowpea landrace widely preferred by the peasants in Burkina Faso. A set of 184 genome-wide SNPs deduced by expressed sequence tags extending over an average span of 2 cM intervals and abutting known annotated loci on either side were used for construing the genotypes of backcross progenies by utilizing the cowpea KASP genotyping platform. This study proclaimed the utility of highly efficient SNPs in performing foreground and background selections under a MABC system for bettering a widely cultivated cowpea variety by introgressing water-stress tolerance and biotic stress resistance genes (Batieno et al. 2016). In black gram, Gupta et al. (2021) used a set of 21 genetic markers for establishing genetic differences between the heat-tolerant and -sensitive lines. VigSatDB, the world's first exhaustive SSR database of genus *Vigna*, comprising more than 875 thousand (772,354 simple and 103,865 compounds) presumed microsatellite markers identified from six genome assemblies belonging to three *Vigna* species, viz. *Vigna radiata* (mung bean), *Vigna angularis* (adzuki bean) and *Vigna unguiculata* (cowpea), could be a treasured tool in legumes for marker discovery and genomic-assisted breeding (Jasrotia et al. 2019).

8.7.7 Transcriptome Analysis

Transcriptome profiling is highly useful in enhancing our understanding of the regulatory mechanisms imparting tolerance to various stresses. Though advanced techniques like RNA-seq have permitted profound expression studies, thereby unravelling several high temperature tolerance-imparting candidate genes in different crops (see Jha et al. 2017), only limited work has been conducted via transcriptome analysis in grain legumes for abiotic stress tolerance. Numerous genes, their pathways and metabolic processes involved in a plant's reaction to many abiotic or biotic stresses have been deciphered through transcriptomics, thus providing avenues for genetic enhancement of stress tolerance. Candidate genes formerly reported to impart water-stress tolerance in other related crops have been used to decipher many drought tolerance-bestowing genes in cowpea, which were later authenticated by differential gene expression studies in response to water stress (Carvalho et al. 2017). In mung bean, Kumar et al. (2020b) carried out transcriptome profiling of contrasting genotypes for water-stress tolerance and identified differentially expressed genes that were mainly mapped to phytohormone signal transduction, carbon metabolism and flavonoid biosynthesis. Tian et al. (2016) reported differential expression of several TFs (MYB, AP2 and NAC), HSPs, late embryogenesis abundant proteins and genes coding methyltransferases and histones in mung bean in response to desiccation. Transcriptome sequencing in black gram has revealed a rich reserve of molecular markers like SSRs and SNPs, which could be exploited for identifying candidate genes for drought and high-temperature tolerance as well as for marker-assisted selection for these traits (Raizada and Souframanien 2019; Souframanien and Reddy 2015).

8.7.8 MicroRNAs (miRNA)

MicroRNAs are small non-coding RNA molecules of 20–24 nucleotides that are post-transcription repressors of genes primarily through recognition, base complementation and cleavage or deadenylation of target RNAs and the genes thereof. Several miRNAs have been reported to have implications in different processes governing plant development and also have definitive roles in a plant's response to various biotic and abiotic stresses. In cowpea, 44 of the 157 miRNAs detected were related to water stress that targeted genes encoding zinc finger family proteins, serine/threonine protein kinases and Kelch repeat-containing F-box proteins (Barrera-Figueroa et al. 2011). Cowpea miRNAs isolated from leaves and roots of plants post-subjection to water stress were corroborated with qPCR studies, and it was observed that the miRNAs exhibited differing tissue-specific responses to water stress treatment (Carvalho et al. 2017). Participation of various miRNAs in water-stress tolerance has been demonstrated in other legumes like chickpea and soybean. Besides miRNAs, there is a rising indication of involvement of long non-coding RNAs (lncRNAs) in retortion to water-stress conditions that have been demonstrated through differential gene expression studies (Jha et al. 2020a).

Recently, miRNAs have also been detected in response to high-temperature stress from several susceptible and tolerant cultivars in different crops, albeit not in legumes. The homeobox leucine-zipper protein and SOD have been reported to be regulated through miRNAs (for details, see Janni et al. 2020). It could be just a matter of time before the role of miRNAs in heat tolerance is elucidated in legumes.

8.8 Metabolite Changes

A comprehensive view of the response of plant metabolism and their regulatory mechanism to various abiotic stresses like water stress has been demonstrated by the rapid progress being made in plant metabolomics. Numerous reviews on techniques determining metabolite variations in retortion to different stresses are available (reviewed in Jha et al. 2020a, b). Significant variations have been reported in different metabolites involved in various pathways in response to water stress. The prime metabolites showing dynamic build-up under water stress include various sugars, proline and γ-aminobutyric acid (GABA) that aid in the maintenance of osmotic potential under water-stress conditions. Plant metabolites including proline, galactinol and quercetin have been found to play prominent roles in acclimatization in response to water stress in cowpea, and their correlation with yield indicated beneficial effects (Goufo et al. 2017). Obata and Fernie (2012) emphasized that integrated comprehensive studies involving metabolite profiling, transcriptomics, genomics and proteomics will help understand the regulation of various metabolic events cascading from the impact of plant's exposure to stresses. Polyamines were also found to protect mung bean plants against drought stress (Sadeghipour 2019). Proline accumulation was found to increase in cowpea, mung bean, black gram and other legumes under water stress (Carvalho et al. 2019a, b; Jha et al. 2020a, b; Pandiyan et al. 2017). Declining levels of GABA (γ-aminobutyric acid) in the cells of heat-stressed mung bean plants have been found to increase the heat sensitivity and external application of GABA served as a thermo-protectant (Priya et al. 2019).

8.9 Genome Editing

Genome editing is a fast-evolving branch of genomics that has solid applications in the evolution of abiotic stress-tolerant cultivars. The genome editing tools such as the CRISPR/Cas9 system, which selectively edits the target genes, have immense potential in breeding better yielding crops under high temperature- and water-stress conditions. Gene editing has proven applications in the manoeuvring of root and nodule traits in cowpea, photoperiod flowering pathway, *GmDrb2a* and *GmDrb2b* genes in *Glycine max*, *SPL9* gene in *Medicago sativa* and Hua enhancer1 (*MtHen1*) gene in *Medicago truncatula* (see Jha et al. 2020a). Therefore, this potential technology could be exploited for custom editing of food legume genomes so as to develop yield-sustaining stress-tolerant genotypes with enhanced ameliorative capabilities under water and high-temperature stresses (Li et al. 2017). However, the rapid generation of novel edited genotypes through CRISPR should also be accompanied with suitable field breeding experiments under stress conditions (Janni et al. 2020).

8.10 Transgenics

Transgenic approaches, albeit controversial, may contribute immensely to the attempts towards developing pulse cultivars with tolerance against abiotic stresses. Transgenics validating cloned genes associated with abiotic stress tolerance could greatly aid functional genomics studies. The legume crops being highly recalcitrant to regeneration, in vitro culture techniques and transformation technologies gained momentum at a slower pace. However, in black gram, the *ALDRXV4* gene was overexpressed through a transgenic approach and engineered multiple stress tolerance. The transgenic lines accumulated more reactive oxygen species and showed increased protection against drought and salinity through sustained photosynthetic efficiency, maintaining increased relative water content and reduced photooxidative damage (Singh et al. 2016). The HSPs and HSFs have been targeted through transgenics for increasing the endurance of crops like wheat, maize, tomato and rice to high temperatures. In cotton, the heat-shock protein, *AtHSP101*, was upregulated in pollen through transgenic events resulting in enhanced pollen germination and improved pollen tube growth under heat stress, consequently leading to significant enhancement in the endurance of high-temperature stress and reduced yield losses (Burke and Chen 2015). Janni et al. (2020) have surveyed lines developed for tolerance towards high-temperature stress through transgenics in several crops. Once the regeneration protocols are standardized in pulses, transgenics could be developed to tackle abiotic stress tolerance.

8.11 Mutation Breeding

Mutational breeding is one of the important tools available to breeders for generating genetic variation in plants and exploiting the created variations for the betterment of agriculture. Mutation breeding has been in vogue since the 1950s and has been variously employed for improving the abiotic stress tolerance of crop plants and also to a limited extent in legumes. The long-root mutant identified in mung bean was found to draw more water compared to its parent and could be utilized under the receding water level, a condition normally experienced under field conditions (Dhole and Reddy 2010). With the advent of TILLING (targeting-induced local lesions in the genome), an innovative approach was introduced to the benefit of breeders, which empowers them to locate mutant lesions within a target locus of known sequence irrespective and independent of their contribution to phenotype (Uauy 2017). The success of TILLING primarily depends on the availability of mutant populations typically generated through chemical mutagenesis such as EMS that generally produce random point mutations across the genome, unlike the physical mutagens that largely produce deletion mutations. In addition, there is also a prerequisite of sequence knowledge about the locus of interest and a well-defined screening protocol for phenotyping the mutants carrying mutations in the region of interest. Unlike genome editing, the induced mutations in TILLING are not targeted and are random and hence do not provide the flexibility envisaged in genome editing. However, the crops developed through TILLING-based chemical or physical mutagenesis do not come under the purview of transgenic regulations. Recently, a set of four novel small *Hsp26* (*sHsp26*) alleles were identified in durum wheat with the potential of augmenting high-temperature tolerance in durum wheat through in silico and in vivo TILLING approaches. Similarly, in tomato also, a mutated HSP identified through TILLING was demonstrated to ameliorate high-temperature tolerance. HSP genes contributing towards heat-stress tolerance were explored for mutations using a TILLING population in *Oryza*, and a number of promising lines exhibiting higher tolerance to heat stress were identified for further breeding (see Janni et al. 2020). A similar application of TILLING to identify HSP genes could be replicated in legumes.

8.12 Next-Generation Platforms

The daunting task of estimating drought tolerance quantitatively is complicated by the intricacy of genetic governance and the prominence of environmental interference. Nevertheless, the advent of new technologies provides us with the sophistication of phenotyping in a non-invasive manner. An array of methodologies in this regard include image analysis software-based platforms, leaf or canopy reflectance spectrometry, analytical instrumentations based on thermal IR-near IR-visible-UV spectrum signatures and satellite-based GIS systems among others. These instruments provide us with the luxury of accurately measuring physiological responses under abiotic stresses over time and space without losing the seed materials. A highly productive phenotyping assay 'legume shovelomics' for examining root traits under water-stress conditions has been established for food legumes including bean and cowpea by integrating visual, manual and image analysis platforms. In *Cicer*, a PVC pipe-based phenotyping protocol has been developed for genotypic identification of water stress tolerance-imparting traits such as increased root biomass and long root lengths. Likewise, therapeutic radiological techniques like magnetic resonance imaging (MRI) and positron-emission tomography (PET) are also rendering enormously in studies related to the dynamics of photo-assimilates under water-stress conditions. Also, nuclear magnetic resonance (NMR) technique is proving to be an immense help in gauging water movement and sucrose apportionment via ^{13}C-labelled sucrose (see Jha et al. 2020a for details).

8.13 Conclusion

Drought and heat tolerance are highly complex traits governed by multiple genes and pose a formidable challenge to the plant breeders to accomplish their task of accurate screening of the constituent traits and induction of resistance against these stresses in plants. Adding to the woes is the climate change and the narrow genetic base of the legumes like cowpea, mung bean and black gram, which further make the improvement of these traits a daunting task through conventional breeding. However, with the advent of modern breeding techniques coupled with high-throughput phenotyping facilities, next-generation sequencing and cues from the accomplishments in cereals like rice and wheat, it is envisaged that the goal of achieving drought- and heat-stress tolerance in legumes is possible in near future. Moreover, the identification and sourcing of untapped adaptive traits from native cultivars and landraces through pre-breeding efforts should be fast-tracked as they are known to house genes against drought and heat stress. The pace of transgenic approaches in legumes should also be accelerated so that the translation genetics could be exploited and transgenics are readily available when the regulations are relaxed. Avenues of improving the genetic gain under drought and heat stress in legumes through physiological trait-based breeding should be explored. The powerful tool of mutation breeding in generating variability for abiotic stress tolerance traits could be deployed in complementation with other breeding techniques. Furthermore, evolving 'omics' sciences, inclusive of genomics, transcriptomics, proteomics and metabolomics, have potential implications in improving our present knowledge on abiotic stress tolerance mechanisms and in furthering the understanding of candidate genes and complex genetic and signalling pathways associated with water and high-temperature stresses in pulse legumes. For enhancing the sustainability of legume production in the context of climate change wherein the crops are exposed to various unpredicted stresses including drought and heat, it becomes imperative to leverage and consolidate the information generated through numerous genetical, physiological, biochemical and 'omics' studies related to stress tolerance.

Conclusively, a cohesive approach integrating genomics with high-throughput phenotyping and genotyping is desired to comprehend the vital mechanisms associated with water and high-temperature stresses, which could eventually augur the development of climate-smart legume cultivars for ensuring food and nutritional security.

References

Agbicodo EM, Fatokun CA, Muranaka S, Visser RGF (2009) Breeding drought tolerant cowpea: constraints, accomplishments, and future prospects. Euphytica 167:353–370

Ahmad F, Gaur PM, Croser J (2005) Chickpea (*Cicer arietinum* L.). Genetic resources, chromosome engineering and crop improvement. Grain Legume 1:187–217

Ahmed FE, Shakoor A, Suliman H (2010) Effect of water stress applied at different stages of growth on seed yield and water-use efficiency of Cowpea. Agric Biol J N Am 1:534–540

Ajayi AT, Gbadamosi AE, Olumekun VO, Omotuyi IO (2021) Variable expression of the candidate gene NCED1 among cowpea accessions under different drought stress conditions. J Genet Resour 7:133–143. https://doi.org/10.22080/jgr.2021.20645.1233

Ali Q, Javed MT, Noman A et al (2018) Assessment of drought tolerance in mung bean cultivars/lines as depicted by the activities of germination enzymes, seedling's antioxidative potential and nutrient acquisition. Arch Agron Soil Sci 64:84–102. https://doi.org/10.1080/03650340.2017.1335393

Anitha Y, Vanaja M, Vijay Kumar G (2016) Identification of attributes contributing to high temperature tolerance in blackgram (*Vigna mungo* L. Hepper) genotypes. Int J Sci Res 5: 1021–1025

Atkinson NJ, Urwin PE (2012) The interaction of plant biotic and abiotic stresses: from genes to the field. J Exp Bot 63:3523–3544

Baloda A, Madanpotra S, Aiwal PK (2017) Transformation of mungbean plants for salt and drought tolerance by introducing a gene for an osmoprotectant glycine betaine. J Plant Stress Physiol 3: 5. https://doi.org/10.19071/jpsp.2017.v3.3148

Bangar P, Chaudhury A, Tiwari B et al (2019) Morphophysiological and biochemical response of mungbean [*Vigna radiata* (L.) Wilczek] varieties at different developmental stages under drought stress. Turkish J Biol 43:58–69. https://doi.org/10.3906/biy-1801-64

Baroowa B, Gogoi N (2014) Biochemical changes in black gram and green gram genotypes after imposition of drought stress. J Food Legume 27:350–353

Barrera-Figueroa BE, Gao L, Diop NN et al (2011) Identification and comparative analysis of drought-associated microRNAs in two cowpea genotypes. BMC Plant Biol 11:127. https://doi.org/10.1186/1471-2229-11-127

Bastos EA, do Nascimento SP, da Silva EM et al (2011) Identification of cowpea genotypes for drought tolerance. Rev Ciência Agronômica 42:100–107. https://doi.org/10.1590/S1806-66902011000100013

Basu PS, Pratap A, Gupta S et al (2019) Physiological traits for shortening crop duration and improving productivity of greengram (*Vigna radiata* L. Wilczek) under high temperature. Front Plant Sci 10:1508. https://doi.org/10.3389/fpls.2019.01508

Batieno BJ, Danquah E, Tignegre J-B et al (2016) Application of marker-assisted backcrossing to improve cowpea (*Vigna unguiculata* L. Walp) for drought tolerance. J Plant Breed Crop Sci 8: 273–286. https://doi.org/10.5897/JPBCS2016.0607

Bharadwaj N, Barthakur S, Biswas AD et al (2019) Transcript expression profiling in two contrasting cultivars and molecular cloning of a SKP-1 like gene, a component of SCF-ubiquitin proteasome system from mungbean *Vigna radiata* L. Sci Rep 9:1–17. https://doi.org/10.1038/s41598-019-44034-4

Bohra A, Singh NP (2015) Whole genome sequences in pulse crops: a global community resource to expedite translational genomics and knowledge-based crop improvement. Biotechnol Lett 37: 1529–1539

Boukar O, Fatokun CA, Huynh BL et al (2016) Genomic tools in cowpea breeding programs: status and perspectives. Front Plant Sci 7:757

Boukar O, Belko N, Chamarthi S et al (2019) Cowpea (*Vigna unguiculata*): genetics, genomics and breeding. Plant Breed 138:415–424

Brachi B, Morris GP, Borevitz JO (2011) Genome-wide association studies in plants: the missing heritability is in the field. Genome Biol 12:232

Briglia N, Montanaro G, Petrozza A et al (2019) Drought phenotyping in *Vitis vinifera* using RGB and NIR imaging. Sci Hortic (Amsterdam) 256:108555. https://doi.org/10.1016/j.scienta.2019.108555

Burke JJ, Chen J (2015) Enhancement of reproductive heat tolerance in plants. PLoS One 10: e0122933. https://doi.org/10.1371/journal.pone.0122933

Carvalho M, Lino-Neto T, Rosa E, Carnide V (2017) Cowpea: a legume crop for a challenging environment. J Sci Food Agric 97:4273–4284. https://doi.org/10.1002/jsfa.8250

Carvalho M, Castro I, Moutinho-Pereira J et al (2019a) Evaluating stress responses in cowpea under drought stress. J Plant Physiol 241:153001. https://doi.org/10.1016/j.jplph.2019.153001

Carvalho M, Matos M, Castro I et al (2019b) Screening of worldwide cowpea collection to drought tolerant at a germination stage. Sci Hortic (Amsterdam). https://doi.org/10.1016/j.scienta.2018.11.082

Cisse N, Ndiaye M, Thiaw S, Hall AE (1995) Registration of 'Mouride' cowpea. Crop Sci 35:1215–1216

Cisse N, Ndiaye M, Thiaw S, Hall AE (1997) Registration of 'Melakh' cowpea. Crop Sci 37:1978

Contour-Ansel D, Torres-Franklin ML, Cruz De Carvalho MH et al (2006) Glutathione reductase in leaves of cowpea: cloning of two cDNAs, expression and enzymatic activity under progressive drought stress, desiccation and abscisic acid treatment. Ann Bot 98:1279–1287

Craufurd PQ, Summerfield RJ, Ellis RH, Roberts EH (1997) Photoperiod, temperature, and the growth and development of cowpea. Adv Cowpea Res 75–86

Cui Q, Xiong H, Yufeng Y et al (2020) Evaluation of drought tolerance in Arkansas Cowpea lines at seedling stage. Hort Sci 55:1132–1143. https://doi.org/10.21273/HORTSCI15036-20

D'Arcy-Lameta A, Ferrari-Iliou R, Contour-Ansel D et al (2006) Isolation and characterization of four ascorbate peroxidase cDNAs responsive to water deficit in cowpea leaves. Ann Bot 97: 133–140

da Silva HAP, Galisa PDS, Oliveira RSDS et al (2012) Expressão gênica induzida por estresses abióticos em nódulos de feijão-caupi. Pesqui Agropecuária Bras 47:797–807. https://doi.org/10.1590/S0100-204X2012000600010

DAC & FW (2020) Agricultural trade. Annual report 2019–2020. Minstry of Agriculture and Farmer's Welfare, Government of India

Dadson RB, Hashem FM, Javaid I et al (2005) Effect of water stress on the yield of cowpea (*Vigna unguiculata* L. Walp.) genotypes in the Delmarva region of the United States. J Agron Crop Sci 191.210–217. https://doi.org/10.1111/j.1439-037X.2005.00155.x

Dash M, Shree D (2013) Temperature effect on morphobiochemical characters in some black gram (*Vigna mungo*) genotypes. ISRN Biotechnol 2013:942868

Deshmukh R, Sonah H, Patil G et al (2014) Integrating omic approaches for abiotic stress tolerance in soybean. Front Plant Sci 5:244

Dhole VJ, Reddy KS (2010) Gamma rays induced moisture stress tolerant long root mutant in mungbean (*Vigna radiata* L. Wilczek). Electron J Plant Breed 1:1299–1305

Diop NN, Kidrič M, Repellin A et al (2004) A multicystatin is induced by drought-stress in cowpea (*Vigna unguiculata* (L.) Walp.) leaves. FEBS Lett 577:545–550

Divyaprasanth R, Ramanathan S, Jeyakumar P, Kokilavani S (2020) Effect of heat stress on the physiological traits and yield of black gram. Int J Agric Sci 12:10206–10209

Ehlers J, Hall A (1998) Heat tolerance of contrasting cowpea lines in short and long days. Field Crops Res 55:11–21

Ehlers JD, Hall AE, Patel PN et al (2000) Registration of 'California Blackeye 27' Cowpea. Crop Sci 40:854

El Maarouf H, Zuily-Fodil Y, Gareil M et al (1999) Enzymatic activity and gene expression under water stress of phospholipase D in two cultivars of *Vigna unguiculata* L. Walp. differing in drought tolerance. Plant Mol Biol 39:1257–1265

El-Nabarawy MAK, El-Kafafi ESH, Helal AG et al (2016) Characterization and evaluation of some Mung bean genotypes for drought tolerance. Egypt J Exp Biol 12:301–312

Farooq M, Gogoi N, Barthakur S et al (2017) Drought stress in grain legumes during reproduction and grain filling. J Agron Crop Sci 203:81–102

Fatima A, Khan MJ, Awan HM et al (2020) Genome-wide identification and expression analysis of SnRK2 gene family in mungbean (*Vigna radiata*) in response to drought stress. Crop Pasture Sci 71:469. https://doi.org/10.1071/CP19392

Fatokun CA, Boukar O, Muranaka S (2012) Evaluation of cowpea (*Vigna unguiculata* (L.) Walp.) germplasm lines for tolerance to drought. Plant Genet Resour Charact Util 10:171–176. https://doi.org/10.1017/S1479262112000214

Feder ME, Hofmann GE (1999) Heat-shock proteins, molecular chaperones, and the stress response: evolutionary and ecological physiology. Annu Rev Physiol 61:243–282

Fischer RA, Rees D, Sayre KD et al (1998) Wheat yield progress associated with higher stomatal conductance and photosynthetic rate, and cooler canopies. Crop Sci 38:1467–1475. https://doi.org/10.2135/cropsci1998.0011183X003800060011x

Fleury D, Jefferies S, Kuchel H, Langridge P (2010) Genetic and genomic tools to improve drought tolerance in wheat. J Exp Bot 61:3211–3222. https://doi.org/10.1093/jxb/erq152

Gazendam I, Oelofse D (2007) Isolation of cowpea genes conferring drought tolerance: construction of a cDNA drought expression library. Water SA 33:387–392

Goufo P, Moutinho-Pereira JM, Jorge TF et al (2017) Cowpea (*Vigna unguiculata* L. Walp.) metabolomics: osmoprotection as a physiological strategy for drought stress resistance and improved yield. Front Plant Sci 8:586. https://doi.org/10.3389/fpls.2017.00586

Gupta DS, Basu PS, Souframanien J et al (2021) Morpho-physiological traits and functional markers based molecular dissection of heat-tolerance in urdbean. Front Plant Sci 12:719381. https://doi.org/10.3389/fpls.2021.719381

Gurumurthy S, Sarkar B, Vanaja M et al (2019) Morpho-physiological and biochemical changes in black gram (*Vigna mungo* L. Hepper) genotypes under drought stress at flowering stage. Acta Physiol Plant 41:42. https://doi.org/10.1007/s11738-019-2833-x

Hall AE (1992) Breeding for heat tolerance. Plant Breed Rev 10:129–168

Hall AE (1993) Physiology and breeding for heat tolerance in cowpea, and comparisons with other crops. In: Kuo CG (ed) Adaptation of food crops to temperature and water stress. Asian Vegetable Research and Development Center, Taipei, Taiwan, pp 271–284

Hall AE (2004) Breeding for adaptation to drought and heat in cowpea. Eur J Agron 21:447–454

Hall AE, Patel PN (1985) Breeding for resistance to drought and heat. In: Singh SR, Rachie KO (eds) Cowpea research, production and utilisation. Wiley, Chichester

Hamidou F, Zombre G, Braconnier S (2007) Physiological and biochemical responses of cowpea genotypes to water stress under glasshouse and field conditions. J Agron Crop Sci 193:229–237

Hanumantharao B, Nair RM, Nayyar H (2016) Salinity and high temperature tolerance in mungbean [*Vigna radiata* (L.) Wilczek] from a physiological perspective. Front Plant Sci 7: 957. https://doi.org/10.3389/fpls.2016.00957

IPCC (2007) Climate change 2007: the physical science basis. In: Solomon S, Qin D, Manning M, Chen Z, Marquis M, Averyt KB, Tignor M, Miller HL (eds) Contribution of Working Group I to the Fourth Assessment Report of the Intergovernmental Panel on Climate Change. Cambridge University Press, p 996

Iseki K, Takahashi Y, Muto C et al (2018) Diversity of drought tolerance in the genus V*igna*. Front Plant Sci 9:729. https://doi.org/10.3389/fpls.2018.00729

Islam MR, Islam MS, Sarker BC et al (2019) Assessment of water stress tolerance in mungbean induced by polyethylene glycol. BioRxiv 2019:872663

Ismail AM, Hall AE (1998) Positive and potential negative effects of heat-tolerance genes in cowpea. Crop Sci 38:381–390

Ismail AM, Hall AE, Ehlers JD (2000) Delayed-leaf-senescence and heat-tolerance traits mainly are independently expressed in cowpea. Crop Sci 40:1049–1055

Iuchi S, Yamaguchi-Shinozaki K, Urao T et al (1996a) Novel drought-inducible genes in the highly drought-tolerant cowpea: cloning of cDNAs and analysis of the expression of the corresponding genes. Plant Cell Physiol 37:1073–1082. https://doi.org/10.1093/oxfordjournals.pcp.a029056

Iuchi S, Yamaguchi-Shinozaki K, Urao T, Shinozaki K (1996b) Characterization of two cDNAs for novel drought-inducible genes in the highly drought-tolerant cowpea. J Plant Res 109:415–424

Iuchi S, Kobayashi M, Yamaguchi-Shinozaki K, Shinozaki K (2000) A stress-inducible gene for 9-cis-epoxycarotenoid dioxygenase involved in abscisic acid biosynthesis under water stress in drought-tolerant cowpea. Plant Physiol 123:553–562

Janni M, Gullì M, Maestri E et al (2020) Molecular and genetic bases of heat stress responses in crop plants and breeding for increased resilience and productivity. J Exp Bot 71:3780–3802

Jasrotia RS, Yadav PK, Iquebal MA et al (2019) VigSatDB: genome-wide microsatellite DNA marker database of three species of *Vigna* for germplasm characterization and improvement. Database 2019:baz055. https://doi.org/10.1093/database/baz055

Jegadeesan S, Raizada A, Dhanasekar P, Suprasanna P (2021) Draft genome sequence of the pulse crop blackgram [*Vigna mungo* (L.) Hepper] reveals potential R-genes. Sci Rep 11:1–10. https://doi.org/10.1038/s41598-021-90683-9

Jha UC, Bohra A, Parida SK, Jha R (2017) Integrated "omics" approaches to sustain global productivity of major grain legumes under heat stress. Plant Breed 136:437–459

Jha UC, Bohra A, Nayyar H (2020a) Advances in "omics" approaches to tackle drought stress in grain legumes. Plant Breed 139:1–27

Jha UC, Nayyar H, Jha R et al (2020b) Heat stress and cowpea: genetics, breeding and modern tools for improving genetic gains. Plant Physiol Rep 25:645–653

Kang YJ, Kim SK, Kim MY et al (2014) Genome sequence of mungbean and insights into evolution within Vigna species. Nat Commun 5:1–9

Khattak SSG, Saeed I, Muhammad T (2006) Breeding for heat tolerance in mungbean (*Vigna radiata* (L.) Wilczek). Pak J Bot 38(5)

Kulkarni A, Gadgil S, Patwardhan S (2016) Monsoon variability, the 2015 Marathwada drought and rainfed agriculture. Curr Sci 111:1182–1193

Kumar S, Kaur R, Kaur N et al (2011) Heat-stress induced inhibition in growth and chlorosis in mungbean (*Phaseolus aureus* Roxb.) is partly mitigated by ascorbic acid application and is related to reduction in oxidative stress. Acta Physiol Plant 33:2091–2101

Kumar S, Kumar A, Jamwal S et al (2019) Divergence studies of blackgram (*Vigna mungo* L) for selection of drought tolerant genotypes under rainfed conditions of North Western Himalayas in J and K, India. Legum Res 44:158–163. https://doi.org/10.18805/lr-4083

Kumar J, Sen Gupta D, Djalovic I (2020a) Breeding, genetics, and genomics for tolerance against terminal heat in lentil: current status and future directions. Legum Sci 2:e38. https://doi.org/10.1002/leg3.38

Kumar S, Ayachit G, Sahoo L (2020b) Screening of mungbean for drought tolerance and transcriptome profiling between drought-tolerant and susceptible genotype in response to drought stress. Plant Physiol Biochem 157:229–238. https://doi.org/10.1016/j.plaphy.2020.10.021

Labbo AM, Mehmood M, Akhtar MN et al (2018) Genome-wide identification of AP2/ERF transcription factors in mungbean (*Vigna radiata*) and expression profiling of the VrDREB subfamily under drought stress. Crop Pasture Sci 69:1009. https://doi.org/10.1071/CP18180

Laing DR, Jones PG, Davis JH (1984) Common bean (Phaseolus vulgaris L). In: Goldsworthy PR, Fisher NM (eds) The physiology of tropical field crops. Wiley, New York, pp 305–351

Lamaoui M, Jemo M, Datla R, Bekkaoui F (2018) Heat and drought stresses in crops and approaches for their mitigation. Front Chem 6:26. https://doi.org/10.3389/fchem.2018.00026

Li M-W, Xin D, Gao Y et al (2017) Using genomic information to improve soybean adaptability to climate change. J Exp Bot 68:1823–1834. https://doi.org/10.1093/jxb/erw348

Li S, Wang R, Jin H et al (2019) Molecular characterization and expression profile analysis of heat shock transcription factors in mungbean. Front Genet 9:736. https://doi.org/10.3389/fgene.2018.00736

Lipiec J, Doussan C, Nosalewicz A, Kondracka K (2013) Effect of drought and heat stresses on plant growth and yield: a review. Int Agrophys 27:463–477

Liu C, Wu J, Wang L et al (2017) Quantitative trait locus mapping under irrigated and drought treatments based on a novel genetic linkage map in mungbean (*Vigna radiata* L.). Theor Appl Genet 130:2375–2393. https://doi.org/10.1007/s00122-017-2965-6

Lonardi S, Muñoz-Amatriaín M, Liang Q et al (2019) The genome of cowpea (*Vigna unguiculata* [L.] Walp.). Plant J 98:767–782

Lopes MS, El-Basyoni I, Baenziger PS et al (2015) Exploiting genetic diversity from landraces in wheat breeding for adaptation to climate change. J Exp Bot 66:3477–3486. https://doi.org/10.1093/jxb/erv122

Lucas MR, Ehlers JD, Huynh BL et al (2013) Markers for breeding heat-tolerant cowpea. Mol Breed 31:529–536. https://doi.org/10.1007/s11032-012-9810-z

Mai-Kodomi Y, Singh B, Terao T et al (1999a) Inheritance of drought tolerance in cowpea. Indian J Genet Plant Breed 59:317–323

Mai-Kodomi Y, Singh BB, Myers O et al (1999b) Two mechanisms of drought tolerance in cowpea. Indian J Genet Plant Breed 59:309–316

Marcel G, Matos A, D'Arcy-Lameta A et al (2000) Two novel plant cDNAs homologous to animal type-2 phosphatidate phosphatase are expressed in cowpea leaves and are differently regulated by water deficits. Biochem Soc Trans 28:915–917

Marfo KO, Hall AE (1992) Inheritance of heat tolerance during pod set in cowpea. Crop Sci 32: 912–918

Matos AR, d'Arcy-Lameta A, França M et al (2001) A novel patatin-like gene stimulated by drought stress encodes a galactolipid acyl hydrolase. FEBS Lett 491:188–192

Matsui T, Singh BB (2003) Root characteristics in cowpea related to drought tolerance at the seedling stage. Exp Agric 39:29–38. https://doi.org/10.1017/S0014479703001108

Meena SK, Pandey R, Sharma S et al (2021) Cross tolerance to phosphorus deficiency and drought stress in mungbean is regulated by improved antioxidant capacity, biological N2-fixation, and differential transcript accumulation. Plant Soil 466:337–356. https://doi.org/10.1007/s11104-021-05062-0

Mitra J (2001) Genetics and genetic improvement of drought resistance in crop plants. Curr Sci 80: 758–763

Muchero W, Ehlers JD, Roberts PA (2008) Seedling stage drought-induced phenotypes and drought-responsive genes in diverse cowpea genotypes. Crop Sci 48:541–552

Muchero W, Ehlers JD, Close TJ, Roberts PA (2009) Mapping QTL for drought stress-induced premature senescence and maturity in cowpea [*Vigna unguiculata* (L.) Walp.]. Theor Appl Genet 118:849–863. https://doi.org/10.1007/s00122-008-0944-7

Muchero W, Roberts PA, Diop NN et al (2013) Genetic architecture of delayed senescence, biomass, and grain yield under drought stress in cowpea. PLoS One 8:e70041. https://doi.org/10.1371/journal.pone.0070041

Muñoz-Amatriaín M, Mirebrahim H, Xu P et al (2017) Genome resources for climate-resilient cowpea, an essential crop for food security. Plant J 89:1042–1054

Nahar K, Hasanuzzaman M, Alam MM, Fujita M (2015) Exogenous glutathione confers high temperature stress tolerance in mung bean (*Vigna radiata* L.) by modulating antioxidant defense and methylglyoxal detoxification system. Environ Exp Bot 112:44–54

Nam NH, Chauhan YS, Johansen C (2001) Effect of timing of drought stress on growth and grain yield of extra-short-duration pigeonpea lines. J Agric Sci 136:179–189. https://doi.org/10.1017/S0021859601008607

Nielsen C, Hall A (1985) Responses of cowpea (*Vigna unguiculata* (L.) Walp.) in the field to high night air temperature during flowering. II. Plant responses. Field Crop Res 10:181–196

Noble T, Douglas C, Williams R et al (2017) Development of the mungbean Nested Association Mapping (NAM) resource. InterDrought V 5:218

Noble TJ, Tao Y, Mace ES et al (2018) Characterization of linkage disequilibrium and population structure in a mungbean diversity panel. Front Plant Sci 8:2102. https://doi.org/10.3389/fpls.2017.02102

Ntare BR (1991) Variation in reproductive efficiency and yield of cowpea under high temperature conditions in a Sahelian environment. Euphytica 59:27–32. https://doi.org/10.1007/BF00025358

Obata T, Fernie AR (2012) The use of metabolomics to dissect plant responses to abiotic stresses. Cell Mol Life Sci 69:3225–3243

Olajide AA, Ilori CO (2018) Genetic analysis of seedlings characters associated with drought tolerance in cowpea under a controlled environment. Plant Genet Resour Charact Util 16: 218–227. https://doi.org/10.1017/S1479262117000235

Olubunmi DI (2015) Genetic analysis of drought tolerance in cowpea [*Vigna unguiculata* (L.) Walp]. University of Ghana

Osmond CB, Austin MP, Berry JA et al (1987) Stress physiology and the distribution of plants. Bioscience 37:38–48. https://doi.org/10.2307/1310176

Pandiyan M, Senthil N, Balaji T et al (2017) Studies on performance of drought tolerant genotypes under drought and normal conditions through morpho, physio and biochemical attributes of blackgram (*Vigna mungo* L.) and green gram (*Vigna radiata* L.). Int J Adv Res 5:489–496. https://doi.org/10.21474/ijar01/3176

Partheeban C, Vijayaraghavan H, Boominathan P (2017) The rapid and reliable method to assess the blackgram (*Vigna mungo* (l.) Hepper) genotypes for thermotolerance. Int J Bioresour Stress Manag 8:6–11. https://doi.org/10.23910/IJBSM/2017.8.1.1710

Patel PN, Hall AE (1988) Inheritance of heat-induced brown discoloration in seed coats of cowpea. Crop Sci 28:929–932

Pootakham W, Nawae W, Naktang C et al (2021) A chromosome-scale assembly of the black gram (*Vigna mungo*) genome. Mol Ecol Resour 21:238–250

Pottorff M, Roberts PA, Close TJ et al (2014) Identification of candidate genes and molecular markers for heat-induced brown discoloration of seed coats in cowpea [*Vigna unguiculata* (L.) Walp]. BMC Genomics 15:328. https://doi.org/10.1186/1471-2164-15-328

Prakash M, Elangaimannan R, Sunilkumar B, Sathiya Narayanan G (2018) Evaluation of blackgram genotypes for drought tolerance based on root dynamics and gas exchange parameters. Legum Res 41:384–391. https://doi.org/10.18805/LR-3702

Priya M, Sharma L, Kaur R et al (2019) GABA (γ-aminobutyric acid), as a thermo-protectant, to improve the reproductive function of heat-stressed mungbean plants. Sci Rep 9:1–14. https://doi.org/10.1038/s41598-019-44163-w

Rainey KM, Griffiths PD (2005) Inheritance of heat tolerance during reproductive development in snap bean (Phaseolus vulgaris L.). J Am Soc Hortic Sci 130:700–706

Raizada A, Souframanien J (2019) Transcriptome sequencing, de novo assembly, characterisation of wild accession of blackgram (*Vigna mungo* var. silvestris) as a rich resource for development of molecular markers and validation of SNPs by high resolution melting (HRM) analysis. BMC Plant Biol 19:358. https://doi.org/10.1186/s12870-019-1954-0

Rao KVM, Raghavendra AS, Reddy KJ (2006) Physiology and molecular biology of stress tolerance in plants. Springer Science & Business Media, New York

Ravelombola W, Shi A et al (2018) Investigation on various aboveground traits to identify drought tolerance in cowpea seedlings. HortScience 53:1757–1765. https://doi.org/10.21273/HORTSCI13278-18

Ravelombola W, Shi A, Huynh BL (2021) Loci discovery, network-guided approach, and genomic prediction for drought tolerance index in a multi-parent advanced generation intercross (MAGIC) cowpea population. Hortic Res 8:24. https://doi.org/10.1038/s41438-021-00462-w

Sadeghipour O (2019) Polyamines protect mung bean [*Vigna radiata* (L.) Wilczek] plants against drought stress. Biol Futur 70:71–78. https://doi.org/10.1556/019.70.2019.09

Sadhukhan A, Kobayashi YY, Kobayashi YY et al (2014) VuDREB2A, a novel DREB2-type transcription factor in the drought-tolerant legume cowpea, mediates DRE-dependent expression of stress-responsive genes and confers enhanced drought resistance in transgenic *Arabidopsis*. Planta 240:645–664. https://doi.org/10.1007/s00425-014-2111-5

Sai CB, Chidambaranathan P (2019) Reproductive stage drought tolerance in blackgram is associated with role of antioxidants on membrane stability. Plant Physiol Rep 24:399–409. https://doi.org/10.1007/s40502-019-00471-x

Salehi-Lisar SY, Bakhshayeshan-Agdam H (2016) Drought stress in plants: causes, consequences, and tolerance. In: Drought stress tolerance in plants, Physiology and biochemistry, vol 1. Springer International Publishing, New York, pp 1–16

Salvi S, Tuberosa R (2005) To clone or not to clone plant QTLs: present and future challenges. Trends Plant Sci 10:297–304

Santos R, Carvalho M, Rosa E et al (2020) Root and agro-morphological traits performance in cowpea under drought stress. Agronomy 10:1604. https://doi.org/10.3390/agronomy10101604
Schmutz J, Cannon SB, Schlueter J et al (2010) Genome sequence of the palaeopolyploid soybean. Nature 463:178–183
Sehgal A, Sita K, Kumar J et al (2017) Effects of drought, heat and their interaction on the growth, yield and photosynthetic function of lentil (*Lens culinaris* Medikus) genotypes varying in heat and drought sensitivity. Front Plant Sci 8:1776. https://doi.org/10.3389/fpls.2017.01776
Sehgal A, Sita K, Bhandari K et al (2019) Influence of drought and heat stress, applied independently or in combination during seed development, on qualitative and quantitative aspects of seeds of lentil (*Lens culinaris* Medikus) genotypes, differing in drought sensitivity. Plant Cell Environ 42:198–211. https://doi.org/10.1111/pce.13328
Shirsath and Bhosale Agro India Ltd (2017) Modern agrotechniques for cultivation of black gram/ urdbean (*Vigna mungo* L.). Shirsath and Bhosale Agro Ltd, Ahmednagar
Siddiqui MH, Al-Khaishany MY, Al-Qutami MA et al (2015) Morphological and physiological characterization of different genotypes of faba bean under heat stress. Saudi J Biol Sci 22:656–663. https://doi.org/10.1016/j.sjbs.2015.06.002
Singh P, Kanemasu ET (1983) Leaf and canopy temperatures of pearl millet genotypes under irrigated and nonirrigated conditions. Agron J 75:497–501. https://doi.org/10.2134/agronj1983.00021962007500030019x
Singh BB, Matsui T (2002) Cowpea varieties for drought tolerance. In: Fatokun C, Tarawali S, Singh B et al (eds) Challenges and opportunities for enhancing sustainable cowpea production. Proceedings of the World Cowpea Conference III held at the International Institute of Tropical Agriculture (IITA), Ibadan, Nigeria, 4–8 September 2000. IITA, Ibadan, Nigeria, pp 287–300
Singh DP, Singh BB (2011) Breeding for tolerance to abiotic stresses in mungbean. J Food Legum 24:83–90
Singh P, Kumar D, Sarin NB (2016) Multiple abiotic stress tolerance in Vigna mungo is altered by overexpression of ALDRXV4 gene via reactive carbonyl detoxification. Plant Mol Biol 91:257–273. https://doi.org/10.1007/s11103-016-0464-9
Singh S, Singh YP, Tomar SS (2018) Review on climatic abnormalities impact on area, productivity of central India and strategies of mitigating technology on yield and benefits of black gram. J Pharmacogn Phytochem 7:1048–1056
Sita K, Sehgal A, Hanumantharao B et al (2017) Food legumes and rising temperatures: effects, adaptive functional mechanisms specific to reproductive growth stage and strategies to improve heat tolerance. Front Plant Sci 8:1658
Slabbert R, Spreeth M, Krüger GHJ, Bornman CH (2004) Drought tolerance, traditional crops and biotechnology: breeding towards sustainable development. South Afr J Bot 70:116–123. https://doi.org/10.1016/S0254-6299(15)30271-4
Sofi PA, Ara A, Gull M, Rehman K (2019) Canopy temperature depression as an effective physiological trait for drought screening. In: Drought-detection and solutions. IntechOpen, London
Souframanien J, Reddy KS (2015) De novo assembly, characterization of immature seed transcriptome and development of genic-SSR markers in black gram [*Vigna mungo* (L.) Hepper]. PLoS One 10:e0128748. https://doi.org/10.1371/journal.pone.0128748
Souframanien J, Saha AJ, Dhole VJ et al (2020) Genetic improvement of pulse crops through induced mutation and biotechnological approaches. IANCAS Bull XV:71–80
Srivastava R, Kumar S, Kobayashi Y et al (2018) Comparative genome-wide analysis of WRKY transcription factors in two Asian legume crops: Adzuki bean and Mung bean. Sci Rep 8:16971. https://doi.org/10.1038/s41598-018-34920-8
Sunayana MSP, Yadav R (2016) Evaluation of mungbean (*Vigna radiata* L. Wilczek) genotypes for high temperature and drought tolerance. Ecoscan 9:875–880
Teulat B, Monneveux P, Wery J et al (1997) Relationships between relative water content and growth parameters under water stress in barley: a QTL study. New Phytol 137:99–107

Thiaw S, Hall AE (2004) Comparison of selection for either leaf-electrolyte-leakage or pod set in enhancing heat tolerance and grain yield of cowpea. Field Crops Res 86:239–253. https://doi.org/10.1016/j.fcr.2003.08.011

Thudi M, Upadhyaya HD, Rathore A et al (2014) Genetic dissection of drought and heat tolerance in chickpea through genome-wide and candidate gene-based association mapping approaches. PLoS One 9:e96758

Tian X, Li S, Liu Y, Liu X (2016) Transcriptomic profiling reveals metabolic and regulatory pathways in the desiccation tolerance of mungbean (*Vigna radiata* [L.] R. Wilczek). Front Plant Sci 7:1921. https://doi.org/10.3389/fpls.2016.01921

Timko M, Singh B (2008) Cowpea, a multifunctional legume. In: Moore P, Ming R (eds) Genomics of tropical crop plants. Springer, New York, pp 227–258

Uauy C (2017) Wheat genomics comes of age. Curr Opin Plant Biol 36:142–148. https://doi.org/10.1016/j.pbi.2017.01.007

Varshney RK, Chen W, Li Y et al (2012) Draft genome sequence of pigeonpea (*Cajanus cajan*), an orphan legume crop of resource-poor farmers. Nat Biotechnol 30:83

Varshney RK, Pandey MK, Bohra A et al (2019) Toward the sequence-based breeding in legumes in the post-genome sequencing era. Theor Appl Genet 132:797–816

Verbree DA, Singh BB, Payne WA (2015) Genetics and heritability of shoot drought tolerance in cowpea seedlings. Crop Sci 55:146–153. https://doi.org/10.2135/cropsci2014.02.0137

Wang L, Zhu J, Li X et al (2018) Salt and drought stress and ABA responses related to bZIP genes from *V. radiata* and *V. angularis*. Gene 651:152–160. https://doi.org/10.1016/j.gene.2018.02.005

Wenham K, Williams A, Rachaputi RCN et al (2020) Critical period of moisture vulnerability in mungbeans. Grain Res Dev Corp

Xu P, Wu X, Muñoz-Amatriaín M et al (2017) Genomic regions, cellular components and gene regulatory basis underlying pod length variations in cowpea (*V. unguiculata* L. Walp). Plant Biotechnol J 15:547–557

Zhu J-K (2002) Salt and drought stress signal transduction in plants. Annu Rev Plant Biol 53:247–273

Current and Future Strategies in Breeding Lentil for Abiotic Stresses

9

Muraleedhar S. Aski, Harsh K. Dikshit, Gyan Prakash Mishra, Prachi S. Yadav, Mir Asif Iquebal, Sarika, Ruchi Bansal, Gayacharan, Akansha Singh, Shiv Kumar, and Sripad Udupa

Abstract

There has been a lot of research on biotic stresses in lentils because they are visible and lead to decline in production and quality losses. Abiotic stresses, on the other hand, are rapidly being identified as key reasons for the low and unpredictable yield of lentils in many regions. Changes in climate, soils, and climate-soil interactions affect lentil productivity and quality directly or indirectly through their influence on foliar and soil-borne diseases, pests, and rhizobia in each growing zone. Furthermore, the relative tolerance of a cultivar and/or the effect of specific cultural control approaches can vary the effects of a specific stress. Salinity, waterlogging, cold, drought, and heat are the key abiotic factors that affect lentil output, and it is critical to produce climate-robust lentil cultivars

M. S. Aski (✉) · H. K. Dikshit · G. P. Mishra · P. S. Yadav
Division of Genetics, ICAR-Indian Agricultural Research Institute, New Delhi, India

M. A. Iquebal · Sarika
Centre for Agricultural Bioinformatics, ICAR-Indian Agricultural Statistics Research Institute, New Delhi, India

R. Bansal
Plant Physiology, ICAR-Indian Agricultural Research Institute, New Delhi, India

Gayacharan
Germplasm Evaluation Division, ICAR-National Bureau of Plant Genetic resources, New Delhi, India

A. Singh
Amity Institute of Organic Agriculture, Amity University, Noida, Uttar Pradesh, India

S. Kumar
International Center for Agricultural Research in the Dry Areas (ICARDA), New Delhi, India

S. Udupa
International Center for Agricultural Research in the Dry Areas (ICARDA), Rabat, Morocco

P. Muthu Arjuna Samy et al. (eds.), *Legumes: Physiology and Molecular Biology of Abiotic Stress Tolerance*, https://doi.org/10.1007/978-981-19-5817-5_9

to address these issues. The implications of several abiotic stresses on lentil production, genetics, and genomics, including mapping of quantitative traits and incorporating the identified genes with the help of marker-assisted selection breeding, and transcriptomics for the advancement of abiotic stress tolerance in lentil are all covered in this chapter. By identifying candidate genes, gene mapping, and marker-assisted selection, advanced genomic tools can supplement traditional breeding procedures to accelerate breeding projects by enhancing accuracy and saving time. There are few reports on lentil resilience to abiotic stress factors, and more work is needed to investigate the inherited biological process. Evaluating germplasm and breeding material for cultivars resistant to abiotic stressors necessitates the use of rigorous and reproducible phenotypic testing approaches. Systemic application of pan-omics with novel omics technologies will fast-track lentil breeding programmes. Additionally, artificial intelligence (AI) algorithms can help in simulating yield under climate change, leading to predicting the genetic gain. Use of machine learning (ML) in quantitative trait locus (QTL) mining will further enhance the understanding of genetic determinants of abiotic stress in lentils.

Keywords

Lentil · Abiotic stress · Stress · Tolerance · Resistance · Breeding · Conventional · Molecular

9.1 Introduction

Lentil is a winter-season, annual food grain legume that is self-pollinating and diploid ($2n = 2x = 14$) and has a bigger genome size of 4 Gbp (Gupta et al. 2011). This crop has been consumed since prehistoric times and originated around 7000–10,000 years ago in the Eastern Mediterranean region (Gupta et al. 2011; Rizvi et al. 2019). This has been produced in Mediterranean regions, southwest Asia, and Australia, as well as South and North America (Hamwieh et al. 2009). Lentil seeds are rich in protein, as well as folate, vitamin C, fibre, carbs, and minerals, and have a low-calorie value (Bari et al. 2009; Choukri et al. 2020). This distinguishes them from other food grain legumes and cereals with lower phytic acid concentration (Thavarajah et al. 2011) and higher total phenolic content (Xu and Chang 2010). Lentils are therefore an excellent source of minerals, amino acids, and high-quality protein (Kahraman 2016; Khazaei et al. 2016). As a result, lentils are particularly good for the general health of both humans and animals (Kumar et al. 2015, Samaranayaka 2017; Kumar et al. 2019a, b). In addition, studies have connected lentil consumption to weight loss, body fat loss (Siva et al. 2018), and antihypertensive function (García-Mora et al. 2017).

The crop has been raised globally, with Canada producing the most lentils, accounting for 48.1% of the world and being the largest exporter of lentils. India is the world's second biggest producer of lentils, but it is also the world's largest importer (Dissanayake et al. 2020). Additionally, the worldwide mean lentil yield is 926 kg/ha, while Asia's average yield is 817 kg/ha, which is much lower than the world average (Singh et al. 2017c). Given the crop's importance for human consumption, animal health, and agricultural systems, it has been largely ignored. To boost resistance to pests and yield, hybridisation with conventional lentil varieties has been used to generate lentil cultivars under production. Notwithstanding these, minimal advancement has been achieved in raising crop yield and productivity. South Asian and Canadian lentil germplasm has a limited genetic background, as per Khazaei et al. (2016), since lentil output is frequently challenged in such locations owing to biotic and abiotic factors. Lentil farmed germplasm has a smaller genetic foundation than its wild equivalents (Duran et al. 2004). As a consequence, there is an urgent need to focus on broadening the genetic base of the lentil cultivar by incorporating commercially relevant genes into the cultivated gene pool. Because it is a climate-resilient and highly nutritious pulse crop, lentil fits well within the target food crops to be turned into the category of a smart crop. As a consequence, scientific engagement is essential in lentil breeding initiatives. The establishment of a breeding approach for lentil genetic improvement has been influenced significantly by recent advancements in current genomic technology.

9.1.1 Nutritional Benefit and Their Health Significance

Lentil has been one of the most essential foods ingested by mankind from the dawn of humanity (Sarker and Erskine 2006; Faris et al. 2013). Its nutritional value has been calculated in order to meet the food and feed requirements of humans and other species. Numerous lentil recipes have been created and enjoyed in South Asia. Inside the Indian subcontinent, it is referred to as "dhal". "Mejadra", a lentil and rice mixture also called as mujaddra, is prepared in Mediterranean regions. Lentil flour will be used to produce purees, soups, and stews, as well as pastries and bread when combined with cereal flour. Lentils cook quickly and efficiently, leading to limited nutrient loss. They are known as "poor man's meat" due to the richness in the nutritional content, 20–36% protein, 60–67% carbohydrate, 2–3% ash on a dry basis, and <4% lipid (Samaranayaka 2017; Johnson et al. 2020). Its dietary qualities are superior to those of other important legumes and grains like chickpeas, soybeans, wheat, and rice (Johnson et al. 2020). Lentils are an excellent source of energy. Because of the low levels of antinutrients, fat, and cholesterol, it is regarded as the best source of protein for human consumption (Sultana and Ghafoor 2008). It is low in fats and abundant in minerals and vitamins, comparative to chickpea and soybean, whereas rice and wheat are low in minerals and vitamins (Johnson et al. 2020). Lentils are poor in cysteine and methionine (sulphur-containing amino acids), as well as tryptophan.

Diets rich in legumes, such as lentils, have numerous health benefits. Lentils have a lower glycaemic index than other legumes. In older persons with a high cardiovascular risk, frequent use of legumes, particularly lentils, in the context of a Mediterranean diet may help avoid type 2 diabetes (Becerra-Tomás et al. 2018, 2020). Red lentils have an extremely low glycaemic index (21%) when compared to other cereal-based meals such as multigrain bread (62%), wholewheat pasta (55%), and basmati rice (69%) (Henry et al. 2005). Lentils cut LDL cholesterol, reduce the risk of cardiovascular problems in persons who eat a lentil-rich diet (Abeysekara et al. 2012), induce satiety (McCrory et al. 2010), and are seen to be a feasible weight management solution (McCrory et al. 2010; Siva et al. 2018). Lentil also contains prebiotic carbohydrates, and variations in these carbs after cooking, cooling, and reheating have been associated with a variety of health benefits (Johnson et al. 2013). According to a study by Rizkalla et al. (2002), lentils could be utilised to treat type 2 diabetes. Fasting blood sugar levels drop considerably if 50 g of lentils are introduced to a diabetic patient's diet (Shams et al. 2010). Several researchers have found that eating lentils improves human health by reducing oxidative stress, adhesion molecules, triglycerides, inflammatory biomarkers, and low-density lipoprotein-cholesterol concentrations and increasing serum antioxidant potential, above all which together slow coronary heart disease progression (Crujeiras et al. 2007; Taku et al. 2007; Azadbakht et al. 2007; Azadbakht and Esmaillzadeh 2012).

9.1.2 Effect of Stress on Quality and Crop Yield

Biological and abiotic factors seriously restrict pulse crops' potential to improve yield qualities genetically (Tomlekova 1998; Yankova and Sovkova-Bobcheva 2009; Muraleedhar et al. 2015; Aski et al. 2021, 2022). Domesticated lentils are exposed to significant biotic and abiotic challenges and have a narrow genetic makeup as compared to their crop wild cousins (Singh et al. 2014). Several traits, in combination with a host of other influences, restrict yield and, as a consequence, lower total productivity, especially in emerging economies (Muehlbauer et al. 2006; Sinclair and Vadez 2012; Tivoli et al. 2006). Abiotic stresses include temperature (both low and high), drought, cold, salt, heat, nutrient scarcity, and toxicity (Yau and Erskine 2000). Abiotic elements like cold, salt, heat, and drought stress are all important on a global scale (Silim and Saxena 1993; Turner et al. 2001). Since lentils seem to be more susceptible to saline condition than other pulses (like soybean and faba bean), salt stress is a serious issue, especially on the Indian subcontinent (Katerji et al. 2005). South Asian countries supply nearly half of the world's lentils, yet they are also the major importers of lentils due to high demand. Lentils grown in these places belong to the Pilosae family, which has a limited genetic makeup. Lentil improvement breeding programmes have been impeded by a paucity of phenological, morphological, and yield-related variation, as well as susceptibility to substantial abiotic stresses. Conventional breeding methods have resulted in variations that are resistant to biotic and abiotic stresses and can handle substantial production constraints (Gahoonia et al. 2005; Materne and Reddy 2007; Muehlbauer et al. 2006). Despite this, breeding operations have been plagued by a paucity of genetic information, a restricted genetic background, and accuracy in selection techniques, which together obstruct a breeder's ultimate breeding goals.

9.1.3 Lentils in the Midst of Climate Change and Rising Population

Climate change is affecting several aspects of agroecosystems. Climate change's cumulative impact on agricultural systems has the potential to cause global shortages of food and mass famine. Leguminous crops are already under strain from a variety of factors, but changing climatic conditions make adaptation even more challenging. According to climate change projections and studies, temperatures will have climbed by 2–4 °C by the late twenty-first century (Girvetz et al. 2019; Tadross et al. 2007), decreasing total growth and productivity of all major crops, notably legumes (Varshney et al. 2011; Zhao et al. 2017). Global warming is expected to have a significant negative impact on farmers' livelihoods, global food security, and agriculture as a result. For improving resilience to climate change in lentils, detecting and introgressing variable attributes or genes useful for widening the genetic basis of existing cultivars are essential. As a result, this chapter will help you understand the abiotic challenges that lentils encounter, as well as significant features, sources of diversity, function of conventional breeding, molecular tools, and genomics, and how to use introgression of such genes in a breeding scheme to generate abiotic-resistant cultivars.

9.2 Major Abiotic Stresses Influencing Lentil Productivity

9.2.1 Heat Stress

Increasing temperatures are one of the big environmental factors impeding the development and growth of economically important crops. To combat heat stress in lentils, high temperature-resistant varieties can be bred since they can maintain production and productivity even in the face of ongoing climate change. Lentils, like all other cool-season legume crops, are vulnerable to temperature increases (Bhandari et al. 2016; Choudhury et al. 2012; Sehgal et al. 2017). The optimal temperature ranges from 18 to 30 °C depending on the stage of development. Throughout the vegetative growth stages, cooler temperatures are required; however, when the plant reaches maturity, warmer temperatures are considered necessary (Choudhury et al. 2012). The crop was grown in India's similarly warmer regions (Northern and Southern) and exposed to temperatures that were above ideal, resulting in decreased grain production (Verma et al. 2014). Moreover, as a result of climate change, the warm phase has become lengthier than the cooling period, exposing crops to significant temperature distress, specifically during the reproductive phase (Hasanuzzaman et al. 2013). In southern Australia, a heatwave of 35 °C for 6 days resulted in a 70% loss in lentil output, according to Delahunty et al. (2015). Temperatures exceeding 32/20 °C (the ratio denotes max/min temperature) can severely degrade grain quality and reduce grain output in lentils throughout the flowering and pod-filling stages (Delahunty et al. 2015). The second most sensitive stage is seed filling, which is impacted by heat stress. The crop is frequently significantly damaged as temperature is high, leading to low grain yields and lower grain quality (Tickoo et al. 2005).

Temperatures above 24.4 °C slow down lentil germination (Covell et al. 1986). Reduced germination percentage, uneven seedling growth, nodule disintegration, deterioration of cell membrane stability, early flowering, reduction in plant biomass, decreased photosynthetic efficiency, and increased lipid peroxidation are all effects of heat stress (Chakraborty and Pradhan 2010; Ellis and Barrett 1994; Muehlbauer et al. 2006; Sehgal et al. 2017). The hampered electron flow, photosystem II (PSII) thermolability, chlorosis, and a lower possibility of assimilation and carbon fixation have all been identified as the most vulnerable stages of photosynthesis in leaves (Sinsawat et al. 2004). As per Chakraborty and Pradhan (2010), greater synthesis of ascorbate peroxidase (APX) in lentils is correlated to heat tolerance. Increased antioxidant contents in the leaf and increased pollen function in lentils are due to heat-stress tolerance (Sita et al. 2017). Responses of plants to heat stress include pollen sterility, flower drop, reduced seed set, pod abortion, and a shortened reproductive period (Bhandari et al. 2016; Gaur et al. 2014). As per Gupta et al. (2019) only very few researchers have examined lentil genotypes for heat tolerance, whereas genetic diversity has indeed been observed within few others. Because a network of genes regulate a variety of physiological, biochemical, and metabolic processes necessary for maintaining plants under heat stress, the genetics of heat-stress tolerance has indeed been examined in different crops and shown to be complex (Kaur et al. 2019).

9.2.2 Cold Stress

The susceptibility of lentils against frost is comparable to that of other pulses. Lentils, for example, are more susceptible than chickpeas but less so than peas (Murray et al. 1988). During the early phases of development, crops recover fast from covered axillary buds. Regardless, the crop dies whenever it reaches maturity or is subjected to frost because the commencement of axillary buds ceases, restricting the plant's capacity to advance to the reproductive phase. The biggest frost damage was caused during the reproductive phase given the availability of flowers to frost and the smaller size of pods. Lentil flower drop and pod abortion, as well as harm to the vegetative organs, are caused by frost injury (Gupta et al. 2019). Frost injury to the seed coat can occur in pod filling and pod formation, impacting the seed's original development. Stem drying and leaf injury have been observed in harsh frost situations. The plant gets vulnerable to pathogenic infiltration as a result of having been exposed to frost, rendering it prone to diseases which include Botrytis grey mould and anthracnose (Gupta et al. 2019). There was a considerable drop in lentil yields. To characterise cold tolerance in lentils, different investigations were performed, along with breeding efforts (Ali et al. 1999; Erskine et al. 1981; Kusmenoglu and Aydin 1995; Murray et al. 1988; Spaeth and Muehlbauer 1991; Summerfield et al. 1985). In documented observations, frost damage and winter tolerance in lentils were studied (Barrios et al. 2007, 2010; Barrios et al. 2016; Kahraman et al. 2004).

9.2.3 Drought Stress

For a number of interesting crops, drought is unquestionably one of the most severe abiotic challenges, resulting in significant production and monetary losses. Crop output is negatively impacted as a result of morphological, physiological, and signalling pathway; transcriptional and post-transcriptional regulation of stress-responsive genes; as well as metabolic alterations. Lentils, more than any other food grain legumes, are drought tolerant to a degree (Abraham 2015). Although lentils may be grown in dry environments and require minimal water for growth and development, harvests are affected by 6–70% in water shortage places, and crop failure happens in extreme cases (Babayeva et al. 2014). Unpredictable yearly variability of rainfall, which contributes to the occurrence of droughts, has an impact on lentil's long-term sustainable production and productivity (Dai 2011). Drought effects differ depending on the level of development; at the pod formation and reproductive phases, 70% and 24% decline were recorded, correspondingly (Shrestha et al. 2006; Allahmoradi et al. 2013). Drought during the flowering or podding phase has a detrimental effect on growth and flowering, leading to a reduction in leaf area (48–55%), flower output (22–55%), pod and seed number (27–66%), and total dry matter (32–50%), and also considerably higher flower fall, pod abortion, and increased number of aborted pods (Shrestha et al. 2006). Drought stress impacts lentil metabolism, osmoregulation, and photosynthetic pigment intensity (Gökçay 2012; Muscolo et al. 2015; Mishra et al. 2016; Biju et al. 2017). Variations in annual precipitation pattern are increasing the frequency of droughts all through the crop growth period, posing a direct danger to the long-term sustainability of lentil cultivation (Dai 2011). Around 90% of the total cultivated area of lentils and other pulses like chickpea depends on conserved, retreating soil water; therefore, yield potential is highly dependent on efficient soil water utilisation (Kumar and Van Rheenen 2000).

Lentil utilises two primary strategies to withstand the effects of drought: drought escape and drought tolerance. Drought tolerance strategies in lentil include regulated stomatal closure, osmotic adjustment, intense pubescence of the leaf, higher antioxidant activities, and improved yield attributes. Early-maturing varieties such as Precoz, Bakaria, BARI M4, BARI M5, BARI M6, and Idlib 3 have drought-escape strategies in the reproductive phase as a reaction to drought conditions (Erskine and Saxena 1993; Hamdi and Erskine 1996; Shrestha et al. 2005). Drought resistance is strongly linked to agromorphological traits, according to Singh et al. (2017a) and Pratap et al. (2019).

The surface area of the leaf, length of the stem, leaf motion, structural canopy, and stomata-related properties all play a critical role in the drought-escape process (Salam and Islam 1994). Drought resistance is mostly determined by root features, which can be exploited to develop drought-resistant cultivars in lentil breeding

(Idrissi et al. 2015; Biju et al. 2017). As a consequence, choosing root-related traits offers a significant opportunity to boost grain output in both perfect nutrient and water circumstances and in environments with limited soil nutrients (Chen et al. 2015; Gahoonia et al. 2005). At above lentils, different approaches are designed to keep plants from drought, including early or late flowering and pubescence. The process of drought resistance in wild lentils is limited as compared to cultivated lentils. Hamdi and Erskine (1996) and Gorim and Vandenberg (2017) examined wild lentil germplasm for variable drought strategies by assessing root-related traits and revealed that wild lentils deploy a variety of drought resistance strategies, including decreased plant height, delaying flowering, and decreased transpiration rates. Despite extensive investigation, the essential strategy that enables wild lentils to adapt to a changing seasonal rainfall situation remained unclear. Plant breeders focused on finding ways to keep productivity going in a water-stress environment (Erskine et al. 2011). A short-term technique for evaluating drought resistance variability in lentil germplasm and a long-term approach for transferring useful traits from wild species to cultivated cultivars are both viable options.

9.2.4 Submergence and Flooding Stress

Flooding and submersion reduce the production of the majority of food grain legumes and subject plants to extensive and unpredictable environmental challenges. Both the issue and the demand for more food due to an expanding human population are getting worse. It is urgently necessary to increase lentil resilience to floods in all of its varied forms due to the contradiction of these conflicting developments. Over the past 25 years or so, crop output losses from flooding amounting to millions of euros per year have been recorded for many countries, including the USA, China, Europe, and Australia (Shi and Tao 2014). Between 2006 and 2016, floods were responsible for two-thirds of agricultural production losses globally (United Nations Food and Agriculture Organization [FAO] 2017 (https://www.fao.org/3/I8656EN/i8656en.pdf). Flooding and submergence have a substantial impact on legume crop quality and productivity (Kang et al. 2017; Solaiman et al. 2007). Waterlogging impedes lentil production in soils with poor drainage, such as duplex types, subsoil compaction, fine-textured soils, or under situations of persistent or severe rainfall (Wiraguna et al. 2017). Waterlogging causes more damage that fluctuates considering the length of the stress, the severity of the stress, and the stage of crop growth, in the worst scenarios culminating in entire crop failure (Toker et al. 2011). Waterlogging is a serious obstacle to lentil productivity, especially in the early vegetative phases of growth (Solaiman et al. 2007). Waterlogging will impede lentil growth at all phases, resulting in decreased yields (Materne and Siddique 2009). Seed germination, root growth, and development are all delayed by waterlogging, and the flowering phase is the most prone to waterlogging, culminating in flower and pod abortion (Jayasundara et al. 1997). Leaf senescence, limited development, wilting, and ultimately death are all symptoms that the plant exhibits. Wiraguna et al. (2017) studied lentil germplasm for waterlogging tolerance and found that varieties from Bangladesh may germinate in wet soil.

To find waterlogging-tolerant cultivars, researchers looked for traits like high root porosity, early flowering and maturity, low biomass, and improved stomatal conductance (Ashraf and Chishti 1993; Erskine et al. 2016; Malik et al. 2015). Waterlogging is now being dealt with by attempting to avoid it. Drainage, seeding rate, sowing time, and paddock selection are among the management strategies that have been shown to help reduce the detrimental effects of waterlogging in lentils (Toker et al. 2011). Breeding for increased tolerance by selecting for more aerenchyma or accidental root formation has been offered as a viable solution (Jayasundara et al. 1997; Materne and Siddique 2009). In recent times, plants have been intensively researched for flood stress and its consequences, such as submergence, waterlogging, hypoxia, and anoxia, especially in rice and *Arabidopsis thaliana*. However, viable solutions to this abiotic stress in lentils need to be explored further in order to understand the underlying molecular basis for flooding or submergence resistance.

9.2.5 Salinity Stress

Modern agriculture faces a serious challenge from salinity, which inhibits and impairs crop growth and development. Water stress, cytotoxicity brought on by the excessive uptake of ions like sodium (Na^+) and chloride (Cl^-), and nutritional imbalance are all effects of salinity that hinder plant growth and development. Saline conditions are harmful to legume plants since this reduces root hair growth and hinders biological nitrogen fixation and nodulation (Rai et al. 1985; Rai and Singh 1999). Lentil roots are impacted by salinity, which inhibits rhizobium infection and root formation (Rai and Singh 1999; Van Hoorn et al. 2001). Plants exposed to salinity stress had an impact on seed germination, growth, survival, and productivity. During salt stress, crops respond in different ways to distinct growth stages (Munns and Tester 2008). Lentils' responses to salinity stress depend greatly on salinity level, growth stage, and environmental variables like relative humidity, soil–water condition, nutrient availability, and temperature (Lachaâl et al. 2002). The reproductive phase is the most vulnerable to salinity stress (Vadez et al. 2007), although the germination stage is less vulnerable than the initial stages of vegetative growth (Sakina et al. 2016). Salinity also increases anthocyanin colouring in leaves and stems, inhibits flower and pod formation (Van Hoorn et al. 2001), and reduces overall growth and development of plants by lowering plant height, biomass production, and biochemical composition (Tewari and Singh 1991). Saline environments also impede lentil growth by affecting biochemical and physiological mechanisms such as photosynthesis (Al-Quraan et al. 2015), ion homeostasis (Hossain et al. 2017), membrane damage (Hossain et al. 2017), oxidative damage (Hossain et al. 2017), and γ-aminobutyric acid (GABA) accumulation (Al-Quraan and Al-Omari 2017). Saline exposure has been shown to reduce lentil production by 90–100% (3 dS/m) and 20% (2 dS/m) at different electrical conductivity levels

(Ghassemi-Golezani and Mahmoodi-Yengabad 2012; Van Hoorn et al. 2001). As part of an ICAR and ICARDA joint research and development programme, some wild lentil accessions (IG 136670, ILWL 297, ILWL 371, ILWL 368, ILWL 417) of *L. culinaris* ssp. *orientalis* were revealed as salt tolerant and promising resources for breeding resistance against salinity in India (DAC-ICAR-ICARDA 2014). Kumawat et al. (2017) reported salt-tolerant genotypes based on their stress-resistant lentil investigations, whereas Sehgal et al. (2017) identified many salinity-tolerant genotypes. New approaches must be devised to prevent the yield loss caused by salinity. Groundwater and soil fertility management techniques can be utilised to minimise soil salinity in salt-affected locations, but these methods are highly costly, and new cost-effective techniques must be devised. As a consequence, establishing a strategy for creating salt-tolerant cultivars is the most effective and long-term technique for stabilising and improving yield in salinity-affected environments. To produce salinity-resistant cultivars, it is critical to develop efficient detection methods. For salinity stress, field and hydroponic screening tests are commonly used. However, screening in the field is problematic due to the absence of homogeneity in the ambient and soil properties. Screening in a hydroponic system helps alleviate the difficulties associated with testing in the fields.

9.3 Crop Wild Relatives (CWRs) of Lentil and Abiotic Stress

Agricultural production and preservation of natural biodiversity face significant problems due to the harmful effects of climate change and human activity as well as a variety of environmental factors. The creation of innovative crop varieties with improved biotic or abiotic resilience that allows them to flourish in marginal soils may be the solution to these problems. However, it is surprising that evolutionary principles have not been fully utilised in tackling these food and environmental concerns given the variety of interactions between crops and environmental elements. Crop wild relatives (CWRs) have faced challenges in their natural habitats for thousands of years and continue to exhibit a far higher amount of genetic variation than domesticated cultivars. CWRs or wild lentil species include a pool of essential abiotic stress tolerance genes. Hamdi et al. (1996) established cold tolerance in *L. culinaris* ssp. *orientalis* and also drought tolerance in *L. ervoides*, *L. odemensis*, and *L. nigricans* (Gupta and Sharma 2006; Hamdi and Erskine 1996). The wild lentil species *L. nigricans*, *L. orientalis*, *L. ervoides*, and *L. odemensis* have been studied in low-precipitation situations (Hamdi and Erskine 1996). Drought tolerance in crop wild relatives of India has been researched, with *L. nigricans* being named one of the most drought-tolerant species (Gupta and Sharma 2006). Several donors were identified for salinity tolerance by Singh et al. (2017a) by screening around 100 accessions of *L. culinaris* subsp. *orientalis* in a hydroponics culture. Root biomass, root dispersal, and other root-related factors varied greatly among wild relatives of the lentil plant (Gorim and Vandenberg 2017). Notwithstanding a reduction in overall pod quantity and yield, a few *L. odemensis* and *L. orientalis* genotypes demonstrated a deep root system, delays in blooming, and comparable stress resistance. Some other *L. lamottei* accession contained many trichomes on the stems, leaves, and pods, whereas *L. tomentosus* transpired slowly (Gorim and Vandenberg 2017).

In recombinant inbred lines generated by crossing lentil genotypes with *L. odemensis*, *L. orientalis*, and *L. ervoides*, drought tolerance features are evaluated and mapped (Sanderson et al. 2019). Drought tolerance in wild lentil species was also studied by Omar et al. (2019). Lentils' heat sensitivity has only been investigated in moderate levels. Heat-stress tolerance in cultivated lentils has only been examined in one investigation by Sita et al. (2017). Using genome-wide transcription, Singh et al. (2019) found heat-responsive genes in the regulatory system of lentil cultivars. More research into the processes of heat tolerance is needed to fully comprehend heat tolerance. A multitude of methods can be utilised to obtain the abiotic stress tolerance or resistance genotype: (a) field phenotyping of accessions based on climate history and (b) GPS data collection according to meteorological data, with curated data examined under frost, heat, and drought stress locations with severe prevalence for various stresses that can be discovered using focused identification of germplasm sets (FIGS) enabling for the identification of acceptable CWRs (Street et al. 2008).

9.3.1 Molecular Genetic Diversity in Lentil

Genetic diversity in cultivated lentils has been studied using a variety of methods (Poyraz 2016). In order to examine the level of genetic diversity in lentils, numerous investigations have used DNA-based markers to categorise a broad range of germplasms (Ferguson et al. 1998; Fikiru et al. 2007; Idrissi et al. 2015; Khazaei et al. 2016; Lombardi et al. 2014; Wong et al. 2015; Yadav et al. 2016). The most prevalent molecular markers detected in the genome are single nucleotide polymorphisms (SNPs) (Agarwal et al. 2008). A high-throughput sequencing technology called genotyping-by-sequencing (GBS) can be used to discover SNPs in the genome (Chung et al. 2017). The expense of sequencing is decreasing because of next-generation sequencing (NGS) technologies, and GBS technologies are rapidly being employed to read large and diverse genomes (Malmberg et al. 2018). So far, the range of SNP-based diversification research has been modest. Many researchers have looked into domesticated and wild lentil accessions from all over the world (Khazaei et al. 2016; Lombardi et al. 2014; Wong et al. 2015).

9.3.2 Next-Generation Technologies

Lentil genomics-aided breeding has been impeded by a paucity of candidate genes, and also a weak genetic background, large genome size, and a low-density linkage map (Kumar et al. 2015). Recent advances in genotyping-by-sequencing and next-generation sequencing (NGS) technology have paved the way for the faster development of sequence-based markers, leading to better lentil genome sequencing around the world. The genome assembly of the CDC Redberry lentil variant has been completed using next-generation DNA sequencing technologies (Bett et al. 2016). The construction of a genomic map enables finding of QTLs/genes associated with the different traits more simpler. Additional genomic resources are required to build a consensus genetic linkage map that will allow the tagging of essential abiotic stress resistance genes in lentil. Bett et al. (2016) emphasised the importance of field phenotyping of lentil germplasms in numerous sites in order to produce strongly related molecular markers for important features. SNPs will be able to detect mutant phenotypes by recognising mutations that occur as a result of chemical and physical techniques. Using sequencing information, reverse genetics can be utilised to dissect the trait functioning. Target-induced local lesions in genomes (TILLING), virus-induced gene silencing (VIGS), and RNAi technologies have all helped researchers better grasp the molecular pathways in lentils. In a lentil breeding programme, such unique approaches have resulted in significant genomic resources for genetic improvement as well as resource utilisation. Lentil breeders used these advanced techniques to add marker-assisted backcrossing and marker-assisted selection into their breeding programmes for genotype and trait selection, as well as trait introgression.

9.3.3 Molecular Mapping of Resistance/Tolerance Genes and QTLs in Lentil

Four and five QTLs for winter damage and survival were found in lentils in 106 RIL populations originating from a hybrid of WA8649090 × Precoz (Kahraman et al. 2004). In this study, experiments were conducted over several locations, and only one of the five QTLs was displayed under all circumstances. In winter-sown lentils, Barrios et al. (2007) identified QTLs for frost resistance, and they also observed that these QTLs are connected to yield. Additional investigation demonstrates a close relationship between the yield and winter hardiness quantitative trait loci in same linkage group (Barrios et al. 2016). In a RIL population produced by hybridising Precoz × WA8649041, Super-SAGE (serial analysis of gene expression) was utilised to find differentially expressed genes of transcripts associated with frost tolerance (Barrios et al. 2010). Singh et al. (2016) found a single major gene Sdt for seedling survival under drought in the lentil F_2 mapping population (JL-3 × PDL-1). Idrissi et al. (2015) reported 18 QTLs associated with root-shoot ratio, LRN, dry root biomass, and specific root length, among other shoot and root characteristics related to drought resistance. Despite the fact that biparental mapping populations have been employed regularly to locate QTLs in lentils, marker-assisted selection has only been able to identify important QTLs. The first linkage map for lentil drought tolerance was published by Singh et al. (2017c), and they also discovered QTL controlling seedling survival under drought tolerance in lentil. The introduction of genes that can withstand drought in cultivated varieties will be facilitated by the molecular markers found via multiple experiments.

A linkage map comprising 291 simple sequence repeat markers and 75 QTLs was created in lentils for attributes related to drought tolerance and yield using an intraspecific RIL mapping population derived from the cross L830 Precoz (Rana et al. 2016). A QTL for heat stress in lentils was found by Singh et al. (2017b). The QTL report showed two important QTLs, qHt ps and qHt ss, accounting for 9.23% and 12.1% of the phenotypic variance, respectively. The identifying of genetic markers connected to phenotypes and the dissecting of potential genes for heat tolerance have both been made possible by the discovery of QTLs. In a mapping population resulting from a Cassab × ILL2024 hybrid, boron tolerance QTLs and genes were found. Linkage map development expedites abiotic stress breeding in molecular breeding programmes, enabling more accurate and precise objective fulfilment. The gene of interest is introduced into elite genetic backgrounds with the help of the known molecular markers. The development of genome-wide SNP markers for biparental and association mapping as well as the acceleration of transcriptome and genome sequencing programmes have all been made possible by the technologies of genotyping-by-sequencing (GBS) and next-generation sequencing (NGS). Molecular breeding that is "omics based" offers more options than traditional breeding to enrich the natural germplasm while also enhancing yield and quality criteria. The approach is currently equipped with cutting-edge "omics" tools like epigenomics, ionomics, RNomics, fluxomics, glycomics, phosphoproteomics, glycoproteomics, regulomics, secretomics, and lipidomics as a result of molecular developments. Pan-omics has been extensively employed to reduce abiotic stress in food legume crops by mRNAs (transcriptomics), identifying genes (genomics), biomolecules (metabolomics), and proteins (proteomics) related with stress control. Programmes for breeding legumes will move more quickly with the integration of pan-omics and innovative omics methods (Singh et al. 2021).

9.3.4 Abiotic Stresses and Transcriptome Analysis in Lentil

A precise representation of the gene expression in a target cell or tissue can be obtained using the potent technique of transcriptomics, which is utilised to measure gene expression. For the purpose of breeding legumes, transcriptomics can identify the gene regulatory networks and candidate genes involved in the development of the abiotic stress response. The development of high-throughput technologies has enabled the extraction of substantial transcriptome data utilising serial analysis of gene expression (SAGE) and microarrays. Ribonucleic acid sequencing (RNA-seq) data can be used to identify the differential expression of genes (DEGs). It is also possible to employ a recently created method for quantitatively estimating gene expression termed digital gene expression (DGE). Huge quantities of transcriptomic data can be analysed using the low-cost, high-throughput sequencing method known as RNA-seq analysis. This method has numerous benefits over microarray technology, including the ability to identify novel transcripts and the lack of a need for genetic information when creating probe sets (Lowe et al. 2017). From a transcriptomic analysis, Singh et al. (2017c) identified putative candidate genes expressed under drought stress at the seedling stage in lentil. Differential gene expression indicated upregulation of reduction of stomatal conductance, correct folding of protein, electron transport chain, oxidation-reduction process, TCA cycle, and organ senescence in drought-tolerant genotypes (PDL-2) in comparison to the sensitive types (JL-3). But negative regulation was observed in GABA synthesis, synthesis of cell wall protein, and abscisic acid that are downregulated in drought-tolerant genotype (PDL-2) in contrast to drought-susceptible JL-3. Barrios et al. (2010) used RILs created from the hybridisation of Precoz (cold tolerant) × WA8649041 to study gene expression in response to cold sensitiveness. Deep Super-SAGE transcriptome sequence analysis on RILs was performed. The discovered sequences encoded proteins related to glycine, proline, proteins regulated by drought and cold, proteins related to dormancy, and other membrane proteins. In the acclimated tolerant lines, they were typically but not always overexpressed.

9.3.5 Marker-Assisted Selection (MAS) in Lentil Improvement

Breeders can choose genotypes with the desired gene combination by using marker-assisted selection, which improves breeding efficiency. This approach to lentil breeding has a number of drawbacks since the genetic resources in lentils develop more slowly than those in other legumes (Kumar et al. 2019a, b). Recombinant inbred lines and close isogenic lines are best suited for closely researching and dissecting the trait of interest since polygenes regulate economic traits and are impacted by both genetic and environmental factors. Lentil linkage mapping was invented by Zamir and Ladizinsky in 1984, and the first lentil linkage map based on DNA-based markers was created by Havey and Muehlbauer in 1989. The creation of lentil linkage maps has been accelerated by the advent of PCR-based markers. For the first time, Eujayl et al. (1998) mapped a population made up of an interspecific hybrid of *Lens* ssp. *culinaris* and ssp. *orientalis* using morphological and molecular markers (RFLP, RAPD, and AFLP) for thorough coverage of the lentil genome and curation of the linkage map. The very first intraspecific lentil map, which includes resistance gene analogue (RGA), 114 RAPD, and inter-simple sequence repeat (ISSR) genetic markers for gene/QTL detection, was reported by Rubeena Ford and Taylor in 2003.

Hamwieh et al. (2005) produced a genomic library of lentils from the cultivar ILL5588 and utilised SSR markers to examine genetic variation in lentils (Hamwieh et al. 2009). In lentils, Tanyolac et al. (2009) found 11 linkage groups, with ISSR, AFLP, and RAPD markers found in each group. Genomic resources (122 SSR functional markers) were generated by Verma et al. (2014) for the improvement of lentils. Due to the limited coverage and larger genome sizes, the information collected from the many linkage maps produced via biparental mapping populations has little practical usefulness (Ates et al. 2018). A high-density consensus linkage map with seven linkage groups that correspond to the seven chromosomes in the lentil genome was produced using Diversity Arrays Technology (DArT) (Ates et al. 2018).

Given recent breakthroughs in molecular innovations and the accessibility of genetic resources, the development of consensus linkage maps has been accelerated by the use of multiple mapping populations instead of a single mapping population. Next-generation sequencing (NGS) methods now use chip-based markers more frequently than PCR-based markers. About 44,879 SNP markers were found by Sharpe et al. (2013), and 50,960 SNPs were found and used to create the lentil linkage map (Temel et al. 2015). The development of marker-assisted selection in lentil breeding and candidate genes for a number of economically beneficial traits were both facilitated by breakthroughs in lentil genome sequencing (Bett et al. 2016). Additionally, crop production under changing environmental conditions can be simulated using artificial intelligence (AI)-based algorithms, which can assist in anticipating the genetic gain. The use of machine learning (ML) in quantitative trait locus (QTL) mining will aid in identifying the genetic factors that influence lentils' ability to withstand abiotic stress.

9.4 Conclusion

Abiotic stressors including drought, salt, and severe temperatures must be endured by plants because they are immobile. These stresses disrupt plant growth and development, severely restrict plant distribution, and lower crop output. In order to improve the resilience of agricultural production systems in the face of changing weather patterns, lentils, an incredibly nutrient-dense and stress-tolerant legume, are absolutely necessary. Initiatives to optimise lentil crops have tremendous potential for even further raising and stabilising lentil yield.

The multidimensional complexity of the molecular mechanisms driving plants' responses to abiotic stresses has recently come to light thanks to recent advances in our knowledge of these processes, which include signalling, transcription, translation, and post-translational protein changes. Through the use of genetic, chemical, and microbiological techniques, this enhanced knowledge can increase crop productivity and agricultural self-sufficiency. Numerous genes linked to stress adaptation have been characterised as a result of research on the physiological and molecular mechanisms of abiotic stress tolerance. The identification of genes associated with stress has been made possible by methods like microarrays. While some of these genes are exclusive to one type of stress, others are common by several types of stresses. It is interesting to note that both biotic and abiotic stress mechanisms share a lot of genes.

Current knowledge of abiotic stress responses is further complicated by the finding that microRNAs control gene expression. Identification of microRNAs linked to the abiotic stress reaction as well as an understanding of how they communicate and how they regulate the action would take a substantial amount of investigation. The introduction of next-generation sequencing methods, which have enabled deep sequencing of mRNAs and microRNAs linked to the abiotic stress reaction, is a healthy sign. Better crop manipulation and increased agricultural output will benefit from a proper study of physiological and molecular dynamics, particularly signalling cascades in relation to abiotic stressors in plant resistance.

References

Abeysekara S, Chilibeck PD, Vatanparast H, Zello GA (2012) A pulse-based diet is effective for reducing total and LDL-cholesterol in older adults. Br J Nutr 108(S1):S103–S110

Abraham R (2015) Lentil (*Lens culinaris* Medikus) Current status and future prospect of production in Ethiopia. Adv Plant Agric Res 2(2):45–53

Agarwal M, Shrivastava N, Padh H (2008) Advances in molecular marker techniques and their applications in plant sciences. Plant Cell Rep 27(4):617–631

Ali A, Johnson DL, Stushnoff C (1999) Screening lentil (*Lens culinaris*) for cold hardiness under controlled conditions. J Agric Sci 133(3):313–319

Allahmoradi P, Mansourifar C, Saidi M, Honarmand SJ (2013) Water deficiency and its effects on grain yield and some physiological traits during different growth stages in lentil (*Lens culinaris* L.) cultivars. Ann. Biol Res 4(5):139–145

Al-Quraan NA, Ghunaim AI, Alkhatib RQ (2015) The influence of chlorsulfuron herbicide on GABA metabolism and oxidative damage in lentil (*Lens culinaris* Medik) and wheat (*Triticum aestivum* L.) seedlings. Acta Physiol Plant 37(11):1–9

Al-Quraan NA, Al-Omari HA (2017) GABA accumulation and oxidative damage responses to salt, osmotic and H_2O_2 treatments in two lentil (*Lens culinaris* Medik) accessions. Plant Biosyst 151(1):148–157

Ashraf M, Chishti SN (1993) Waterlogging tolerance of some accessions of lentil (*Lens culinaris* Medic.). Trop Agric 70(1):60–67

Aski MS, Rai N, Reddy VRP, Dikshit HK, Mishra GP, Singh D et al (2021) Assessment of root phenotypes in mungbean mini-core collection (MMC) from the World Vegetable Center (AVRDC) Taiwan. PLoS One 16(3):e0247810

Aski M, Mehra R, Mishra GP, Singh D, Yadav P, Rai N et al (2022) Genotypic variation in root architectural traits under contrasting phosphorus levels in Mediterranean and Indian origin lentil genotypes. PeerJ 10:e12766

Ates D, Aldemir S, Alsaleh A, Erdogmus S, Nemli S, Kahriman A, Ozkan H, Vandenberg A, Tanyolac B (2018) A consensus linkage map of lentil based on DArT markers from three RIL mapping populations. PLoS One 13(1):e0191375

Azadbakht L, Esmaillzadeh A (2012) Dietary patterns and attention deficit hyperactivity disorder among Iranian children. Nutrition 28(3):242–249

Azadbakht L, Kimiagar M, Mehrabi Y, Esmaillzadeh A, Hu FB, Willett WC (2007) Soy consumption, markers of inflammation, and endothelial function: a cross-over study in postmenopausal women with the metabolic syndrome. Diabetes Care 30(4):967–973

Babayeva S, Akparov Z, Damania A, Izzatullayeva V, Aslanova G, Abbasov M (2014) Genetic diversity for drought tolerance in lentils from Central Asia and the Caucasus: CACLentil. Albanian J Agric Sci 13(2):1–8

Bari ML, Nei D, Enomoto K, Todoriki S, Kawamoto S (2009) Combination treatments for killing *Escherichia coli* O157: H7 on alfalfa, radish, broccoli, and mung bean seeds. J Food Prot 72(3): 631–636

Barrios A, Kahraman A, Aparicio T, Rodríguez M, Mosquera P, García P, McPhee K, Pérez de la Vega M, Caminero C (2007) Preliminary identification of QTLS for winter hardiness, frost tolerance and other agronomic characters in lentil (*Lens culinaris* Medik.) for Castilla y León (SPAIN) region. In: Proceedings of the 6th European Conference on Grain Legumes, Lisbon, 12–16

Barrios A, Martin-Sanz A, Ramos S, Rotter B, Horres R, Plotner A, Kretzdorn N, Constantino Saldaña C, Pérez de la Vega M, Winter P (2010) Allele-specific differential expression of transcripts potentially involved in cold tolerance QTL in lentil revealed by bulked extremes SuperTag digital gene expression (BE-STDGE) profiling and qRT-PCR. In: Proceedings of the 5th International Food Legumes Research Conference and 7th European Conference on Grain Legumes, Antalya, 26–30

Barrios A, Aparicio T, Rodríguez MJ, de la Vega MP, Caminero C (2016) Winter sowing of adapted lines as a potential yield increase strategy in lentil (*Lens culinaris* Medik.). Span J Agric Res 14(2):e0702

Becerra-Tomás N, Díaz-López A, Rosique-Esteban N, Ros E, Buil-Cosiales P, Corella D, Estruch R, Fitó M, Serra Majem L, Arós F (2018) Legume consumption is inversely associated with type 2 diabetes incidence in adults: a prospective assessment from the PREDIMED study. Clin Nutr 37(3):906–913

Becerra-Tomás N, Blanco Mejía S, Viguiliouk E, Khan T, Kendall CW, Kahleova H et al (2020) Mediterranean diet, cardiovascular disease and mortality in diabetes: a systematic review and meta-analysis of prospective cohort studies and randomized clinical trials. Crit Rev Food Sci Nutr 60(7):1207–1227

Bett K, Ramsay L, Chan C, Sharpe AG, Cook DR, Varma R (2016) OP06: the lentil genome–from the sequencer to the field. In: PAG XXIV: plant and animal genomics conference, San Diego, California. https://mel.cgiar.org/reporting/download/hash/gJFbVXUc

Bhandari K, Siddique KHM, Turner NC, Kaur J, Singh S, Agrawal SK, Nayyar H (2016) Heat stress at reproductive stage disrupts leaf carbohydrate metabolism, impairs reproductive function, and severely reduces seed yield in lentil. J Crop Improv 30(2):118–151

Biju S, Fuentes S, Gupta D (2017) Silicon improves seed germination and alleviates drought stress in lentil crops by regulating osmolytes, hydrolytic enzymes and antioxidant defense system. Plant Physiol Biochem 119:250–264

Chakraborty U, Pradhan D (2010) Biochemical responses of lentil (*Lens culinaris* Medik.) to elevated temperature stress. Res J Pharm Biol Chem Sci 1(3):575–585

Chen H, Wang L, Wang S, Liu C, Blair MW, Cheng X (2015) Transcriptome sequencing of mung bean (*Vigna radiate* L.) genes and the identification of EST-SSR markers. PLoS One 10(4): e0120273

Choudhury DR, Tarafdar S, Das M, Kundagrami S (2012) Screening lentil (*Lens culinaris* Medik.) germplasms for heat tolerance. Trends Biosci 5(2):143–146

Choukri H, Hejjaoui K, El-Baouchi A, El Haddad N, Smouni A, Maalouf F et al (2020a) Heat and drought stress impact on phenology, grain yield, and nutritional quality of lentil (*Lens culinaris* Medikus). Front Nutr 7:596307

Chung YS, Choi SC, Jun T-H, Kim C (2017) Genotyping-by-sequencing: a promising tool for plant genetics research and breeding. Hortic Environ Biotechnol 58(5):425–431

Covell S, Ellis RH, Roberts EH, Summerfield RJ (1986) The influence of temperature on seed germination rate in grain legumes: I. A comparison of chickpea, lentil, soyabean and cowpea at constant temperatures. J Exp Bot 37(5):705–715

Crujeiras AB, Parra D, Abete I, Martínez JA (2007) A hypocaloric diet enriched in legumes specifically mitigates lipid peroxidation in obese subjects. Free Radic Res 41(4):498–506

DAC-ICAR-ICARDA (2014) Annual progress report: International centre for agricultural research in dry areas. South Asia and China Regional Programme

Dai A (2011) Drought under global warming: a review. Wiley Interdiscip Rev Clim Chang 2(1): 45–65

Delahunty A, Nuttall J, Nicolas M, Brand J (2015) Genotypic heat tolerance in lentil. In: Proceedings of the 17th ASA Conference, 20–24

Dissanayake R, Braich S, Cogan NOI, Smith K, Kaur S (2020) Characterization of genetic and allelic diversity amongst cultivated and wild lentil accessions for germplasm enhancement. Front Genet 11:546

Duran Y, Fratini R, Garcia P, Perez De La Vega M (2004) An intersubspecific genetic map of Lens. Theor Appl Genet 108(7):1265–1273

Ellis RH, Barrett S (1994) Alternating temperatures and rate of seed germination in lentil. Ann Bot 74(5):519–524

Erskine W, Saxena MC (1993) Problems and prospects of stress resistance breeding in lentil. Wiley, Chichester

Erskine W, Myveci K, Izgin N (1981) Screening a world lentil collection for cold tolerance. Lens Newsl 8:5–8

Erskine W, Sarker A, Kumar S (2011) Crops that feed the world 3. Investing in lentil improvement toward a food secure world. Food Secur 3(2):127–139

Erskine W, Sarker A, Kumari S (2016) Lentil breeding. In: Wrigley CW, Corke H, Seetharaman K, Faubion JM (eds) Encyclopedia of food grains, 2nd edn. Elsevier, Oxford, pp 317–324

Eujayl I, Baum M, Powell W, Erskine W, Pehu E (1998) A genetic linkage map of lentil (*Lens* sp.) based on RAPD and AFLP markers using recombinant inbred lines. Theor Appl Genet 97(1): 83–89

Faris MEAIE, Takruri HR, Issa AY (2013) Role of lentils (*Lens culinaris* L.) in human health and nutrition: a review. Mediterr. J Nutr Metab 6(1):3–16

Ferguson ME, Ford-Lloyd BV, Robertson LD, Maxted N, Newbury HJ (1998) Mapping the geographical distribution of genetic variation in the genus Lens for the enhanced conservation of plant genetic diversity. Mol Ecol 7(12):1743–1755

Fikiru E, Tesfaye K, Bekele E (2007) Genetic diversity and population structure of Ethiopian lentil (*Lens culinaris* Medikus) landraces as revealed by ISSR marker. Afr J Biotechnol 6(12): 1460–1468

Ford R, Taylor PWJ (2003) Construction of an intraspecific linkage map of lentil (*Lens culinaris* ssp. culinaris). Theor Appl Genet 107(5):910–916

Gahoonia TS, Ali O, Sarker A, Rahman MM (2005) Root traits, nutrient uptake, multi-location grain yield and benefit–cost ratio of two lentil (*Lens culinaris*, *Medikus*.) varieties. Plant Soil 272(1):153–161

García-Mora P, Martín-Martínez M, Bonache MA, González-Múniz R, Peñas E, Frias J, Martinez-Villaluenga C (2017) Identification, functional gastrointestinal stability and molecular docking studies of lentil peptides with dual antioxidant and angiotensin I converting enzyme inhibitory activities. Food Chem 221:464–472

Gaur PM, Samineni S, Krishnamurthy L, Varshney RK, Kumar S, Ghanem ME, Beebe SE, Rao IM, Chaturvedi SK, Basu PS (2014) High temperature tolerance in grain legumes. Legum Perspect 7:23

Ghassemi-Golezani K, Mahmoodi-Yengabad F (2012) Physiological responses of lentil (*Lens culinaris* Medik.) to salinity. Int J Agric Crop Sci 4(20):1531–1535

Girvetz E, Ramirez-Villegas J, Claessens L, Lamanna C, Navarro-Racines C, Nowak A, Thornton P, Rosenstock TS (2019) Future climate projections in Africa: where are we headed? In: The climate-smart agriculture papers. Springer, Cham, pp 15–27

Gökçay D (2012) Physiological and biochemical screeing of different Turkish Lentil (*Lens culinaris* M.) cultivars under drought stress condition, Master's thesis. Middle East Technical University

Gorim LY, Vandenberg A (2017) Evaluation of wild lentil species as genetic resources to improve drought tolerance in cultivated lentil. Front Plant Sci 8:1129

Gupta D, Sharma SK (2006) Evaluation of wild Lens taxa for agro-morphological traits, fungal diseases and moisture stress in North Western Indian Hills. Genet Resour Crop Evol 53(6): 1233–1241

Gupta D, Ford R, Taylor PWJ (2011) Lens. In: Kole C (ed) Wild crop relatives: genomic and breeding resources legume crops and forages. Springer, Berlin

Gupta D, Dadu RHR, Sambasivam P, Bar I, Singh M, Biju S (2019) Toward climate-resilient lentils: challenges and opportunities. In: Genomic designing of climate-smart pulse crops. Springer, Cham, pp 165–234

Hamdi A, Erskine W (1996) Reaction of wild species of the geneus Lens to drought. Euphytica 91(2):173–179

Hamdi A, Küsmenoĝlu I, Erskine W (1996) Sources of winter hardiness in wild lentil. Genet Resour Crop Evol 43(1):63–67

Hamwieh A, Udupa SM, Choumane W, Sarker A, Dreyer F, Jung C, Baum M (2005) A genetic linkage map of Lens sp. based on microsatellite and AFLP markers and the localization of fusarium vascular wilt resistance. Theor Appl Genet 110(4):669–677

Hamwieh A, Udupa SM, Sarker A, Jung C, Baum M (2009) Development of new microsatellite markers and their application in the analysis of genetic diversity in lentils. Breed Sci 59(1): 77–86

Hasanuzzaman M, Nahar K, Alam M, Roychowdhury R, Fujita M (2013) Physiological, biochemical, and molecular mechanisms of heat stress tolerance in plants. Int J Mol Sci 14(5):9643–9684

Havey MJ, Muehlbauer FJ (1989) Linkages between restriction fragment length, isozyme, and morphological markers in lentil. Theor Appl Genet 77(3):395–401

Henry CJK, Lightowler HJ, Strik CM, Renton H, Hails S (2005) Glycaemic index and glycaemic load values of commercially available products in the UK. Br J Nutr 94(6):922–930

Hossain MS, Alam MU, Rahman A, Hasanuzzaman M, Nahar K, Al Mahmud J, Fujita M (2017) Use of iso-osmotic solution to understand salt stress responses in lentil (*Lens culinaris* Medik.). S Afr J Bot 113:346–354

Idrissi O, Udupa SM, Houasli C, De Keyser E, Van Damme P, De Riek J (2015) Genetic diversity analysis of Moroccan lentil (*Lens culinaris* Medik.) landraces using Simple Sequence Repeat and Amplified Fragment Length Polymorphisms reveals functional adaptation towards agro-environmental origins. Plant Breed 134(3):322–332

Jayasundara HPS, Thomson BD, Tang C (1997) Responses of cool season grain legumes to soil abiotic stresses. Adv Agron 63:77–151

Johnson CR, Combs GF Jr, Thavarajah P (2013) Lentil (*Lens culinaris* L.): a prebiotic-rich whole food legume. Food Res Int 51(1):107–113

Johnson N, Johnson CR, Thavarajah P, Kumar S, Thavarajah D (2020) The roles and potential of lentil prebiotic carbohydrates in human and plant health. Plants People Planet 2(4):310–319

Kahraman A (2016) Nutritional components and amino acids in lentil varieties. Selcuk J Agric Food Sci 30(1):34–38

Kahraman A, Kusmenoglu I, Aydin N, Aydogan A, Erskine W, Muehlbauer FJ (2004) Genetics of winter hardiness in 10 lentil recombinant inbred line populations. Crop Sci 44(1):5–12

Kang U-G, Choi J-S, Kim J-J, Cho J-S (2017) Yield potentials of rice and soybean as affected by cropping systems in mid-mountainous paddy soils of Korea. Korean J Soil Sci Fertilizer 50(4): 259–274

Katerji N, Van Hoorn JW, Hamdy A, Mastrorilli M, Oweis T (2005) Salt tolerance analysis of chickpea, faba bean and durum wheat varieties: I. Chickpea and faba bean. Agric Water Manag 72(3):177–194

Kaur R, Sinha K, Bhunia RK (2019) Can wheat survive in heat? Assembling tools towards successful development of heat stress tolerance in *Triticum aestivum* L. Mol Biol Rep 46(2): 2577–2593

Khazaei H, Caron CT, Fedoruk M, Diapari M, Vandenberg A, Coyne CJ, McGee R, Bett KE (2016) Genetic diversity of cultivated lentil (*Lens culinaris* Medik.) and its relation to the world's agro-ecological zones. Front Plant Sci 7:1093

Kumar J, Van Rheenen HA (2000) Brief communication. A major gene for time of flowering in chickpea. J Hered 91(1):67–68

Kumar S, Rajendran K, Kumar J, Hamwieh A, Baum M (2015) Current knowledge in lentil genomics and its application for crop improvement. Front Plant Sci 6:78

Kumar H, Singh A, Dikshit HK, Mishra GP, Aski M, Meena MC, Kumar S (2019a) Genetic dissection of grain iron and zinc concentrations in lentil (*Lens culinaris* Medik.). J Genet 98 (3):1–14

Kumar J, Kumar S, Gupta DS, Dubey S, Gupta S, Gupta P (2019b) Molecular marker assisted gene pyramiding. In: Lentils. Elsevier, Amsterdam, pp 125–139

Kumawat KR, Gothwal DK, Singh D (2017) Salinity tolerance of lentil genotypes based on stress tolerance indices. J Pharmacogn Phytochem 6:1368–1372

Kusmenoglu I, Aydin N (1995) The current status of lentil germplasm exploitation for adaptation to winter sowing in the Anatolian highlands. In: Autumn-sowing of lentil in the highlands of West Asia and North Africa. Central Research Institute for Field Crops (CRIFC), Ankara, Turkey, pp 63–71

Lachaâl M, Grignon C, Hajji M (2002) Growth rate affects salt sensitivity in two lentil populations. J Plant Nutr 25(12):2613–2625

Lombardi M, Materne M, Cogan NOI, Rodda M, Daetwyler HD, Slater AT, Forster JW, Kaur S (2014) Assessment of genetic variation within a global collection of lentil (*Lens culinaris* Medik.) cultivars and landraces using SNP markers. BMC Genet 15(1):1–10

Lowe R, Shirley N, Bleackley M, Dolan S, Shafee T (2017) Transcriptomics technologies. PLoS Comput Biol 13(5):e1005457

Malik AI, Ailewe TI, Erskine W (2015) Tolerance of three grain legume species to transient waterlogging. AoB Plants 7:plv040

Malmberg MM, Pembleton LW, Baillie RC, Drayton MC, Sudheesh S, Kaur S, Shinozuka H, Verma P, Spangenberg GC, Daetwyler HD (2018) Genotyping-by-sequencing through transcriptomics: implementation in a range of crop species with varying reproductive habits and ploidy levels. Plant Biotechnol J 16(4):877–889

Materne M, Reddy AA (2007) Commercial cultivation and profitability. In: Lentil. Springer, New York, pp 173–186

Materne M, Siddique KHM (2009) Agroecology and crop adaptation. In: Erskine W, Muehlbauer FJ, Sarker A, Sharma B (eds) The lentil: botany, production and uses. CABI, pp 47–63

McCrory MA, Hamaker BR, Lovejoy JC, Eichelsdoerfer PE (2010) Pulse consumption, satiety, and weight management. Adv Nutr 1(1):17–30

Mishra BK, Srivastava JP, Lal JP, Sheshshayee MS (2016) Physiological and biochemical adaptations in lentil genotypes under drought stress. Russ J Plant Physiol 63(5):695–708

Muehlbauer FJ, Cho S, Sarker A, McPhee KE, Coyne CJ, Rajesh PN, Ford R (2006) Application of biotechnology in breeding lentil for resistance to biotic and abiotic stress. Euphytica 147(1): 149–165

Munns R, Tester M (2008) Mechanisms of salinity tolerance. Annu Rev Plant Biol 59:651–681

Muraleedhar A, Dikshit HK, Singh D, Singh A, Prapti P (2015) Inheritance of resistance to Mungbean Yellow Mosaic India Virus (MYMIV) in mungbean (Vigna radiata). Indian J Agric Sci 85(9):1144–1147

Murray GA, Eser D, Gusta LV, Eteve G (1988) Winterhardiness in pea, lentil, faba bean and chickpea. In: World crops: cool season food legumes. Springer, New York, pp 831–843

Muscolo A, Junker A, Klukas C, Weigelt-Fischer K, Riewe D, Altmann T (2015) Phenotypic and metabolic responses to drought and salinity of four contrasting lentil accessions. J Exp Bot 66(18):5467–5480

Omar I, Ghoulam SB, Abdellah EA, Sahri A (2019) Evaluation and utilization of lentil crop wild relatives for breeding in Morocco: towards development of drought and herbicide tolerant varieties. In: First international experts workshop on pre-breeding utilizing crop wild relatives

Poyraz I (2016) Comparison of ITS, RAPD and ISSR from DNA-based genetic diversity techniques. C R Biol 339(5–6):171–178

Pratap A, Gupta S, Basu PS, Tomar R, Dubey S, Rathore M, Prajapati US, Singh P, Kumari G (2019) Towards development of climate smart mungbean: challenges and opportunities. In: Genomic designing of climate-smart pulse crops. Springer, New York, pp 235–264

Rai R, Singh RP (1999) Effect of salt stress on interaction between lentil (*Lens culinaris*) genotypes and Rhizobium spp. strains: symbiotic N2 fixation in normal and sodic soils. Biol Fertil Soils 29(2):187–195

Rai R, Nasar SKT, Singh SJ, Prasad V (1985) Interactions between Rhizobium strains and lentil (*Lens culinaris* Linn.) genotypes under salt stress. J Agric Sci 104(1):199–205

Rana D, Dass A, Rajanna G, Kaur R (2016) Biotic and abiotic stress management in pulses. Indian J Agron 61:238–248. https://doi.org/10.1007/978-981-13-6986-5_12

Rizkalla SW, Bellisle F, Slama G (2002) Health benefits of low glycaemic index foods, such as pulses, in diabetic patients and healthy individuals. Br J Nutr 88(S3):255–262

Rizvi AH, Aski M, Sarker A, Dikshit HK, Yadav P (2019) Origin, distribution, and gene pools. In: Singh M (ed) Lentils. Academic, Burlington, pp 7–19. ISBN: 978-0-12-813522-8

Sakina A, Ahmed I, Shahzad A, Iqbal M, Asif M (2016) Genetic variation for salinity tolerance in Pakistani rice (*Oryza sativa* L.) germplasm. J Agron Crop Sci 202(1):25–36

Salam MA, Islam MT (1994) Growth, yield and leaf-water attributes of some advanced mutant lentil lines under different soil moisture regimes [*Lens culinaris*]. Lens Newsl 21(1):32

Samaranayaka A (2017) Lentil: revival of poor man's meat. In: Sustainable protein sources. Academic, San Diego, pp 185–196

Sanderson L-A, Caron CT, Tan R, Shen Y, Liu R, Bett KE (2019) KnowPulse: a web-resource focused on diversity data for pulse crop improvement. Front Plant Sci 10:965

Sarker A, Erskine W (2006) Recent progress in the ancient lentil. J Agric Sci 144(1):19–29

Sehgal A, Sita K, Kumar J, Kumar S, Singh S, Siddique KHM, Nayyar H (2017) Effects of drought, heat and their interaction on the growth, yield and photosynthetic function of lentil (*Lens culinaris* Medikus) genotypes varying in heat and drought sensitivity. Front Plant Sci 8:1776

Shams H, Tahbaz F, Hassan Entezari M, Abadi A (2010) Effects of cooked lentils on glycemic control and blood lipids of patients with type 2 diabetes. ARYA Atherosclerosis 4(1)

Sharpe AG, Ramsay L, Sanderson L-A, Fedoruk MJ, Clarke WE, Li R, Kagale S, Vijayan P, Vandenberg A, Bett KE (2013) Ancient orphan crop joins modern era: gene-based SNP discovery and mapping in lentil. BMC Genomics 14(1):1–13

Shi W, Tao F (2014) Spatio-temporal distributions of climate disasters and the response of wheat yields in China from 1983 to 2008. Nat Hazards 74(2):569–583

Shrestha R, Siddique KHM, Turner NC, Turner DW, Berger JD (2005) Growth and seed yield of lentil (*Lens culinaris* Medikus) genotypes of West Asian and South Asian origin and crossbreds between the two under rainfed conditions in Nepal. Aust J Agric Res 56(9):971–981

Shrestha R, Turner NC, Siddique KHM, Turner DW, Speijers J (2006) A water deficit during pod development in lentils reduces flower and pod numbers but not seed size. Aust J Agric Res 57(4):427–438

Silim SN, Saxena MC (1993) Adaptation of spring-sown chickpea to the Mediterranean basin. II. Factors influencing yield under drought. Field Crops Res 34(2):137–146

Sinclair TR, Vadez V (2012) The future of grain legumes in cropping systems. Crop Pasture Sci 63(6):501–512. https://doi.org/10.1071/CP12128

Singh M, Bisht IS, Kumar S, Dutta M, Bansal KC, Karale M et al (2014) Global wild annual Lens collection: a potential resource for lentil genetic base broadening and yield enhancement. PLoS One 9(9):e107781

Singh D, Singh CK, Taunk J, Tomar RSS (2016) Genetic analysis and molecular mapping of seedling survival drought tolerance gene in lentil (*Lens culinaris* Medikus). Mol Breed 36(5): 1–12

Singh D, Singh CK, Kumari S, Singh Tomar RS, Karwa S, Singh R, Singh RB, Sarkar SK, Pal M (2017a) Discerning morpho-anatomical, physiological and molecular multiformity in cultivated and wild genotypes of lentil with reconciliation to salinity stress. PLoS One 12(5):e0177465

Singh D, Singh CK, Singh Tomar RS, Pal M (2017b) Genetics and molecular mapping of heat tolerance for seedling survival and pod set in lentil. Crop Sci 57(6):3059–3067

Singh D, Singh CK, Taunk J, Tomar RSS, Chaturvedi AK, Gaikwad K, Pal M (2017c) Transcriptome analysis of lentil (*Lens culinaris* Medikus) in response to seedling drought stress. BMC Genomics 18:206

Singh D, Singh CK, Taunk J, Jadon V, Pal M, Gaikwad K (2019) Genome wide transcriptome analysis reveals vital role of heat responsive genes in regulatory mechanisms of lentil (*Lens culinaris* Medikus). Sci Rep 9(1):1–19

Singh D, Chaudhary P, Taunk J, Singh CK, Singh D, Tomar RSS et al (2021) Fab advances in Fabaceae for abiotic stress resilience: from 'omics' to artificial intelligence. Int J Mol Sci 22(19): 10535

Sinsawat V, Leipner J, Stamp P, Fracheboud Y (2004) Effect of heat stress on the photosynthetic apparatus in maize (*Zea mays* L.) grown at control or high temperature. Environ Exp Bot 52(2): 123–129

Sita K, Sehgal A, Kumar J, Kumar S, Singh S, Siddique KHM, Nayyar H (2017) Identification of high-temperature tolerant lentil (*Lens culinaris* Medik.) genotypes through leaf and pollen traits. Front Plant Sci 8:744

Siva N, Johnson CR, Richard V, Jesch ED, Whiteside W, Abood AA, Thavarajah P, Duckett S, Thavarajah D (2018) Lentil (*Lens culinaris* Medikus) diet affects the gut microbiome and obesity markers in rat. J Agric Food Chem 66(33):8805–8813

Solaiman Z, Colmer TD, Loss SP, Thomson BD, Siddique KHM (2007) Growth responses of cool-season grain legumes to transient waterlogging. Aust J Agric Res 58(5):406–412

Spaeth SC, Muehlbauer FJ (1991) Registration of three germplasms of winter hardy lentil. Crop Sci 31(5):1395

Street K, Mackay M, Zuev E, Kaul N, El Bouhssini M, Konopka J, Mitrofanova O (2008) Swimming in the genepool-a rational approach to exploiting large genetic resource collections. Sydney University Press

Sultana T, Ghafoor A (2008) Genetic diversity in ex-situ conserved Lens culinaris for botanical descriptors, biochemical and molecular markers and identification of landraces from indigenous genetic resources of Pakistan. J Integr Plant Biol 50(4):484–490

Summerfield RJ, Roberts EH, Erskine W, Ellis RH (1985) Effects of temperature and photoperiod on flowering in lentils (*Lens culinaris* Medic.). Ann Bot 56(5):659–671

Tadross M, Suarez P, Lotsch A, Hachigonta S, Mdoka M, Unganai L, Lucio F, Kamdonyo D, Muchinda M (2007) Changes in growing-season rainfall characteristics and downscaled scenarios of change over southern Africa: implications for growing maize. In: IPCC Regional Expert Meeting on regional impacts, adaptation, vulnerability, and mitigation, Nadi, Fiji, pp 193–204

Taku K, Umegaki K, Sato Y, Taki Y, Endoh K, Watanabe S (2007) Soy isoflavones lower serum total and LDL cholesterol in humans: a meta-analysis of 11 randomized controlled trials. Am J Clin Nutr 85(4):1148–1156

Tanyolac B, Ozatay S, Kahraman A, Muehlbauer F (2009) Linkage mapping of Lentil (*Lens culinaris* L.) genome using recombinant inbred lines revealed by AFLP, ISSR, RAPD and some morphologic markers. J Appl Biol Sci 3(2):179–185

Temel HY, Göl D, Akkale HBK, Kahriman A, Tanyolac MB (2015) Single nucleotide polymorphism discovery through Illumina-based transcriptome sequencing and mapping in lentil. Turk J Agric For 39(3):470–488

Tewari TN, Singh BB (1991) Stress studies in lentil (*Lens esculenta* Moench). Plant Soil 136(2): 225–230

Thavarajah D, Thavarajah P, Wejesuriya A, Rutzke M, Glahn RP, Combs GF, Vandenberg A (2011) The potential of lentil (*Lens culinaris* L.) as a whole food for increased selenium, iron, and zinc intake: preliminary results from a 3 year study. Euphytica 180(1):123–128

Tickoo JL, Sharma B, Mishra SK, Dikshit HK (2005) Lentil (*Lens culinaris*) in India: present status aaad future perspectives. Indian J Agric Sci 75(9)

Tivoli B, Baranger A, Avila CM, Banniza S, Barbetti M, Chen W, Davidson J, Lindeck K, Kharrat M, Rubiales D (2006) Screening techniques and sources of resistance to foliar diseases caused by major necrotrophic fungi in grain legumes. Euphytica 147(1):223–253

Toker C, Mutlu N, Pratap A, Kumar J (2011) Breeding for abiotic stress. In: Pratap A, Kumar J (eds) Biology and breeding of food legumes. CABI

Tomlekova N (1998) Mechanism of resistance against bruchid beetle (A. obtectus) (Say) in common bean (P. vulgaris L.). Agric Sci 5:19–21

Turner NC, Wright GC, Siddique KHM (2001) Adaptation of grain legumes (pulses) to water-limited environments. Adv Agron 71:193–231

Vadez V, Krishnamurthy L, Serraj R, Gaur PM, Upadhyaya HD, Hoisington DA, Varshney RK, Turner NC, Siddique KHM (2007) Large variation in salinity tolerance in chickpea is explained by differences in sensitivity at the reproductive stage. Field Crop Res 104(1–3):123–129

Van Hoorn JW, Katerji N, Hamdy A, Mastrorilli M (2001) Effect of salinity on yield and nitrogen uptake of four grain legumes and on biological nitrogen contribution from the soil. Agric Water Manag 51(2):87–98

Varshney RK, Bansal KC, Aggarwal PK, Datta SK, Craufurd PQ (2011) Agricultural biotechnology for crop improvement in a variable climate: hope or hype? Trends Plant Sci 16(7):363–371

Verma P, Sharma TR, Srivastava PS, Abdin MZ, Bhatia S (2014) Exploring genetic variability within lentil (*Lens culinaris* Medik.) and across related legumes using a newly developed set of microsatellite markers. Mol Biol Rep 41(9):5607–5625

Wiraguna E, Malik AI, Erskine W (2017) Waterlogging tolerance in lentil (*Lens culinaris* Medik. subsp. culinaris) germplasm associated with geographic origin. Genet Resour Crop Evol 64(3): 579–586

Wong MML, Gujaria-Verma N, Ramsay L, Yuan HY, Caron C, Diapari M, Vandenberg A, Bett KE (2015) Classification and characterization of species within the genus Lens using genotyping-by-sequencing (GBS). PLoS One 10(3):e0122025

Xu B, Chang SKC (2010) Phenolic substance characterization and chemical and cell-based antioxidant activities of 11 lentils grown in the Northern United States. J Agric Food Chem 58(3): 1509–1517

Yadav NK, Ghimire SK, Shakya SM, Sah SK, Sah BP, Sarker A, Kushwaha UKS (2016) Genetic diversity analysis of lentil (*Lens culinaris* L.) germplasm using DNA based SSR markers. Am J Food Sci Health 2(3):18–24

Yankova V, Sovkova-Bobcheva S (2009) Studying of bean varieties (*Phaseolus vulgaris* L.) reaction to bean weevil infestation (Acanthoscelides obtectus Say). In: Reports of bean improvement cooperative and national dry bean council research conference, pp 144–145

Yau SK, Erskine W (2000) Diversity of boron-toxicity tolerance in lentil growth and yield. Genet Resour Crop Evol 47(1):55–62

Zamir D, Ladizinsky G (1984) Genetics of allozyme variants and linkage groups in lentil. Euphytica 33(2):329–336

Zhao C, Liu B, Piao S, Wang X, Lobell DB, Huang Y, Huang M, Yao Y, Bassu S, Ciais P (2017) Temperature increase reduces global yields of major crops in four independent estimates. Proc Natl Acad Sci 114(35):9326–9331

Molecular and Physiological Approaches for Effective Management of Drought in Black Gram

10

M. Pandiyan, M. Sivaji, M. Yuvaraj, A. Krishnaveni, C. Sivakumar, and E. Jamuna

Abstract

Black gram (*Vigna mungo*) is one of the major pulse crops cultivated in more areas almost all over the world, particularly in India. This pulse crop is cultivated in both kharif and rabi seasons. Recent statistical data revealed that black gram yields are depleting drastically year by year in both India and the rest of the world. Drought stress is one of the primary causes for this yield reduction, and this stress happens due to lack of sufficient and improper distribution of rainfall. Black gram is extremely sensitive to drought, especially in both vegetative and reproductive stages, and witnesses 20–30% yield reduction. Drought stress is the reason for improper germination, reduced growth, injuring of the photosynthetic machinery, and declining of net photosynthesis and nutrient uptake, thereby causing yield reduction in black gram. Since stress sensing, signal transduction, and various adoption mechanisms are highly complex networks, to understand more about this network system, one should know more about physiological, biochemical, and molecular level changes that occurred in the plant system during stress and stress-responding mechanisms. Advance molecular technologies can be used to limelight the different gene regulation patterns and adoptive mechanisms concerning drought resistance, and this will be useful to develop drought stress-tolerant or -resistant black gram variety with higher yield potential under water-stress conditions. This chapter mainly concentrates on the impact of drought, morphological and physiochemical changes, stress-adaptive mechanisms, important drought-resistant traits, and physiological and molecular methods to manage drought stress in black gram.

M. Pandiyan (✉) · M. Sivaji · M. Yuvaraj · A. Krishnaveni · C. Sivakumar · E. Jamuna
Agricultural College and Research Institute, Tamil Nadu Agricultural University, Tiruvannamalai, Tamil Nadu, India

P. Muthu Arjuna Samy et al. (eds.), *Legumes: Physiology and Molecular Biology of Abiotic Stress Tolerance*, https://doi.org/10.1007/978-981-19-5817-5_10

Keywords

Black gram · Drought stress · Stress response · Physiochemical modifications · Drought-tolerant traits · Physiological and molecular approach

10.1 Introduction

Abiotic stresses are the major reason for decrease in the crop yield of many agriculture crops up to 80%. Among various abiotic stresses, drought is the most prevalent stress and causes a major threat to agricultural crop production and productivity worldwide (Anjum et al. 2017; Golldack et al. 2014; Hussain et al. 2018). The conditions, when water availability in soil is not enough for the active growth of the plant or transpiration loss is more when compared to the water absorbed by the roots; generating drought stress in the plants (Anjum et al. 2011). Majorly drought affects the morphological features of plant organs, enzyme activity, osmotic balance, source, and sink relationship and causes a substantial decline in crop yields (Barnabás et al. 2008; Hussain et al. 2008; Yordanov et al. 2000). Plant sensitivity to drought varies based upon plant age, genetic nature, duration, and intensity of water-deficit condition (Zhu 2002). Legumes are known to be a major source of low-cost protein that is needed to maintain better human health and also contribute more to soil fertility through fixing atmospheric nitrogen into the soil (Rubiales and Mikic 2015; Siddique et al. 1999). Among many cultivated pulse crops, black gram is ranked fourth in both cultivating areas and production in India. Chickpea, pigeon pea, and green gram are the other pulse crops that account for the first three places, respectively (Singh and Ahlawat 2005). Even though black gram is cultivated in more areas in India, average productivity is still very low (550 kg ha^{-1}). Like other pulse crops, black gram is also highly susceptible to many abiotic stresses, especially drought that significantly affects crop yield (Fang et al. 2010; Farooq et al. 2017; Micheletto et al. 2007). Drought will affect and reduce crop performance at any of its growth stages, but severe yield reduction occurs when black gram faces drought during the grain-filling and reproductive stage (Cortes and Sinclair 1986; Delmer 2005; Pushpavalli et al. 2014; Uprety and Bhatia 1989). So the development of drought-tolerant varieties or improving drought-tolerant mechanisms in black gram is one of the important approaches to improve yield, production, and productivity under water-deficit conditions. Developing black gram cultivars with drought-tolerant characteristics like high water-use efficiency (WUE), deep and profuse root growth, stomatal conductivity, and osmolyte production will be highly useful to increase the productivity in drought-prone areas (Miyauchi et al. 2012; Ulemale et al. 2013). To make significant progress in the development of drought-resistant cultivars, one should know the modification that occurred in plant physiological, cellular, and molecular levels due to environmental stress factors and differential gene expression patterns. At present, this deep molecular analysis and interpretation of data in a positive way are possible only because of the advancement of "omic" technologies. These technologies allow scientists to study and understand the interconnected mechanisms of gene expression and various plant metabolic responses towards drought stress. Although adverse effects of drought and detailed molecular analysis of plants to manage the stress have been studied and reviewed on several pulse crops, i.e., soybean, red gram, and chickpea (Bechtold 2018; Golldack et al. 2014; Kamanga et al. 2018; Shavrukov et al. 2017), in black gram no such type of molecular analysis exists for drought stress and its management. This chapter is mainly aimed to collect and review all works previously done on other pulse crops to alleviate the water deficit, and it also explains how this can be utilized in black gram crops to improve the performance of crops in terms of production and productivity under water-stress conditions.

10.2 Different Mechanisms of Plants to Manage Drought Stress

Plants have adapted dynamic responses, helping them continue to exist in unfavorable climate conditions (Huber and Bauerle 2016). Various inherent mechanisms are used by plants to manage water-deficit conditions that are grouped into three categories: drought escape, avoidance, and tolerance (Turner et al. 2001).

10.2.1 Drought Escape

This is one type of mechanism adopted by crop plants, in which the whole life cycle of the plant is completed before the commencement of drought. So, enough water would be available to all stages of the crop, which leads to better vegetative growth, reproduction, and yield. Agronomic and cultural practices such as showing time, plant geometry, fertilizer management, and selection of short-duration, suitable black gram variety will give promising yield. Even though yield reduction is commonly based upon the severity of the drought, complete crop failure will not occur under this mechanism. Unfortunately, if drought commences earlier during the crop growth period itself, then drought-escape plants will slowly switch over to the drought avoidance mechanism (Liu et al. 2005; Sicher et al. 2012; Waraich et al. 2011).

10.2.2 Drought Avoidance

Drought avoidance mechanism is mainly adopted by plants which undergo full or any part of the life cycle in the water-deficit condition. During this period, the plant changes various morphological and physiological features to tackle the stress situation. Some of the important modifications are more root production and deep penetration into the soil, reduction in stomatal conductance, small and thick leaf formation to reduce evapotranspiration, and production of various osmolytes (Goufo et al. 2017; Sicher et al. 2012). Biosynthesis of cuticular wax on leaf surfaces is also one of the avoidance mechanisms of drought (Lee and Suh 2013).

10.2.3 Drought Tolerance

Any external condition or an influence, either biotic or abiotic, that changes the regular homeostatic state of the plant is called stress (Zhu 2002). In general, whatever the stress may be, to attain tolerance, there are three important interrelated activities in plants that are necessary. The first one is that the harmfulness of stress needs to be prevented by activation of stress-responsive mechanisms. The next one is the establishment of homeostatic conditions in the current stressful situation faced by plants through the action of stress-responsive mechanisms. The third one is that plant growth may return to the original state, even though at a reduced rate at the end of the stress (Negrao et al. 2017; Zhu 2002). This tolerance-adoptive mechanism allows the plant to maintain turgor pressure and continue normal function even at low water potential, e.g., by protoplasmic tolerance or synthesis of osmoprotectants, osmolytes, or compatible solutes (Blum 2017; Khan et al. 2015; Nguyen et al. 1997).

10.3 Drought Tolerance Mechanism in Legumes

Various drought tolerance mechanisms are adopted by legume plants. It includes physio-morphological changes and cellular adaptations in plants based on the different stages or nature of the drought stress. Osmotic imbalance is the first symptom developed in plants by drought stress. Thereafter, it leads to damage of cell membranes, macromolecules like proteins, deoxyribonucleic acid, and lipids and finally affects the metabolic function of the cell. The cells with osmotic imbalance produce a hyperosmotic signal that in turn increases the accumulation of abscisic acid (ABA). ABA accumulation induces many drought-adaptive responses in whole plants (Negrao et al. 2017). Drought stress activates several biomolecules like reactive oxygen species (ROS), abscisic acid (ABA), Ca^{2+}, and jasmonic acid (JA). Accumulation of ABA and JA increases the expression (transcription) of ion transporter genes. WRKY, DREB, ZIP, AP2/ERF, MYB, etc. are some of the transcription factors (TF) that are overexpressed during drought stress (Fig. 10.1 [adopted from Nadeem et al. 2019]).

10.4 Compatible Solute Accumulation

Polyhydroxy alcohols, carbohydrates (oligosaccharides and disaccharides), and amino acids such as prolines and polyamines are some of the examples of compatible solutes. Overproduction or accumulation of these solutes is a basic strategy to maintain osmoprotectant and thereby an osmotic balance in the cell. Osmoprotection is a mechanism to protect cell components (important macromolecules such as proteins and enzymes) from the accumulation of higher concentration toxic ions like Na^{+}, which is the result of dehydration. Upon dissolving of compatible solutes in the cell solvent, it decreases the osmotic potential, which is more important to retain the water content and sustain the turgor pressure of the cell. This process is

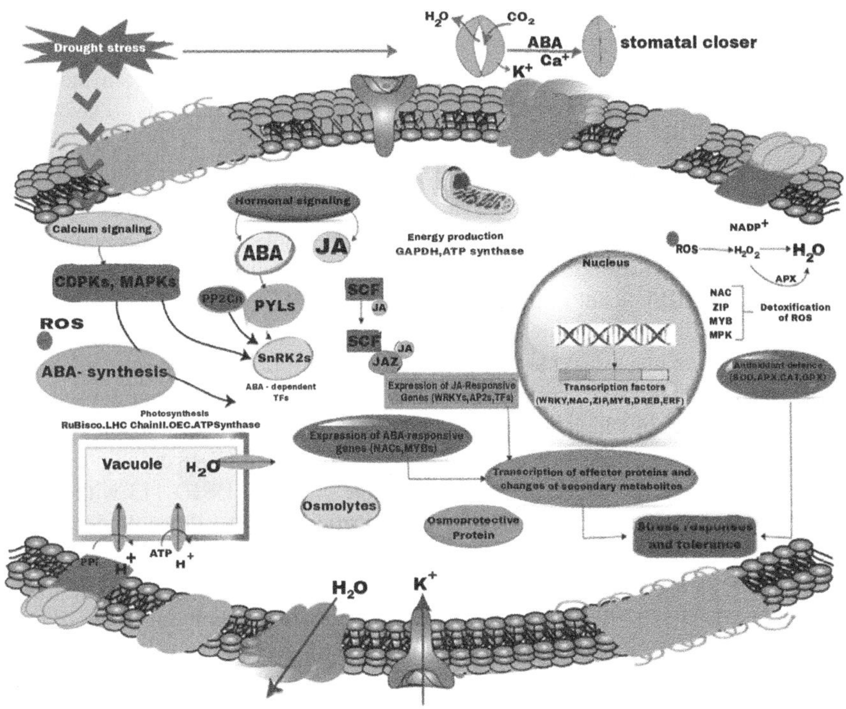

Fig. 10.1 Drought tolerance mechanism in legumes

called osmotic regulation (Majumdar et al. 2016; Slama et al. 2015; Solanki and Sarangi 2014). Compatible solutes accumulate mainly in drought-stressed cells to prevent the negative effect caused by drought stress (Majumdar et al. 2016). Accumulation of proline, sucrose, trehalose, mannitol, and glycine betaine helps in stabilizing membrane integrity by protecting the major functional molecules like proteins and lipid molecules, thereby increasing the photosynthetic activity during water-deficit conditions (Amede et al. 2003; Bhauso et al. 2014; Ibrahim and Abdellatif 2016; Khater et al. 2018; Ramanjulu and Bartels 2002; Shen et al. 1997; Shinde et al. 2016). Under moisture-stress conditions, stress-tolerant black gram varieties' proline content increases when compared to irrigated control (Yohan 2017).

10.5 Antioxidant Defense

At lower concentrations, reactive oxygen species will turn on several adaptive mechanisms in plants. But increasing drought stress leads to a higher accumulation of ROS, causing damage to macromolecules (Choudhury et al. 2017; Farnese et al. 2016; Kurutas 2016; Smirnoff 1993). So, ROS in lower concentration helps the plant cell by way of inducing adaptive responses; contrastingly, higher concentration causes the negative effect; this negative effect is nullified when cells produce antioxidants. Some of the examples of antioxidants are catalase, glutathione peroxidase, reductase, carotenoids, ascorbic acid, etc. Hence, maintaining an increased level of antioxidants in plant cells is one of the best mechanisms to avoid the negative effect of ROS (Al Hassan et al. 2017; Chakrabarty et al. 2016; Farooq et al. 2009; Sahitya et al. 2018). It has been witnessed in many legumes like soybean and green gram that during stress conditions plants contain low levels of antioxidants, whereas the quantity increases when plants are in the drought stress recovery phase (Guler and Pehlivan 2016; Mittler and Zilinskas 1994; Noctor et al. 2000; Osman 2015; Patel et al. 2016; Yasar et al. 2013; Zoz et al. 2013). Under drought-stress conditions, stress-resistant black gram cultivars accumulate more amount of antioxidant enzymes POD and SOD than the control (Jyoti and Yadav 2012; Saglam et al. 2011; Yohan 2017). The flowering stage is one of the sensitive stages for drought in which crop yield is highly reduced. Drought-tolerant genotype T9 experienced yield reduction of around 12.10–33.91% with 7.48 drought tolerance index when compared to the drought-susceptible genotype USJD 113 with a yield loss of 26.48–60.99% and 6.07 drought tolerance index value. Physiological analysis of these genotypes revealed that during stress conditions, genotype T9 has severalfold increase in the antioxidant enzyme activity when compared to the genotype USJD 113 (Baroowa and Gogoi 2017). A higher amount of antioxidant accumulation helps the varieties VBN4 and K1 protect themselves from drought stress. Hence, this trait of black gram can be used to study its molecular and physiological response to drought stress (Sai and Chidambaranathan 2019).

10.6 Hormone Regulation

Plant hormones are more important for good plant growth and its good establishment. Among many, auxins, cytokinins, and gibberellins are important plant hormones, regulating plant growth through cell division, cell differentiation, cell elongation, and organogenesis in a normal stress-free environment. ABA and ethylene are also phytohormones but are mainly involved in the control of plant metabolic function under stress conditions (Bielach et al. 2017; Ullah et al. 2018). During drought conditions, growth hormones such as auxins, cytokinins, and gibberellins are at a low level in plants, whereas ABA and ethylene are observed at a higher level (Weyers and Paterson 2001). This raise under stress triggers ABA entry to root xylem, actively involving reduction of stomatal conductance and increasing the hydraulic conductivity of roots (Hartung et al. 2002; Miyashita et al. 2005); thereby, water uptake is increased in plants (Merilo et al. 2015; Park et al. 2017) and it increases the production of superoxide radicals and H_2O_2; by this way, ABA increases the antioxidant enzyme activities. Many drought-responsive gene regulations are induced by ABA and jasmonic acid (JA) and play an important role in drought resistance in plants (Mohamed and Latif 2017; Ullah et al. 2018). In addition, black gram varieties, namely VBN4 and K1, have fivefold of ABA, 4.5-fold of proline, and fivefold of lipid peroxidase activity that will protect these varieties from drought stress (Sai and Chidambaranathan 2019).

10.7 Important Traits for Managing or Adopting Drought Stress in Black Gram

The ever-increasing water shortage, mainly because of insufficient rainfall, improper rainfall distribution, and climate change, is highly prevalent in the agricultural ecosystem and creates severe yield reduction or at the maximum crop failure. Identifying important drought-tolerant traits in black gram is the major goal for developing drought-tolerant black gram varieties, through the conventional or molecular breeding approach. Once drought-tolerant traits are identified, then it is easy to understand the stress-alleviating mechanism that is connected to that particular trait; then this may be effectively utilized to develop efficient drought-resistant cultivars through various crop improvement programs. However, accumulated knowledge about drought resistance in black gram is quite limited, so the drought-resistant traits which are studied with other legumes might be utilized in the screening process of drought-tolerant traits in black gram. Water-use efficiency, size of leaf area, leaf area maintenance, osmotic balance, shoot and root biomass, number of pods per plant, 100-seed weight, membrane stability index (MSI), proline content of leaves, relative water content (RWC), flowering time, height of the plant, early flowering and maturity time, seedling vigor, chlorophyll content, biological yield, harvest index, and plant yield are few of the physiological related drought-resistant traits that could be used in screening black gram (Yohan 2017). Varieties with more RWC, MSI, proline content, leaf area, plant height, and yield have high tolerance to drought (Bangar et al. 2019; Kumar et al. 2012; Upadhyaya et al. 2012). Among many, some of the important traits in legumes are discussed below.

10.7.1 Root Morphology and Plasticity

In plants, the root is the first organ to come across and act in response to water stress (Khodarahmpour 2011; Xiong et al. 2006). Plants possessing more roots and more root length will increase the absorption of water from the surrounding area, and a deep layer of soil in turn gives more tolerance to the drought condition (Bibi et al. 2010; Khayatnezhad et al. 2010; Okcu et al. 2005; Taylor et al. 1978; Vadez 2014; Yadav et al. 2013). Black gram varieties (T9 and CBG-09-13) with ideal root and shoot performed well under moisture-stress conditions (Kaydan and Yagmur 2008; Leishman and Westoby 1994; Prakash et al. 2018; Price et al. 2002; Sinclair Thomas 2011). Therefore, among several crop characters, more number of roots and more root length are targeted to increase crop yields, such as chickpea (Silim and Saxena 1993), wheat (Reynolds et al. 2007), and rice (Price et al. 2002; Yadav et al. 1997) under rainfed conditions. Drought stress reduces the root depth up to 14%, root mass up to 29%, and root length and root volume up to 35% and 41%, respectively. Reductions in the root parameters invariably decline the pod setting and pod weight up to 53% and 43%, respectively (Sofi et al. 2018).

10.7.2 Stomatal Conductance

During water deficit or drought period, plants completely depend upon soil water for their whole water requirement, but in normal conditions, plants' water requirements for their metabolic functions are not wholly dependent on the soil water. During this phase of transition, average stomatal conductance is reduced, and it directly affects the source and sink relationship, thereby decreasing the yield (Sinclair and Ludlow 1986). Genotypes capable of stomata opening despite internal water stress will perform well in drought-stress conditions. In water-loss conditions, many genotypes of black gram (*Vigna mungo*) show good stomatal control compared with other legumes (Bennett et al. 1987; Liu et al. 2005).

10.7.3 Slow Canopy Wilting (SW)

Slow canopy wilting is one of the water-conserving characters or traits, saving water through limiting maximum transpiration rates. So, under drought conditions, slow canopy wilting provides higher drought tolerance in early-maturating soybean varieties (Abdel-Haleem et al. 2012). This trait is governed by single quantitative trait loci (QTL); hence, this QTL can be used to select the drought-tolerant genotype of black gram and legume crops through molecular assisted selection (Charlson et al. 2009).

10.7.4 Epidermal Conductance

In normal stress-free conditions, epidermal conductance of water vapor (water vapor conductance through cuticle diffusion) is in negligible fraction because all the stomata will be in open condition (Charlson et al. 2009). But in water-stressed conditions, the maximum number of stomata present in the leaves is likely to be closed; at this time, stomatal conductance will be lower than the epidermal conductance (Gardingen and Grace 1992). In critical drought, epidermal conductance will be the major part of water loss in leaves (Boyer et al. 1997). Crops with lower epidermal conductance will have more survival ability; hence, this trait can be used to identify the drought-tolerant black gram varieties from the field level (Bennett et al. 1987; Riederer and Schreiber 2001).

10.7.5 Leaf Pubescence Density

More density of leaf pubescence increases the reflectance and reduces the temperatures of the leaf. So, more leaf pubescence density crop variety has more vigor; more and deep roots will help strengthen tolerance to the water-deficit condition (Garay and Wilhelm 1983; Jovanovi 1996).

10.7.6 Water-Use Efficiency

The quantity of biomass produced for the utilization of one unit of water is defined as water-use efficiency (WUE). It is an important factor deciding the production and productivity under limited water availability or water scarcity condition (Specht and Williams 2022). So, the positive correlation of high WUE with more crop yield will be used to select black gram variety with more drought tolerance (Mian et al. 1996; Wright 1996; Yohan 2017).

10.7.7 Osmotic Adjustment

Accumulation of more amounts of compatible solutes in the plant tissue during water scarcity is called osmotic adjustment. At the time of water loss, cells' osmotic potential is decreased because of the accumulation of harmful solutes. Decreased cell osmotic potential could help plants to absorb more water from the soil, but the harmful solutes will affect the protein structure and enzyme function, leading to collapsing of major cell organelle, and finally cell becomes functionless. At drought stress, mannitol, proline, sucrose, trehalose, and glycine betaine like compatible solutes are produced at a higher level to protect the proteins and enzymes that keep cells working even in stress conditions (Fried et al. 2019). Drought-tolerant black gram varieties showed more amount of proline accumulation during water-stress conditions (Turner et al. 2001; Yohan 2017).

10.8 Various Strategies of Drought Stress Management

10.8.1 Physiological Approach

10.8.1.1 Exogenous Application of Growth-Regulating Chemicals

Water deficit seriously affects physiological processes such as cell division, cell differentiation, cell elongation, and reduction in growth. Application of growth-regulating chemicals with clear growth regulatory actions like auxins, cytokinins, gibberellins, and ethylene can redefine the growth and development of plants undergoing drought stress by overcoming the stress damage or by inducing quick stress responses (Bangar et al. 2019). Compatible solutes and hormones such as polyamines, proline, glycine betaine, gibberellic acid, and salicylic acid enhance the drought-tolerant process in plants through better osmotic adjustment, detoxification of ROS, and protecting the integrity of cell membranes and macromolecules (Upreti and Sharma 2016). So, external application of these compounds to crops will change the metabolic activity and biomolecule production within the plant and modify it to adapt to the water-stress condition (Merilo et al. 2015; Park et al. 2017).

10.8.1.2 Hydrogels

Hydrogels are one type of polymers that absorb more quantity of water used to increase the water-holding ability of the soil. Carboxymethyl cellulose, pectin, cellulose, and chitin are some of the examples of polymers to form hydrogels (Jyoti and Yadav 2012; Rathinasabapathi 2000). Plants' performance can be improved by applying hydrogels through enhancing soil permeability, decreasing soil erosion, and lessening water loss. Black gram crop height, pod numbers, number of seeds in a single pod, and yield increase through hydrogel application (Ullah et al. 2018).

10.8.1.3 Application of Fertilizer

Fertilizer application to plants is more important because this will supply critically needed nutrients to the growing plants. Among many nutrients, potassium (K) is considered to play a significant role in drought-stress tolerance. This nutrient influences metabolism, growth, and development (Suriyaprakash et al. 2019). Application of K fertilizer to the plant improves the root growth, membrane stability, stomata regulation, total biomass, leaf photosynthetic activity, and water uptake and enhances downstream carbohydrate metabolism. In this way, through the application of appropriate K fertilizer at moisture-stress conditions, plant growth and yield improve when compared to the control conditions (Hartung et al. 2002).

10.8.2 Molecular Approaches for the Development of DS-Tolerant Legumes

The development of drought-tolerant genotypes is the most important strategy to increase the black gram productivity and for efficient utilization of available water. For this purpose, modern and conventional approaches and breeding methods will be more helpful for efficient variety development within a short period.

10.8.2.1 Breeding Approach

Conventional breeding will be a success only when existing enormous genetic variations that contain all the possible traits are involved in drought-stress mitigation (Farooq et al. 2014; Frahm et al. 2004; Miyashita et al. 2005). Limited resource, inappropriate information about crop yield during water stress, and limited existence of genetic variability hamper the drought resistance breeding progress in legumes (Beebe et al. 2008; Torres et al. 2010). Generally, legumes are narrow genetic-based crops, so black gram belonging to legume crops also has limited variation in the primary gene pool. This is the reason for the unavailability of improved variety for drought (Mir et al. 2012).

Root-related traits, such as root length, root depth and density, root architecture, and root biomass, are potential breeding traits used for identifying black gram cultivar with drought stress avoidance mechanism (Hall 2012; Thamodharan et al. 2016), and traits such as early maturity, 50% flowering, and podding provide mass screening traits for selecting the black gram cultivar with drought escape mechanism

(Mir et al. 2012). Cooler canopies, high stomatal conductance, slow canopy wilting, and WUE are some of the major traits useful for screening black gram genotypes for drought tolerance (Duc et al. 2015; Khan et al. 2010; Umamahesh et al. 2017). Wide hybridization is another important breeding method to achieve certain desirable traits within or between species (Chapman 2008; Hou et al. 2018; Martynenko et al. 2016; Muchero et al. 2013). Nine well-adopted drought-tolerant genotypes of chickpea were developed through transferring of genes from wild chickpea (*C. reticulatum*) into cultivated chickpea (Chapman 2008; Hou et al. 2018; Kashiwagi et al. 2005; Martynenko et al. 2016; Muchero et al. 2013). MCV-1, PLM-32, LGG-407, LGG-450, TM-96-2, and Sattyaare, the wild relative of black gram, *Vigna sublobata*, possess more drought tolerance since these wild relatives contain elevated amount of protein and proline, higher values of RWC and MSI, and more leaf area, plant height, and yield traits. Better physiological drought tolerance traits of these varieties might be useful to develop drought stress-tolerant black gram variety in breeding programs (Turner et al. 2001).

10.8.2.2 Quantitative Trait Loci (QTL) and Molecular Assisted Breeding

The traditional breeding methods require large investments in land, labor, and capital, and it requires quite a long time for varietal development with limited success. In this context, the DNA marker and QTL-based improvement method displays the high potential regarding legume crop improvement. All the drought-tolerant traits, either physiological or morphological traits, are governed by genes, and these genes are quantitatively inherited from one generation to the next. So, identification of QTL which is responsible for drought-tolerant traits makes it easy to track and identify desirable plants through marker-assisted breeding or marker-assisted selection (Miyashita et al. 2005).

QTL discovery for drought tolerance-related traits involves (1) development of segregating mapping populations for water stress-associated traits, (2) selection of polymorphic markers, (3) genotyping of the mapping populations with selected markers, (4) creation of genetic maps, (5) accurate phenotyping for drought resistance traits, and (6) QTL mapping using both genotypic and phenotypic data. This is called linkage mapping/linkage analysis-based QTL mapping (Basu et al. 2007; Cattivelli et al. 2008; Chamarthi et al. 2011). Twelve QTLs (NCPGR-50, TR-50, SCEA19, TAA-58, H6C-07, H5E-02, H5G-01, H6C-07, H1B-04, TA-113, H6C-07, H1F-21) responsible for seedling drought resistance and one QTL *Qncl.Sw1* related to grain yield were identified in chickpea (Fleury et al. 2010; Hamwieh et al. 2013). In cowpea (Muchero et al. 2009; Radhika et al. 2007), there are seven markers ACC-3, VuPAT1-2, CPRD8-1, CPRD14-2, CPRD14-3, CPRD22-2, and CPRD22-4 associated with Dro-1, Dro-2, Dro-3, Dro-3, Dro-4, Dro-5, and Dro-5, respectively. Scientists (Carpentieri-Pipolo et al. 2012; Muchero et al. 2009) identified four QTLs*qSV_Gm03, qSV_Gm05, qSV_Gm10, and qSV_Gm12* connected to drought tolerance in soybean. Two QTLs were identified for both leaf ash and WUE, influencing root architecture, a significant trait for tolerating water-deficit stress (Carpentieri-Pipolo et al. 2012). Various researchers (Abdel-Haleem et al. 2011; Manavalan et al. 2009) identified five QTLs Gm01, Gm02,

Gm03, Gm04, and Gm20 related to fibrous roots in soybean. These QTL-related information and available knowledge can be utilized for breeding drought-tolerant black gram cultivars through marker-assisted selection.

10.8.2.3 Transgenic Approach

With the advancement of bioinformatics and rapid development of "omics" technology, at present, it is easy to identify and predict the genes and respective proteins produced by those genes after expression. With the breakthrough of molecular biology technique and various tools, identification of complete pathway and enzymes involved in the specific pathway of any organism is also possible. Advancement of bioinformatics and molecular biology tools, opened the ways to trace out the genes from various organism; predicting, comprehensive analysis of gene functions and those genes can be integrated into the targeted plant to develop desirable character in that plant through the transgenic approach (Abdel-Haleem et al. 2011). Transgenic legume plants developed with the osmoprotectant gene and lipid membrane modification gene showed increased growth under drought stress (Pushpavalli et al. 2014). An enzyme ALDRXV4 involved in carbonyl metabolite reduction in cells has an important role as osmoprotectant and detoxicant of the reactive carbonyl species. Transgenic black gram plants integrated with ALDRXV4 showed multiple tolerance to abiotic stresses such as water deficit, oxidative, and salt stress. Since the transgenic lines have higher aldose reductase activity, they accumulate more amount of sorbitol and less amount of toxic metabolites such as methylglyoxal; which leads to an increase the photosynthetic efficiency, higher relative water content (RWC), and low photooxidative damage (Kishor et al. 1995). *P5CSF129A* gene is responsible for the biosynthesis of amino acid proline. Transgenic chickpea with the *P5CSF129A* gene accumulates more proline, thereby decreasing malonaldehyde and free radical levels (Singh et al. 2016). Transgenic chickpea plants with overexpression of *DREB1A* gene face 50% moisture reduction and display increased efficiency in drought tolerance compared to non-transgenic controls by effective control of the stomatal response, modified root architecture, improved water uptake, and transpiration efficiency (Bhatnagar-Mathur et al. 2009). Overexpression of *LOSS/ABA3* gene in soybean and cowpea synthesizes more ABA, which activates stress-upregulated gene expression and enhances drought tolerance (Anbazhagan et al. 2015; Li et al. 2013).

10.8.2.4 Genome Editing (GE) by CRISPR/Cas9

Genome editing is a process in which modification of genes is possible in a precise manner at a specific location. CRISPR/Cas9 is one of the highly researched and utilized modern genome editing (GE) tools. Many drought-resistant varieties were developed from economically important crops like wheat (Iuchi et al. 2000), rice (Sanchez Leon et al. 2018), barley (Huang et al. 2018; Lawrenson et al. 2015), maize (Kapusi et al. 2017; Zhu et al. 2016), and potato (Agarwal et al. 2018) by utilization of CRISPR/Cas9. Since it is in the beginning stage, very little research has been done in legumes related to editing of drought-tolerant genes, but this has a great future for developing drought-tolerant black gram varieties (Abdelrahman et al. 2018; Cai et al. 2015, 2018; Nakayasu et al. 2018).

10.9 Conclusions and Future Research Perspectives

Research by UNO predicts that water crisis is the major menace to mankind in the twenty-first century, leading to severe drought spells all over the world. Therefore, to avoid nutritional imbalance and maintain food security (yield improvement and stability in drought conditions) is the best option to develop water stress-resistant plant variety. Stress adaptation in plants is multigene regulated, and coordinated expression of many genes will bring about the desired level of tolerance. One of the important challenges here is to understand the signals, signaling cascade, transcription factor, and various gene networks working against water stress and plant survival. Many strategies are involved in the understanding of stress-responsive mechanisms and the development of stress-tolerant varieties. Modern breeding approaches, molecular approaches like genome editing, transgenic crop techniques, etc. will make it easy to complete the understanding and speed up the development of drought stress-tolerant black gram varieties. Even though many promising molecular techniques are available and so much progress has been made in many crops in the development of drought resistance cultivar, very few works have been done in legume crops such as soybean, chickpea, red gram, pigeon pea, and green gram. As of now, there are no or negligible documentary reports on the molecular level study of drought stress-related genes and transgenic study in black gram. Importance of such knowledge in improving the yield of black gram demands the need to research the abovesaid areas. In the future, it is necessary to initiate various research programs, which lead to producing highly promising black gram varieties that perform well in water-stress conditions.

References

Abdel-Haleem H, Lee GJ, Boerma RH (2011) Identification of QTL for increased fibrous roots in soybean. Theor Appl Genet 122:935–946

Abdel-Haleem H, Carter TE, Purcell LC, King CA, Ries LL, Chen P, Schapaugh W, Sinclair TR, Boerma HR (2012) Mapping of quantitative trait loci for canopy-wilting trait in soybean (*Glycine max* L. Merr). Theor Appl Genet 125:837–846

Abdelrahman M, Al-Sadi AM, Pour-Aboughadareh A, Burritt DJ, Tran LS (2018) Genome editing using CRISPR/Cas9–targeted mutagenesis: an opportunity for yield improvements of crop plants grown under environmental stresses. Plant Physiol Biochem 131:31–36

Agarwal A, Yadava P, Kumar K, Singh I, Kaul T, Pattanayak A, Agrawal PK (2018) Insights into maize genome editing via CRISPR/Cas9. Physiol Mol Biol Plants 24:175–183

Al Hassan M, Chaura J, Donat-Torres MP, Boscaiu M, Vicente O (2017) Antioxidant responses under salinity and drought in three closely related wild monocots with different ecological optima. AoB Plants 9:plx009

Amede T, Schubert S, Stahr K (2003) Mechanisms of drought resistance in grain legumes I: Osmotic adjustment. SINET 26:37–46

Anbazhagan K, Bhatnagar-Mathur P, Vadez V, Dumbala SR, Kishor PB, Sharma KK (2015) DREB1A overexpression in transgenic chickpea alters key traits influencing plant water budget across water regimes. Plant Cell Rep 34:199–210

Anjum SA, Xie X, Wang L, Saleem MF, Man C, Lei W (2011) Morphological, physiological and biochemical responses of plants to drought stress. Afr J Agric Res 6:2026–2032

Anjum SA, Ashraf U, Zohaib A, Tanveer M, Naeem M, Ali I, Tabassum T, Nazir U (2017) Growth and development responses of crop plants under drought stress: a review. Zemdirbyste 104:267–276

Bangar P, Chaudhury A, Tiwari B, Kumar S, Kumari R, Bhat KV (2019) Morphophysiological and biochemical response of mungbean [*Vigna radiata* (L.) Wilczek] varieties at different developmental stages under drought stress. Turk J Biol 43:58–69

Barnabás B, Jäger K, Fehér A (2008) The effect of drought and heat stress on reproductive processes in cereals. Plant Cell Environ 31:11–38

Baroowa B, Gogoi N (2017) The effect of osmotic stress on anti-oxidative capacity of black gram (*Vigna mungo* L.). Exp Agric 53:84–99

Basu PS, Berger JD, Turner NC, Chaturvedi SK, Ali M, Siddique KHM (2007) Osmotic adjustment of chickpea (*Cicer arietinum*) is not associated with changes in carbohydrate composition or leaf gas exchange under drought. Ann Appl Biol 150:217–225

Bechtold U (2018) Plant life in extreme environments: how do you improve drought tolerance? Front Plant Sci 9:543. https://doi.org/10.3389/fpls.2018.00543

Beebe SE, Rao IM, Cajiao C, Grajales M (2008) Selection for drought resistance in common bean also improves yield in phosphorus limited and favorable environments. Crop Sci 48:582–592

Bennett JM, Sinclair TR, Muchow RC, Costello SR (1987) Dependence of stomatal conductance on leaf water potential, turgor potential, and relative water content in field grown soybean and maize. Crop Sci 27:984–990

Bhatnagar-Mathur P, Vadez V, Jyostna Devi M, Lavanya M, Vani G, Sharma KK (2009) Genetic engineering of chickpea (*Cicer arietinum* L.) with the P5CSF129A gene for osmoregulation with implications on drought tolerance. Mol Breed 23:591–606

Bhauso TD, Radhakrishnan T, Kumar A, Mishra GP, Dobaria JR, Patel K, Rajam MV (2014) Overexpression of bacterial mtlD gene in peanut improves drought tolerance through accumulation of mannitol. Sci World J 2014:125967

Bibi A, Sadaqat HA, Akram HM, Mohammed MI (2010) Physiological markers for screening sorghum (*Sorghum bicolor*) germplasm under water stress condition. Int J Agric Biol 12:451–455

Bielach A, Hrtyan M, Tognetti VB (2017) Plants under stress: involvement of auxin and cytokinin. Int J Mol Sci 18:1427

Blum A (2017) Osmotic adjustment is a prime drought stress adaptive engine in support of plant production. Plant Cell Environ 40:4–10

Boyer JS, Wong SC, Farquhar GD (1997) CO_2 and water vapor exchange across leaf cuticle (epidermis) at various water potentials. Plant Physiol 114:185–191

Cai Y, Chen L, Liu X, Sun S, Wu C, Jiang B, Han T, Hou W (2015) CRISPR/Cas9-mediated genome editing in soybean hairy roots. PLoS One 10:e0136064. https://doi.org/10.1371/journal.pone.0136064

Cai Y, Chen L, Liu X, Guo C, Sun S, Wu C, Jiang B, Han T, Hou W (2018) CRISPR/Cas9 mediated targeted mutagenesis of GmFT2a delays flowering time in soya bean. Plant Biotechnol J 16:176–185

Carpentieri-Pipolo V, Pipolo AE, Abdel-Haleem H, Boerma HR, Sinclair TR (2012) Identification of QTLs associated with limited leaf hydraulic conductance in soybean. Euphytica 186:679–686

Cattivelli L, Rizza F, Badeck FW, Mazzucotelli E, Mastrangelo AM, Francia E, Mare C, Tondelli A, Stanca AM (2008) Drought tolerance improvement in crop plants: an integrated view from breeding to genomics. Field Crop Res 105:1–14

Chakrabarty A, Aditya M, Dey N, Banik N, Bhattacharjee S (2016) Antioxidant signaling and redox regulation in drought- and salinity stressed plants. Drought Stress Tolerance Plants 1:465–498

Chamarthi SK, Kumar A, Vuong TD, Blair MW, Gaur PM, Nguyen HT, Varshney RK (2011) 19 Trait mapping and molecular breeding. Biology and breeding of food legumes. CAB International, Wallingford, UK, pp 296–313

Chapman SC (2008) Use of crop models to understand genotype by environment interactions for drought in real-world and simulated plant breeding trials. Euphytica 161:195–208

Charlson DV, Bhatnagar S, King CA, Ray JD, Sneller CH, Carter TE, Purcell LC (2009) Polygenic inheritance of canopy wilting in soybean [*Glycine max* (L.) Merr.]. Theor Appl Genet 119:587–594

Choudhury FK, Rivero RM, Blumwald E, Mittler R (2017) Reactive oxygen species, abiotic stress and stress combination. Plant J 90:856–867

Cortes PM, Sinclair TR (1986) Gas exchange of field grown soybean under drought. Agron J 78:454–458

Delmer DP (2005) Agriculture in the developing world: connecting innovations in plant research to downstream applications. Proc Natl Acad Sci 102:15739–15746

Duc G, Agrama H, Bao S, Berger J, Bourion V, De Ron AM, Gowda CLL, Mikic A, Millot D, Singh KB (2015) Breeding annual grain legumes for sustainable agriculture: new methods to approach complex traits and target new cultivar ideotypes. Crit Rev Plant Sci 34:381–411

Fang X, Turner NC, Yan G, Li F, Siddique KHM (2010) Flower numbers, pod production, pollen viability, and pistil function are reduced and flower and pod abortion increased in chickpea (*Cicer arietinum* L.) under terminal drought. J Exp Bot 61:335–345

Farnese FS, Menezes-Silva PE, Gusman GS, Oliveira JA (2016) When bad guys become good ones: the key role of reactive oxygen species and nitric oxide in the plant responses to abiotic stress. Front Plant Sci 7:471

Farooq M, Wahid A, Kobayashi N, Fujita D, Basra SMA (2009) Plant drought stress: effects, mechanisms and management. In: Sustainable agriculture. Springer, Netherlands, pp 153–188

Farooq M, Hussain M, Siddique KHM (2014) Drought stress in wheat during flowering and grain-filling periods. Crit Rev Plant Sci 33:331–349. https://doi.org/10.1080/07352689.2014.875291

Farooq M, Gogoi N, Barthakur S, Baroowa B, Bharadwaj N, Alghamdi SS, Siddique KHM (2017) Drought stress in grain legumes during reproduction and grain filling. J Agron Crop Sci 203:81–102

Fleury D, Jefferies S, Kuchel H, Langridge P (2010) Genetic and genomic tools to improve drought tolerance in wheat. J Exp Bot 61:3211–3222

Frahm MA, Rosas JC, Mayek-Perez N, Lopez-Salinas E, Acosta-Gallegos JA, Kelly JD (2004) Breeding beans for resistance to terminal drought in the lowland tropics. Euphytica 136:223–232

Fried HG, Narayanan S, Fallen B (2019) Evaluation of soybean [*Glycine max* (L.) Merr.] genotypes for yield, water use efficiency, and root traits. PLoS One 14:e0212700. https://doi.org/10.1371/journal.pone.0212700. eCollection 2019

Garay AF, Wilhelm W (1983) Root system characteristics of two soybean isolines undergoing water stress condition. Agron J 75:973–977

Gardingen PRV, Grace J (1992) Vapour pressure deficit response of cuticular conductance in intact leaves of Fagus sylvatica L. J Exp Bot 43:1293–1299

Golldack D, Li C, Mohan H, Probst N (2014) Tolerance to drought and salt stress in plants: unraveling the signaling networks. Front Plant Sci 5:151

Goufo P, Moutinho-Pereira JM, Jorge TF, Correia CM, Oliveira MR, Rosa EAS, Antonio C, Trindade H (2017) Cowpea (*Vigna unguiculata* L. Walp.) metabolomics: osmoprotection as a physiological strategy for drought stress resistance and improved yield. Front Plant Sci 8:586

Guler NS, Pehlivan N (2016) Exogenous low-dose hydrogen peroxide enhances drought tolerance of soybean (*Glycine max* L.) through inducing antioxidant system. Acta Biol Hung 67:169–183

Hall A (2012) Phenotyping cowpeas for adaptation to drought. Front Physiol 3:1–8. https://doi.org/10.3389/fphys.2012.00155

Hamwieh A, Imtiaz M, Malhotra RS (2013) Multi-environment QTL analyses for drought-related traits in a recombinant inbred population of chickpea (*Cicer arientinum* L.). Theor Appl Genet 126:1025–1038

Hartung W, Sauter A, Hose E (2002) Abscisic acid in the xylem: where does it come from, where does it go to? J Exp Bot 53:27–32

Hou M, Tian F, Zhang L, Li S, Du T, Huang M, Yuan Y (2018) Estimating crop transpiration of soybean under different irrigation treatments using thermal infrared remote sensing imagery. Agronomy 9:1–18. https://doi.org/10.3390/agronomy9010008

Huang L, Zhang R, Huang G, Li Y, Melaku G, Zhang S, Chen H, Zhao Y, Zhang J, Zhang Y (2018) Developing superior alleles of yield genes in rice by artificial mutagenesis using the CRISPR/Cas9 system. Crop J 6:475–481

Huber AE, Bauerle TL (2016) Long-distance plant signaling pathways in response to multiple stressors: the gap in knowledge. J Exp Bot 67:2063–2079

Hussain M, Malik MA, Farooq M, Ashraf MY, Cheema MA (2008) Improving drought tolerance by exogenous application of glycine betaine and salicylic acid in sunflower. J Agron Crop Sci 194:193–199

Hussain M, Farooq S, Hasan W, Ul-Allah S, Tanveer M, Farooq M, Nawaz A (2018) Drought stress in sunflower: physiological effects and its management through breeding and agronomic alternatives. Agric Water Manag 201:152–166

Ibrahim HA, Abdellatif YMR (2016) Effect of maltose and trehalose on growth, yield and some biochemical components of wheat plant under water stress. Ann Agric Sci 61:267–274

Iuchi S, Kobayashi M, Yamaguchi-Shinozaki K, Shinozaki K (2000) A stress-inducible gene for 9-cis-epoxycarotenoid dioxygenase involved in abscisic acid biosynthesis under water stress in drought-tolerant cowpea. Plant Physiol 123:553–562

Jovanovi Z (1996) Cuticular characteristics in maize lines differing in drought resistance and maturity grouping. J Exp Bot 47:59–60

Jyoti B, Yadav SK (2012) Comparative study on biochemical parameters and antioxidant enzymes in a drought tolerant and a sensitive variety of horsegram (*Macrotyloma uniflorum*) under drought stress. Am J Plant Physiol 7:17–29

Kamanga RM, Mbega E, Ndakidemi P (2018) Drought tolerance mechanisms in plants: physiological responses associated with water deficit stress in Solanum lycopersicum. Adv Crop Sci Technol 6:1–8

Kapusi E, Corcuera-Gomez M, Melnik S, Stoger E (2017) Heritable genomic fragment deletions and small indels in the putative ENGase gene induced by CRISPR/Cas9 in barley. Front Plant Sci 25:8–540. https://doi.org/10.3389/fpls.2017.00540. eCollection 2017

Kashiwagi J, Krishnamurthy L, Upadhyaya HD, Krishna H, Chandra S, Vadez V, Serraj R (2005) Genetic variability of drought-avoidance root traits in the mini-core germplasm collection of chickpea (Cicer arietinum L.). Euphytica 146:213–222

Kaydan D, Yagmur M (2008) Germination, seedling growth and relative water content of shoot in different seed sizes of triticale under osmotic stress of water and NaCl. Afr J Biotechnol 7:2862–2868

Khan HR, Paull JG, Siddique KHM, Stoddard FL (2010) Faba bean breeding for drought-affected environments: a physiological and agronomic perspective. Field Crop Res 115:279–286

Khan MS, Kanwal B, Nazir S (2015) Metabolic engineering of the chloroplast genome reveals that the yeast ArDH gene confers enhanced tolerance to salinity and drought in plants. Front Plant Sci 6:725. https://doi.org/10.3389/fpls.2015.00725

Khater MA, Dawood MG, Sadak MS, Shalaby MAF, El-Awadi ME, El-Din KG (2018) Enhancement the performance of cowpea plants grown under drought conditions via trehalose application. Middle East J 7:782–800

Khayatnezhad M, Gholamin R, Jamaatie-Somarin SH, Zabihi-Mahmoodabad R (2010) Effects of peg stress on corn cultivars (*Zea mays* L.) at germination stage. World Appl Sci J 11:504–506

Khodarahmpour Z (2011) Effect of drought stress induced by polyethylene glycol (PEG) on germination indices in corn (*Zea mays* L.) hybrids. Afr J Biotechnol 10:18222–18227

Kishor PBK, Hong Z, Miao GH, Hu CAA, Verma DPS (1995) Overexpression of [delta]-pyrroline-5-carboxylate synthetase increases proline production and confers osmotolerance in transgenic plants. Plant Physiol 108:1387–1394

Kumar J, Basu PS, Srivastava E, Chaturvedi SK, Nadarajan N, Kumar S (2012) Phenotyping of traits imparting drought tolerance in lentil. Crop Pasture Sci 63:547–554

Kurutas EB (2016) The importance of antioxidants which play the role in cellular response against oxidative nitrosative stress: current state. Nutr J 15:1–22. https://doi.org/10.1186/s12937-016-0186-5

Lawrenson T, Shorinola O, Stacey N, Li C, Ostergaard L, Patron N, Uauy C, Harwood W (2015) Induction of targeted, heritable mutations in barley and Brassica oleracea using RNA-guided Cas9 nuclease. Genome Biol 16:1–13

Lee SB, Suh MC (2013) Recent advances in cuticular wax biosynthesis and its regulation in Arabidopsis. Mol Plant 6:246–249

Leishman MR, Westoby M (1994) The role of seed size in seedling establishment in dry soil conditions—experimental evidence from semi-arid species. J Ecol 82:249–258

Li Y, Zhang J, Zhang J, Hao L, Hua J, Duan L, Zhang M, Li Z (2013) Expression of an Arabidopsis molybdenum cofactor sulphurase gene in soybean enhances drought tolerance and increases yield under field conditions. Plant Biotechnol J 11:747–758

Liu F, Andersen MN, Jacobsen SE, Jensen CR (2005) Stomatal control and water use efficiency of soybean (*Glycine max* L. Merr.) during progressive soil drying. Environ Exp Bot 54:33–40

Majumdar R, Barchi B, Turlapati SA, Gagne M, Minocha R, Long S, Minocha SC (2016) Glutamate, ornithine, arginine, proline, and polyamine metabolic interactions: the pathway is regulated at the post-transcriptional level. Front Plant Sci 7:78. https://doi.org/10.3389/fpls.2016.00078

Manavalan LP, Guttikonda SK, Phan Tran LS, Nguyen HT (2009) Physiological and molecular approaches to improve drought resistance in soybean. Plant Cell Physiol 50:1260–1276

Martynenko A, Shotton K, Astatkie T, Petrash G, Fowler C, Neily W, Critchley AT (2016) Thermal imaging of soybean response to drought stress: the effect of Ascophyllum nodosum seaweed extract. Springerplus 5:1–14

Merilo E, Jalakas P, Laanemets K, Mohammadi O, Horak H, Kollist H, Brosche M (2015) Abscisic acid transport and homeostasis in the context of stomatal regulation. Mol Plant 8:1321–1333

Mian MA, Bailey MA, Ashley DA, Wells R, Carter TE Jr, Parrott WA, Boerma HR (1996) Molecular markers associated with water use efficiency and leaf ash in soybean. Crop Sci 36: 1252–1257

Micheletto S, Rodriguez-Uribe L, Hernandez R, Richins RD, Curry J, O'Connell MA (2007) Comparative transcript profiling in roots of Phaseolus acutifolius and P. vulgaris under water deficit stress. Plant Sci 173:510–520

Mir RR, Zaman-Allah M, Sreenivasulu N, Trethowan R, Varshney RK (2012) Integrated genomics, physiology and breeding approaches for improving drought tolerance in crops. Theor Appl Genet 125:625–645

Mittler R, Zilinskas BA (1994) Regulation of pea cytosolic ascorbate peroxidase and other antioxidant enzymes during the progression of drought stress and following recovery from drought. Plant J 5:397–405

Miyashita K, Tanakamaru S, Maitani T, Kimura K (2005) Recovery responses of photosynthesis, transpiration, and stomatal conductance in kidney bean following drought stress. Environ Exp Bot 53:205–214

Miyauchi Y, Isoda A, Li Z, Wang P (2012) Soybean cultivation on desert sand using drip irrigation with mulch. Plant Prod Sci 15:310–316

Mohamed HI, Latif HH (2017) Improvement of drought tolerance of soybean plants by using methyl jasmonate. Physiol Mol Biol Plants 23:545–556

Muchero W, Ehlers JD, Close TJ, Roberts PA (2009) Mapping QTL for drought stress-induced premature senescence and maturity in cowpea [Vigna unguiculata (L.) Walp.]. Theor Appl Genet 118:849–863

Muchero W, Roberts PA, Diop NN, Drabo I, Cisse N, Close TJ, Muranaka S, Boukar O, Ehlers JD (2013) Genetic architecture of delayed senescence, biomass, and grain yield under drought stress in cowpea. PLoS One 8:e70041. https://doi.org/10.1371/journal.pone.0070041

Nadeem M, Li J, Yahya M, Sher A, Ma C, Wang X, Qiu L (2019) Research progress and perspective on drought stress in legumes: a review. Int J Mol Sci 20:2541. https://doi.org/10.3390/ijms20102541

Nakayasu M, Akiyama R, Lee HJ, Osakabe K, Osakabe Y, Watanabe B, Sugimoto Y, Umemoto N, Saito K, Muranaka T (2018) Generation of α-solanine-free hairy roots of potato by CRISPR/Cas9 mediated genome editing of the St16DOX gene. Plant Physiol Biochem 131:70–77

Negrao S, Schmockel SM, Tester M (2017) Evaluating physiological responses of plants to salinity stress. Ann Bot 119:1–11

Nguyen HT, Babu RC, Blum A (1997) Breeding for drought resistance in rice: physiology and molecular genetics considerations. Crop Sci 37:1426–1434

Noctor G, Veljovic-Jovanovic S, Foyer CH (2000) Peroxide processing in photosynthesis: antioxidant coupling and redox signalling. Philos Trans R Soc Lond B Biol Sci 355:1465–1475

Okcu G, Kaya MD, Atak M (2005) Effects of salt and drought stresses on germination and seedling growth of pea (*Pisum sativum* L.). Turk J Agric For 29:237–242

Osman HS (2015) Enhancing antioxidant–yield relationship of pea plant under drought at different growth stages by exogenously applied glycine betaine and proline. Ann Agric Sci 60:389–402

Park J, Lee Y, Martinoia E, Geisler M (2017) Plant hormone transporters: what we know and what we would like to know. BMC Biol 15:1–15

Patel KG, Mandaliya VB, Mishra GP, Dobaria JR, Thankappan R (2016) Transgenic peanut overexpressing mtlD gene confers enhanced salinity stress tolerance via mannitol accumulation and differential antioxidative responses. Acta Physiol Plant 38:1–14

Prakash M, Elangaimannan R, Sunilkumar B, Narayanan GS (2018) Evaluation of blackgram genotypes for drought tolerance based on root dynamics and gas exchange parameters. Legum Res 41:384–391

Price AH, Cairns JE, Horton P, Jones HG, Griffiths H (2002) Linking drought-resistance mechanisms to drought avoidance in upland rice using a QTL approach: progress and new opportunities to integrate stomatal and mesophyll responses. J Exp Bot 53:989–1004

Pushpavalli R, Zaman-Allah M, Turner NC, Baddam R, Rao MV, Vadez V (2014) Higher flower and seed number leads to higher yield under water stress conditions imposed during reproduction in chickpea. Funct Plant Biol 42:162–174

Radhika P, Gowda SJM, Kadoo NY, Mhase LB, Jamadagni BM, Sainani MN, Chandra S, Gupta VS (2007) Development of an integrated intraspecific map of chickpea (*Cicer arietinum* L.) using two recombinant inbred line populations. Theor Appl Genet 115:209–216

Ramanjulu S, Bartels D (2002) Drought and desiccation induced modulation of gene expression in plants. Plant Cell Environ 25:141–151

Rathinasabapathi B (2000) Metabolic engineering for stress tolerance: installing osmoprotectant synthesis pathways. Ann Bot 86:709–716

Reynolds M, Dreccer F, Trethowan R (2007) Drought-adaptive traits derived from wheat wild relatives and landraces. J Exp Bot 58:177–186

Riederer M, Schreiber L (2001) Protecting against water loss: analysis of the barrier properties of plant cuticles. J Exp Bot 52:2023–2032

Rubiales D, Mikic A (2015) Introduction: legumes in sustainable agriculture. Crit Rev Plant Sci 34:1–8

Saglam A, Saruhan N, Terzi R, Kadioglu A (2011) The relations between antioxidant enzymes and chlorophyll fluorescence parameters in common bean cultivars differing in sensitivity to drought stress. Russ J Plant Physiol 58:60–68

Sahitya UL, Krishna MSR, Deepthi R, Prasad GS, Kasim D (2018) Seed antioxidants interplay with drought stress tolerance indices in chilli (*Capsicum annuum* L) seedlings. Biomed Res Int. https://doi.org/10.1155/2018/1605096

Sai CB, Chidambaranathan P (2019) Reproductive stage drought tolerance in blackgram is associated with role of antioxidants on membrane stability. Plant Physiol Rep 24:399–409

Sanchez Leon S, Gil Humanes J, Ozuna CV, Gimenez MJ, Sousa C, Voytas DF, Barro F (2018) Low gluten, nontransgenic wheat engineered with CRISPR/Cas9. Plant Biotechnol J 16:902–910

Shavrukov Y, Kurishbayev A, Jatayev S, Shvidchenko V, Zotova L, Koekemoer F, De Groot S, Soole K, Langridge P (2017) Early flowering as a drought escape mechanism in plants: how can it aid wheat production? Front Plant Sci 8(8):1950. https://doi.org/10.3389/fpls.2017.01950

Shen B, Jensen RG, Bohnert HJ (1997) Mannitol protects against oxidation by hydroxyl radicals. Plant Physiol 115:527–532

Shinde S, Villamor JG, Lin W, Sharma S, Verslues PE (2016) Proline coordination with fatty acid synthesis and redox metabolism of chloroplast and mitochondria. Plant Physiol 172:1074–1088

Sicher RC, Timlin D, Bailey B (2012) Responses of growth and primary metabolism of water-stressed barley roots to rehydration. J Plant Physiol 169:686–695

Siddique KHM, Loss SP, Regan KL, Jettner RL (1999) Adaptation and seed yield of cool season grain legumes in Mediterranean environments of south-western Australia. Aust J Agric Res 50: 375–388

Silim SN, Saxena MC (1993) Adaptation of spring-sown chickpea to the Mediterranean basin. II. Factors influencing yield under drought. Field Crop Res 34:137–146

Sinclair TR, Ludlow MM (1986) Influence of soil water supply on the plant water balance of four tropical grain legumes. Funct Plant Biol 13:329–341

Sinclair Thomas R (2011) Challenges in breeding for yield increase for drought. Trends Plant Sci 16:289–293

Singh DP, Ahlawat IPS (2005) Greengram (*Vigna radiata*) and blackgram (*V. mungo*) improvement in India: past, present and future prospects. Indian J Agric Sci 75:243–250

Singh P, Kumar D, Sarin NB (2016) Multiple abiotic stress tolerance in *Vigna mungo* is altered by overexpression of ALDRXV4 gene via reactive carbonyl detoxification. Plant Mol Biol 91:257–273

Slama I, Abdelly C, Bouchereau A, Flowers T, Savoure A (2015) Diversity, distribution and roles of osmoprotective compounds accumulated in halophytes under abiotic stress. Ann Bot 115: 433–447

Smirnoff N (1993) Tansley Review No. 52. The role of active oxygen in the response of plants to water deficit and desiccation. New Phytol 125:27–58

Sofi PA, Djanaguiraman M, Siddique KHM, Prasad PVV (2018) Reproductive fitness in common bean (*Phaseolus vulgaris* L.) under drought stress is associated with root length and volume. Indian J Plant Physiol 23:796–809

Solanki JK, Sarangi SK (2014) Effect of drought stress on proline accumulation in peanut genotypes. Int J Adv Res 2:301–309

Specht JE, Williams JH (2022) Breeding for drought and heat resistance: prerequisites and examples. In: World Soybean Research Conference III: Proceedings, pp 468–475

Suriyaprakash R, Sivakumar T, KrishnaSurendar K, Senthivelu M (2019) Morpho-physiological studies on response of blackgram (*Vigna mungo* L.) under soil water deficit amended with hydrogel. Int J Pure Appl Biosci 7:498–502

Taylor HM, Burnett E, Booth GD (1978) Taproot elongation rates of soybeans. Zeitschrift Fuer Acker Und Pflanzenbau

Thamodharan G, Geetha S, Ushakumari ARR (2016) Studies on heterosis in blackgram (*Vigna mungo* (L.) Hepper). Electronic Journal of Plant Breeding 7:670–676

Torres AM, Avila CM, Gutierrez N, Palomino C, Moreno MT, Cubero JI (2010) Marker-assisted selection in faba bean (*Vicia faba* L.). Field Crop Res 115:243–252

Turner NC, Wright GC, Siddique KHM (2001) Adaptation of grain legumes (pulses) to water-limited environments. Adv Agron 71:123–231

Ulemale CS, Mate SN, Deshmukh DV (2013) Physiological indices for drought tolerance in chickpea (*Cicer arietinum* L.). World J Agric Sci 9:123–131

Ullah A, Manghwar H, Shaban M, Khan AH, Akbar A, Ali U, Ali E, Fahad S (2018) Phytohormones enhanced drought tolerance in plants: a coping strategy. Environ Sci Pollut Res 25:33103–33118. https://doi.org/10.1007/s11356-018-3364-5

Umamahesh YYPSV, Reddy DM, Sumathi V (2017) Screening of blackgram genotypes for total drymatter and yield components under imposition of moisture stress

Upadhyaya HD, Kashiwagi J, Varshney RK, Gaur PM, Saxena KB, Krishnamurthy L, Gowda CLL, Pundir RPS, Chaturvedi SK, Basu PS (2012) Phenotyping chickpeas and pigeonpeas for adaptation to drought. Front Physiol 3:1–10
Upreti KK, Sharma M (2016) Role of plant growth regulators in abiotic stress tolerance. In: Abiotic stress physiology of horticultural crops. Springer, New Delhi, pp 19–46. https://doi.org/10.1007/978-81-322-2725-0_2
Uprety DC, Bhatia A (1989) Effect of water stress on the photosynthesis, productivity and water status of mung bean (*Vigna radiata* L. Wilczek). J Agron Crop Sci 163:115–123
Vadez V (2014) Root hydraulics: the forgotten side of roots in drought adaptation. Field Crop Res 165:15–24
Waraich EA, Ahmad R, Ashraf MY (2011) Role of mineral nutrition in alleviation of drought stress in plants. Aust J Crop Sci 5:764–777
Weyers JDB, Paterson NW (2001) Plant hormones and the control of physiological processes. New Phytol 152:375–407
Wright G (1996) Review of ACIAR selection for water use efficiency in legumes project recommends further research. ACIAR Food Legum Newslett:2–3
Xiong L, Wang RG, Mao G, Koczan JM (2006) Identification of drought tolerance determinants by genetic analysis of root response to drought stress and abscisic acid. Plant Physiol 142:1065–1074
Yadav R, Courtois B, Huang N, McLaren G (1997) Mapping genes controlling root morphology and root distribution in a doubled-haploid population of rice. Theor Appl Genet 94:619–632. https://doi.org/10.1007/s001220050459
Yadav SK, Jyothi Lakshmi N, Singh V, Patil A, Tiwari YK, Nagendram E, Sathish P, Vanaja M, Maheswari M, Venkateswarlu B (2013) In vitro screening of *Vigna mungo* genotypes for PEG induced moisture deficit stress. Indian J Plant Physiol 18:55–60
Yasar F, Uzal O, Yasar TO (2013) Investigation of the relationship between the tolerance to drought stress levels and antioxidant enzyme activities in green bean (*Phaseolus Vulgaris* L.) genotypes. Afr J Agric Res 8:5759–5763
Yohan Y (2017) Screening of blackgram (*Vigna mungo* L. Hepper) genotypes for physiological efficiency and drought tolerance. Acharya N.G. Ranga Agricultural University
Yordanov I, Velikova V, Tsonev T (2000) Plant responses to drought, acclimation, and stress tolerance. Photosynthetica 38:171–186
Zhu JK (2002) Salt and drought stress signal transduction in plants. Annu Rev Plant Biol 53:247–273
Zhu J, Song N, Sun S, Yang W, Zhao H, Song W, Lai J (2016) Efficiency and inheritance of targeted mutagenesis in maize using CRISPR-Cas9. J Genet Genomics 43:25–36
Zoz T, Steiner F, Guimaraes VF, Castagnara DD, Meinerz CC, Fey R (2013) Peroxidase activity as an indicator of water deficit tolerance in soybean cultivars. Biosci J 29:1664–1677

Abiotic Stress Responses in Groundnut (*Arachis hypogaea* L.): Mechanisms and Adaptations

11

M. K. Kalarani, A. Senthil, S. Punitha, S. Sowmyapriya, M. Umapathi, and V. Geethalakshmi

Abstract

The world population is rising at a fast pace and may reach 6–9.3 billion by 2050, but food production is rapidly declining due to the detrimental effects of numerous environmental stresses. According to climate change predictions, extreme weather events will become more frequent in tropical and subtropical areas. Extreme weather events cause abiotic stresses such as water stress, temperature stress, radiation stress, and salt stress, all of which have a significant influence on the productivity of crops such as groundnut (*Arachis hypogaea* L.). Globally, groundnut is a significant oil and food crop, ranking third and fourth in terms of protein and edible oil, respectively. Recent advances in groundnut physiology, plant phenotyping, and genomics have resulted in new insights into the abiotic stress tolerance mechanisms in groundnut, providing breeders with a better understanding of the gene networks involved in stress tolerance as well as newer tools for genetic improvement of groundnut for higher yield under stress conditions. This chapter discusses the abiotic stresses that impact groundnut production, as well as recent advances in employing physiological and genetic methods to increase abiotic stress tolerance in groundnut.

M. K. Kalarani (✉) · A. Senthil · S. Punitha · S. Sowmyapriya
Directorate of Crop Management, Tamil Nadu Agricultural University, Coimbatore, Tamil Nadu, India
e-mail: kalarani.mk@tnau.ac.in

M. Umapathi
Department of Crop Physiology, Tamil Nadu Agricultural University, Coimbatore, Tamil Nadu, India

V. Geethalakshmi
Tamil Nadu Agricultural University, Coimbatore, Tamil Nadu, India

P. Muthu Arjuna Samy et al. (eds.), *Legumes: Physiology and Molecular Biology of Abiotic Stress Tolerance*, https://doi.org/10.1007/978-981-19-5817-5_11

Keywords

Oil seed · Moisture stress · Antioxidants · Reactive oxygen species · Yield

11.1 Introduction

Next to soybean, groundnut (*Arachis hypogaea* L.) is the world's most important legume in terms of production and is often produced as a rain-fed crop. India is the world's second largest producer of groundnut (68.57 lakh tonnes), after China (166.24 lakh tonnes) (Groundnut Outlook 2018). The area under groundnut cultivation in India has decreased from 41.35 lakh ha in 2017–2018 to 40.12 lakh ha in 2018–2019 due to various reasons. Hence, both the production and yield decreased from 52.75 lakh tonnes and 1269 kg ha^{-1} in 2017–2018 to 37.70 lakh tonnes and 931 kg ha^{-1} in 2018–2019, respectively. Both biotic and abiotic stresses have contributed to groundnut's declining production and productivity in recent years, and among these, abiotic stresses have emerged as a major cause for this.

Among the abiotic stresses, drought, high-temperature, and salinity stresses are the significant abiotic factors that cause limitations to groundnut production. More than 70% of groundnut cultivation in Asia is in arid or semiarid regions, where the crop is subjected to drought regularly with varying intensity and duration and 15% of this land is also subject to salt stress (Reddy et al. 2003). Because it is a rain-fed crop, groundnut is particularly vulnerable to drought, which is brought on by climate change, which reduces yields dramatically. However, groundnut is a moderately salt-sensitive plant, but the amount of salinity of the soil or water might restrict the yield. Because of climate change and global warming, groundnut yields in India are expected to decline by 23–36% with a mean air temperature increase of 2–3 °C (Hoegh-Guldberg et al. 2018). High temperatures at crucial growth phases have an impact on the pod yield and also intensify the moisture stress, both of which contribute to additional yield losses (Prasad et al. 1999).

As the crop progresses throughout its life cycle, the plant's response to various environmental stresses will differ depending upon the genotype. The response to environmental stresses is more prominent at some phenological stages than at different phenophases. As a result, field tolerance of a genotype to any abiotic stress is assessed throughout multiple development stages. Several studies have found that the sensitivity of groundnut genotypes to abiotic stresses varies during the vegetative stage and pod growth stage, although the reproductive stage is more vulnerable to such challenges (Hamidou et al. 2013). There are just a few studies on screening groundnut genotypes for high-temperature and salt stress under controlled conditions and in the field conditions. To establish a breeding program and create better cultivars for abiotic stress environments, it is critical to know how plants respond under stress and in non-stress situations. This review addresses the groundnut plant's responses to numerous abiotic stresses, as well as the mechanism of tolerance and adaptability at various growth stages.

11.2 Abiotic Stress Responses in Groundnut

Abiotic stresses have different effects on plants, and the plants which have inherent tolerance strive to change their morphophysiological and biochemical characteristics to combat these stresses, whereas vulnerable plants acquire symptoms (Zafar et al. 2018) that lead to a reduction in growth and development.

11.2.1 Morphological Responses

Drought stress has a major influence on morphological characteristics, particularly the roots. The root is the plant's primary organ, which responds and perceives the drought signals and maintains its growth and development during drought. Under drought conditions, phenotyping of roots demonstrates the relevance of root characteristics as a screening tool for drought resistance. Phenotyping helps to understand how various root characteristics contribute to drought tolerance by drawing soil resources from deeper soil layers and using them to carry out many metabolic processes in the plant system. Deeper rooted plants can draw water from the soil at a deeper level, which protects them from drought stress (Bao et al. 2014).

Under water stress, root traits are critical for maintaining crop yields because they affect water and nutrient absorption (Narayanan et al. 2014). Plants with a larger main root have more growth potential since it is directly related to water absorption and has a greater ability to penetrate the compacted soil. Particularly, herbaceous plants have thin roots, which are more permeable and better able to absorb water. Due to increased climate variability under current cropping systems, this role is considered even more important in soils with low water and nutrient availability. Root architecture has a substantial influence on nitrogen-usage efficiency. As the roots grow deeper and faster, more adventitious roots are formed in the upper soil layer, which increases nutrient and water uptake and reduces soil surface evaporation losses (Sinclair 1994). The initiation, branching, and turnover of new roots are all regulated by soil temperature. The increase in root characteristics such as root length and volume, during a water shortage and after recovery, is linked to greater drought tolerance. Water may be conserved by increasing partitioning to root mass, which would result in a tendency to allocate less assimilates to other parts of the plant. Rapid root development into the surrounding soil would be an adaptive benefit in using the soil water more completely. Several root traits, such as root tissue density, specific surface area, and specific root length are also linked to higher crop yield under drought. The diameter of the roots and the density of the root tissue determine the surface area and length of the roots, respectively. A bigger root system and deeper rooting ability will support the acquisition of required soil water under stress conditions, where deeper soil water is accessible and contributes to sustaining yield under terminal drought conditions.

11.2.2 Reproductive Responses

When drought is induced shortly before the reproductive growth of groundnut, a large burst in flowering may represent a distinctive characteristic in the pattern of flowering under moisture stress. The main flush of flowers generated up to 45 days after planting does not form pegs when stress is imposed (30–45 days after sowing); however, this loss is compensated by the flowers produced after rewatering (Gowda and Hegde 1986). The ultimate groundnut yield is determined by the flowering pattern or the number of flowers produced at various stages of the reproductive cycle. After 75–90 days of post-flowering water stress, groundnut plants act like watered ones and produce more flowers. The flowering time of groundnuts is around 41–60 days; however, plants exposed to water stress during this period have fewer flowers, because the plants have followed a 51–70 days' flowering pattern. Water-stressed plants had more early-formed flowers, which determine the optimum flowering pattern of 31–50 days, even if they had fewer flowers overall than plants that were exposed to stress after flowering (Kalarani et al. 2018). Drought stress slows groundnut peg elongation, which is turgor dependent and delays pod and seed development due to dry soil at the pegging zone. Soil water deficiencies in the pegging and root zone reduced pod and seed development rates by around 30% and lowered the weight per seed by 428–563 mg.

Studies showed that at higher air temperatures, pod yield reduced due to fewer pegs and pods as a result of less fruit set. Ketring (1984) found that exposure to day temperature of 35 °C reduced the number of pegs and pods by 33% when compared to 30 °C under a controlled environment. When flowering and pod development occurred at a temperature of 40 °C, pod yield dropped by more than 50% of its potential yield. Heat stress significantly reduced kernel weight by 45–46% compared to non-stressed environments.

11.2.3 Physiological Responses

There is a negative impact of abiotic stress on groundnut, which is plant water relations and mineral nutrition as well as metabolism and photosynthesis (Suthar and Patel 1992). Under abiotic stress conditions, biomass output and pod yield are used as selection standards for resistant groundnut genotypes. Many factors, including relative water content (RWC), leaf water potential (WP), stomatal resistance, transpiration rate, leaf temperature, and canopy temperature, affect the groundnut water relations under drought. Plants that are under stress have lower RWC values than plants that are not under stress. The radiation and vapor pressure deficits are low in the morning, and the groundnut leaves have high RWC values. By midday, low values were observed, and after midday, again gradual increase in RWC values was observed (Erickson and Ketring 1985).

Water-deficit plants lose moisture from pods, which reduces the physiological activity of seeds, and so inevitably both yield and nutritious quality are affected. The characteristics related to drought tolerance and pod production exhibit significant relationships between drought tolerance index (DTI), pod yield, root length density (RLD), and harvest index (HI), demonstrating that RLD in deeper soil contributes to pod yield and HI under drought circumstances. In the pegging and root zones, a lack of water can reduce pod and seed output by 30% (Kambiranda et al. 2011). As previously reported in groundnut (Sheshshayee et al. 2006; Songsri et al. 2008), a strong direct association between water-usage efficiency (WUE) and chlorophyll index as well as an indirect correlation between leaf area and SPAD chlorophyll meter readings (SCMR) have been found. Maheswari et al. (2016) found that drought had a beneficial effect on plant osmotic adjustment and photosynthetic rate during the pre-flowering period when plants were exposed to drought. Drought stress kills the chlorophyll and inhibits its production in groundnut. A higher chlorophyll a/b ratio with a decreasing total chlorophyll content indicates more damage to chlorophyll b than chlorophyll a (Mafakheri et al. 2010).

Temperature-induced adaptations in plants include long-term evolutionary changes in phenology and molecular structure as well as short-term avoidance or acclimation strategies involving leaf orientation and transpiration cooling. High temperature has detrimental effects on plants, such as lower leaf water potential, reduced leaf area, and premature leaf senescence. These factors all have an impact on the plant's photosynthetic ability (Greer and Weedon 2012). Drought and salt stress are known to be harsher on shoot growth than on root growth. Plants might preserve soil moisture by using less water if their leaf area expansion was reduced in relation to their root growth. Water absorption is restricted in salt stressed organisms, and this causes osmotic stress. Salt stress also increases the buildup of Na^+ and Cl^- ions, which can cause cytotoxicity, impede enzyme activity, and lead to other elements being unbalanced. Salt stress impairs cellular metabolism and photosynthetic activity.

11.2.4 Biochemical and Molecular Responses

When plants are under abiotic stress, their cellular biochemistry is altered such as protein content, ion transporters, signal molecules, free radical scavengers, and other biochemical reactions. The stress caused by drought and high temperatures promotes the formation of reactive oxygen species (ROS), which inactivate enzymes, damage cellular components, and decrease the defense capability of the plants. High temperatures have a significant impact on starch and sucrose synthesis because they lower the activity of enzymes including sucrose phosphate synthase, ADP-glucose pyrophosphorylase, and invertase. Under conditions of sufficient water supply, the transpiration rate is often correlated with the incident radiation. Drought-stressed plants lose more water through transpiration than healthy ones. As well as rendering groundnuts more prone to aflatoxin contamination (Cole et al. 1989), this makes them unsuitable for human consumption.

When membranes are subjected to high temperatures, they experience lipid bilayer stress, which allows membrane proteins to displace and solutes to leak, also leading to deterioration of membrane selectivity (Du et al. 2011). Physiological characteristics, such as leaf area and chlorophyll concentration, were linked with membrane damage in groundnut under high-temperature stress. Free radical production compromises a plant's defense capabilities, resulting in oxidative stress. Additionally, the Fenton/Haber-Weiss pathway generates a harmful hydroxyl radical ($^{\bullet}OH$) inside the plants, which destabilizes the membrane lipids through lipid peroxidation, resulting in membrane damage. Damage to cellular membranes and chlorophyll is a good measure of how much oxidative stress has damaged the plant.

There are numerous ways in which plants alter their metabolism in response to abiotic stress, including producing compatible solutes that can organize proteins and cellular structure to keep the cell turgor, as well as making changes to the antioxidant system to restore cell redox balance and maintain homeostasis (Janská et al. 2010). Due to abiotic stress, modification of physiological and biochemical processes by gene expression changes gradually leads to the development of tolerance in the form of acclimation or, in the ideal case, to adaptation (Mirza et al. 2010).

11.3 Tolerance Mechanisms and Adaptation

11.3.1 Morphophysiological Mechanisms

Understanding the abiotic stress tolerance mechanism through the physiological and genetic processes of plants helps in the development of the newest varieties with abiotic stress tolerance. Plants have numerous adjustment responses to abiotic stress, including stomatal behavior and osmotic changes. The adaptive response of the plant to survive under long-lasting drought is in relation to decreased oxidative damage to cells (Azevedo Neto et al. 2010). Groundnut crop might undergo closure of stomata during drought in semiarid regions. Prolonged water shortages combined with high temperatures can limit active gas exchange duration gradually by stomatal behavior, which impacts plant growth and development processes. But the quick and complete recovery from severe stress after rewatering leads to normal stomatal conductivity, which is frequently recorded in groundnut. This ability to quickly recover to normal transpiration and CO_2 assimilation is a key mechanism for the adaptive reactions of the groundnut. Higher palmitic and stearic acid accumulations in groundnut at high temperatures improve membrane stability.

Cellular activities of plants are altered differently when exposed to a particular abiotic stress or when combined with specific stress environments. In addition to these changes, the production and accumulation of highly soluble, low-molecular-weight, electrically neutral, nontoxic compounds, generally known as osmolytes or osmoprotectants (Behelgardy et al. 2012), are important due to their protective role against the damage that was done by the stress, which could affect the cellular machinery. These compounds directly scavenge toxic ROS and protect antioxidant enzymes, thereby improving plants' antioxidant defense system. As a result, osmoprotectants function to activate genes associated with defense mechanism under a various stress. Thus, to survive under hostile conditions, plants have evolved the osmoprotectants as an important evolutionary strategy. Plant cells are protected against the damaging effects of diverse environmental stresses by amino alkanoic acid proline, which functions as a molecular chaperone.

An increased accumulation of soluble sugars (beta-D-galactofuranoside, hexopyranose, D-glucopyranose) and osmoprotectors (D-mannitol and pentitol) in the groundnut may have played a key role in regulating osmotic changes and in protecting diverse cell structures from temperature stress via maintaining cellular water balance and membrane stability, as well as buffering the cellular redox potential. The increased availability of carbohydrates during heat stress is an essential physiological feature associated with stress tolerance and acclimatization processes.

Antioxidant enzymes such as superoxide dismutase (SOD), catalase (CAT), peroxidase (POD), ascorbate peroxidase (APX), glutathione reductase (GR), and other associated enzymes involved in the cellular detoxification are considered as the primary critical defense mechanism towards abiotic stresses. Drought-induced damage was mitigated with an increase in SOD and POD activities. It is the tolerance capacity, mainly dependent on genotypes. POD and SOD activities are increased with the progression of water stress to a certain degree and subsequently stabilized in certain genotypes and have very little activity in certain genotypes, which are sensitive genotypes (Maheswari et al. 2018). The initial line of defense is SOD, which detoxifies the superoxide radicals into hydrogen peroxide (H_2O_2); CAT and POD break down H_2O_2. The conversion of H_2O_2 to the nontoxic water molecule is via either CAT activity or ascorbate-glutathione cycle. This process limits the cellular buildup of H_2O_2. Finally, glutathione reductase performs NADPH-dependent reduction of oxidized glutathione (GSSG) to reconstitute the cellular pool of reduced glutathione (GSH) (Noctor et al. 2002).

11.3.2 Molecular Mechanisms

The use of diverse methodologies has assisted scientists in drawing a global picture of how plants perceive environmental stress signals, transmit the stress signals to the nucleus, and then regulate gene expression to create appropriate responses when stressed. Although insights into plant tolerance mechanisms, as well as particular activities of various participants and their interactions with other members within the network, have not yet been precisely characterized, several different network components have been targeted and utilized for genetic modification.

Transcription factors (TFs), kinases, phosphatases, microRNAs (miRNAs), and two-component systems (TCSs) have all been discovered as members of genes that encode regulatory proteins that play important roles in influencing plant behavior in response to abiotic stresses (Hoang et al. 2014). The role of regulatory members is to transfer stress signals from the external environment to the nucleus (such as TFs and TCSs) and to directly control the gene expression (like TFs) by interacting with gene promoters. In eukaryotes, small ubiquitin-like modifier (SUMO) peptide binding to protein substrates (SUMOylation) is a key posttranslational regulation mechanism. SUMOylation was shown to play a vital role in pod formation and abiotic stress.

The expressional control of auxin transporters, including ABCB transporters, is crucial in drought response. Plant response to abiotic stress might be improved by engineering hormone signaling. Exogenous brassinolide (BR) increased cucumber photosynthesis through influencing photosynthetic enzymes such as ribulose 1,5 bisphosphate carboxylase/oxygenase (Rubisco). To control photosynthesis, BR signaling might move from brassinosteroid insensitive (BRI) to transcription factors brassinazole resistant 1 (BZR) and phytochrome interacting factors (PIFs) (Osakabe et al. 2014). As a result, BR priming may be able to control photosynthesis transcriptionally during drought stress. Exogenous BR signal promotes auxin biosynthesis and expression of the transcription factor genes, such as small auxin upregulated 15 (SAUR15). As a result, the enhanced growth might be due to the cross talk between the BR signal and the auxin signal during BR priming. Plant defense against stress is regulated by salicylic acid (SA), jasmonic acid (JA), and abscisic acid (ABA).

Protein analysis is a straightforward way to determine the function of their linked genes. Proteome analysis linked with genome sequence data is important for functional genomics. However, the data on protein expression that is now accessible is inadequate. Spectroscopy study of changes in seed protein composition in response to drought stress indicated that methionine-rich proteins (MRPs) and arachin proteins were downregulated in drought-susceptible (DS) genotypes, but not in drought-tolerant (DT) genotypes. The upregulation of mRNA transcripts in DT genotypes showed a link to stress tolerance. Continued production of those proteins appears to improve drought tolerance; it lowers the aflatoxin levels and increases the nutritional value of groundnuts (Basha et al. 2007).

Metabolomics is an interdisciplinary branch of study that focuses on the metabolomes of biological systems. Metabolomics, as a high-throughput technology, has the fundamental goal of providing a comprehensive view of all metabolites participating in cellular processes, which necessitate the use of nonselective, universally applicable, and comprehensive analytical approaches for metabolite identification and quantification. Relative electrolyte leakage (REC) fluctuation in reaction to salt stress, particularly during recovery, was demonstrated by metabolite changes, with 92 metabolites, out of a total of 391 detected, varying in response to salt and 42 metabolites responding specifically to recovery. Transcriptomics data revealed that 1742 transcripts in shoots and 3281 transcripts in roots changed in response to salt stress, whereas 372 transcripts in shoots and 1386 transcripts in roots responded particularly to recovery but not salt stress. Finally, 95 transcripts and 1 metabolite were identified as potential candidates for REC, photosynthesis, transpiration, and variation in Na^+ accumulation (Cui et al. 2018).

Dehydration-responsive element binding (DREB) (a member of the ethylene-responsive element-binding factor (ERF) family) has been shown to efficiently alter the expression of several stress-inducible genes in groundnut, therefore conferring drought tolerance. Groundnut transgenic lines had a higher proline content (30–40%) and higher levels of pyrroline-5-carboxylate synthase (P5CS) than non-transgenic plants during drought and salt stress. Transgenic groundnut plants not only pile up a lot of solutes, but they also had better membrane integrity under stress. Groundnut transgenics overexpressing AtNAC2 (Arabidopsis NAM, ATAF1, 2, and CUC2) demonstrated better drought and salinity tolerance as well as yield under water-stressed circumstances.

Compared to the wild-type plants, transgenic groundnut plants containing a novel stress-inducible WRKY transcription factor, MuWRKY3 (*Macrotyloma uniflorum* Lam. Verdc) gene isolated from horse gram, improved drought tolerance by slowing down the wilting; upregulating stress-inducible genes; accumulating higher proline, total sugars, and antioxidant enzymes; and lowering malondialdehyde, hydrogen peroxide, and superoxide anion. Arabidopsis homeodomain-leucine zipper transcription factor (AtHDG11) was overexpressed under stress-inducible promoter desiccation-responsive protein 29A (rd29A), which increased drought and salt tolerance in transgenic groundnut plants by upregulating stress-responsive genes, antioxidative enzymes, and free proline. Furthermore, the transgenic plants had longer roots, lower stomatal density, higher chlorophyll content, greater specific leaf area, and better photosynthetic rates (Banavath et al. 2018). At NAC2, groundnut transgenics showed a lower rate of water loss and higher RWC than wild type, indicating that transgenics had a greater ability to retain water and maintain a higher leaf water status. One of the major factors that result in the preservation of a greater canopy photosynthetic rate during stress is the maintenance of higher chlorophyll content, which indicates the stay-green nature of AtNAC2 transgenics under stress conditions. Transgenics have also shown less membrane damage when stressed. When the PDH45 gene (pea DNA helicase) was overexpressed in genetically modified groundnut, it showed about 10% higher yield under salt stress.

11.4 Strategies for Improving Abiotic Stress Tolerance

Complete groundnut sequencing will be too expensive and labor consuming to undertake with existing resources. The cultivated form of groundnut is an amphidiploid with $2n = 4x = 40$ chromosomes. The study of molecular features of the groundnut genome began in the 1980s, when protein and isozyme variation in groundnut was shown to be ineffective for defining variation in the cultivated groundnut. Although many polymorphisms were found in other species within the genus, the number of markers was too small to be used in breeding programs on a regular basis.

Gene knockout and knockdown strategies based on ethyl methane sulfonate, T-DNA insertion mutations, transposable element insertion, target-induced local lesions in genomes, and gene silencing by RNAi have been used, but these methods have drawbacks such as the need to screen large populations of mutants and transgenic lines. These issues may be avoided by using virus-induced gene silencing (VIGS), and a strong functional genomics method will be employed to create multi-stress-resistant groundnut plants. However, the creation of VIGS procedures to examine the activity of a single gene under many stressful situations in a given time frame is required.

Seed priming has the potential to enhance seed germination, produce anti-stress chemicals, and acclimatize groundnut to abiotic stresses. Breeding for abiotic stress resistance has been a critical technique used by researchers to relieve abiotic stress issues and ensure to meet out the abiotic stress-prone environments (Pereira et al. 2012). Understanding physiology and genetics may lead to a better understanding of stress response and assist in the creation of new stress-tolerant cultivars. However, due to the number and order of genes regulating quantitative characteristics, the transmission of features associated with abiotic stress adaptation is likely to be genetically complicated.

Although there is significant phenotypic diversity for yield-related characteristics in groundnut, the variability for vitamin (primarily vitamin E) and micronutrient (particularly Fe and Zn) levels, as well as resistance to aflatoxin, insect, and disease, is extremely low in the cultivated groundnut. Because of the minimal diversity of the abovementioned characteristics, genetic improvement of groundnut by traditional and marker-assisted breeding is limited.

Stress-inducible expression of AtHDG11 in three independent homozygous transgenic groundnut lines improved the drought and salt tolerance by upregulating known stress-responsive genes (LEA, HSP70, Cu/Zn SOD, APX, P5CS, NCED1, RRS5, ERF1, NAC4, MIPS, aquaporin, TIP, ELIP) in the stress gene network, antioxidative enzymes, and free proline and enhanced water-use efficiency traits such as longer root system, reduced stomatal density, higher chlorophyll content, increased specific leaf area, improved photosynthetic rates, and increased intrinsic and instantaneous WUE. Transgenic groundnut plants outperformed non-transgenic plants in terms of yield under both drought- and salt-stress conditions. Transgenic methods would aid in the introduction of those genes into groundnut to improve mineral content, vitamin E content, and aflatoxin resistance.

Direct irradiation of groundnut seed or seeds of mutant(s) produced from it will result in a modified salt tolerance level. The clustered regularly interspaced short palindromic repeats and CRISPR-associated protein 9 (CRISPR–Cas9) system has great potential for evaluating gene/genome function and developing abiotic stress tolerance in a variety of plants. It is a low-cost, simple, user-friendly, and fast-accepted genome editing technology for generating genome-edited crops to meet rising food demands in the face of climate change. To better understand the underlying metabolism, it is critical to identify and classify the individual genes linked with the complicated processes of tolerance. Plant tissue culture system is an efficient and dependable approach for studying salt tolerance in groundnut. The method is simple to operate, allowing the tolerance potential of the plants to be accurately measured.

11.5 Conclusion

So far, detailed investigations have substantially contributed to a better knowledge of groundnut plant responses in terms of morphophysiological, biochemical, and molecular features under drought, high-temperature, and, to a lesser extent, salt stress. Several researchers have reported on the underlying adaptation and tolerance mechanisms in groundnut against abiotic stresses such as drought, salt, and high temperature. In addition, numerous drought tolerance techniques have been identified. However, in addition to drought and high-temperature stresses, an in-depth study of groundnut responses to different abiotic stresses such as waterlogging, salt and sodic conditions, low temperature, and low light intensity is required, under changing climatic circumstances. Furthermore, the adaptive characteristics and tolerance mechanisms against each of these stresses must be elucidated in order to be used in breeding programs to create tolerant genotypes. Proteomics, metabolomics, marker-assisted selection, gene editing methods, and transgenic technologies must be used successfully to produce novel groundnut genotypes tolerant to diverse abiotic stresses to preserve agricultural community output and livelihood security.

References

Azevedo Neto AD, Nogueira RJMC, Melo Filho PA, Santos RC (2010) Physiological and biochemical responses of peanut genotypes to water deficit. J Plant Interact 5(1):1–10

Banavath JN, Chakradhar T, Pandit V, Konduru S, Guduru KK, Akila CS, Podha S, Puli COR (2018) Stress inducible overexpression of AtHDG11 leads to improved drought and salt stress tolerance in peanut (*Arachis hypogaea* L.). Front Chem 6:34

Bao Y, Aggarwal P, Robbins NE, Sturrock CJ, Thompson MC, Tan HQ, Tham C, Duan L, Rodriguez PL, Vernoux T (2014) Plant roots use a patterning mechanism to position lateral root branches toward available water. Proc Natl Acad Sci 111(25):9319–9324

Basha SM, Katam R, Naik KSS (2007) Differential response of peanut genotypes to water stress. Peanut Sci 34(2):96–104

Behelgardy MF, Motamed N, Jazii FR (2012) Expression of the P5CS gene in transgenic versus nontransgenic olive (*Olea europaea*) under salinity stress. World Appl Sci J 18(4):580–583

Cole RJ, Sanders TH, Dorner JW, Blankenship PD (1989) Environmental conditions required to induce preharvest aflatoxin contamination of groundnuts: summary of six years' research. In: International workshop on aflatoxin contamination of groundnut, Patancheru, AP (India), 6–9 Oct 1987

Cui F, Sui N, Duan G, Liu Y, Han Y, Liu S, Wan S, Li G (2018) Identification of metabolites and transcripts involved in salt stress and recovery in peanut. Front Plant Sci 9:217

Du H, Wang Z, Yu W, Liu Y, Huang B (2011) Differential metabolic responses of perennial grass Cynodon transvaalensis × Cynodon dactylon (C4) and Poa Pratensis (C3) to heat stress. Physiol Plant 141(3):251–264

Erickson PI, Ketring DL (1985) Evaluation of peanut genotypes for resistance to water stress in situ 1. Crop Sci 25(5):870–876

Gowda A, Hegde BR (1986) Moisture stress and hormonal influence on the flowering behavior and yield of groundnut. Plant Physiol 66:835–837

Greer DH, Weedon MM (2012) Modelling photosynthetic responses to temperature of grapevine (*Vitis vinifera* cv. Semillon) leaves on vines grown in a hot climate. Plant Cell Environ 35 (6):1050–1064

Groundnut Outlook (2018) Groundnut Outlook, PJTSAU. pp 1–3

Hamidou F, Halilou O, Vadez V (2013) Assessment of groundnut under combined heat and drought stress. J Agron Crop Sci 199(1):1–11

Hoang XLT, Thu NBA, Thao NP, Tran LSP (2014) Transcription factors in abiotic stress responses: their potentials in crop improvement. In: Improvement of crops in the era of climatic changes. Springer, New York, NY, pp 337–366

Hoegh-Guldberg O, Jacob D, Bindi M, Brown S, Camilloni I, Diedhiou A et al (2018) Impacts of 1.5 C global warming on natural and human systems. Global warming of 1.5 C. An IPCC special report

Janská A, Maršík P, Zelenková S, Ovesná J (2010) Cold stress and acclimation–what is important for metabolic adjustment? Plant Biol 12(3):395–405

Kalarani MK, Maheswari P, Senthil A, Umapathi M (2018) Influence of pre flowering drought on physiological parameters and yield in groundnut. Madras Agric J 105:378–380

Kambiranda DM, Vasanthaiah HKN, Katam R, Ananga A, Basha SM, Naik K (2011) Impact of drought stress on peanut (*Arachis hypogaea* L.) productivity and food safety. Plants Environ:249–272

Ketring DL (1984) Temperature effects on vegetative and reproductive development of peanut. Crop Sci 24(5):877–882

Mafakheri A, Siosemardeh AF, Bahramnejad B, Struik PC, Sohrabi Y (2010) Effect of drought stress on yield, proline and chlorophyll contents in three chickpea cultivars. Aust J Crop Sci 4(8):580–585

Maheswari P, Kalarani MK, Senthil A, Sowmiyapriya S (2016) Physiological and yield responses of groundnut (*Arachis hypogaea* L.) genotypes to drought. Madras Agric J 103:309–313

Maheswari P, Kalarani MK, Senthil A, Umapathi M (2018) Effect of pre flowering drought on flowering behaviour of yield in groundnut. Madras Agric J 105:368–369

Mirza H, Hossain MA, Fujita M (2010) Physiological and biochemical mechanisms of nitric oxide induced abiotic stress tolerance in plants. Am J Plant Physiol 5(6):295–324

Narayanan S, Mohan A, Gill KS, Prasad PVV (2014) Variability of root traits in spring wheat germplasm. PLoS One 9(6):e100317

Noctor G, Veljovic-Jovanovic S, Driscoll S, Novitskaya L, Foyer CH (2002) Drought and oxidative load in the leaves of C3 plants: a predominant role for photorespiration? Ann Bot 89(7):841–850

Osakabe Y, Osakabe K, Shinozaki K, Tran L-SP (2014) Response of plants to water stress. Front Plant Sci 5:86

Pereira J, MeloFilho P, Albuquerque M (2012) Mudanças bioquímicasemgenótipos de amendoimsubmetidos a déficithídricomoderado. Rev Cienc Agron 43:766–773

Prasad PVV, Craufurd PQ, Summerfield RJ (1999) Fruit number in relation to pollen production and viability in groundnut exposed to short episodes of heat stress. Ann Bot 84(3):381–386

Reddy TY, Reddy VR, Anbumozhi V (2003) Physiological responses of groundnut (*Arachis hypogea* L.) to drought stress and its amelioration: a critical review. Plant Growth Regul 41 (1):75–88

Sheshshayee MS, Bindumadhava H, Rachaputi NR, Prasad TG, Udayakumar M, Wright GC, Nigam SN (2006) Leaf chlorophyll concentration relates to transpiration efficiency in peanut. Ann Appl Biol 148(1):7–15

Sinclair TR (1994) Limits to crop yield? In: Physiology and determination of crop yield, pp 509–532

Songsri P, Jogloy S, Vorasoot N, Akkasaeng C, Patanothai A, Holbrook CC (2008) Root distribution of drought-resistant peanut genotypes in response to drought. J Agron Crop Sci 194(2): 92–103

Suthar DM, Patel MS (1992) Yield and nutrient absorption by groundnut and iron availability in soil as influenced by lime and soil water. J Indian Soc Soil Sci 40(3):594–596

Zafar SA, Noor MA, Waqas MA, Wang X, Shaheen T, Raza M, Rahman MU (2018) Temperature extremes in cotton production and mitigation strategies. Past, present and future trends in cotton breeding. IntechOpen, London, pp 65–91

Molecular Mechanisms of Nutrient Deficiency Stress Tolerance in Legumes

12

Sandeep Sharma, Neha Anand, Krishnapriya Vengavasi, and Renu Pandey

Abstract

Legumes, which are an essential source of plant proteins and dietary fibre, are the most valued diet for humans after cereals. Globally, legume is commonly grown in the arid and semi-arid tropics. Legumes play an important role in the effective management of fertilisers and improve soil fertility, thereby sustaining agriculture. Improved nutrient absorption, translocation, and cellular homeostasis are essential for optimum plant growth and development. Legumes have evolved strategies to adapt to nutritional deprivation at both physiological and molecular levels. High-throughput sequencing as well as other recent advancements in molecular biology techniques have allowed researchers to investigate the molecular basis of nutrient deficiency tolerance in legume crops. In this chapter, we attempt to present various physiological and molecular mechanisms, specific to legumes wherever available, assisting in adaptation to nutrient-deficient conditions. However, increased efforts are needed on food and feed legumes in the area of mineral nutrition covering physiology and molecular aspects.

Keywords

Legume · Mineral nutrients · Transporters · Nutrient stress · miRNA

S. Sharma · N. Anand · R. Pandey (✉)
Mineral Nutrition Lab, Division of Plant Physiology, ICAR-Indian Agricultural Research Institute, New Delhi, India
e-mail: renu_pphy@iari.res.in

K. Vengavasi
Plant Physiology Section, Division of Crop Production, ICAR-Sugarcane Breeding Institute, Coimbatore, Tamil Nadu, India

P. Muthu Arjuna Samy et al. (eds.), *Legumes: Physiology and Molecular Biology of Abiotic Stress Tolerance*, https://doi.org/10.1007/978-981-19-5817-5_12

12.1 Introduction

Legumes, belonging to the Fabaceae family, consist of more than 700 genera, including 20,000 species that comprise the second major group of food and fodder crops cultivated globally (Iantcheva et al. 2013). Around 250 Mt of grain and legumes are produced annually accounting for 27% of the global primary food output and 33% of human dietary protein requirements (Hussain et al. 2018). Global malnutrition is a serious threat to nutritional security that leads to a high mortality rate due to the emerging non-communicable diseases (Roorkiwal et al. 2021). Legumes are considered an economical source of nutrition with a high percentage of protein (20–25%) and fibre (8–27%) and a low glycaemic index (Sánchez-Chino et al. 2015). A cup of cooked dried legume contains 6–8 g of fibre and 14–16 g of protein. The majority of legume grains are storage protein, which consists primarily of globulin (70%), albumin (10–20%), and glutelins (10–20%) (Sharif et al. 2018). Protein quality is determined by its amino acid composition, and a protein containing all the essential amino acids (EAA) is called a 'complete protein'. Most of the proteins in legumes are deficient in EAA and considered 'incomplete proteins', whereas proteins from eggs, meat, and milk products are categorised as 'complete proteins'. Usually, legumes contain low fat (<5%) except for soybean (*Glycine max*), lupin (*Lupinus albus*), and chickpea (*Cicer arietinum*) (15–47%). Besides, legumes also contain substantial amounts of nutritionally important minerals as well as vitamins (B1, B2, B3, B6, and B9) (Rebello et al. 2013; Roorkiwal et al. 2021).

Legumes constitute a major part of sustainable agriculture as it improves soil fertility through symbiotic association with beneficial rhizobia and mycorrhizal fungi (Abdelrahman et al. 2018). The interaction of plant roots with soil and water influences nutrient availability in soil and their uptake, leading to a significant role in the growth and productivity of plants. Plants require 17 nutrients for completing their life cycle, which is grouped as macro- and micronutrients based on the quantity required by plants. The macronutrients include carbon (C), hydrogen (H), oxygen (O), nitrogen (N), phosphorus (P), potassium (K), calcium (Ca), magnesium (Mg), and sulphur (S). Out of these, C, H, and O constitute 90–95% of the total biomass and are available to plants from carbon dioxide and water. Other macronutrients which make up 0.2–4.0% of plant dry weight are divided into two categories: primary (N, P, K) and secondary (Ca, Mg, S). Micronutrients, although required in very less amounts, constitute only 0.002% of the total plant dry weight, but they are indispensable for plant growth. Micronutrients are divided into two groups: positively charged (iron (Fe), copper (Cu), manganese (Mn), zinc (Zn), and nickel (Ni)) and negatively charged (boron (B), chlorine (Cl), and molybdenum (Mo)) (Singh et al. 2013).

The root of legume crops forms two types of symbiotic association with soil microorganisms: rhizobial symbiosis, responsible for atmospheric N fixation, and arbuscular mycorrhizal (AM) symbiosis, which enhances plant P uptake (Püschel et al. 2017). The AM fungi colonise roots, and its hyphae spread over the surrounding soil, forming enormous mycelium networks, which enhance P and Zn uptake by improving root-soil interaction (Kiers et al. 2011; Püschel et al. 2017).

This chapter deals with the physiological and biochemical adaptation strategies with a focus on molecular mechanisms that allow legumes to tolerate nutritional deprivation.

12.2 Physiological Tolerance Mechanisms to Nutrient Deficiency in Legumes

The atmospheric N_2-fixing ability not only benefits the legume crop but is also useful for succeeding crops or main crops with the former as an intercrop. However, several factors influence the process of symbiotic N_2 fixation, including crop growth stage, soil water status, soil temperature, N concentration in the rhizosphere, and presence of other nutrients in the soil (Garg and Geetanjali 2006). In legumes, N deficiency is less common, and the mechanism of symbiotic N_2 fixation is another vast topic and therefore not covered in this chapter. The biological role and the physiological tolerance mechanisms specific to legumes for different nutrients' stress are summarised in Table 12.1. Under low-P conditions, legumes adopt many physiological strategies for mitigating P starvation by adjusting their external and internal P demand. The first strategy involves an improved root-soil interaction by increasing root surface area through alteration in root architecture like an increased number of secondary roots with more root hairs and nodules (Lazali and Bargaz 2017; Meena et al. 2021; Ramtekey et al. 2021; Reddy et al. 2020; Richardson et al. 2011). In addition to altered root morphology, other changes include rhizosphere acidification, root exudation of low-molecular-weight organic acids and acid phosphatase, and symbiotic association with microorganisms including fungi and bacteria (Meena et al. 2021; Singh and Pandey 2003; Smith and Read 2010; Vengavasi et al. 2016; Vengavasi and Pandey 2018). K plays a vital role in CO_2 assimilation, and under its deficiency, the rate of photosynthesis drastically reduces due to a reduction in the leaf size, leaf number, leaf sunlight interception, stomatal conductance, increased mesophyll resistance, and reduced Rubisco (ribulose 1,5-*bis*phosphate carboxylase/oxygenase) activity in plants (Liu et al. 2008; Pettigrew 2008; Zhao et al. 2001). The physiological influence of S starvation is a reduction in root hydraulic conductivity, which is the first response that signals nutrient hunger through root to shoot (Hawkesford and De Kok 2006). Availability of S in soil determines the relative status of reduced sulphate pools as a means of mobilising S within the plant. If vacuolar sulphate pools are small due to S starvation, the involvement of reduced S compounds translocated through phloem increases dramatically and plays an important role in delivering sulphur to sink tissues like developing seed (Hawkesford and De Kok 2006). The most common adaptation of plants under Mg starvation is starch accumulation in chloroplast at the source leaves. The de-chelating of Mg^{2+} ion from chlorophyll molecules during chlorophyll catabolism is a defence strategy of plants experiencing Mg starvation. Mg is relatively a phloem-mobile element; thus, the regenerated Mg is transported in favour of growth of young tissues (Ceppi et al. 2012; Yang et al. 2012). In legumes, during the early stage of infection in nodule development process, the rhizobia invade plants through a transcellular tunnel and

Table 12.1 A summary of biological roles and physiological tolerance mechanisms developed by legumes in response to various nutritional stresses

Nutrient	Biological roles in plants	Physiological tolerance mechanisms	References
Phosphorus	Constituents of ATP, phospholipids, and nucleic acids, important for root growth and nodule development	Improved root-soil interaction by improved root surface area; exudation of low-molecular-weight organic acids and acid phosphatase	Lazali and Bargaz (2017), Meena et al. (2021)
Potassium	Osmoregulator and involved in ROS detoxification	Increased mesophyll resistance; lowered the Rubisco activity in leaf; altered root gravitropic behaviour	Pettigrew (2008)
Sulphur	Major constituent of cysteine (C) and methionine (M) and vitamins like biotin and thiamine, promotes nodule formation in legumes	Reduction in root hydraulic conductivity; reduces translocation of S towards seeds	Afzal et al. (2015)
Magnesium	Central atom of chlorophyll molecule, involved in protein synthesis, N uptake, and assimilation	Starch accumulation in chloroplast; increased mobilization of photosynthates towards root for nodule development	Peng et al. (2018, 2020), Yang et al. (2012)
Calcium	Secondary messenger; involved in cell division and cell wall strengthening	Reduces the passive flow of monovalent ions, which decreases membrane fluidity	De Freitas et al. (2016)
Iron	Cofactor, structural constituent of many antioxidative enzymes, involved in lipid peroxidation	Improved root growth, root-tip swelling, increased ferric reductase activity in the root, release of phyto-siderophore	Hindt and Guerinot (2012), Sharma et al. (2019)
Zinc	Regulates activities of all six classes of enzymes, involved in transcriptional control of the Ros-type regulator MucR in legumes	Increased length and number of root hairs; release of phyto-siderophore	Lurthy et al. (2020)
Manganese	Acts as a cofactor, component of antioxidant enzyme, oxygen-evolving complex of photosystem II	Lignin concentration decreased in the root	Socha and Guerinot (2014)

embed in the plant matrix glycoprotein (MGP), secreted by host plants. Ca with B plays an essential role in these stages to modulate plant-rhizobia interaction at the cell surface. The degree of attachment and cell invasion by *Rhizobium* in the root is regulated by both Ca and B nutrition, so the deficiency of both elements reduces the induction capability of *nod* genes (Redondo-Nieto et al. 2003).

Plants respond to Fe deficiency by exhibiting morphological changes, including increased root surface area, enhanced root hair development and branching, root-tip swelling, and increased lateral root formation for Fe reduction and uptake (Hindt and Guerinot 2012; Muller and Schmidt 2004). Previous studies on legumes showed that increased Fe^{3+} reductase activity in the root of soybean (*Glycine max*) and lotus (*Lotus japonicus*) provided higher tolerance to cope with Fe deficiency (Klein et al. 2012; Li et al. 2011). To manage Zn deficiency, a tightly regulated network of coordinated expression of Zn transporters for acquisition from the soil, translocation between tissues, and intracellular sequestration has been evolved in plants (Kabir et al. 2017). The cellular utilisation of Zn is considered as a potential Zn efficiency mechanism (Rengel and Graham 1995). Importantly, the activity of carbonic anhydrase, a metallo-enzyme that catalyses the conversion of CO_2 to HCO_3^-, is associated with cellular Zn concentration. In black gram (*Vigna mungo*), the activity of carbonic anhydrase and Cu/ZnSOD enzymes significantly correlated with Zn supply, which can be used as a marker for Zn deficiency (Pandey et al. 2002). Under Mn deficiency, Mg replaces Mn, which could have a detrimental effect on the cellular process such as lignin synthesis that involves Mg. The lignin concentration was found to decrease significantly in root tissue under Mn deficiency because Mn is a cofactor of phenylalanine ammonia lyase (PAL) enzyme, which is involved in the phenylpropanoid pathway to produce monolignols (Socha and Guerinot 2014).

12.3 Molecular Basis of Nutrient Uptake Under Starvation Conditions

All efforts have been made to present the information available up to date with particular reference to legumes wherever available for each nutrient element in the subsequent paragraphs. The transporters characterised for each nutrient element and their regulation are presented briefly in Table 12.2.

12.3.1 Phosphorus

12.3.1.1 Uptake and Transport

Plant roots absorb inorganic P (Pi) from the soil as $H_2PO_4^-$ or HPO_4^{2-} ions depending on soil pH; however, these ionic forms are present in the soil solution at very low concentrations, usually at micromolar (<10 μM) levels (Hinsinger 2001). Phosphate transporters (PTs) are localised in the plasma membrane of root cells and play a major role in the acquisition of soluble Pi from the soil solution against the concentration gradient. Plants possess two nutrient transport systems: (1) high-affinity transport system (HATS), regulated by P concentration in the media, and (2) low-affinity transport system (LATS) which is constitutively expressed. The plants' internal Pi status adjusts their P uptake, especially by raising I_{max} (maximum influx), while changes in K_m are insignificant in this process (Muchhal and Raghothama 1999; Pandey et al. 2018). According to the protein sequence, location, and structure, plants have a wide variety of Pi transporter families such as *Pht1*, *Pht2*, *Pht3*, *Pht4*, and *Pht5* (Guo et al. 2008; Liu et al.

Table 12.2 List of transporter and regulatory genes involved in the uptake and homeostasis of different nutrient elements in legume crops

Nutrients	Crop	Transporter/ regulatory gene	Description	References
Phosphorus	*Phaseolus vulgaris*	*PvPHR1*, *PvmiR399*	Positive regulator of genes implicated in P transport, remobilization, and homeostasis	Valdés-López et al. (2008)
	Medicago truncatula	*MtPT1*, *MtPT2*, *MtPT3*	Low-affinity P uptake	Cao et al. (2021), Liu et al. (2008)
		MtPT5	High-affinity P uptake	Liu et al. (2008)
	Glycine max	*GmPT5*	High-affinity P uptake and homeostasis	Qin et al. (2012)
	Cicer arietinum	Putative *CaPHO1*, *CaPHO2*, *CaPHT1;4*, *CaPAP17*, *CaPPase4*, CaDGD1	P uptake, transport, and mobilization from roots and leaves to nodules	Esfahani et al. (2016)
Potassium	*Lotus japonicus*	*LjKUP*	K transport across plasma membrane	Desbrosses et al. (2004)
	Glycine max	*GmKEA2* to *6*	Cation/proton antiporter involved in K accumulations	Chen et al. (2015)
	Cicer arietinum	K^+ efflux antiporter (KEA)	Accumulation of K	Azeem et al. (2018)
Calcium	*Medicago truncatula*	*MCA8*	Involved in calcium signalling during symbiotic contacts	Capoen et al. (2011)
	Medicago, *Lupinus luteus*, *Vicia faba*	Ca^{2+}/ATPases	Ca absorption into symbiosomes	Andreev et al. (1997, 1998), Benedito et al. (2010), Kataoka et al. (2004)
Sulphur	*Lotus japonicus*	Homolog of *AtSultr3.5*	Essential for S supply to the bacteroides	Kataoka et al. (2004)
	Medicago truncatula	*MtSULTR*	High-affinity sulphate transporter	Casieri et al. (2012)
	Glycine max	*GmSULTR1;2b*	High-affinity sulphate transporter	Ding et al. (2016)
Iron	*Phaseolus vulgaris*	Phvul.005G130500/ FIT1-like, Phvul.002G099700/ IRT1-like	Fe uptake	Castro-Guerrero et al. (2016)
		Phvul.003G086500/ OPT3-like	Fe signalling	Castro-Guerrero et al. (2016)

(continued)

Table 12.2 (continued)

Nutrients	Crop	Transporter/ regulatory gene	Description	References
	Melilotus japonicus	*MtIRT* and *MtFRD3*	Fe uptake and transport	Li et al. (2014)
	Medicago truncatula	*MtNRAMP1*	Fe uptake and transport, expressed in roots and nodules	Tejada-Jiménez et al. (2015)
	Glycine max, Medicago	*DMT1* (divalent metal transporter1)	Ferrous transporter in symbiosome membrane	Benedito et al. (2010), Kaiser et al. (2003)
	Glycine max	Glyma03g28610, Glyma03g28630	Fe acquisition	Peiffer et al. (2012)
Zinc	*Glycine max*	*GmZIP1*	Zn uptake and transport	Moreau et al. (2002)
	Phaseolus vulgaris	*PvZIP12, PvZIP13, PvZIP16, PvbZIP1*	Zn uptake and transport	Astudillo et al. (2013)
	Medicago truncatula	*MtZIP1, MtZIP3, MtZIP4, MtZIP5, MtZIP6, MtZIP7*	Zn uptake and transport	Lopez-Millan et al. (2004)
	Arachis hypogaea	*AhNRAMP1*	Zn, Fe, and Mn transport	Wang et al. (2019), Xiong et al. (2012)
Manganese	*Medicago truncatula*	*MtZIP4, MtZIP7*	Mn uptake and transport	Socha and Guerinot (2014)
	Pisum sativum	*PsIRT1*	Mn uptake and transport	Socha and Guerinot (2014)
Molybdenum	*Medicago truncatula*	*MtMOT1.3*	Mo transport to nodule cells	Tejada-Jiménez et al. (2015)
	Lotus japonicus	*LjMOT1*	Mo uptake and translocation to shoots	Gao et al. (2016)

2011; Qin et al. 2012; Raghothama 1999; Rausch and Bucher 2002; Schachtman et al. 1998). The *Pht1* family belonging to HATS is responsible for P absorption from rhizospheres and its transport to the xylem (Gu et al. 2016), while the families of *Pht2*, *Pht3*, *Pht4*, and *Pht5* are organelle transporters responsible for the transport of P across the plastid (*Pht2/4*), mitochondrial (*Pht3*), Golgi membrane (*Pht4*), and vacuole (*Pht5*) (Huang et al. 2019; Liu et al. 2016). The *Pht1* family has received utmost attention among the Pi transporter families, and the members of *Pht1* were identified and functionally validated from a wide range of plant species including *Arabidopsis*, tomato (*Lycopersicum esculentum*), rice (*Oryza sativa*), maize (*Zea mays*), soybean, *Medicago truncatula*, and lotus (Bulgarelli et al. 2020; Liu et al. 2008, 2011; Maeda et al. 2006; Nagy et al. 2006; Paszkowski et al. 2002). All the members of *Pht1* family are $H_2PO_4^-/H^+$ symporters with a similar structure containing 12 membrane-spanning domains with hydrophilic N- and C-terminals. A putative glycosylation site is present in transmembrane domain 10, while a hydrophilic loop is located between transmembrane domains six and seven (Karandashov and Bucher 2005; Smith and Read 2010).

Major transcripts of high-affinity transporter are strongly induced by P starvation and are preferentially expressed in the epidermal cells of root hairs and cortical cells, while a few are expressed in various aerial parts like stems, leaves, flowers, and grains (Ai et al. 2009; Qin et al. 2012). In soybean, 14 members of *Pht1* family, namely *GmPht1;1–14*, as well as one pseudogene (Glyma13g18420) have been identified. *GmPht1* transporters are distributed unevenly on soybean chromosomes ($2n = 20$); however, these transporters are located only on 8 chromosomes out of 20. Among 14 *GmPht1* transporters, maximum four (*GmPht1;4* to *GmPht1;7*) are located on chromosome 10, three (*GmPht1;12* to *GmPht1;14*) are on chromosome 20, two (*GmPht1;9* and *GmPht1;10*) on chromosome 14, while *GmPht1;1, 2, 3, 8,* and *11* are located on chromosomes 2, 3, 7, 13, and 19, respectively (Qin et al. 2012). Except for *GmPht1;8*, which is located in the endoplasmic reticulum, all other *GmPht1* transporters are located in the plasma membrane (Fan et al. 2013). Similar to other Pi transporters, *GmPht1* transporters were significantly upregulated by P deficiency, with the exception of *GmPht1;10*. Among *GmPht1* transporters, seven, including *GmPht1;1, 2, 3, 4, 7, 8,* and *12*, are expressed only in root tissues. *GmPht1;9* and *GmPht1;13* were strongly induced in roots and stems as well as in immature leaves and roots, while flowers and stems were the primary sites for the expression of *GmPht1;5* and *GmPht1;14* (Gu et al. 2016; Qin et al. 2012). The β-glucuronidase staining of transgenic soybean roots showed expression of *GmPht1;5* predominantly in the junction region of roots and young nodules as well as in nodule vascular bundles, suggesting its function in Pi transport from root vascular system into nodules. In *M. truncatula*, four *Pht1* members, *MtPT1*, *MtPT2*, *MtPT3*, and *MtPT5*, were identified which showed significant expression in root tissue under P starvation (Cao et al. 2021; Liu et al. 1998, 2008).

12.3.1.2 Regulation of Pi Transporters

The Pi trafficking across the plasma membrane is coordinated among different cellular organelles and regulated by cytosolic Pi homeostasis (Pratt et al. 2009). Under P deficiency, the expression of genes involved in C metabolism (glyceraldehyde 3-phosphate dehydrogenase), N assimilation (glutamine synthetase and glutamate synthase), phospholipid biosynthesis (phosphoethanolamine *N*-methyl transferase), photosynthesis, and mitochondrial electron transport (ferredoxin NADPH reductase) is suppressed in response to cytosolic P and maintains cytosolic Pi homeostasis (Misson et al. 2005; Valdés-López et al. 2008). Proteins containing SPX domain at the N-terminal have been linked to Pi sensing and transport. The SPX-domain proteins (SPX1 and SPX2) function as intracellular Pi sensors and, when bound to PHR1 (PHOSPHORUS STARVATION RESPONSE1), suppress P starvation response under P-depleted condition (for details, see Wang et al. 2021). Inositol polyphosphate (InsP) is an intracellular P signalling molecule that binds with the SPX domain affecting the PHR1-SPX1 interaction. Inactivating the redundant genes, *VIH1* (*VIP1 HOMOLOG1*) and *VIH2*, which encode *PPIP5K* (diphosphoinositol pentakisphosphate kinase), limits InsP8 production and induces the expression of *PHT1* genes causing excessive Pi accumulation (Yan Wang et al. 2021; Zhu et al. 2019). Several genes with consensus *cis*-acting DNA sequences such as W-box, G(E)-box, TATA-box, P1BS (PHR1-binding sequence), MBS (MYB-binding site), helix-loop-helix, and PHO have been associated to the responsiveness of Pi transporters and other P starvation-responsive genes (details in Gu et al. 2016; He et al. 2019). The expression of most of the Pi transporter genes was induced by P starvation, while some of them are controlled by P starvation response transcriptional factors (TFs) such as MYB-coiled coil (MYB-CC), WRKY, and C2H2-type zinc finger protein. The transcription factor belonging to MYB-CC family regulates the transcription of P starvation-induced (PSI) genes by binding to their proximal promoter regions with the imperfect palindromic sequences (GNATATNC) (Baek et al. 2017; Bustos et al. 2010; Gu et al. 2016; Guo et al. 2015). The members of WRKY (WRKY6, WRKY42, WRKY45, and WRKY75) and C2H2 (ZAT6) families are involved in Pi starvation signalling in *Arabidopsis*, bean (*Phaseolus vulgaris*), soybean, *Medicago*, and lupin (Devaiah et al. 2007a, b; Graham et al. 2006). These TFs are localised in the nucleus and overexpressed under P starvation to regulate root architectural modifications. The WRKY75 recognises W-box ((T)TGAC(C/T)), DNA *cis*-regulatory elements, and a region of genes involved in P homeostasis and remobilisation, while ZAT6 regulates the expression of several genes of WRKY75 pathway (Devaiah et al. 2007a, b; Su et al. 2015; Valdés-López et al. 2008). Recently, two new transcriptional factors, namely, OsbHLH6 (He et al. 2021) and RLI1/HINGE1 (Zhang et al. 2021), were identified in rice, which regulates the expression of PHT1 family genes.

The regulation of Pi transporter genes at post-transcriptional level has been reported in plants. Small regulatory RNAs, microRNAs (miRNA), and small interfering RNAs (siRNAs) are considered the most ubiquitous molecules that regulate post-transcriptional gene expression (Bartel 2004). The expression profiles of various miRNAs in legumes under P starvation have been reported earlier. In lupin and soybean, 167 and 57 miRNAs, respectively, showed significant alteration in their expression (Zeng et al. 2010; Zhu et al. 2010). The role of miR399 during P deficiency is well characterised in plants; however, P deprivation alters the expression of some other miRNAs such as miR827, miR2111, miR778, miR169, and miR395 (Franco-Zorrilla et al. 2007; Fujii et al. 2005; Hsieh et al. 2009; Pant et al. 2008). In *Arabidopsis*, miR399 binds to the five complementary bases of the PHOSPHATE OVER ACCUMULATOR2 (PHO2) transcripts and inhibits internal Pi mobilisation from older to new leaves (Chiou et al. 2006; Fujii et al. 2005). miR399 also influences the PSI signalling in the roots of *Phaseolus vulgaris* and phloem sap of *Brassica napus* and *Cucurbita moschata* (Pant et al. 2008; Ramírez et al. 2013; Valdés-López et al. 2008). miR211 accumulates in the phloem sap only under low-P conditions, targets the F-box protein in soybean and *Arabidopsis*, and regulates the protein abundance under P starvation (Hsieh et al. 2009; Xu et al. 2013). Besides miRNAs, long non-coding RNAs (lncRNAs) are also expressed in response to P starvation, which plays a significant role in the regulation of P uptake. The well-studied *IPS1* (induced by P starvation1) acts as a ribo-regulator rather than the target of miR399 and functions as an endogenous target mimic (eTM) of *PHO2* in *Arabidopsis* (for details, see Franco-Zorrilla et al. 2007). The ribo-regulators *At4* and *Mt4* were induced by *IPS1* in *Arabidopsis* and *Medicago*, respectively. Further, in *Medicago*, three PHOSPHORUS DEFICIENCY INDUCED lncRNAs (*PDILs*) were characterised under P starvation, out of which *PDIL1* suppresses the degradation of *MtPHO2* transcripts (Wang et al. 2017). Borah et al. (2018) identified putative lncRNAs for nitrogen and P starvation in soybean and *Arabidopsis*, respectively, which can act as eTMs. They showed computationally that miR827 (P starvation induced) and miR169 (N starvation induced) could be sponged by two and three eTMs, respectively, thereby regulating nutrient uptake through the regulatory module of 'eTM-miRNA-mRNA'.

12.3.1.3 Regulation of Pi Transporters by Arbuscular Mycorrhizal Fungi

Legumes establish root symbiosis not only with rhizobia but also with AM fungi, which significantly influences the expression of Pi transporters. Only *Pht1*, high-affinity H^+/Pi symporters, have been identified which are involved in mycorrhizal driven P acquisition among different Pi transporter families. The mycorrhiza-specific *Pht1* transporters are grouped into two subgroups, namely, subfamilies I and III. During AM symbiosis, most of the subfamily I transporters are expressed only in the arbuscule-containing cortical cells, while subfamily III *Pht1* genes are expressed in plant roots but specifically induced in cortical cells (Bucher 2007; Harrison et al. 2002; Javot et al. 2007). The upregulation of AM-inducible Pi transporter generally suppresses the expression of other Pi transporters, specifically those involved in direct P uptake from the rhizosphere. This interaction between *Pht1* transporters could indicate the association between mycorrhizal and direct Pi uptake routes. However, it is still unclear whether the downregulation of other Pi transporters is caused by a direct plant response to symbiosis or is caused by an enhancement in Pi acquisition (Garcia-Brugger et al. 2006; Paszkowski et al. 2002). The AM symbiosis-inducible *PHT1* subfamily I transporters were identified in a few plant species such as *M. truncatula* (*MtPT4*), rice (*OsPT11*), and *Astragalus sinicus* (*AsPT4*) (Breuillin-Sessoms et al. 2015; Xie et al. 2013; Yang et al. 2012). Generally, P starvation induced the expression of most of the *Pht1* family transporters in soybean, but AM symbiosis suppressed the expression of *GmPht1;6*, *7*, and *10* in root tissues, while the expression of *GmPht1;1*, *7*, and *11* was significantly induced (Bulgarelli et al. 2020; Tamura et al. 2012).

12.3.2 Potassium

12.3.2.1 K Uptake and Transport

Plant roots acquire potassium ion (K^+) from soil solution, which is derived from several sources such as potassium chloride (KCl), potassium nitrate (KNO_3), potassium carbonate (K_2CO_3), and potassium sulphate (K_2SO_4) present in soil or applied as chemical fertilisers. A wide variety of K transporters and channels are involved in the uptake of K by roots and its mobilisation throughout the plant. The transporter proteins have a high affinity for K^+ and are active at low K concentrations, whereas the channels have a low affinity for K and are active only at high K concentrations (>300 μM external K) (Wang and Wu 2013). The K transporters are grouped into five different classes: (1) shakers/voltage-gated channels, (2) non-voltage-gated channels/tandem pores, (3) HAK/KT/KUP high-affinity transporter family, (4) HAT high-affinity family, and (5) KEA family of antiporter (Gomez-Porras et al. 2012; He et al. 2012; Rehman et al. 2017).

The members of the shaker family, which controls membrane conductance in most plant cell types, are further classified into three groups: *inwards-rectifying* (K_{in}), activated by membrane hyperpolarisation and mediates K uptake; *outward-rectifying* (K_{out}), activated by membrane depolarisation and facilitates K efflux; and *weakly-inward rectifying* (K_{weak}) that mediates K efflux and influx based on the electrochemical gradient due to K^+ (Shabala and Pottosin 2010; Véry et al. 2014; Yi Wang and Wu 2013). All the voltage-gated K^+ channels contained a conserved amino acid motif (TVGYGD) and were widely expressed in plant tissues, allowing a fast K distribution across various parts of the plant and cellular compartments (Kuang et al. 2015; Rehman et al. 2017). In soybean, 16 genes encode voltage-gated K^+ channels, and all of them have a highly conserved gene structure with varying lengths, 57, 98, and 185 bp exons (Rehman et al. 2017). According to Damiani et al. (2016), a candidate gene implicated in membrane repolarisation, movements of stomata, and K^+ extrusion into the xylem sap of *M. truncatula* belongs to this family. The non-voltage-gated K^+ channels, also known as tandem pore channels (TPKs), contained two pore loops per subunit and four transmembrane domains (TM domain). There are six members of the non-voltage-gated K^+ channel family, including a single subunit channel and five tandem pore channels. With the exception of TPK3 and TPK4, voltage-gated channels are located at the plasma membrane in plants, whereas non-voltage-gated channels are located on the endomembrane of several organelles (Pandey and Mahiwal 2020).

The HAK/KT/KUP family plays a critical role in K acquisition from soil and is assumed to function as H^+/K^+ symporters (Véry et al. 2014). The HAK/KT/KUP transporter families have a wide range of subcellular localisation, including the plasma membrane, tonoplast, and another endomembrane, while its transcript is expressed in diverse plant tissues such as guard cells, vascular tissues, root meristems, and fruits (Scherzer et al. 2015). A large number of HAK/KUP/KT genes have been found in different plant species such as 17 in *Vitis vinifera*, 13 in *Arabidopsis*, 20 in *Medicago*, and 29 in soybean and poplar (*Populus alba*) (Davies et al. 2006; Nieves-Cordones et al. 2016; Rehman et al. 2017). Among legumes, *LjKUP* was the first KUP family high-affinity K transporter and was identified in *L. japonicus* with maximum expression in nodules under K stress (Desbrosses et al. 2004). The transcriptomic profiling of soybean showed that 22 HAK/KUP/KT genes were differentially expressed during nodulation, wherein *GmHAK5*, *GmKUP8*, and *GmKUP8* recorded higher expression in root hairs during nodulation (Clarke et al. 2014; Rehman et al. 2017).

The HKT family belonging to the high-affinity K transporters has been widely studied after the cloning of *TaHKT2;1* from *Triticum aestivum*, the first member of HKT gene family (Schachtman et al. 1992). Based on the presence of Gly or Ser residue in P loop, the selectivity pore-forming area, the members of this family were categorised into two subfamilies: subfamily I has a Ser residue (SGGG type) in the P-loop region that is thought to be linked to the specialised Na^+ transport. The subfamily II has only Gly residues (GGGG-type) in the P loop, which mediate the transport of both K^+ and Na^+ (Horie et al. 2009; Huang et al. 2020; Platten et al. 2006). These transporters are still very poorly characterised in legumes; *GmHKT1* and *GmHKT1;4* are two soybean genes that have been identified and functionally validated as participating in salt tolerance (Chen et al. 2014). Only 4 out of 70 potential K^+ transporters identified in soybean belong to the HKT family (Rehman et al. 2017).

The KEA (K^+ efflux antiporter) belongs to the cation/proton antiporter family-2 (CPA2 family) and is responsible for the active accumulation of K in plants. The first KEA was identified in gram-negative bacteria involved in a mechanism for cytosol acidification as a defence against harmful electrophiles (Munro et al. 1991). In plants, KEAs are located in tonoplast, plasma membrane, and membranes of mitochondria and chloroplast (Sze et al. 2004; Walker et al. 1996). Till date, six KEA genes have been identified in *Arabidopsis* genome (*AtKEA1* to *6*). The mutation in *KEA1* and *KEA2* gene in *Arabidopsis* showed that they have diverse effects on leaf development and photosynthetic rate (Dana et al. 2016). Chen et al. (2015) identified 12 members of a novel KEA gene family in soybean, which was divided into five subgroups based on their similarity with the *Arabidopsis* KEA gene family as *GmKEA2* to *6*, whereas the *KEA1*-type gene was not found in the entire genome of soybean. Recently, 23 K channels and transporter genes were identified by genome-wide analysis in chickpea (Azeem et al. 2018). Among 23 genes, only 6 belonged to KEA family, while 2 and 15 genes belonged to HKT and KUP/HAK/KT family, respectively.

12.3.2.2 Regulation of K Transporters

In most of the plant species, transcriptional regulation of K transporter is a ubiquitous mechanism to cope with K starvation conditions (Wang and Wu 2013). When high concentration of K is available in soil solution, most channels are employed to transport K through the membrane along with the concentration gradient, while under K starvation conditions, an active or energy-driven transport system is required to pull K inside the cell (Ragel de la Torre 2019; Rubio et al. 2010). In a few higher plants, the activity of K transporter and channels is regulated by external NH_4^+ concentration. K absorption is competitively reduced by NH_4^+ uptake via these K transporters and channels at high NH_4^+ concentrations (Wang and Wu 2013). The sensitivity to NH_4^+ is a key feature of carrier protein-mediated K^+ uptake. Several NH_4^+-sensitive or -insensitive high-affinity K uptake systems have been found in plants such as *Arabidopsis* (Nieves-Cordones et al. 2007), rice (Chen et al. 2015), and barley (Santa-Marıa et al. 2000). The $NH4^+/K^+$ channels mediate the trafficking of K across the symbiosome membrane (SM) of soybean, faba bean (*Vicia faba*), and *L. japonicus*; however, the identity of these transporters is unknown (Andreev et al. 2005). The interaction of CBL (calcineurin B-like proteins, major Ca^{2+} sensor in plants) proteins with CIPK (CBL-interacting protein kinase) plays a key role in regulating K acquisition in plants in response to K-starved conditions (Xu et al. 2006). In *Arabidopsis*, 26 CIPK and 10 CBL proteins have been identified that control multiple signalling pathways in response to many abiotic stresses (see review Ragel de la Torre 2019; Wang and Wu 2013). In soybean, the upregulated expression of CBL1/9 (Glyma05g05580), CIPK23 (Glyma14g04430), and CDPKs (Glyma14g02680, Glyma04g38150, and Glyma14g00320) genes under K starvation revealed that they probably play a vital role in adaptation to low-K stress (Wang et al. 2012). Earlier studies found that the phosphorylation and dephosphorylation processes regulate the activity of K channels (Chérel et al. 2002; Hashimoto et al. 2012; Lee et al. 2007; Xu et al. 2006). Most of the K channels have cytoplasmic regulatory domains and therefore could be regulated by many cytoplasmic regulatory domains, viz. trafficking proteins (such as SYP121), 14-3-3 proteins (such as GF14-6), and K channel β-subunits (Honsbein et al. 2009; Sottocornola et al. 2006; Sutter et al. 2006). Another gene family possessing BURP-domain protein (a plant-specific protein with a conserved C-terminal domain named after four common members: *B*NM2, *U*SP, *R*D22, and *P*G1) might be significant for plants' responses to stresses. Wang et al. (2012) identified 23 members of BURP gene family in soybean, which exhibited an alteration in expression under K stress, implying that they are involved in K uptake. Further, the expression of a few jasmonic acid biosynthesis-related genes (allene oxide synthase, allene oxide cyclase, and lipoxygenase) was found to be significantly induced by K starvation, but K resupply downregulated their expression indicating that jasmonic acid plays a prominent role in K starvation signalling (Armengaud et al. 2004). Many transcription factor genes such as GATA transcription factors (Glyma11g04060, Glyma07g05960) of the MYB family are thought to have a role in low K tolerance in soybean (Wang et al. 2012).

A few miRNAs have been characterised as post-transcriptional regulators in response to K starvation such as miR319 and miR396 in barley (Zeng et al. 2019), miR399 in rice (Hu et al. 2015), and miR168 in tomato (Zeng et al. 2019). A recent study on cotton showed that the expression of miR165, miR166, and miR390 was inhibited in cotton after 8 days of K starvation, leading to increased expression of their target genes (ADF3 and HD-Zip) indicating their probable role in the K deficiency-regulating mechanism (Fontana et al. 2020). However, studies related to the regulation of K transporters by non-coding RNA in legumes are lacking.

12.3.3 Sulphur

12.3.3.1 S Uptake and Transport

Sulphate (SO_4^{2-}) is the predominant inorganic S form acquired by roots from the soil solution. Sulphate content in the cytoplasm is relatively constant, and the excess sulphate is stored in the vacuole. Once inside the cytoplasm, it travels through plasmodesmata from cell to cell and reaches the distant leaf chloroplast, where it is converted from sulphate to sulphide and subsequently assimilated into amino acids or other metabolites (Mitra 2015). A large family of sulphate transporters

(SULTRs) are employed in sulphate absorption from the soil solutions. The majority of the SULTR proteins are expressed in the root cell plasma membrane and are made up of a polypeptide chain of ~70 kDa. Sulphate transport through the plasma membrane is most likely a pH-dependent H^+-linked cotransport including $3H^+/SO_4^{2-}$ stoichiometry (Hawkesford and De Kok 2006). According to their function and location, SULTRs are categorised into five groups. The transporters of Group 1 and Group 2 are located in the plasma membrane; the former includes high-affinity while the latter includes low-affinity S transporters. Group 1 SULTRs are predominantly expressed in root tissue, while Group 2 transporters are expressed in vascular tissues (Buchner et al. 2004; Smith et al. 1997). Group 1 SULTRs were first identified in *Stylosanthes hamata*, a tropical legume (Smith et al. 1995), followed by characterisation in the many other plant species such as rice (Godwin et al. 2003), chickpea (Tabe et al. 2003), *Arabidopsis*, *Brassica oleracea* (Buchner et al. 2004), *L. japonicus* (Krusell et al. 2005), *Zea mays* (Nocito et al. 2006), and *T. aestivum* (Shinmachi et al. 2010). Group 3 SULTRs are also localised in the plasma membrane and associated with heterodimer association with unknown function. In *Arabidopsis*, one isoform, *AtSultr3.5*, failed to mediate sulphate transport itself but, after forming heterodimer with *AtSultr2.1*, catalysed sulphate transportation. In *L. japonicus*, a homolog of *AtSultr3.5* was identified, which was localised on the symbiosome membrane of nodules and indispensable for S mobilisation to the bacteroides (Kataoka et al. 2004). The SULTRs belonging to Group 4 mediate the efflux of sulphate from the vacuole to cytoplasm. Group 5 SULTRs, like Group 4, are located in tonoplast and are thought to be important in the absorption of molybdenum (Mo) and selenium (Se) (Shinmachi et al. 2010). Group 5 SULTRs, similar to Group 4, are located in the tonoplast and are thought to be important in the absorption of molybdenum (Mo) and selenium (Se).

Some SULTRs mediate the mobilisation of sulphate from plant cells to rhizobia and play an essential role in the establishment of symbiotic association (Frendo et al. 2013). The *Sst1* gene in *L. japonicus* is expressed in the symbiosome membrane of root nodules and encodes a SULTR protein, which mediates the transport of sulphate from plant cytoplasm to bacteroides, thus playing a vital role in symbiotic N_2 fixation (Krusell et al. 2005). Casieri et al. (2012) identified eight putative *MtSULTR* genes in *M. truncatula* belonging to four SULTR groups, expressed differentially in leaves and root tissue, and their transcript levels were affected by S concentration. Although SULTR genes have been characterised in many crops, only a few were reported in soybean. Ding et al. (2016) isolated and characterised a high-affinity sulphate transporter gene (*GmSULTR1;2b*) from soybean that was extensively expressed in root tissues and induced by S starvation.

12.3.3.2 Regulation of S Transporter

The regulation of sulphate uptake is well coordinated with the transcript levels of the SULTRs, which are mostly higher under low S supply and are rapidly reduced after resupplying of sulphate to S-starved plants (Koralewska et al. 2009; Rouached et al. 2008; Smith et al. 1997). The transcript levels of *AtSULTR1;1*, *1;2*, *2;1*, *4;1*, and *4;2* were induced by S starvation in *Arabidopsis*, and the same was true for wheat,

Medicago, and *Brassica* (Gigolashvili and Kopriva 2014). Uptake of nitrate influences the sulphate uptake; with low nitrate concentrations, sulphate acquisition was also suppressed (Kopriva et al. 2002). An intermediary metabolite, *O*-acetylserine (OAS), acts as a sulphate starvation regulatory signal, which accumulates under S starvation-induced expression of SULTR genes at low or even at sufficient S conditions (Hopkins et al. 2005). A *cis*-element characterised in *Arabidopsis*, known as S-responsive element (SURE), regulates the S response in plants under S starvation. SURE is a 16-base pair sequence found in the promoter region of several S starvation-inducible genes (Maruyama-Nakashita et al. 2006). SURE21A and SURE21B are present in the 3′-flanking region of *SULTR2;1* (low-affinity sulphate transporter), which is required for the transcriptional activation of these low-affinity SULTRs and is essential for enhancing the rate of root-to-shoot sulphate mobilisation under S starvation (Maruyama-Nakashita et al. 2006). Additionally, transcriptional regulator, Sulfur LIMitation1 (SLIM1), was found to induce the expression of many SULTRs under S starvation (Maruyama-Nakashita et al. 2006). SLIM1, also known as ETHYLENE-INSENSITIVE3-LIKE3 (EIL3), is a member of the transcription factor family that controls the ethylene response. It could be hypothesised that ethylene regulates sulphate absorption and metabolism; however, the effect of ethylene on S metabolism remains unknown (Takahashi 2019). The soybean embryo factors (SEFs) 3 and 4 are also known to be S-responsive factors that bind to the 235 bp region of β-conglycinin promoter (Awazuhara et al. 2002). A similar component has also been reported in the promoter region of serine acetyltransferase in *Citrullus vulgaris* (Saito et al. 1997).

The post-transcriptional regulation of SULTRs by microRNAs such as miR395 induced by a S starvation regulates several genes of sulphur assimilation pathway, including *SULTR2;1* and two chloroplast-localising ATP sulphurylases (*APS1* and *APS4*) (Jones-Rhoades and Bartel 2004; Kawashima et al. 2009). Furthermore, Kawashima et al. (2009) found that the transcription factor SLIM1 regulates the accumulation of miR395 in addition to directing the expression of protein-coding genes involved in sulphur metabolism. Li et al. (2017) identified five novel miRNAs and 27 conserved miRNAs whose accumulation was altered under S starvation in *Arabidopsis*. Among five novel miRNAs, two (miR66 and miR67) were upregulated, while the other three (miR14, miR20, and miR43) were downregulated under S starvation condition.

12.3.4 Magnesium

12.3.4.1 Mg Uptake and Transport

The Mg content in the soil is usually very low because it binds weakly with soil particles and could be leached out by rainwater. The Mg homeostasis in various plant tissues is maintained by a very efficient transporter system, which is involved in acquiring Mg^{2+} from the soil and their allocation throughout the plants. The majority of Mg transporters are members of a single protein family belonging to bacterial CorA Mg^{2+} transporter (MGTs) (Li et al. 2001). The first family of MGTs in the plant was reported in *Arabidopsis*, *AtMGT* and *AtMRS2* (Li et al. 2001). According to the cellular localisation and tissue of expression, ten members of MGT were identified in *Arabidopsis*. Their structural analysis revealed that they possessed two transmembrane domains with a conserved amino acid (GMN) motif (Tang and Luan 2017). The molecular mechanisms of Mg acquisition in plants are poorly understood, and members of the MGT family have been identified in only a few crops including rice (Chen and Ma 2013), *Brassica napus* (Zhang et al. 2019), and maize (Li et al. 2016). To the best of our knowledge, no Mg transporter has yet been identified and characterised in legume crops.

12.3.5 Calcium

12.3.5.1 Ca Uptake and Transport

Plants absorb Ca as a divalent cation (Ca^{2+}) from the soil solution and its uptake by roots against the electrochemical potential gradient. Inside the root cells, Ca can be mobilised via symplast or apoplast, thereby maintaining a low Ca concentration in root cells and preventing its toxicity in the shoot (Marschner 2011). Plastids, endoplasmic reticulum, and mitochondria have the ability to store Ca, but vacuoles serve as the principal Ca storage organelle with a concentration 10,000 times more than the cytoplasm. The cytosolic Ca concentration is almost 0.1 μM during the resting phase of cells, while it rises to 1 μM when Ca participates in any signalling process (Dodd et al. 2010). Ca channels are located in the plasma membrane, and according to their voltage dependence, they are classified into two groups: (1) voltage-dependent cation channels (VDCCs) and (2) voltage-independent cation channels (VICCs) (Sanders et al. 2002). The VDCCs are further divided into two subgroups: (a) *depolarisation-activated cation channels* (DACC), permeable to both mono- and divalent cations and contributing to only short and transient Ca influx, and (b) *hyperpolarisation-activated cation channels* (HACC), permeable for sustained Ca influx and playing a key role in stomatal closure under drought condition. VICCs located at the plasma membrane can be constitutively opened, so are permeable to both mono- and di-valent cations and play a vital role in maintaining cytosolic Ca level (González-Fontes et al. 2017; Tang and Luan 2017).

Ca^{2+}/ATPases and H^+/Ca^{2+} antiporters actively regulate the trafficking of Ca between cytosol and apoplast or vacuoles against the electrochemical potential gradient. Previous studies proposed that the Ca^{2+}/ATPases possessing a higher affinity (K_m = 0.4–10 μM) but lower Ca transport capacity are essential for maintaining cytosolic Ca homeostasis in resting cells (Hayter and Peterson 2004; Hirschi 2001). Two major families of Ca^{2+}/ATPases that are identified in plants include (a) P-type ATPase II A family and (b) P-type ATPase II B family (details in González-Fontes et al. 2017; Tuteja and Mahajan 2007). The H^+/Ca^{2+} antiporters have lower affinities (K_m = 10–15 μM) but a higher efficiency for Ca transport. They function to withdraw Ca from the cytosol during signalling events and control cytosolic Ca concentration fluctuations (Hirschi 2001; Pittman and Hirschi 2016; Shabala and Palmgren 2011; Sze et al. 2000).

The first H^+/Ca^{2+} antiporter was characterised in yeast followed by *Arabidopsis*, oat (*Avena sativa*), barley (*Hordeum vulgare*), maize, rice, mung bean, soybean, and *Medicago* (Chanson 1991; Charpentier et al. 2016; Cunningham and Fink 1996; DuPont et al. 1990; Hirschi et al. 1996; Schumaker and Sze 1986; Ueoka-Nakanishi et al. 1999; Zeng et al. 2020). In *Medicago*, the calcium ATPase (MCA8) was identified, which was localised in the nuclear envelope; however, in the endoplasmic reticulum, MCA8 was necessary for nuclear calcium signalling during symbiotic contacts (Capoen et al. 2011). A Ca^{2+}/ATPase-driven Ca absorption into symbiosomes has been reported in yellow lupin and broad bean (Andreev et al. 1997, 1998), while NH_4^+/K^+ channels mediate Ca transport in the symbiosome membrane of *L. japonicus*. Out of the 15 Ca^{2+}/ATPases characterised in *Medicago*, only one showed a >150-fold increase in expression during the late stages of nodule growth (Benedito et al. 2010).

12.3.5.2 Regulation of Ca Transporters

The perturbations in cytosolic Ca concentration in response to a specific environmental challenge or developmental signal are referred to as the 'Ca^{2+} signature' that is unique to each response. An increase in cytosolic Ca concentration measured by an array of Ca sensors is a common response to stress (Tuteja and Mahajan 2007). Calmodulin (CaMs), calmodulin-like proteins (CMLs), Ca-dependent protein kinase (CDPKs), and calcineurin B-like proteins (CBLs) are the major families of plant Ca sensors whose conformation or catalytic activity changes after Ca^{2+} binding (González-Fontes et al. 2017).

CaMs are usually located in the cytosol; however, they have also been found in the nucleus, endoplasmic reticulum, and plasma membrane. CaMs/Ca complex regulates the expression of genes for several plant responses through post-translational modification of transcription factors (Rudd and Franklin-Tong 2001). Members of the CAMTA (calmodulin-binding transcription activator), bZIP, CBP60, MYB, MADS-box, NAC, and WRKY transcription factor families bind to CaM and control gene expression in response to light, mechanical stress, heat shock, and osmotic stress in plants (Kim et al. 2009; Reddy et al. 2011). Wang et al. (2015) identified 15 CAMTA proteins in soybean, all expressed in root tissues and induced by several stresses (dehydration, cold, H_2O_2) and hormone signals (abscisic acid, methyl jasmonate, and salicylic acid). Although all *GmCAMTAs* express constitutively in root and leaf tissue, a recent study found that five of them (*GmCAMTA2*, *4*, *5*, *11*, and *12*) were upregulated under drought indicating their contribution to the drought tolerance of soybean (Noman et al. 2021). An increased Ca influx and Ca accumulation in cells enhanced phytase (PA) and acid phosphatase (PAP) activity by increasing the expression of *PA*, *PAP*, and alkaline phosphatase (ALP) gene in the mung bean sprouts (Zhou et al. 2018). The Ca signature was also triggered by a variety of elicitors (either a group of compounds secreted or constituents of pathogens) including protein, oligogalacturonides, β-heptaglucosans, lipopolysaccharides, and xylanases. The perception of elicitors significantly increases Ca influx through various Ca channels such as cyclic nucleotide-gated channels (CNGC) and activated multiple protein kinases (Garcia-Brugger et al.

2006; Reddy et al. 2011). Recently, a few miRNAs were identified that target the sites within putative Ca transporter genes (gma-miR156b target sites on *GmACA22*, gma-miR156b target sites on *GmMCA13* and *GmMCA14*, and gma-miR9750 target sites on *GmMCA3* and *GmMCA4*), indicating that miRNA may be involved in Ca homeostasis and signalling (Zeng et al. 2020).

12.3.6 Metal Divalent Cations: Fe, Zn, and Mn

12.3.6.1 Uptake, Transport, and Regulation of Metal Divalent Cations

The transport of metal divalent cations is mostly mediated by similar transporter families such as zinc-regulated transporter/iron-regulated transporter [ZRT/IRT1]-related protein (ZIP), natural resistance-associated macrophage protein (NRAMP), yellow stripe-like (YSL), P-type ATPases, and vacuolar iron transporter (VIT) (Guerinot 2000; Socha and Guerinot 2014). Plants have a limited number of 'Mn-only' transporter because most of the divalent cation (Fe and Zn) transporters such as NRAMP, YSL, zinc-regulated transporter/iron-regulated transporter-related protein (ZIP), and cation exchanger (CAX) are involved in Mn transport (for details, see review Socha and Guerinot 2014).

In legumes, Strategy I which is a reduction-based mechanism is operational to acquire insoluble Fe^{3+} from the rhizosphere into the root cells. The enzymes, iron-regulated transporter (IRT) and ferric chelate reductase (FCR), are required for the uptake of the reduced form of ferric by the roots (White 2012). The gene encoding IRT belongs to the ZIP family (ZRT-IRT-like protein), and the FCR enzyme belongs to the ferric reductase oxidase (FRO) family (Wu et al. 2005). The IRT is a major Fe importer expressed in the root tissue and located in the plasma membrane, which contains eight transmembrane domains. In soybean, homologs of *Arabidopsis* IRT (*AtIRT1*) and FRO (*AtFRO2*) were identified, which showed an increased transcript level under Fe starvation in root tissue (Stribe 2012). Later, *MtIRT* and *MtFRD3* genes from *Melilotus japonicus* (Li et al. 2014) and homologs of IRT and FRO genes were characterised from *L. japonicus* (Campestre et al. 2016), which showed enhanced expression under low-Fe conditions. Besides Strategy I, the NRAMP family is another Fe transporter family with a highly conserved domain that mediates the trafficking of a divalent metal ion such as Mn and Fe across cellular membranes (Thomine and Vert 2013). The members of NRAMP gene family have been characterised in several plant species such as *Arabidopsis*, barley, rice, and mustard (Qin et al. 2017; Yamaji et al. 2013). Recently, various NRAMP genes have been characterised in legumes. For example, the *AhNRAMP1* gene in groundnut (*Arachis hypogaea*) was expressed in roots and leaves (Xiong et al. 2012), while the *MtNRAMP1* in *M. truncatula* was expressed in roots and nodules under low-Fe stress (Tejada-Jiménez et al. 2015). Further, Qin et al. (2017) identified 17 NRAMP genes in soybean that are differentially regulated by deficiencies of several nutrient elements such as N, P, K, S, and Fe. In contrast to soil conditions where Fe is present in ferric form, the nodule cytosol maintains Fe in its reduced form; hence, the absorption of ferrous is faster than ferric in the nodules (Moreau et al. 1995). The members of NRAMP, vacuolar iron transporter (VIT), yellow stripe-like (YSL), and ZIP transporter family are overexpressed in nodules and thereby may be involved in iron transport across symbiotic membranes (Brear et al. 2013). A ferrous transporter, *GmDMT1* (divalent metal transporter 1), has been identified in soybean symbiosome membrane and showed maximum similarity with NRAMP transporter family (Kaiser et al. 2003). Similarly, the homologs of *GmDMT1* were identified in *Medicago*, which were expressed specifically in the nodules (Benedito et al. 2010). The release of citrate by *LjMATE1* (multidrug and toxic compound extrusion 1) increased Fe transport into rhizobia-infected cells of *L. japonicus*, resulting in enhanced leghaemoglobin concentration and nitrogenase activity in nodules (Takanashi et al. 2013).

The Zrt and Irt-like proteins (ZIP) and bZIP families of transporters are involved in Zn absorption, and its mobilisation to shoot, developing embryo and seeds (Eide et al. 1996). The ZIP family is highly conserved in prokaryotes and eukaryotes, and it is thought to have eight transmembrane domains with a histidine motif (Chen et al. 2008; Eng et al. 1998). Members of the ZIP family transporter have been identified in several plant species, including *Arabidopsis* (15 members), rice (17 members), and wheat (14 members), demonstrating a wide range of localisation and function (Evens et al. 2017; Milner et al. 2012). Moreau et al. (2002) discovered that a member of the ZIP family, *GmZIP1*, was highly selective for zinc uptake in soybean nodules. *VvZIP3*, a member of ZIP family, identified in *Vitis vinifera* showed higher expression in flower tissue under Zn deficiency (Gainza-Cortés et al. 2012). Lopez-Millan et al. (2004) identified six genes in *M. truncatula*, namely, *MtZIP1*, *3*, *4*, *5*, *6*, and *7*, all of which contained a conserved Zn motif with eight transmembrane domains. They showed that *MtZIP1*, *5*, and *6* transporters restored yeast growth in Zn-deficient media; *MtZIP3*, *5*, and *6* proteins restored yeast growth in Fe-limited media; while *MtZIP4* and *7* proteins restored yeast growth in Mn-deficient media. Astudillo et al. (2013) identified and characterised a large family of Zn transporters in *Phaseolus vulgaris*, 23 of which belonged to the Zip family and three to the bZIP family.

The regulation of uptake and translocation of most divalent cations and their deficiency responses are controlled by the master regulator, *FER* transcription factor, which belongs to the bHLH transcription factor family and was first cloned from tomato (Ling et al. 2002). Its homolog, *AtFIT* (FER-like iron deficiency-induced transcription factor), was later found in Arabidopsis (Colangelo and Guerinot 2004; Yuan et al. 2008). Similar to *IRT1* and *FRO2*, the expression of *FIT* is also induced by Fe starvation in root tissue, where it upregulates the expression of *IRT1* and *FRO2*. Two soybean genes, Glyma03g28610 and Glyma03g26830, showed homology with *AtFIT* and upregulated the Fe acquisition genes *IRT* and *FRO2* under Fe starvation (Yuan et al. 2008). Another member of bHLH family transcription factor, *POPEYE* (*PYR*), controls the internal mobilisation of Fe or Zn by regulating the activity of *FRO6*, *ZIF1* (zinc-induced facilitator 1), and *NAS4* (nicotianamine synthase 4) (Long et al. 2010). The impact of phytohormones on Fe uptake has also been studied; auxin and ethylene positively control the Fe starvation response (Romera et al. 2011; Zuchi et al. 2009), while cytokinin and jasmonate act as a negative regulator of Fe acquisition by decreasing the expression of FRO2 and IRT1 (for details, see review, Hindt and Guerinot 2012).

12.4 Conclusions

Protein calorie malnutrition is a prevalent nutritional disorder, especially among children, in underdeveloped nations. The lower income populations are particularly vulnerable because they cannot afford to buy conventional protein sources like milk and meat. The high protein content in legumes makes them a viable replacement for more energy-dense animal protein sources. The availability of several mineral nutrients may influence legume productivity and N_2 fixation. Due to symbiotic nitrogen fixation, the demand for other nutrients is higher for legumes as compared to other non-legume crops. Among nutrient elements, P is a common limiting factor for nodulation in legume crops because of the energy-intensive N_2-fixation reaction. Similarly, Ca is significantly important for early symbiotic activities. On the other hand, S and K are not a major bottleneck for nodulated legumes, but the K supplement for osmoadaptation is necessary for the development of legume crops. Due to the anaerobic and acidic environment inside the nodule, Fe is more or less deficient in legume crops even though the soil contains sufficient Fe concentration.

Available literature showed that the characterisation of transporters and identification of their regulatory genes in legumes have been accomplished for a few nutrient elements. However, such studies for the majority of essential nutrient elements in legumes are still in the primitive stage. This chapter has outlined the various physiological and molecular mechanisms which assist in the adaptation of legumes to nutrient-deficient conditions. Future efforts should be directed to determine the molecular basis of nutrient absorption, translocation, and cellular homeostasis in legume crops.

References

Abdelrahman M, El-Sayed MA, Hashem A, Abd Allah EF, Alqarawi AA, Burritt DJ, Tran LSP (2018) Metabolomics and transcriptomics in legumes under phosphate deficiency in relation to nitrogen fixation by root nodules. Front Plant Sci 9:922

Afzal F, Khan T, Khan A, Khan S, Raza H, Ihsan A, Ahanger MA, Kazi AG (2015) Nutrient deficiencies under stress in legumes: an overview. In: Legumes under environmental stress: yield, improvement and adaptations. Wiley, New York, pp 53–65

Ai P, Sun S, Zhao J, Fan X, Xin W, Guo Q, Yu L, Shen Q, Wu P, Miller AJ (2009) Two rice phosphate transporters, OsPht1; 2 and OsPht1; 6, have different functions and kinetic properties in uptake and translocation. Plant J 57(5):798–809

Andreev I, Dubrovo P, Krylova V, Andreeva IN, Koren'kov V, Sorokin EM, Izmailov SF (1997) Characterization of ATP-hydrolyzing and ATP-driven proton-translocating activities associated with the peribacteroid membrane from root nodules of *Lupinus luteus* L. J Plant Physiol 151(5): 563–569

Andreev IM, Dubrovo PN, Krylova VV, Izmailov SF (1998) Calcium uptake by symbiosomes and the peribacteroid membrane vesicles isolated from yellow lupin root nodules. J Plant Physiol 153(5–6):610–614

Andreev I, Krylova V, Dubrovo P, Izmailov S (2005) Passive potassium transport by symbiosomes from broad bean root nodules. Plant Sci 168(4):1005–1010

Armengaud P, Breitling R, Amtmann A (2004) The potassium-dependent transcriptome of Arabidopsis reveals a prominent role of jasmonic acid in nutrient signaling. Plant Physiol 136(1):2556–2576

Astudillo C, Fernandez A, Blair MW, Cichy KA (2013) The *Phaseolus vulgaris* ZIP gene family: identification, characterization, mapping, and gene expression. Front Plant Sci 4:286

Awazuhara M, Kim H, Goto DB, Matsui A, Hayashi H, Chino M, Kim SG, Naito S, Fujiwara T (2002) A 235-bp region from a nutritionally regulated soybean seed-specific gene promoter can confer its sulfur and nitrogen response to a constitutive promoter in aerial tissues of *Arabidopsis thaliana*. Plant Sci 163(1):75–82

Azeem F, Ahmad B, Atif RM, Ali MA, Nadeem H, Hussain S, Manzoor H, Azeem M, Afzal M (2018) Genome-wide analysis of potassium transport-related genes in chickpea (*Cicer arietinum* L.) and their role in abiotic stress responses. Plant Mol Biol Rep 36(3):451–468

Baek D, Chun HJ, Yun DJ, Kim MC (2017) Cross-talk between phosphate starvation and other environmental stress signaling pathways in plants. Mol Cells 40(10):697

Bartel DP (2004) MicroRNAs: genomics, biogenesis, mechanism, and function. Cell 116(2): 281–297

Benedito VA, Li H, Dai X, Wandrey M, He J, Kaundal R, Torres-Jerez I, Gomez SK, Harrison MJ, Tang Y (2010) Genomic inventory and transcriptional analysis of *Medicago truncatula* transporters. Plant Physiol 152(3):1716–1730

Borah P, Das A, Milner MJ, Ali A, Bentley AR, Pandey R (2018) Long non-coding RNAs as endogenous target mimics and exploration of their role in low nutrient stress tolerance in plants. Gene 9(9):459

Brear EM, Day DA, Smith PMC (2013) Iron: an essential micronutrient for the legume-rhizobium symbiosis. Front Plant Sci 4:359

Breuillin-Sessoms F, Floss DS, Gomez SK, Pumplin N, Ding Y, Levesque-Tremblay V, Noar RD, Daniels DA, Bravo A, Eaglesham JB (2015) Suppression of arbuscule degeneration in *Medicago truncatula* phosphate transporter4 mutants is dependent on the ammonium transporter 2 family protein AMT2; 3. Plant Cell 27(4):1352–1366

Bucher M (2007) Functional biology of plant phosphate uptake at root and mycorrhiza interfaces. New Phytol 173(1):11–26

Buchner P, Prosser IM, Hawkesford MJ (2004) Phylogeny and expression of paralogous and orthologous sulphate transporter genes in diploid and hexaploid wheats. Genome 47(3): 526–534

Bulgarelli RG, De Oliveira VH, de Andrade SAL (2020) Arbuscular mycorrhizal symbiosis alters the expression of PHT1 phosphate transporters in roots and nodules of P-starved soybean plants. Theor Exp Plant Physiol 32(3):243–253

Bustos R, Castrillo G, Linhares F, Puga MI, Rubio V, Pérez-Pérez J, Solano R, Leyva A, Paz-Ares J (2010) A central regulatory system largely controls transcriptional activation and repression responses to phosphate starvation in Arabidopsis. PLoS Genet 6(9):e1001102

Campestre MP, Antonelli C, Calzadilla PI, Maiale SJ, Rodríguez AA, Ruiz OA (2016) The alkaline tolerance in *Lotus japonicus* is associated with mechanisms of iron acquisition and modification of the architectural pattern of the root. J Plant Physiol 206:40–48

Cao Y, Liu J, Li Y, Zhang J, Li S, An Y, Hu T, Yang P (2021) Functional analysis of the phosphate transporter gene MtPT6 from *Medicago truncatula*. Front Plant Sci 11:2313

Capoen W, Sun J, Wysham D, Otegui MS, Venkateshwaran M, Hirsch S, Miwa H, Downie JA, Morris RJ, Ané JM (2011) Nuclear membranes control symbiotic calcium signaling of legumes. Proc Natl Acad Sci 108(34):14348–14353

Casieri L, Gallardo K, Wipf D (2012) Transcriptional response of *Medicago truncatula* sulphate transporters to arbuscular mycorrhizal symbiosis with and without sulphur stress. Planta 235(6): 1431–1447

Castro-Guerrero NA, Isidra-Arellano MC, Mendoza-Cozatl DG, Valdés-López O (2016) Common bean: a legume model on the rise for unraveling responses and adaptations to iron, zinc, and phosphate deficiencies. Front Plant Sci 7:600

Ceppi MG, Oukarroum A, Çiçek N, Strasser RJ, Schansker G (2012) The IP amplitude of the fluorescence rise OJIP is sensitive to changes in the photosystem I content of leaves: a study on plants exposed to magnesium and sulfate deficiencies, drought stress and salt stress. Physiol Plant 144(3):277–288

Chanson AA (1991) Ca^{2+}/H^+ antiport system driven by the tonoplast pyrophosphate-dependent proton pump from maize roots. J Plant Physiol 137(4):471–476

Charpentier M, Sun J, Martins TV, Radhakrishnan GV, Findlay K, Soumpourou E, Thouin J, Véry AA, Sanders D, Morris RJ (2016) Nuclear-localized cyclic nucleotide–gated channels mediate symbiotic calcium oscillations. Science 352(6289):1102–1105

Chen ZC, Ma JF (2013) Magnesium transporters and their role in Al tolerance in plants. Plant Soil 368(1):51–56

Chen WR, Feng Y, Chao YE (2008) Genomic analysis and expression pattern of OsZIP1, OsZIP3, and OsZIP4 in two rice (*Oryza sativa* L.) genotypes with different zinc efficiency. Russ J Plant Physiol 55(3):400–409

Chen H, Chen GH, Wu B, Zhang H, Yuan X, Cui X (2014) GmHKT1; 4, a novel soybean gene regulating Na^+/K^+ ratio in roots enhances salt tolerance in transgenic plants. Plant Growth Regul 73(3):299–308

Chen G, Hu Q, Luo LE, Yang T, Zhang S, Hu Y, Yu L, Xu G (2015) Rice potassium transporter O s HAK 1 is essential for maintaining potassium-mediated growth and functions in salt tolerance over low and high potassium concentration ranges. Plant Cell Environ 38(12):2747–2765

Chérel I, Michard E, Platet N, Mouline K, Alcon C, Sentenac H, Thibaud JB (2002) Physical and functional interaction of the Arabidopsis K^+ channel AKT2 and phosphatase AtPP2CA. Plant Cell 14(5):1133–1146

Chiou TJ, Aung K, Lin SI, Wu CC, Chiang SF, Su C (2006) Regulation of phosphate homeostasis by microRNA in Arabidopsis. Plant Cell 18(2):412–421

Clarke VC, Loughlin PC, Day DA, Smith PMC (2014) Transport processes of the legume symbiosome membrane. Front Plant Sci 5:699

Colangelo EP, Guerinot ML (2004) The essential basic helix-loop-helix protein FIT1 is required for the iron deficiency response. Plant Cell 16(12):3400–3412

Cunningham KW, Fink GR (1996) Calcineurin inhibits VCX1-dependent H^+/Ca^{2+} exchange and induces Ca^{2+} ATPases in *Saccharomyces cerevisiae*. Mol Cell Biol 16(5):2226–2237

Damiani I, Drain A, Guichard M, Balzergue S, Boscari A, Boyer JC, Brunaud V, Cottaz S, Rancurel C, Da Rocha M (2016) Nod factor effects on root hair-specific transcriptome of *Medicago truncatula*: focus on plasma membrane transport systems and reactive oxygen species networks. Front Plant Sci 7:794

Dana S, Herdean A, Lundin B, Spetea C (2016) Retracted: Each of the chloroplast potassium efflux antiporters affects photosynthesis and growth of fully developed Arabidopsis rosettes under short-day photoperiod. Physiol Plant 158(4):483–491

Davies C, Shin R, Liu W, Thomas MR, Schachtman DP (2006) Transporters expressed during grape berry (*Vitis vinifera* L.) development are associated with an increase in berry size and berry potassium accumulation. J Exp Bot 57(12):3209–3216

De Freitas S, Amarante C, Mitcham E (2016) Calcium deficiency disorders in plants. Postharvest Biol Technol:477–502

Desbrosses G, Kopka C, Ott T, Udvardi MK (2004) Lotus japonicus LjKUP is induced late during nodule development and encodes a potassium transporter of the plasma membrane. Mol Plant Microbe Interact 17(7):789–797

Devaiah BN, Karthikeyan AS, Raghothama KG (2007a) WRKY75 transcription factor is a modulator of phosphate acquisition and root development in Arabidopsis. Plant Physiol 143(4):789–1801
Devaiah BN, Nagarajan VK, Raghothama KG (2007b) Phosphate homeostasis and root development in Arabidopsis are synchronized by the zinc finger transcription factor ZAT6. Plant Physiol 145(1):147–159
Ding Y, Zhou X, Zuo L, Wang H, Yu D (2016) Identification and functional characterization of the sulfate transporter gene GmSULTR1; 2b in soybean. BMC Genet 7(1):1–19
Dodd AN, Kudla J, Sanders D (2010) The language of calcium signaling. Annu Rev Plant Biol 61: 593–620
DuPont FM, Bush DS, Windle JJ, Jones RL (1990) Calcium and proton transport in membrane vesicles from barley roots. Plant Physiol 94(1):179–188
Eide D, Broderius M, Fett J, Guerinot ML (1996) A novel iron-regulated metal transporter from plants identified by functional expression in yeast. Proc Natl Acad Sci 93(11):5624–5628
Eng BH, Guerinot ML, Eide D, Saier MH (1998) Sequence analyses and phylogenetic characterization of the ZIP family of metal ion transport proteins. J Membr Biol 166(1):1–7
Esfahani MN, Kusano M, Nguyen KH, Watanabe Y, Van Ha C, Saito K, Sulieman S, Herrera-Estrella L, Tran LSP (2016) Adaptation of the symbiotic Mesorhizobium–chickpea relationship to phosphate deficiency relies on reprogramming of whole-plant metabolism. Proc Natl Acad Sci 113(32):E4610–E4619
Evens NP, Buchner P, Williams LE, Hawkesford MJ (2017) The role of ZIP transporters and group F bZIP transcription factors in the Zn-deficiency response of wheat (*Triticum aestivum*). Plant J 92(2):291–304
Fan C, Wang X, Hu R, Wang Y, Xiao C, Jiang Y, Zhang X, Zheng C, Fu YF (2013) The pattern of Phosphate transporter 1 genes evolutionary divergence in *Glycine max* L. BMC Plant Biol 13(1):1–16
Fontana JE, Wang G, Sun R, Xue H, Li Q, Liu J, Pan X (2020) Impact of potassium deficiency on cotton growth, development and potential microRNA-mediated mechanism. Plant Physiol Biochem 153:72–80
Franco-Zorrilla JM, Valli A, Todesco M, Mateos I, Puga MI, Rubio-Somoza I, Leyva A, Weigel D, García JA, Paz-Ares J (2007) Target mimicry provides a new mechanism for regulation of microRNA activity. Nat Genet 39(8):1033–1037
Frendo P, Matamoros MA, Alloing G, Becana M (2013) Thiol-based redox signaling in the nitrogen-fixing symbiosis. Front Plant Sci 4:376
Fujii H, Chiou TJ, Lin SI, Aung K, Zhu JK (2005) A miRNA involved in phosphate-starvation response in Arabidopsis. Curr Biol 15(22):2038–2043
Gainza-Cortés F, Pérez-Dïaz R, Pérez-Castro R, Tapia J, Casaretto JA, González S, Peña-Cortés H, Ruiz-Lara S, González E (2012) Characterization of a putative grapevine Zn transporter, VvZIP3, suggests its involvement in early reproductive development in *Vitis vinifera* L. BMC Plant Biol 12(1):1–13
Gao J, Wu F, Shen Z, Meng Y, Cai Y, Lin Y (2016) A putative molybdate transporter LjMOT1 is required for molybdenum transport in *Lotus japonicus*. Physiol Plant 158(3):331–340
Garcia-Brugger A, Lamotte O, Vandelle E, Bourque S, Lecourieux D, Poinssot B, Wendehenne D, Pugin A (2006) Early signaling events induced by elicitors of plant defenses. Mol Plant Microbe Interact 19(7):711–724
Garg N, Geetanjali KA (2006) Arbuscular mycorrhiza: nutritional aspects: (Arbusculäre Mykorrhiza: Nährstoffaspekte). Arch Agron Soil Sci 52(6):593–606
Gigolashvili T, Kopriva S (2014) Transporters in plant sulfur metabolism. Front Plant Sci 5:442
Godwin RM, Rae AL, Carroll BJ, Smith FW (2003) Cloning and characterization of two genes encoding sulfate transporters from rice (*Oryza sativa* L.). Plant Soil 257(1):113–123
Gomez-Porras JL, Riaño Pachó DM, Benit B, Haro R, Sklodowski K, Rodríguez-Navarro A, Dreyer I (2012) Phylogenetic analysis of K^+ transporters in bryophytes, lycophytes, and flowering plants indicates a specialization of vascular plants. Front Plant Sci 3:167

González-Fontes A, Navarro-Gochicoa MT, Ceacero CJ, Herrera-Rodríguez MB, Camacho-Cristóbal JJ, Rexach J (2017) Understanding calcium transport and signaling, and its use efficiency in vascular plants. In: Plant macronutrient use efficiency. Elsevier, Amsterdam, pp 165–180

Graham MA, Ramírez M, Valdés-López O, Lara M, Tesfaye M, Vance CP, Hernandez G (2006) Identification of candidate phosphorus stress induced genes in *Phaseolus vulgaris* through clustering analysis across several plant species. Funct Plant Biol 33(8):789–797

Gu M, Chen A, Sun S, Xu G (2016) Complex regulation of plant phosphate transporters and the gap between molecular mechanisms and practical application: what is missing? Mol Plant 9(3): 396–416

Guerinot ML (2000) The ZIP family of metal transporters. Biochim Biophys Acta Biomembr 1465(1–2):190–198

Guo B, Jin Y, Wussler C, Blancaflor EB, Motes CM, Versaw WK (2008) Functional analysis of the Arabidopsis PHT4 family of intracellular phosphate transporters. New Phytol 177(4):889–898

Guo M, Ruan W, Li C, Huang F, Zeng M, Liu Y, Yu Y, Ding X, Wu Y, Wu Z (2015) Integrative comparison of the role of the PHOSPHATE RESPONSE1 subfamily in phosphate signaling and homeostasis in rice. Plant Physiol 168(4):762–1776

Harrison MJ, Dewbre GR, Liu J (2002) A phosphate transporter from *Medicago truncatula* involved in the acquisition of phosphate released by arbuscular mycorrhizal fungi. Plant Cell 14(10):2413–2429

Hashimoto K, Eckert C, Anschütz U, Scholz M, Held K, Waadt R, Reyer A, Hippler M, Becker D, Kudla J (2012) Phosphorylation of calcineurin B-like (CBL) calcium sensor proteins by their CBL-interacting protein kinases (CIPKs) is required for full activity of CBL-CIPK complexes toward their target proteins. J Biol Chem 287(11):7956–7968

Hawkesford MJ, De Kok LJ (2006) Managing sulphur metabolism in plants. Plant Cell Environ 29(3):382–395

Hayter ML, Peterson CA (2004) Can Ca^{2+} fluxes to the root xylem be sustained by Ca^{2+}-ATPases in exodermal and endodermal plasma membranes? Plant Physiol 136(4):4318–4325

He C, Cui K, Duan A, Zeng Y, Zhang J (2012) Genome-wide and molecular evolution analysis of the P oplar KT/HAK/KUP potassium transporter gene family. Ecol Evol 2(8):1996–2004

He Q, Wang F, Wang Y, Lu H, Yang Z, Lv Q (2019) Molecular control and genetic improvement of phosphorus use efficiency in rice. Mol Breed 39(12):1–15

He Q, Lu H, Guo H, Wang Y, Zhao P, Li Y, Wang F, Xu J, Mo X, Mao C (2021) OsbHLH6 interacts with OsSPX4 and regulates the phosphate starvation response in rice. Plant J 105(3): 649–667

Hindt MN, Guerinot ML (2012) Getting a sense for signals: regulation of the plant iron deficiency response. Biochim Biophys Acta Mol Cell Res 1823(9):1521–1530

Hinsinger P (2001) Bioavailability of soil inorganic P in the rhizosphere as affected by root-induced chemical changes: a review. Plant Soil 237(2):173–195

Hirschi K (2001) Vacuolar H^+/Ca^{2+} transport: who's directing the traffic? Trends Plant Sci 6(3): 100–104

Hirschi KD, Zhen RG, Cunningham KW, Rea PA, Fink GR (1996) CAX1, an $H^{+/}Ca^{2+}$ antiporter from Arabidopsis. Proc Natl Acad Sci 93(16):8782–8786

Honsbein A, Sokolovski S, Grefen C, Campanoni P, Pratelli R, Paneque M, Chen Z, Johansson I, Blatt MR (2009) A tripartite SNARE-K^+ channel complex mediates in channel-dependent K^+ nutrition in Arabidopsis. Plant Cell 21(9):2859–2877

Hopkins L, Parmar S, Błaszczyk A, Hesse H, Hoefgen R, Hawkesford MJ (2005) O-acetylserine and the regulation of expression of genes encoding components for sulfate uptake and assimilation in potato. Plant Physiol 138(1):433–440

Horie T, Hauser F, Schroeder JI (2009) HKT transporter-mediated salinity resistance mechanisms in Arabidopsis and monocot crop plants. Trends Plant Sci 14(12):660–668

Hsieh LC, Lin SI, Shih ACC, Chen JW, Lin WY, Tseng CY, Li WH, Chiou TJ (2009) Uncovering small RNA-mediated responses to phosphate deficiency in Arabidopsis by deep sequencing. Plant Physiol 151(4):2120–2132

Hu B, Wang W, Deng K, Li H, Zhang Z, Zhang L, Chu C (2015) MicroRNA399 is involved in multiple nutrient starvation responses in rice. Front Plant Sci 6:188

Huang KL, Wang H, Wei YL, Jia HX, Zha L, Zheng Y, Ren F, Li XB (2019) The high-affinity transporter BnPHT1;4 is involved in phosphorus acquisition and mobilization for facilitating seed germination and early seedling growth of Brassica napus. BMC Plant Biol 19(1):1–13

Huang L, Wu D, Zhang G (2020) Advances in studies on ion transporters involved in salt tolerance and breeding crop cultivars with high salt tolerance. J Zhejiang Univ Sci B 21(6):426–441

Hussain SS, Hussain M, Irfan M, Siddique KHM (2018) Legume, microbiome, and regulatory functions of miRNAs in systematic regulation of symbiosis. In: Plant microbiome: stress response. Springer, New York, pp 255–282

Iantcheva A, Mysore KS, Ratet P (2013) Transformation of leguminous plants to study symbiotic interactions. Int J Dev Biol 57:577–586

Javot H, Penmetsa RV, Terzaghi N, Cook DR, Harrison MJ (2007) A *Medicago truncatula* phosphate transporter indispensable for the arbuscular mycorrhizal symbiosis. Proc Natl Acad Sci 104(5):1720–1725

Jones-Rhoades MW, Bartel DP (2004) Computational identification of plant microRNAs and their targets, including a stress-induced miRNA. Mol Cell 14(6):787–799

Kabir AH, Hossain MM, Khatun MA, Sarkar MR, Haider SA (2017) Biochemical and molecular mechanisms associated with Zn deficiency tolerance and signaling in rice (*Oryza sativa* L.). J Plant Interact 12(1):447–456

Kaiser BN, Moreau S, Castelli J, Thomson R, Lambert A, Bogliolo S, Puppo A, Day DA (2003) The soybean NRAMP homologue, GmDMT1, is a symbiotic divalent metal transporter capable of ferrous iron transport. Plant J 35(3):295–304

Karandashov V, Bucher M (2005) Symbiotic phosphate transport in arbuscular ycorrhizas. Trends Plant Sci 10(1):22–29

Kataoka T, Watanabe-Takahashi A, Hayashi N, Ohnishi M, Mimura T, Buchner P, Hawkesford MJ, Yamaya T, Takahashi H (2004) Vacuolar sulfate transporters are essential determinants controlling internal distribution of sulfate in Arabidopsis. Plant Cell 16(10):2693–2704

Kawashima CG, Yoshimoto N, Maruyama-Nakashita A, Tsuchiya YN, Saito K, Takahashi H, Dalmay T (2009) Sulphur starvation induces the expression of microRNA-395 and one of its target genes but in different cell types. Plant J 57(2):313–321

Kiers ET, Duhamel M, Beesetty Y, Mensah JA, Franken O, Verbruggen E, Fellbaum CR, Kowalchuk GA, Hart MM, Bago A (2011) Reciprocal rewards stabilize cooperation in the mycorrhizal symbiosis. Science 333(6044):880–882

Kim MC, Chung WS, Yun DJ, Cho MJ (2009) Calcium and calmodulin-mediated regulation of gene expression in plants. Mol Plant 2(1):13–21

Klein MA, López-Millán AF, Grusak MA (2012) Quantitative trait locus analysis of root ferric reductase activity and leaf chlorosis in the model legume, *Lotus japonicus*. Plant Soil 351(1): 363–376

Kopriva S, Suter M, von Ballmoos P, Hesse H, Krähenbühl U, Rennenberg H, Brunold C (2002) Interaction of sulfate assimilation with carbon and nitrogen metabolism in *Lemna minor*. Plant Physiol 130(3):1406–1413

Koralewska A, Buchner P, Stuiver CEE, Posthumus FS, Kopriva S, Hawkesford MJ, De Kok LJ (2009) Expression and activity of sulfate transporters and APS reductase in curly kale in response to sulfate deprivation and re-supply. J Plant Physiol 166(2):168–179

Krusell L, Krause K, Ott T, Desbrosses G, Krämer U, Sato S, Nakamura Y, Tabata S, James EK, Sandal N (2005) The sulfate transporter SST1 is crucial for symbiotic nitrogen fixation in *Lotus japonicus* root nodules. Plant Cell 17(5):1625–1636

Kuang Q, Purhonen P, Hebert H (2015) Structure of potassium channels. Cell Mol Life Sci 72(19): 3677–3693

Lazali M, Bargaz A (2017) Examples of belowground mechanisms enabling legumes to mitigate phosphorus deficiency. In: Legume nitrogen fixation in soils with low phosphorus availability. Springer, New York, pp 135–152

Lee SC, Lan WZ, Kim BG, Li L, Cheong YH, Pandey GK, Lu G, Buchanan BB, Luan S (2007) A protein phosphorylation/dephosphorylation network regulates a plant potassium channel. Proc Natl Acad Sci 104(40):15959–15964

Li L, Tutone AF, Drummond RSM, Gardner RC, Luan S (2001) A novel family of magnesium transport genes in Arabidopsis. Plant Cell 13(12):2761–2775

Li LY, Cai QY, Yu DS, Guo CH (2011) Overexpression of AtFRO6 in transgenic tobacco enhances ferric chelate reductase activity in leaves and increases tolerance to iron-deficiency chlorosis. Mol Biol Rep 38(6):3605–3613

Li G, Wang B, Tian Q, Wang T, Zhang WH (2014) Medicago truncatula ecotypes A17 and R108 differed in their response to iron deficiency. J Plant Physiol 171(8):639–647

Li H, Du H, Huang K, Chen X, Liu T, Gao S, Liu H, Tang Q, Rong T, Zhang S (2016) Identification, and functional and expression analyses of the CorA/MRS2/MGT-type magnesium transporter family in maize. Plant Cell Physiol 57(6):1153–1168

Li L, Yi H, Xue M, Yi M (2017) miR398 and miR395 are involved in response to SO_2 stress in Arabidopsis thaliana. Ecotoxicology 26(9):1181–1187

Ling HQ, Bauer P, Bereczky Z, Keller B, Ganal M (2002) The tomato fer gene encoding a bHLH protein controls iron-uptake responses in roots. Proc Natl Acad Sci 99(21):13938–13943

Liu H, Trieu AT, Blaylock LA, Harrison MJ (1998) Cloning and characterization of two phosphate transporters from *Medicago truncatula* roots: regulation in response to phosphate and to colonization by arbuscular mycorrhizal (AM) fungi. Mol Plant Microbe Interact 11(1):14–22

Liu J, Versaw WK, Pumplin N, Gomez SK, Blaylock LA, Harrison MJ (2008) Losely related members of the *Medicago truncatula* PHT1 phosphate transporter gene family encode phosphate transporters with distinct biochemical activities. J Biol Chem 283(36):24673–24681

Liu F, Chang XJ, Ye Y, Xie WB, Wu P, Lian XM (2011) Comprehensive sequence and whole-life-cycle expression profile analysis of the phosphate transporter gene family in rice. Mol Plant 4(6):1105–1122

Liu TY, Huang TK, Yang SY, Hong YT, Huang SM, Wang FN, Chiang SF, Tsai SY, Lu WC, Chiou TJ (2016) Identification of plant vacuolar transporters mediating phosphate storage. Nat Commun 7(1):1–11

Long TA, Tsukagoshi H, Busch W, Lahner B, Salt DE, Benfey PN (2010) The bHLH transcription factor POPEYE regulates response to iron deficiency in Arabidopsis roots. Plant Cell 22(7): 2219–2236

Lopez-Millan AF, Ellis DR, Grusak MA (2004) Identification and characterization of several new members of the ZIP family of metal ion transporters in *Medicago truncatula*. Plant Mol Biol 54(4):583–596

Lurthy T, Cantat C, Jeudy C, Declerck P, Gallardo K, Barraud C, Leroy F, Ourry A, Lemanceau P, Salon C (2020) Impact of bacterial siderophores on iron status and ionome in pea. Front Plant Sci 11:730

Maeda D, Ashida K, Iguchi K, Chechetka SA, Hijikata A, Okusako Y, Deguchi Y, Izui K, Hata S (2006) Knockdown of an arbuscular mycorrhiza-inducible phosphate transporter gene of *Lotus japonicus* suppresses mutualistic symbiosis. Plant Cell Physiol 47(7):807–817

Marschner H (2011) Marschner's mineral nutrition of higher plants. Academic, San Diego

Maruyama-Nakashita A, Nakamura Y, Tohge T, Saito K, Takahashi H (2006) Arabidopsis SLIM1 is a central transcriptional regulator of plant sulfur response and metabolism. Plant Cell 18(11): 3235–3251

Meena SK, Pandey R, Sharma S, Vengavasi K, Dikshit HK, Siddique KHM, Singh MP (2021) Cross tolerance to phosphorus deficiency and drought stress in mungbean is regulated by improved antioxidant capacity, biological N_2-fixation, and differential transcript accumulation. Plant Soil 466(1):337–356

Milner MJ, Craft E, Yamaji N, Koyama E, Ma JF, Kochian LV (2012) Characterization of the high affinity Zn transporter from Noccaea caerulescens, NcZNT1, and dissection of its promoter for its role in Zn uptake and hyperaccumulation. New Phytol 195(1):113–123

Misson J, Raghothama KG, Jain A, Jouhet J, Block MA, Bligny R, Ortet P, Creff A, Somerville S, Rolland N (2005) A genome-wide transcriptional analysis using Arabidopsis thaliana Affymetrix gene chips determined plant responses to phosphate deprivation. Proc Natl Acad Sci 102(33):11934–11939

Mitra GN (2015) Regulation of nutrient uptake by plants, vol 10. Springer, New Delhi, pp 978–981

Moreau S, Meyer JM, Puppo A (1995) Uptake of iron by symbiosomes and bacteroids from soybean nodules. FEBS Lett 361:225–228

Moreau S, Thomson RM, Kaiser BN, Trevaskis B, Guerinot ML, Udvardi MK, Puppo A, Day DA (2002) GmZIP1 encodes a symbiosis-specific zinc transporter in soybean. J Biol Chem 277(7): 4738–4746

Muchhal US, Raghothama KG (1999) Transcriptional regulation of plant phosphate transporters. Proc Natl Acad Sci 96(10):5868–5872

Muller M, Schmidt W (2004) Environmentally induced plasticity of root hair development in Arabidopsis. Plant Physiol 134(1):409–419

Munro AW, Ritchie GY, Lamb AJ, Douglas RM, Booth IR (1991) The cloning and DNA sequence of the gene for the glutathione-regulated potassium-efflux system KefC of Escherichia coli. Mol Microbiol 5(3):607–616

Nagy R, Vasconcelos MJV, Zhao S, McElver J, Bruce W, Amrhein N, Raghothama KG, Bucher M (2006) Differential regulation of five Pht1 phosphate transporters from maize (*Zea mays* L.). Plant Biol 8(02):186–197

Nieves-Cordones M, Martínez-Cordero MA, Martínez V, Rubio F (2007) An NH_4^+-sensitive component dominates high-affinity K^+ uptake in tomato plants. Plant Sci 172(2):273–280

Nieves-Cordones M, Ródenas R, Chavanieu A, Rivero R, Martinez V, Gaillard I, Rubio F (2016) Uneven HAK/KUP/KT protein diversity among angiosperms: species distribution and perspectives. Front Plant Sci 7:127

Nocito FF, Lancilli C, Crema B, Fourcroy P, Davidian JC, Sacchi GA (2006) Heavy metal stress and sulfate uptake in maize roots. Plant Physiol 141(3):1138–1148

Noman M, Aysha J, Ketehouli T, Yang J, Du L, Wang F, Li H (2021) Calmodulin binding transcription activators: an interplay between calcium signalling and plant stress tolerance. J Plant Physiol 256:153327

Pandey GK, Mahiwal S (2020) Potassium uptake and transport system in plant. In: Role of potassium in plants. Springer, New York, pp 19–28

Pandey N, Pathak GC, Singh AK, Sharma CP (2002) Enzymic changes in response to zinc nutrition. J Plant Physiol 159(10):1151–1153

Pandey R, Lal MK, Vengavasi K (2018) Differential response of hexaploid and tetraploid wheat to interactive effects of elevated [CO_2] and low phosphorus. Plant Cell Rep 37(9):1231–1244

Pant BD, Buhtz A, Kehr J, Scheible W (2008) MicroRNA399 is a long-distance signal for the regulation of plant phosphate homeostasis. Plant J 53(5):731–738

Paszkowski U, Kroken S, Roux C, Briggs SP (2002) Rice phosphate transporters include an evolutionarily divergent gene specifically activated in arbuscular mycorrhizal symbiosis. Proc Natl Acad Sci 99(20):13324–13329

Peiffer GA, King KE, Severin AJ, May GD, Cianzio SR, Lin SF, Lauter NC, Shoemaker RC (2012) Identification of candidate genes underlying an iron efficiency quantitative trait locus in soybean. Plant Physiol 158(4):1745–1754

Peng WT, Zhang LD, Zhou Z, Fu C, Chen ZC, Liao H (2018) Magnesium promotes root nodulation through facilitation of carbohydrate allocation in soybean. Physiol Plant 163(3):372–385

Peng WT, Qi WL, Nie MM, Xiao YB, Liao H, Chen ZC (2020) Magnesium supports nitrogen uptake through regulating NRT2. 1/2.2 in soybean. Plant Soil 457(1):97–111

Pettigrew WT (2008) Potassium influences on yield and quality production for maize, wheat, soybean and cotton. Physiol Plant 133(4):670–681

Pittman JK, Hirschi KD (2016) Phylogenetic analysis and protein structure modelling identifies distinct Ca^{2+}/Cation antiporters and conservation of gene family structure within Arabidopsis and rice species. Rice 9(1):1–6

Platten JD, Cotsaftis O, Berthomieu P, Bohnert H, Davenport RJ, Fairbairn DJ, Horie T, Leigh RA, Lin HX, Luan S (2006) Nomenclature for HKT transporters, key determinants of plant salinity tolerance. Trends Plant Sci 11(8):372–374

Pratt J, Boisson AM, Gout E, Bligny R, Douce R, Aubert S (2009) Phosphate (Pi) starvation effect on the cytosolic Pi concentration and Pi exchanges across the tonoplast in plant cells: an in vivo 31P-nuclear magnetic resonance study using methylphosphonate as a Pi analog. Plant Physiol 151(3):1646–1657

Püschel D, Janoušková M, Voříšková A, Gryndlerová H, Vosátka M, Jansa J (2017) Arbuscular mycorrhiza stimulates biological nitrogen fixation in two *Medicago spp.* through improved phosphorus acquisition. Front Plant Sci 8:390

Qin L, Guo Y, Chen L, Liang R, Gu M, Xu G, Zhao J, Walk T, Liao H (2012) Functional characterization of 14 Pht1 family genes in yeast and their expressions in response to nutrient starvation in soybean. PLoS One 7(10):e47726

Qin L, Han P, Chen L, Walk TC, Li Y, Hu X, Xie L, Liao H, Liao X (2017) Genome-wide identification and expression analysis of NRAMP family genes in soybean (*Glycine max* L.). Front Plant Sci 8:1436

Ragel de la Torre P (2019) Regulation of K^+ nutrition in plants. Front Plant Sci 10:281

Raghothama KG (1999) Phosphate acquisition. Annu Rev Plant Biol 50(1):665–693

Ramírez CM, Goedeke L, Rotllan N, Yoon JH, Cirera-Salinas D, Mattison JA, Suárez Y, de Cabo R, Gorospe M, Fernández-Hernando C (2013) MicroRNA 33 regulates glucose metabolism. Mol Cell Biol 33(15):2891–2902

Ramtekey V, Bansal R, Aski MS, Kothari D, Singh A, Pandey R, Tripathi K, Mishra GP, Kumar S, Dikshit HK (2021) Genetic variation for traits related to phosphorus use efficiency in lens species at the seedling stage. Plants 10(12):2711

Rausch C, Bucher M (2002) Molecular mechanisms of phosphate transport in plants. Planta 216(1): 23–37

Rebello CJ, Liu AG, Greenway FL, Dhurandhar NV (2013) Dietary strategies to increase satiety. Adv Food Nutr Res 69:105–182

Reddy ASN, Ali GS, Celesnik H, Day IS (2011) Coping with stresses: roles of calcium- and calcium/calmodulin-regulated gene expression. Plant Cell 23:2010–2032

Reddy VRP, Aski MS, Mishra GP, Dikshit HK, Singh A, Pandey R, Singh MP, Ramtekey V, Rai N, Nair RM (2020) Genetic variation for root architectural traits in response to phosphorus deficiency in mungbean at the seedling stage. PLoS One 15(6):e0221008

Redondo-Nieto M, Wilmot AR, El-Hamdaoui A, Bonilla I, Bolaños L (2003) Relationship between boron and calcium in the N_2-fixing legume–rhizobia symbiosis. Plant Cell Environ 26(11): 1905–1915

Rehman HM, Nawaz MA, Shah ZH, Daur I, Khatoon S, Yang SH, Chung G (2017) In-depth genomic and transcriptomic analysis of five K^+ transporter gene families in soybean confirm their differential expression for nodulation. Front Plant Sci 8:804

Rengel Z, Graham RD (1995) Importance of seed Zn content for wheat growth on Zn-deficient soil. Plant Soil 173(2):259–266

Richardson AE, Lynch JP, Ryan PR, Delhaize E, Smith FA, Smith SE, Harvey PR, Ryan MH, Veneklaas EJ, Lambers H (2011) Plant and microbial strategies to improve the phosphorus efficiency of agriculture. Plant Soil 349(1):121–156

Romera FJ, García MJ, Alcántara E, Pérez-Vicente R (2011) Latest findings about the interplay of auxin, ethylene and nitric oxide in the regulation of Fe deficiency responses by Strategy I plants. Plant Signal Behav 6(1):167–170

Roorkiwal M, Pandey S, Thavarajah D, Hemalatha R, Varshney RK (2021) Molecular mechanisms and biochemical pathways for micronutrient acquisition and storage in legumes to support biofortification for nutritional security. Front Plant Sci 12:682842

Rouached H, Wirtz M, Alary R, Hell R, Arpat AB, Davidian JC, Fourcroy P, Berthomieu P (2008) Differential regulation of the expression of two high-affinity sulfate transporters, SULTR1. 1 and SULTR1. 2, in Arabidopsis. Plant Physiol 147(2):897–911

Rubio F, Alemán F, Nieves-Cordones M, Martínez V (2010) Studies on Arabidopsis athak5, atakt1 double mutants disclose the range of concentrations at which AtHAK5, AtAKT1 and unknown systems mediate K^+ uptake. Physiol Plant 139(2):220–228

Rudd JJ, Franklin-Tong VE (2001) Unravelling response-specificity in Ca^{2+} signalling pathways in plant cells. New Phytol 151(1):7–33

Saito K, Inoue K, Fukushima R, Noji M (1997) Genomic structure and expression analyses of serine acetyltransferase gene in *Citrullus vulgaris* (watermelon). Gene 189(1):57–63

Sánchez-Chino X, Jiménez-Martínez C, Dávila-Ortiz G, Álvarez-González I, Madrigal-Bujaidar E (2015) Nutrient and nonnutrient components of legumes, and its chemopreventive activity: a review. Nutr Cancer 67(3):401–410

Sanders D, Pelloux J, Brownlee C, Harper JF (2002) Calcium at the crossroads of signaling. Plant Cell 14(Suppl):S401–S417

Santa-Maria GE, Danna CH, Czibener C (2000) High-affinity potassium transport in barley roots. Ammonium-sensitive and -insensitive pathways. Plant Physiol 123(1):297–306

Schachtman DP, Schroeder J, Lucas WJ, Anderson JA, Gaber RF (1992) Expression of an inward-rectifying potassium channel by the Arabidopsis KAT1 cDNA. Science 258(5088):1654–1658

Schachtman DP, Reid RJ, Ayling SM (1998) Phosphorus uptake by plants: from soil to cell. Plant Physiol 116(2):447–453

Scherzer S, Böhm J, Krol E, Shabala L, Kreuzer I, Larisch C, Bemm F, Al-Rasheid KAS, Shabala S, Rennenberg H (2015) Calcium sensor kinase activates potassium uptake systems in gland cells of *Venus flytraps*. Proc Natl Acad Sci 112(23):7309–7314

Schumaker KS, Sze H (1986) Calcium transport into the vacuole of oat roots. Characterization of H^+/Ca^{2+} exchange activity. J Biol Chem 261(26):12172–12178

Shabala SS, Palmgren MG (2011) Calcium efflux systems in stress signalling and adaptation in plants. Front Plant Sci 2:121–129

Shabala S, Pottosin II (2010) Potassium and potassium-permeable channels in plant salt tolerance. In: Ion channels and plant stress responses, pp 87–110

Sharif HR, Williams PA, Sharif MK, Abbas S, Majeed H, Masamba KG, Safdar W, Zhong F (2018) Current progress in the utilization of native and modified legume proteins as emulsifiers and encapsulants–a review. Food Hydrocoll 76:2–16

Sharma S, Malhotra H, Borah P, Meena MK, Bindraban P, Chandra S, Pande V, Pandey R (2019) Foliar application of organic and inorganic iron formulation induces differential detoxification response to improve growth and biofortification in soybean. Plant Physiol Rep 24(1):119–128

Shinmachi F, Buchner P, Stroud JL, Parmar S, Zhao FJ, McGrath SP, Hawkesford MJ (2010) Influence of sulfur deficiency on the expression of specific sulfate transporters and the distribution of sulfur, selenium, and molybdenum in wheat. Plant Physiol 153(1):327–336

Singh B, Pandey R (2003) Differences in root exudation among phosphorus-starved genotypes of maize and green gram and its relationship with phosphorus uptake. J Plant Nutr 26(12): 2391–2401

Singh J, Singh M, Jain A, Bhardwaj S, Singh A, Singh D, Bhushan B, Dubey S (2013) An introduction of plant nutrients and foliar fertilization: a review "Precision farming: a new approach". Daya Publishing Co., New Delhi

Smith SE, Read DJ (2010) Mycorrhizal symbiosis. Academic, San Diego

Smith FW, Ealing PM, Hawkesford MJ, Clarkson DT (1995) Plant members of a family of sulfate transporters reveal functional subtypes. Proc Natl Acad Sci 92(20):9373–9377

Smith FW, Hawkesford MJ, Ealing PM, Clarkson DT, Vanden Berg PJ, Belcher AR, Warrilow AGS (1997) Regulation of expression of a cDNA from barley roots encoding a high affinity sulphate transporter. Plant J 12(4):875–884

Socha AL, Guerinot ML (2014) Mn-euvering manganese: the role of transporter gene family members in manganese uptake and mobilization in plants. Front Plant Sci 5:106

Sottocornola B, Visconti S, Orsi S, Gazzarrini S, Giacometti S, Olivari C, Camoni L, Aducci P, Marra M, Abenavoli A (2006) The potassium channel KAT1 is activated by plant and animal 14-3-3 proteins. J Biol Chem 281(47):35735–35741

Stribe DR (2012) Analysis of iron transporters in the soybean (*Glycine max* (L.) Merr.) genome. Iowa State University

Su J, Ding L, Xue K, Yao H, Quensen J, Bai S, Wei W, Wu J, Zhou J, Tiedje JM (2015) Long-term balanced fertilization increases the soil microbial functional diversity in a phosphorus-limited paddy soil. Mol Ecol 24(1):136–150

Sutter JU, Campanoni P, Tyrrell M, Blatt MR (2006) Selective mobility and sensitivity to SNAREs is exhibited by the Arabidopsis KAT1 K^+ channel at the plasma membrane. Plant Cell 18(4): 935–954

Sze H, Liang F, Hwang I, Curran AC, Harper JF (2000) Diversity and regulation of plant Ca^{2+} pumps: insights from expression in yeast. Annu Rev Plant Biol 51(1):433–462

Sze H, Padmanaban S, Cellier F, Honys D, Cheng NH, Bock KW, Conéjéro G, Li X, Twell D, Ward JM (2004) Expression patterns of a novel *AtCHX* gene family highlight potential roles in osmotic adjustment and K^+ homeostasis in pollen development. Plant Physiol 136(1): 2532–2547

Tabe LM, Venables I, Grootemaat A, Lewis D (2003) Sulfur transport and assimilation in developing embryos of chickpea (*Cicer arietinum*). Sulfur transport and assimilation in plants. Backhuys, Leiden, The Netherlands, pp 335–337

Takahashi H (2019) Sulfate transport systems in plants: functional diversity and molecular mechanisms underlying regulatory coordination. J Exp Bot 70(16):4075–4087

Takanashi K, Yokosho K, Saeki K, Sugiyama A, Sato S, Tabata S, Ma JF, Yazaki K (2013) LjMATE1: a citrate transporter responsible for iron supply to the nodule infection zone of *Lotus japonicus*. Plant Cell Physiol 54(4):585–594

Tamura Y, Kobae Y, Mizuno T, Hata S (2012) Identification and expression analysis of arbuscular mycorrhiza-inducible phosphate transporter genes of soybean. Biosci Biotechnol Biochem 76(2):309–313

Tang RJ, Luan S (2017) Regulation of calcium and magnesium homeostasis in plants: from transporters to signaling network. Curr Opin Plant Biol 39:97–105

Tejada-Jiménez M, Castro-Rodríguez R, Kryvoruchko I, Lucas MM, Udvardi M, Imperial J, González-Guerrero M (2015) *Medicago truncatula* natural resistance-associated macrophage protein1 is required for iron uptake by rhizobia-infected nodule cells. Plant Physiol 168(1): 258–272

Thomine S, Vert G (2013) Iron transport in plants: better be safe than sorry. Curr Opin Plant Biol 16(3):322–327

Tuteja N, Mahajan S (2007) Calcium signaling network in plants: an overview. Plant Signal Behav 2(2):79–85

Ueoka-Nakanishi H, Nakanishi Y, Tanaka Y, Maeshima M (1999) Properties and molecular cloning of Ca^{2+}/H^+ antiporter in the vacuolar membrane of mung bean. Eur J Biochem 262(2):417–425

Valdés-López O, Arenas-Huertero C, Ramirez M, Girard L, Sanchez F, Vance CP, Luis Reyes J, Hernandez G (2008) Essential role of MYB transcription factor: PvPHR1 and microRNA: PvmiR399 in phosphorus-deficiency signalling in common bean roots. Plant Cell Environ 31(12):1834–1843

Vengavasi K, Pandey R (2018) Root exudation potential in contrasting soybean genotypes in response to low soil phosphorus availability is determined by photo-biochemical processes. Plant Physiol Biochem 124:1–9

Vengavasi K, Kumar A, Pandey R (2016) Transcript abundance, enzyme activity and metabolite concentration regulates differential carboxylate efflux in soybean under low phosphorus stress. Indian J Plant Physiol 21(2):179–188

Véry AA, Nieves-Cordones M, Daly M, Khan I, Fizames C, Sentenac H (2014) Molecular biology of K^+ transport across the plant cell membrane: what do we learn from comparison between plant species? J Plant Physiol 171(9):748–769

Walker DJ, Leigh RA, Miller AJ (1996) Potassium homeostasis in vacuolate plant cells. Proc Natl Acad Sci 93(19):10510–10514

Wang Y, Wu WH (2013) Potassium transport and signaling in higher plants. Annu Rev Plant Biol 64:451–476

Wang C, Chen H, Hao Q, Sha A, Shan Z, Chen L, Zhou R, Zhi H, Zhou X (2012) Transcript profile of the response of two soybean genotypes to potassium deficiency. PLoS One 7(7):e39856

Wang G, Zeng H, Hu X, Zhu Y, Chen Y, Shen C, Wang H, Poovaiah BW, Du L (2015) Identification and expression analyses of calmodulin-binding transcription activator genes in soybean. Plant Soil 386(1):205–221

Wang T, Zhao M, Zhang X, Liu M, Yang C, Chen Y, Chen R, Wen J, Mysore K, Zhang W (2017) Novel phosphate deficiency-responsive long non-coding RNAs in the legume model plant *Medicago truncatula*. J Exp Bot 68:5937–5948

Wang N, Qiu W, Dai J, Guo X, Lu Q, Wang T, Li S, Liu T, Zuo Y (2019) AhNRAMP1 enhances manganese and zinc uptake in plants. Front Plant Sci 10:415

Wang YWF, Lu H, Liu Y, Mao C (2021) Phosphate uptake and transport in plants: an elaborate regulatory system. Plant Cell Physiol 62(4):564–572

White PJ (2012) Ion uptake mechanisms of individual cells and roots: short-distance transport. In: Marschner's mineral nutrition of higher plants. Elsevier, Amsterdam, pp 7–47

Wu H, Li L, Du J, Yuan Y, Cheng X, Ling HQ (2005) Molecular and biochemical characterization of the Fe (III) chelate reductase gene family in *Arabidopsis thaliana*. Plant Cell Physiol 46(9): 1505–1514

Xie X, Huang W, Liu F, Tang N, Liu Y, Lin H, Zhao B (2013) Functional analysis of the novel mycorrhiza-specific phosphate transporter As PT 1 and PHT 1 family from Astragalus sinicus during the arbuscular mycorrhizal symbiosis. New Phytol 198(3):836–852

Xiong H, Kobayashi T, Kakei Y, Senoura T, Nakazono M, Takahashi H, Nakanishi H, Shen H, Duan P, Guo X (2012) AhNRAMP1 iron transporter is involved in iron acquisition in peanut. J Exp Bot 63(12):4437–4446

Xu J, Li HD, Chen LQ, Wang Y, Liu LL, He L, Wu WH (2006) A protein kinase, interacting with two calcineurin B-like proteins, regulates K^+ transporter AKT1 in Arabidopsis. Cell 125(7): 1347–1360

Xu F, Liu Q, Chen L, Kuang J, Walk T, Wang J, Liao H (2013) Genome-wide identification of soybean microRNAs and their targets reveals their organ-specificity and responses to phosphate starvation. BMC Genomics 14(1):1–30

Yamaji N, Sasaki A, Xia JX, Yokosho K, Ma JF (2013) A node-based switch for preferential distribution of manganese in rice. Nat Commun 4(1):1–11

Yang GH, Yang LT, Jiang HX, Li Y, Wang P, Chen LS (2012) Physiological impacts of magnesium-deficiency in Citrus seedlings: photosynthesis, antioxidant system and carbohydrates. Trees 26(4):1237–1250

Yuan Y, Wu H, Wang N, Li J, Zhao W, Du J, Wang D, Ling HQ (2008) FIT interacts with AtbHLH38 and AtbHLH39 in regulating iron uptake gene expression for iron homeostasis in Arabidopsis. Cell Res 18(3):385–397

Zeng HQ, Zhu YY, Huang SQ, Yang ZM (2010) Analysis of phosphorus-deficient responsive miRNAs and cis-elements from soybean (*Glycine max* L.). J Plant Physiol 167(15):1289–1297

Zeng J, Ye Z, He X, Zhang G (2019) Identification of microRNAs and their targets responding to low-potassium stress in two barley genotypes differing in low-K tolerance. J Plant Physiol 234: 44–53

Zeng H, Zhao B, Wu H, Zhu Y, Chen H (2020) Comprehensive in silico characterization and expression profiling of nine gene families associated with calcium transport in soybean. Agronomy 10(10):1539

Zhang L, Wen A, Wu X, Pan X, Wu N, Chen X, Chen Y, Mao D, Chen L, Luan S (2019) Molecular identification of the magnesium transport gene family in *Brassica napus*. Plant Physiol Biochem 136:204–214

Zhang Z, Li Z, Wang W, Jiang Z, Guo L, Wang X, Qian Y, Huang X, Liu Y, Liu X (2021) Modulation of nitrate-induced phosphate response by the MYB transcription factor RLI1/HINGE1 in the nucleus. Mol Plant 14(3):517–529

Zhao D, Oosterhuis DM, Bednarz CW (2001) Influence of potassium deficiency on photosynthesis, chlorophyll content, and chloroplast ultrastructure of cotton plants. Photosynthetica 39(1): 103–109

Zhou T, Wang P, Yang R, Wang X, Gu Z (2018) Ca^{2+} influxes and transmembrane transport are essential for phytic acid degradation in mung bean sprouts. J Sci Food Agric 98(5):1968–1976

Zhu YY, Zeng HQ, Dong CX, Yin XM, Shen QR, Yang ZM (2010) microRNA expression profiles associated with phosphorus deficiency in white lupin (*Lupinus albus* L.). Plant Sci 178(1):23–29

Zhu J, Lau K, Puschmann R, Harmel RK, Zhang Y, Pries V, Gaugler P, Broger L, Dutta AK, Jessen HJ (2019) Two bifunctional inositol pyrophosphate kinases/phosphatases control plant phosphate homeostasis. Elife 8:e43582

Zuchi S, Cesco S, Varanini Z, Pinton R, Astolfi S (2009) Sulphur deprivation limits Fe-deficiency responses in tomato plants. Planta 230(1):85–94

Stress Memory and Its Mitigation via Responses Through Physiological and Biochemical Traits in Mung Bean Under Moisture Stress

13

B. Sunil Kumar, K. R. Saravanan, P. Sudhakar, R. Anandan, G. Sathiyanarayanan, J. Gokulakrishnan, and M. Prakash

Abstract

The experience of stressful events can alter the plant structures involved in memory encoding, storage, and its retrieval. Stress may sometimes improve memories while sometimes diminishing them. Some contributing traits like the type of stress (stressor) involved, the plant's nature in reacting to the stress, and the interplay of various environmental factors appear to be important in stress memory. Both modes of action including the stimulation of various receptors and the activation of related biomolecules in the plant system play a pivotal role. A unified conceptual framework is required to identify scenarios in which these contradicting stress effects are likely to occur and further determine the downstream mechanisms that govern these stress-related responses that improve or hinder plant stress memory.

Keywords

Mung bean · Moisture stress · Stress memory · Mitigation · Physiological · Biochemical · Molecular levels

B. Sunil Kumar (✉) · K. R. Saravanan · R. Anandan · G. Sathiyanarayanan · J. Gokulakrishnan · M. Prakash
Department of Genetics and Plant Breeding, Faculty of Agriculture, Annamalai University, Annamalai Nagar, Tamil Nadu, India

P. Sudhakar
Department of Agronomy, Faculty of Agriculture, Annamalai University, Annamalai Nagar, Tamil Nadu, India

P. Muthu Arjuna Samy et al. (eds.), *Legumes: Physiology and Molecular Biology of Abiotic Stress Tolerance*, https://doi.org/10.1007/978-981-19-5817-5_13

13.1 Introduction

Pulses are recognised for their high protein content, especially when compared to cereal crops. Mung bean is the cheapest source of plant protein (Chandrashekhar 2019), with an estimated protein content of 22–27%. Calcium and sodium are also abundant in mung bean. Vitamins A and B are plentiful in dried mung bean, but calcium, phosphorous, potassium, and other minerals are plentiful in sprouting mung bean. Similarly, qualities like palatability, digestibility, and a low proportion of flatulence factors increase their value among diverse pulse crops (Bangar et al. 2019; Markam et al. 2018). India is the world's largest producer, buyer, and importer of pulses, allowing it to influence the decisions of all other pulse markets (Chandrashekhar 2016).

Plant stress responses are multifaceted and rely on an initial alarm phase that activates multiple stress resistance/avoidance/escape mechanisms before slowing down growth-related processes to endure the stress and repair any damage that has already occurred. However, in the event of severe stress, the damage is permanent and the plant dies. However, if the plant recovers physically from drought stress during the recovery phase, the first stress will have an impact and leave an imprint that will help it respond to future challenges. The structural changes that occur under moisture stress range from the whole plant (e.g. changes in the number of leaves, the area of the plant, or the thickness of the plant) to the genetic level (e.g. histone modification). The root and shoot biomass ratio, number of leaves, leaf area, leaf mass per area ratio (LMA), leaf size, and photosynthetic system, including chloroplast shape and organisation, are among the structural alterations (Aroca 2013).

Plant growth and development are slowed down by drought stress, resulting in a loss of economic output (Basu et al. 2016; Ciais et al. 2005). Many elements connected to droughts, such as their intensity, duration, plant genotype, and stress imprint left on the plant, have been documented to influence the impacts of drought stress. This imprint/stress memory can be divided into genetic, metabolic, and structural changes that happened because of stress exposure, allowing the plant to withstand future stress. However, resistance may lower the economic output in the short term by reducing photosynthesis, but raising the plant's tolerance to subsequent stress may improve its efficiency in the long run (Fahad et al. 2017; Bruce et al. 2007). However, if the stress is too extreme and lasts too long, it will harm productivity in the short and long terms.

Photosynthesis inhibition is one of the plant's responses to drought stress, and it directly leads to lower production output. Because of the excess energy in chloroplasts, this reduction disturbs the electron transport chain, resulting in an increase in the generation of reactive oxygen species (ROS) (Ghosh and Xu 2014). Drought-stressed plants seal their stomata, preventing dehydration by photoinhibition caused by high ABA levels. Similarly, abscisic acid creates protective biomolecules such as osmolytes, which aid in maintaining membrane structure integrity (Verslues et al. 2006).

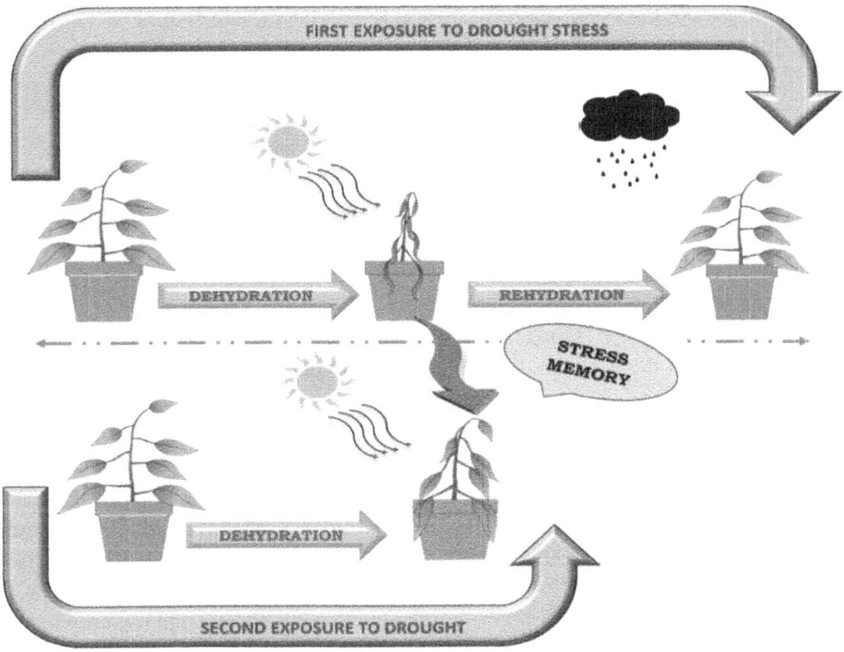

Fig. 13.1 Stress memory—dehydration and hydration mechanism

Drought cues are essential for the gene's expression, according to Campos and Reinberg (2009). It was discovered that variations in gene expression patterns are linked to changes in chromatin status, resulting in the modification of histone tails in drought stress-responsive genes in response to dehydration. Plants have extensive defensive systems for stress memory (Kinoshita and Seki 2014). If plant wilts owing to drought stress, it subsequently rebounds to complete its life cycle due to rehydration. But if the plant is exposed to second drought stress, the plant 'remembers' the previous drought experience, and as a result, the plant can perform better during dehydration and increase its chances of survival when compared to plants that have not been exposed to drought stress. These multiple drought exposures have enabled the plants to respond more quickly to fresh stress by generating adaptive changes to gene expression at a faster rate (Fig. 13.1).

Mung bean has traditionally been grown as a rainfed crop with high temperatures ranging from 27 to 30 °C, low humidity, and 60–80 cm of rainfall. As a result, it is subjected to drought stress throughout critical periods of development (Nair et al. 2019). Mung bean thrives in areas with little or no water. Moisture stress affects mung bean varieties differently depending on their duration, growth stage, and variety (Dutta and Bera 2008; Ahmad et al. 2015). The heterogeneity in morphophysiological features for drought tolerance during important developmental phases of growth has been noted by several researchers (Naresh et al. 2013; Uddin et al. 2013).

13.1.1 Drought Stress

Drought is one of the most common abiotic pressures on the planet. Drought is currently the world's main food supply constraint, and it has been difficult to attain tolerance since it causes a wide range of phenotypic responses in plants, depending on the severity of the stress, the environment, the species, and the stage of development (Tuberosa 2012). Mung bean diversity grown in different climate zones demonstrates that it has evolved to withstand drought stress through morphological, physiological, and biochemical modifications (Bharadwaj et al. 2018; Bohnert et al. 1995). Droughts have become more common as climatic conditions have changed, and basic plant growth processes such as seed germination, enzyme activities, turgor pressure, cell division and elongation, photosynthesis, source-sink relationships, and secondary metabolite production have been harmed (Yordanov et al. 2000; Flexas et al. 2004; Farooq et al. 2009; Kinoshita and Seki 2014; Basu et al. 2016; Tietjen et al. 2017).

Legumes have a significant impact on the photosynthetic rate and total dry matter production during the vegetative stage (Jain 1975). Jain (1975) also stated that when the pod began to mature, the photosynthetic rate dropped dramatically, which was accompanied by a decrease in total nitrogen content. Similarly, other biophysical parameters like photosynthetic and transpiration rates, as well as stomatal conductance, were higher during early pod growth and thereafter decreased as pod size increased (Venkateswarlu and Subramanian 1993; Maiti and Kumari 2016).

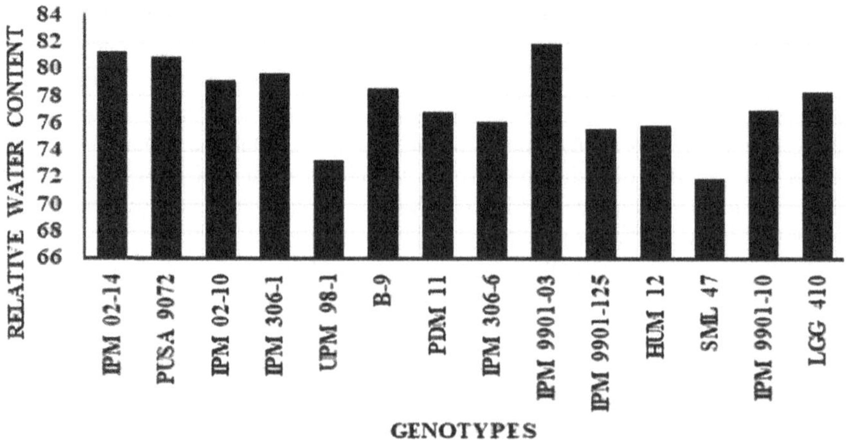

Relative water content among various genotypes under drought stress

13.1.2 Hormonal Profiling Reveals Stress Memory

Abscisic acid (ABA), cytokinin (CK), gibberellic acid (GA_3), auxin, and ethylene are phytohormones that regulate a variety of plant activities, allowing plants to respond to drought stress (Wilkinson et al. 2012). Drought causes abscisic acid (ABA) to be triggered and produced in the roots, and then translocated to the leaves. It starts the drought adaption process by closing the stomata and reducing plant growth. However, because of the invariable drop in CO_2 intake due to stomatal closure and ABA-related senescence during the reproductive stage, using the drought ABA-induced reaction model for higher economic production is still difficult (Ji et al. 2011). Several ABA signalling genes, such as Osan, OsNAC5, and DSM2, have been demonstrated to improve economic production under crucial drought-stress circumstances (Chen et al. 2014). Under reproductive drought, these ABA-induced non-stomatal genes can be used and exploited to increase grain yield.

Similarly, the role of GA_3, which is naturally present in young leaves and germinating embryos in regulating crop productivity, needs to be investigated further in plants because it improves shoot elongation (Ratnasekera and Subhashi 2015; Keykha et al. 2014) and plant growth and development (Kundu et al. 2017; Rahman et al. 2018) and enhances enzymes such as carbonic anhydrase (CA) (El Karamany et al. 2019). GA_3 has also been shown to aid in the mitigation of water balance in drought-stressed plants. When compared to controls, GA_3 treatments significantly increase photosynthetic pigments such as chlorophyll a, b, and carotenoids (Mubeen et al. 2015; Baliah et al. 2018; El Karamany et al. 2019). Similarly, a phenolic phytohormone called salicylic acid promotes plant growth and development, photosynthesis, transpiration, ion uptake, and transport at lower concentrations, but is determinantal and inhibitive at greater doses. Salicylic acid at 150 ppm boosted the structural component of amino acid RNA molecules, as well as DNA, RNA, and protein production in the ribosome, according to Raj Kumar et al. (2018). Under drought stress, ethylene inhibits leaf senescence, imminent root growth and development, shoot/leaf expansion, and photosynthesis (Sharp 2002). Other hormones, such as brassinosteroids, jasmonic acid (JA), salicylic acid (SA), and strigolactone, have also been discovered to be as important in plant growth and development. Even though their function is well understood, little is known about their role in mung bean drought stress.

13.1.3 Leaf Water Contents, Gas Exchange, and Chlorophyll Fluorescence

When plants are stressed by drought, the initial response is the closure of the stomatal pore. However, closing stomata not only limits water loss in the form of water vapour, but also reduces CO_2 intake and increases O_2, resulting in O_2 saturation, oxidation of the leaf, and altered metabolic pathways (Xiong and Zhu 2002). Plants have evolved xeromorphic features to limit water loss through reduced transpiration and plant adaptations such as sunken stomata, thick cuticle, presence of thorns, powdery coating, and others to withstand drought stress over time. Plants achieve transpirational loss under stress by losing leaves, reducing leaf number, size, and branching behaviour. Sclerophylly, on the other hand, is an adaptation in which the plant leaves thicken to resist withering during extreme drought stress and to revert to normal functioning once the stress is relieved (Micco Veronica De and Giovanna 2016). Researchers discovered that under drought stress, stomatal conductance is affected not only by reduced expression of aquaporin genes but also by the intercellular leaf chloroplast surface area. The developmental stage of the leaf and the leaf angle for light collecting, which influence conductance and photosynthetic capacity, are further elements that interfere with mesophyll and chloroplast differentiation modelling. According to research, stomatal density increases under mild drought stress and decreases completely during severe drought. Plant adaptations not only reduce the negative consequences of photosynthesis but also improve water-use efficiency (WUE), resulting in higher economic yields (Blum 2005).

13.1.4 Photosynthetic Pigments and Antioxidants

Drought stress is known to decrease both leaf area and photosynthetic rate per unit leaf area. Stomatal closure or impairment of metabolic activity is caused by a decrease in photosynthetic rate (Tezara et al. 1999; Flexas et al. 2004). Photosynthesis is the most widely studied for its role in enhancing economic yields (Foyer et al. 2017). Photosynthesis is described to improve efficiency by linking yield production with different transposing mechanisms (Long et al. 2015). Ainsworth and Bush (2011) supported and emphasised the positive correlation between enhanced photosynthesis and yield under elevated (CO_2) conditions, which showed that increased source strength is needed to improve yields. Similarly, the continued photosynthetic light reactions under drought stress result in impeded photosynthetic electron transport, which in turn prevents the provision of reactive oxygen species (ROS). ROS has been found to severely impair the photosynthetic apparatus (Lawlor 2002). Some of the adaptive responses in plants to reduce induced damage to photosynthesis include certain pathways like the xanthophyll cycle, thermal dissipation of light energy, water-water cycle, and damaged light harvesting complexes from photosynthetic reaction centres (Kinoshita and Seki 2014; Jahns and Holzwarth 2012). Niyogi (1999) revealed a positive correlation with photosynthetic carbon metabolism during drought stress. Similarly, the biochemical efficiency of drought-based photosynthesis depends on the first carbon compound of the C_3 and C_4 cycles, namely ribulose-1,5-bisphosphate (RuBP) and ribulose-1,5-bisphosphate carboxylase/oxygenase (RuBisCo) (Lawlor 2002). Of late, considerable progress has been reported in improving the stoma centre diffusion, photosynthetic light reaction, and metabolic changes in the expression of regulatory genes under drought (Chaves et al. 2009).

Genotypes having superior means under drought stress for biophysical traits

Characters	Range	Genotypes having superior means under drought
Photosynthetic rate (Pn)	10.31–39.42	IPM 02-14, IPM 306-1, UPM 98-1, B-9, PDM 11, IPM 9901-125, HUM 12, IPM 9901-10
Transpiration rate (Tr)	3.68–7.64	IPM 02-14, IPM 02-10, IPM 306-1, UPM 98-1, PDM 11, IPM 306-6, IPM 9901-03, IPM 9901-10
Stomatal conductance (Cs)	0.22–0.79	IPM 02-14, IPM 02-10, IPM 306-1, IPM 9901-03, IPM 9901-125, HUM 12
Leaf temperature (*T* ° C)	31.90–40.48	PUSA 9072, IPM 02-10, IPM 306-1, UPM 98-1, PDM 11, SML 47, IPM 9901-10
Intercellular CO_2 concentration ratio (Ci)	224.30–269.48	PUSA 9072, IPM 02-10, SML 47

Under constrained water stress, Flexas et al. (2012) emphasised the relevance of photosynthetic responses. They shed light on the mechanisms that limit photosynthesis during drought and reduce CO_2 diffusion during the early stages of stress, a topic that is becoming increasingly crucial in the context of climate change.

Under drought stress, Ivlev (2017) stressed the importance of carbon assimilation pathways, stating that C_4 is more effective than C_3 in terms of water loss, photorespiration, and photosynthetic efficiency. Even if transferring the C_3 pathway into C_4 crops is a far-fetched fantasy for improving grain yields (Gowik and Westhoff 2011), combining all computational models to integrate physiological and metabolic processes with genetic/gene expression data is a pressing necessity. Similarly, contemporary breeding techniques utilising transgenic technology may open the way for improved photosynthesis and productivity in drought-stricken areas.

Under constrained water stress, Flexas et al. (2012) emphasised the relevance of photosynthetic responses. They shed light on the mechanisms that limit photosynthesis during drought and reduce CO_2 diffusion during the early stages of stress, a topic that is becoming increasingly crucial in the context of climate change.

Under drought stress, Edwards and Walker (1983) stressed the importance of carbon assimilation pathways, stating that C_4 is more effective than C_3 in terms of water loss, photorespiration, and photosynthetic efficiency. Even if transferring the C_3 pathway into C_4 crops is a far-fetched fantasy for improving grain yields (Gowik and Westhoff 2011), combining all computational models to integrate physiological and metabolic processes with genetic/gene expression data is a pressing necessity. Similarly, contemporary breeding techniques utilising transgenic technology may open the way for improved photosynthesis and productivity in drought-stricken areas.

13.1.5 Source-Sink Relationships

The relationship between changing climatic conditions and movement of photosynthates from leaves/green parts to competitive sinks, which is then used to build new plant biomass and carbohydrate remobilisation into reproductive structures, is well understood. To assess increased crop productivity, strong source-sink interactions are required. Depending on the stage of growth and development of the plant, the plant sections contribute to the source and sink change. During the grain-filling stage, for example, the growing grains serve as primary sinks, while the top two leaves serve as key sources. With a restricted supply of water and nutrients, an increase in photosynthetic rates is attributed to an increase in photosynthate production rather than photosynthetic performance. Drought stress impairs the source-sink interaction by impounding or lowering photosynthate production at the source, resulting in a drop in economic yield.

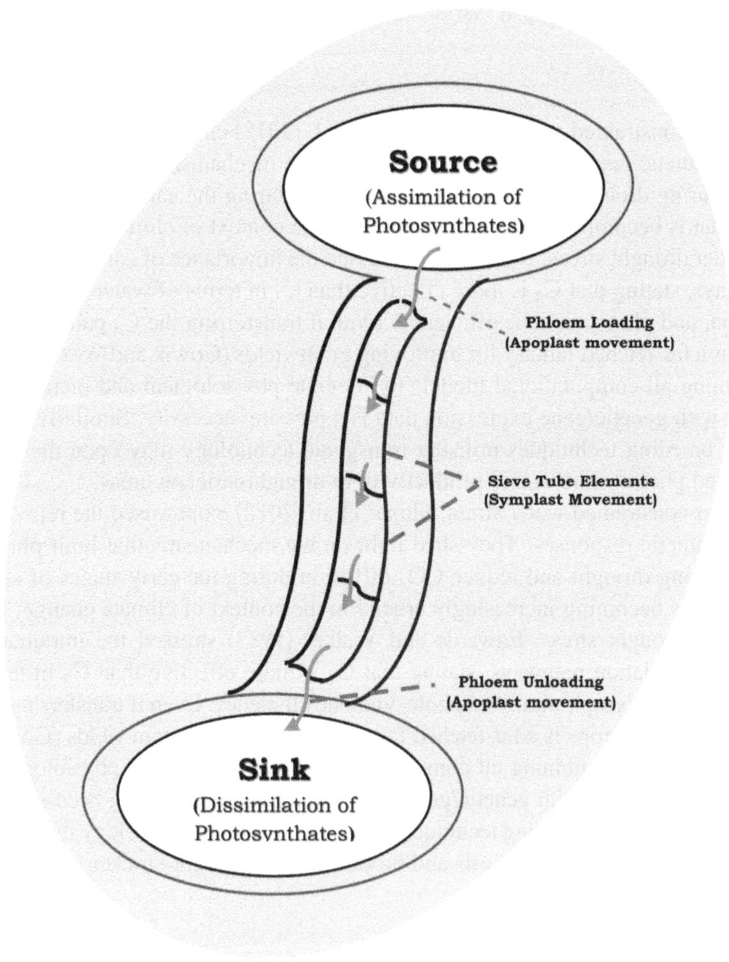

Diagrammatic representation of source-sink relationship

Drought stress causes aberrant starch to build up in pollens, reducing pollen viability significantly.

The interchange of photosynthates/sugars from leaves and the demand for photosynthates from competing by sinks can be discussed and investigated further to improve productivity (Ainsworth and Bush 2011; Lemoine et al. 2013; White et al. 2016). Under drought stress, phloem loading and unloading of metabolites, as well as their transport, are considered essential factors regulating yield (Turgeon 1996).

Photosynthate metabolism and storage occur at near distances in the circuit between source and sink, according to Bihmidine et al. (2013), and their participation in partitioning could be a major yield determiner. As a result, we must increase our current knowledge of the effect of the environment on the source and sink, their route linkages, and the mechanisms that obstruct photosynthate transfer between source and sink.

According to research experts, the flow of materials from source to sink may be managed by a highly evolved regulated signalling network that determines the most potent and competitive sink based on available photosynthates (Paul and Foyer 2001; Rossi et al. 2015). Manipulation of source-sink connections may help boost overall light penetration into the plant canopy while also lowering the competition between vegetative and reproductive sinks during important plant phases like seed filling, resulting in higher yields. Increased photosynthate accumulation from the source to the reproductive sink could lead to more seeds being produced (Howlader 1995).

The exchange of photosynthates/sugars from leaves and the demand for photosynthates from the competitive sinks can be further deliberated and explored for enhancing productivity (Ainsworth and Bush 2011; Lemoine et al. 2013; White et al. 2016). Phloem loading and unloading of metabolites and their transport are considered as central mechanisms influencing yield under drought stress (Turgeon 1996).

13.1.6 Biometric Traits

Drought-stressed plants have been shown to lose 50–60% of their economic output. According to Nadeem et al. (2019), drought stress has a less damaging effect on the number of pods per plant than seed weight or biomass per plant. They also reported that drought stress is more harmful to seed production than filling (Kumar and Sharma 2009).

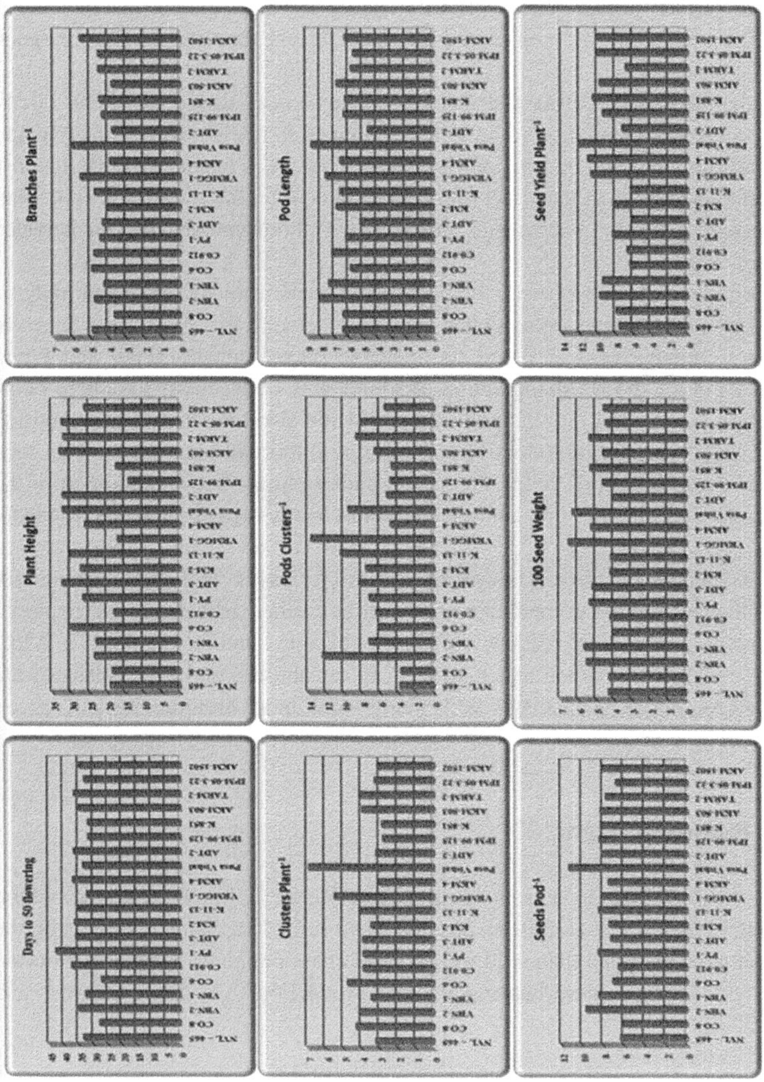

Mean performance of mung bean genotypes for growth parameters and yield-attributing traits under drought

Reduced total plant dry weight and harvest index are substantially linked to lower seed yield in mung bean due to drought stress (Robertson et al. 2004; Sadasivan et al. 1988). According to Begg (1980), the primary factors under drought stress during the blooming and pod-filling stages were pod initiation and pod development rates. Water stress during flowering has been shown to reduce production due to flower abscission (Chauhan and Williams 2018). According to Kumar and Sharma (2009), water potential in leaves and biomass partitioning were contributing aspects of drought-stress tolerance in mung bean. Drought stress has been reported to impair mung bean economic output by 31–57% during blooming and 26% after flowering/podding (Nadeem et al. 2019). Drought has also been discovered to cause an electron imbalance in photosynthesis, resulting in damaging superoxide molecules, which have been identified as the principal source of cellular damage. They also claimed that lowering ascorbic acid and glutathione levels, as well as oxidative phases, helped to relieve drought-related oxidative stress (Sharma et al. 2020). Lower nitrogenase and glutamate activity was also connected to decreased leaf water potential.

Genotypes having superior means under drought stress for biometric traits

Characters	Range	Genotypes having superior means under drought
Plant height (cm)	25.67–39.00	IPM 02-14, UPM 98-1, B-9, IPM 306-6, IPM 9901-03, IPM 9901-10, LGG 410
Number of branches/plant	3.00–5.00	PUSA 9072, IPM 02-10, IPM 306-1, PDM 11, IPM 306-6, IPM 9901-10, IPM 9901-10, LGG 410
Number of pods/plant	9.00–12.00	IPM 02-10, IPM 306-1, UPM 98-1, PDM 11, IPM 9901-10, LGG 410
100-seed weight (g)	6.00–8.94	PUSA 9072, PDM 11, IPM 306-6, IPM 9901-03, IPM 9901-125, IPM 9901-10, LGG 410
Seed yield/plant (g)	6.66–12.36	IPM 306-1, IPM 306-6, IPM 9901-03, IPM 9901-125, IPM 9901-10, LGG 410

13.1.7 Biochemical Traits

Sowmiya Selvanayagi (2020) investigated the average performance of 20 distinct mung bean genotypes in terms of growth characteristics and yield determinants. The genotypes VRMGG 1, VBN 2, VBN 1, and Pusa Vishal, according to her research, performed well for both yield and drought tolerance traits and might be used in future hybridisation programmes. Varma et al. (2018) reported on the biochemical composition and storage protein profiling of mung bean (*Vigna radiata* L. Wilczek) cultivars. According to their data, the methionine content of mung bean cultivars ranged from 0.79 g/16 gN to 1.76 g/16 gN, with an overall mean of 1.26 g/16 gN. Under drought-stress circumstances, the cultivars HUM-2 (1.76 g/16 gN) and Pairy Mung (1.72 g/16 gN) had the greatest amounts of methionine, whereas MUM-2

(0.79 g/16 gN) had the least. The physiological characteristics of green gram (*Vigna radiata* L.) variants for drought tolerance were examined by Krishna et al. (2018). When compared to the other genotypes, the KM-1423 (20.133) and IPM 02-3 (19.037) genotypes were reported to have the highest protein content. It thrives in both irrigated and rainfed settings, making it an ideal crop for severe drought zones.

Inamul Hasan Madar et al. (2017) identified the nutritional and biochemical alterations in *Vigna radiata* seeds by germination. The results showed that pre-germinated seeds had lower protein content (0.027 mg/g), whereas protein content increased dramatically after germination to 0.207 mg/g. There was a significant difference between the dry and germinated seeds at $P < 0.05$. Thus, a considerable increase in protein content was observed post-germination in the seeds. As a result, it is hypothesised that following germination, the nutritional content and quality of *Vigna radiata* seeds improve.

Anandhi Lavanya and Vanniarajan (2014) reported the biochemical characterisation of elite green gram (*Vigna radiata* (L.) Wilczek) genotypes. The results indicated that the protein content of the genotypes ranged from 12.2 g/100 g to 24 g/100 g in all environments. The maximum amount of protein was observed in Co (Gg) 7 E3 (22.8 g/100 g). The genotypes, viz. K. Pudur 2, CO(Gg) 7, NM65, NM67, K. Pudur3, 76-47/1, 76-43, CO6, M986, and Samrat, were stable for seed yield per plant in average environmental conditions, whereas K. Pudur 2 was found to be good for high Fe and protein content. Hence, these genotypes could be used in the next breeding effort.

Genotypes having superior means under drought stress for biochemical traits

Characters	Range	Genotypes having superior means under drought
Sugar (mg/g)	0.20–0.34	NVL 465, Pusa Vishal, ADT 2, IMP-99-125, IMP-05-22
Protein (g/100 g)	6.00–25.00	NVL 465, VBN 1, CO 6, ADT 3, KM 2, Pusa Vishal, AKM 503, TARM 2
Methionine (mg/g)	0.70–1.52	NVL 465, VBN-2, VBN 1, CO 6, K 11-13, VRMGG 1, Pusa Vishal, ADT 2, AKM 503, IPM-05-3-22
Albumin (%)	12.00–78.00	VBN 1, CO 912, VRMGG 1, ADT 2, IMP-99-125
Globulin (%)	5.00–18.00	VBN 2, CO 912, KM 2, VRMGG 1, Pusa Vishal
Phytic acid (mg/g)	1.20–8.63	NVL 465, CO 6, CO 912, ADT 3, VRMGG 1, Pusa Vishal

Tajoddin et al. (2011) investigated the levels of phytic acid and minerals in various mung bean cultivars. The phytate content of yellow and green mung bean cultivars was found to differ substantially. The phytate P and phytic acid contents of mung bean cultivars were found to be in the range of 1.74–2.79 mg/g and 6.17–9.90 mg/g, respectively. The cultivar 'TAP-7' had the highest phytic acid concentration (9.90 mg/g), whereas the cultivar 'ALM-3' had the lowest phytate level (6.17 mg/g). Yellow mung bean cultivars had a lower average phytate concentration (7.38 mg/g) than green cultivars (9.02 mg/g). The mineral contents of mung bean cultivars should be assessed simultaneously with their phytate content to obtain a better appraisal of their potential as a mineral source.

Biochemical components of mung bean cultivars grown throughout the summer and kharif seasons were studied by Sital et al. (2011). Seed cotyledonary biomolecule levels, such as proteins, total sugars, starch, and lipids, were lower in kharif than in summer, affecting seed grain quality. Growing seeds of both sorts sown in the summer had higher crude protein, soluble protein, sulphur-containing amino acids, albumins, and storage protein globulins than seeds sown in the kharif season, demonstrating that seed protein quality changes by season. Even among variations, SML668 had more of these features than ML1333.

Souframanien et al. (2021) discovered genetic variation in phytic acid content in mung bean that ranged from 6.17 to 12 mg/g. Mung bean VC-6379 (5.74 mg/g) and YBSM (5.85 mg/g) both had low PA content. It has been reported that dominant alleles control phytic acid accumulation at two independent loci of major genes with duplicate recessive epistasis. Two major QTLs, SDPAP4.1 and SDPAP11.1, were also found on linkage groups 4A and 11A in interval markers CEDG139-MBSSR179 and BM141-VR222, respectively. In many cases, phytic acid content varies depending upon the cultivars, climatic conditions, locations, irrigation conditions, type of soil, and year during which they are grown. The level of phytic acid in the tested mung bean cultivars was lower than 1%, thus suggesting that the nutritive value of mung bean seeds would be impaired to a lesser extent, which is relevant for the selection of low phytate cultivars to improve mineral bioavailability and for preparation of weaning foods.

13.1.8 Seedling Traits

Sowmiya Selvanayagi (2020) experimented under in vitro conditions and concluded that the rate of germination varied and decreased in the genotypes as the concentration of PEG increased. The decrease in the water potential gradient between seeds and their surrounding media by the effect of PEG impacted the seed germination and its related seedling growth-related characters due to limited water uptake by the seeds. The genotypes, namely VRMGG 1, VBN 2, VBN 1, and Pusa Vishal, were found to be drought tolerant.

Jincy et al. (2021) revealed that the genotypes COGG 1332, VGG 16069, VGG 17003, VGG 17004, VGG 17009, VGG 17019, and VGG 17045 exhibited high tolerance levels to moisture stress during the seedling stage. The physiological characterisation of green gram genotypes (*Vigna radiata* L.) for drought tolerance was examined by Nithila et al. (2019) who reported that an increased PEG concentrations had a negative impact on seedling growth. Seed germination and seedling growth play a key role in the establishment of stressed crops. VBN 2 has the highest stress tolerance index among these types, followed by ADT 3 and CO 8.

Swathi et al. (2017) indicated that the genotypes ML-267, MH-565, MGG-350, and MGG-347 were determined to be stress tolerant at high (−0.9 MPa) concentrations of PEG due to their good germination percentage paired with increased root and shoot lengths. KM-122, EC-396117, LGG-460, LGG-450, LGG-407, MH-3-18, and PM-110, on the other hand, were shown to be particularly susceptible to water stress. Rajwinder Kaur et al. (2017) investigated and showed that the genotypes experienced a linear decrease in germination, shoot and root length, and their corresponding fresh and dry weight as the PEG concentration was raised.

In order to screen for drought tolerance, Sunil Kumar et al. (2016) studied drought tolerance at the seedling stage. Significant differences were found among the accessions, treatments, and their interactions when plant characteristics were evaluated, showing that drought tolerance in mung bean is very variable. PEG-induced water stress generated differences in sensitivity across seedling features. Shoot-related traits, on the other hand, reacted to water stress the most.

In the lab, Dutta and Bera (2008) tested 15 mung bean genotypes using PEG 6000 at a water potential of −0.3 bar. They discovered that when genotypes were exposed to water stress, all development indices except radicle dry weight decreased in some genotypes. Among the genotypes tested, Pusa 105 and Khargoan 1 showed the biggest and smallest percent reduction in overall seedling length, respectively. Under drought stress, hypocotyl extension is inhibited more in separated cotyledons than in intact cotyledons. As a result, more research is needed to discover if plants with larger cotyledons can help plants establish more quickly in drought-stricken areas. In the low leaf water loss (LWL) genotype, Raina et al. (2016) reported a twofold increase in physiological and molecular responses to drought, as well as efficient stomatal management. Drought-induced stomatal closure was associated with downregulation of the farnesyl transferase gene in this genotype, which was accompanied by a cooler canopy temperature and a branching root system for scavenging available soil moisture (Raina et al. 2016). These mung bean plant adaptations, mechanisms, and traits are appropriate for harsh environments, but they necessitate extensive knowledge based on drought stress levels, crop stage, and agroecological factors. Other significant physiological measures that can be used to screen mung bean for drought tolerance include water consumption efficiency, root growth/biomass, carbon isotope discrimination (13 °C), and leaf temperature (canopy temperature differential).

Genotypes having superior means under various drought stress for seedling traits

		Superior genotypes at different levels of PEG concentrations/drought induction			
S. No.	Character	Control	−0.3 MPa	−0.6 MPa	−0.9 MPa
1.	Germination percentage	–	CO 8, VBN I, PY I, KM 2, VRMGG I. Pusa Vishal, IPM-99-125	NVL 465, CO 8, VBN 1, CO 6, C0 912, PY 1, KM 2, K 11-13, VRMGG 1, AKM 4, Pusa Vishal, ADT	CO 8, VBN 1, KM 2, VRMGG 1, Pusa Vishal, IPM-99-125, TARM 2, IPM-05-3-22

(continued)

S. No.	Character	Superior genotypes at different levels of PEG concentrations/drought induction			
		Control	−0.3 MPa	−0.6 MPa	−0.9 MPa
				2, IPM-99-125, K 851, AKM 503, TARM 2, IPM-05-3-22, AKM 1502	
2.	Plumule length	VBN 2, VBN 1, KM 2, VRMGG 1, AKM 4, ADT 2, K 851	VBN 2, VBN 1, CO 912, KM 2, VRMGG 1, AKM 4, ADT 2, K 851, TARM 2	VBN 2, VBN 1, CO 912, KM 2, VRMGG 1, AKM 4, ADT 2, K 851, TARM 2	VBN 2, VBN 1, CO 912, KM 2, VRMGG 1, AKM 4, ADT 2, K 851, AKM 503, TARM 2
3.	Radicle length	VBN 2, PY 1, ADT 3, VRMGG 1, Pusa Vishal, IPM-05-3-22	NVL 465, VBN 2, ADT 3, K 11-13, VRMGG 1, Pusa Vishal, IPM-05-3-22	NVL 465, VBN 2, PY 1, ADT 3, K 11-13, VRMGG 1, Pusa Vishal, ADT 2, IPM-05-3-22	NVL 465, VBN 2, PY 1, ADT 3, K 11-13, VRMGG 1, Pusa Vishal, ADT 2, IPM-05-3-22
4.	Plumule fresh weight	VBN 2, VBN 1, KM 2, K 11-13, VRMGG 1, Pusa Vishal, ADT 2, IPM-99-125, AKM 503, IPM-05-3-22	VBN 2, VBN 1, KM 2, K 11-13, VRMGG 1, Pusa Vishal, ADT 2, IPM-99-125, AKM 503, IPM-05-3-22	VBN 2, VBN 1, KM 2, K 11-13, VRMGG 1, Pusa Vishal, IPM-99-125, AKM 503, IPM-05-3-22	VBN 2, VBN 1, KM 2, K 11-13, VRMGG 1, Pusa Vishal, ADT 2, IPM-99-125, AKM 503, IPM-05-3-22
5.	Radicle fresh weight	VBN 2, VBN 1, KM 2, K 11-13, VRMGG 1, Pusa Vishal, ADT 2, IPM-99-125, AKM 503, IPM-05-3-22	VBN 2, VBN 1, KM 2, K 11-13, VRMGG 1, Pusa Vishal, ADT 2, IPM-99-125	VBN 2, VBN 1, KM 2, K 11-13, VRMGG 1, Pusa Vishal, IPM-99-125	NVL 465, VBN 2, KM 2, K 11-13, VRMGG 1, Pusa Vishal, ADT 2, IPM-99-125, AKM 503, AKM 1502
6.	Plumule dry weight	CO 912, KM 2, K 11-13, VRMGG	CO 912, KM 2, K 11-13, VRMGG	CO 912, KM 2, K 11-13, VRMGG	CO 912, KM 2, K 11-13, VRMGG

(continued)

		Superior genotypes at different levels of PEG concentrations/drought induction			
S. No.	Character	Control	−0.3 MPa	−0.6 MPa	−0.9 MPa
		1, AKM 4, ADT 2, AKM 503, IPM-05-3-22	1, AKM 4, ADT 2, AKM 503, IPM-05-3-22, AKM 1502	1, AKM 4, ADT 2, K 851, IPM-05-3-22, AKM 1502	1, IPM-99-125, K 851, IPM-05-3-22
7.	Radicle dry weight	VBN 2, VBN 1, CO 912, VRMGG 1, Pusa Vishal, ADT 2, IPM-99-125, AKM 503, IPM-05-3-22	VBN 2, VBN 1, CO 912, VRMGG 1, AKM 4, Pusa Vishal, ADT 2, IPM-99-125, AKM 503, IPM-05-3-22	VBN 2, VBN 1, CO 6, CO 912, ADT 3, VRMGG 1, AKM 4, IPM-99-125, IPM-05-3-22	VBN 2, VBN 1, CO 6, CO 912, ADT 3, VRMGG 1, IPM-99-125, IPM-05-3-22, AKM 1502

13.2 Conclusion

It is very crucial to breed mung bean lines that can withstand harsh conditions. While stress dominates a population of habitats, especially when agroecology is characterised by various stresses, each agroecology is distinct, necessitating the use of systemised solutions. It is critical to understand the core mechanism for stress tolerance from intrinsic physiological and biochemical perspectives to make the ideal combination of abiotic stress and the traits to incorporate. We want to create root systems that can help plants cope with water shortages by extracting water from deeper soils. To find robust donors for these features, screening for diverse abiotic stressors must be more exact and stricter. The breeders must utilise the identified donors as quickly as possible. Increasing photosynthetic rates with depleted or current water and nutrient supplies would be a critical goal and a game changer for agriculture in the twenty-first century (Foyer et al. 2017). To boost yields and crop nutritional value, more efficient photosynthesis must be developed with sustainable and climate-resilient cropping methods. Plants with a deep root system, an early maturity span, tall stature, sympodial pod bearing, many pods per cluster, and longer pods with many nodes and shorter internodes will be better able to endure heat and drought stress.

Hence, there is an urgent need for a unified and multidisciplinary conceptual framework to identify the stress scenarios and the involvement of various plant biomolecules which help in stress memory–mitigation mechanisms. Further downstream mechanisms that govern these stress-related responses that improve or hinder plant stress memory have to be studied. Lately, combining various modern technologies such as infrared thermography, automated robotics, camera images, and computational algorithms, all of which are components of high-throughput phenotyping facilities (phenomics and phenospex), has made it possible to conduct high-throughput phenotyping for stress tolerance (Pratap et al. 2019). However, for establishing a relationship between known difficult-to-measure features and surrogate parameters obtained from photos, which indicate plant responses to abiotic stresses, non-destructive approaches used for particular regions or environments need to be optimised. These phenomics techniques can aid in the accurate quantification of plant shoot architectural responses to pressures such as soil moisture deficits, salt, and high temperatures, among others. More than a dozen picture parameters have been given to depict plant responses to stress, which can help in identifying key features and the methodology for screening many breeding lines or mapping the population to find stress-tolerant genes.

References

Ahmad A, Selim MM, Alderfasi AA, Afzal M (2015) Effect of drought stress on mungbean (*Vigna radiata* L.) under arid climatic conditions of Saudi Arabia. Ecosyst Sustain Dev X 1(December 2016):185–193. https://doi.org/10.2495/eco150171

Ainsworth EA, Bush DR (2011) Carbohydrate export from the leaf: a highly regulated process and target to enhance photosynthesis and productivity. Plant Physiol 155(1):64–69

Anandhi Lavanya S, Vanniarajan C (2014) Biochemical characterisation of elite green gram (*Vigna radiata* (L.) Wilczek) genotypes. Int J Sci Res 3(3):1–2

Aroca R (2013) Plant responses to drought stress: from morphological to molecular features. Springer, Berlin, pp 1–466. https://doi.org/10.1007/978-3-642-32653-0

Baliah NT, Sheeba PC, Mallika S (2018) Encouraging effect of gibberellic acid on the growth and biochemical characters of green gram (*Vigna radiata L.*). J Glob Biosci 7(8):5522–5529

Bangar P, Chaudhury A, Tiwari B, Kumar S, Kumari R, Bhat KV (2019) Morphophysiological and biochemical response of mungbean [*Vigna radiata* (L.) Wilczek] varieties at different developmental stages under drought stress. Turk J Biol 43(1):58–69

Basu S, Ramegowda V, Kumar A, Pereira A (2016) Plant adaptation to drought stress [version 1; referees: 3 approved]. F1000Research 5:1–10. https://doi.org/10.12688/F1000RESEARCH.7678.1

Begg JE (1980) Morphological adaptations of leaves to water stress. In: Adaptation of plants to water and high temperature stress. Wiley, Chichester

Bharadwaj N, Gogoi N, Barthakur S, Basumatary N (2018) Morphophysiological responses in different mungbean genotypes under drought stress. Res J Recent Sci 7(7):1–5

Bihmidine S, Hunter CT III, Johns CE, Koch KE, Braun DM (2013) Regulation of assimilate import into sink organs: update on molecular drivers of sink strength. Front Plant Sci 4:177

Blum A (2005) Drought resistance, water-use efficiency, and yield potential - are they compatible, dissonant, or mutually exclusive? Aust J Agric Res 56(11):1159–1168. https://doi.org/10.1071/AR05069

Bohnert HJ, Nelson DE, Jensen RG (1995) Adaptations to environmental stresses. Plant Cell 7(7):1099–1111. https://doi.org/10.1105/tpc.7.7.1099

Bruce TJA, Matthes MC, Napier JA, Pickett JA (2007) Stressful "memories" of plants: evidence and possible mechanisms. Plant Sci 173(6):603–608

Campos EI, Reinberg D (2009) Histones: annotating chromatin. Annu Rev Genet 43:559–599

Chandrashekhar G (2016) Rebound in Indian pulse production and softer prices will boost domestic consumption. Pulse Market Report. Saskatchewan Pulse Growers

Chandrashekhar G (2019) Indian pulse market turning slowly. Pulse Market Report. Saskatchewan Pulse Growers. saskpulse.com
Chauhan YS, Williams R (2018) Physiological and agronomic strategies to increase mungbean yield in climatically variable environments of Northern Australia. Agronomy 8(6):83
Chaves MM, Flexas J, Pinheiro C (2009) Photosynthesis under drought and salt stress: regulation mechanisms from whole plant to cell. Ann Bot 103(4):551–560. https://doi.org/10.1093/aob/mcn125
Chen X, Wang Y, Lv B, Li J, Luo L, Lu S, Zhang X, Ma H, Ming F (2014) The NAC family transcription factor OsNAP confers abiotic stress response through the ABA pathway. Plant Cell Physiol 55(3):604–619
Ciais P, Reichstein M, Viovy N, Granier A, Ogée J, Allard V, Aubinet M, Buchmann N, Bernhofer C, Carrara A, Chevallier F, De Noblet N, Friend AD, Friedlingstein P, Grünwald T, Heinesch B, Keronen P, Knohl A, Krinner G et al (2005) Europe-wide reduction in primary productivity caused by the heat and drought in 2003. Nature 437(7058):529–533. https://doi.org/10.1038/nature03972
Dutta P, Bera AK (2008) Screening of mungbean genotypes for drought tolerance. Legum Res 31(2):145–148
Edwards G, Walker D (eds) (1983) C3, C4: mechanisms, and cellular and environmental regulation, of photosynthesis. Blackwell Scientific Publications, Oxford
El Karamany MF, Sadak MS, Bakry BA (2019) Synergistic effect of indole acetic acid and gibberellic acid on mung bean grown under sandy soil conditions. J Appl Sci 19(7):718–724
Fahad S, Bajwa AA, Nazir U, Anjum SA, Farooq A, Zohaib A, Sadia S, Nasim W, Adkins S, Saud S (2017) Crop production under drought and heat stress: plant responses and management options. Front Plant Sci 8:1147
Farooq M, Wahid A, Kobayashi N, Fujita D, Basra SMA (2009) Plant drought stress: effects, mechanisms and management. Agron Sustain Dev 29(1):185–212
Flexas J, Bota J, Loreto F, Cornic G, Sharkey T (2004) Diffusive and metabolic limitations to photosynthesis under drought. Plant Biol 6:269–279
Flexas J, Gallé A, Galmés J, Ribas-Carbo M, Medrano H (2012) The response of photosynthesis to soil water stress. In: Plant responses to drought stress. Springer, New York, pp 129–144
Foyer CH, Ruban AV, Nixon PJ (2017) Photosynthesis solutions to enhance productivity. Philos Trans R Soc B Biol Sci 372(1730):20160374
Ghosh D, Xu J (2014) Abiotic stress responses in plant roots: a proteomics perspective. Front Plant Sci 5:6
Gowik U, Westhoff P (2011) The path from C3 to C4 photosynthesis. Plant Physiol 155(1):56–63
Howlader M (1995) Influence of source manipulation on photosynthesis and dry matter partitioning in mungbean. Bangladesh Agricultural University, Mymonsingh
Ivlev AA (2017) Global photosynthesis and its regulatory role in natural carbon cycle. J Ecosyst Ecograph 07(02). https://doi.org/10.4172/2157-7625.1000233
Jahns P, Holzwarth AR (2012) The role of the xanthophyll cycle and of lutein in photoprotection of photosystem II. Biochim Biophys Acta Bioenergetics 1817(1):182–193
Jain SK (1975) Genetic analysis of yield and protein content in green gram (Phaseolus Aureus Roxb.). IARI, Division of Genetics, New Delhi
Ji X, Dong B, Shiran B, Talbot MJ, Edlington JE, Hughes T, White RG, Gubler F, Dolferus R (2011) Control of abscisic acid catabolism and abscisic acid homeostasis is important for reproductive stage stress tolerance in cereals. Plant Physiol 156(2):647–662
Jincy M, Prasad VBR, Jeyakumar P, Senthil A, Manivannan N (2021) Evaluation of green gram genotypes for drought tolerance by PEG (polyethylene glycol) induced drought stress at seedling stage. Legum Res 44(6):684–691. https://doi.org/10.18805/LR-4149
Kaur R, Kaur J, Bains TS (2017) Screening of mungbean genotypes for drought tolerance using different water potential levels. J Adv Agric Technol 4(2):159–164
Keykha M, Ganjali HR, Mobasser HR (2014) Effect of salicylic acid and gibberellic acid on some characteristics in mungbean (*Vigna radiata*). Int J Biosci 5:70–75

Kinoshita T, Seki M (2014) Epigenetic memory for stress response and adaptation in plants. Plant Cell Physiol 55(11):1859–1863

Krishna DV, Rao KP, Lal EP (2018) Physiological characterisation of green gram (*Vigna radiata* L.) genotypes for drought tolerance. J Pharmacogn Phytochem 6(5):589–593

Kumar A, Sharma KD (2009) Physiological responses and dry matter partitioning of summer mungbean (*Vigna radiata* L.) genotypes subjected to drought conditions. J Agron Crop Sci 195(4):270–277

Kumar R, Yadav RK, Sharma N, Yadav A, Nehal N (2018) Influence of plant growth regulators on biochemical changes of mungbean (*Vigna radiata* L. Wilczek). J Pharmacogn Phytochem 2007:386–389

Kundu S, Das B, Yonzone R (2017) Effect of different hydro and osmopriming materials on germination and seedling growth of mung bean. Int J Chem Stud 5:99–103

Lawlor DW (2002) Limitation to photosynthesis in water-stressed leaves: stomata vs. metabolism and the role of ATP. Ann Bot 89(SPEC. ISS):871–885. https://doi.org/10.1093/aob/mcf110

Lemoine R, La Camera S, Atanassova R, Dédaldéchamp F, Allario T, Pourtau N, Bonnemain JL, Laloi M, Coutos-Thévenot P, Maurousset L, Faucher M, Girousse C, Lemonnier P, Parrilla J, Durand M (2013) Source-to-sink transport of sugar and regulation by environmental factors. Front Plant Sci 4(JUL):1–21. https://doi.org/10.3389/fpls.2013.00272

Long SP, Marshall-Colon A, Zhu XG (2015) Meeting the global food demand of the future by engineering crop photosynthesis and yield potential. Cell 161(1):56–66. https://doi.org/10.1016/j.cell.2015.03.019

Madar IH, Asangani AH, Srinivasan S, Tayubi IA, Ogu GI (2017) Nutritional and biochemical alterations in *Vigna radiata* (Mung Bean) seeds by germination. Int J Curr Microbiol App Sci 6 (9):3307–3313

Maiti R, Kumari A (2016) Physiological basis of crop productivity. In: Bioresource and stress management, pp 71–85

Markam NK, Nair SK, Saxena RR, Nanda HC (2018) Assessment of genetic diversity using SSR marker in moonbeam [*Vigna radiata* (L.) Wilczek] genotypes. J Pharmacogn Phytochem 7 (2):2750–2754

Micco Veronica De, Giovanna (2016) Plant adaptation to drought stress [version 1; referees: 3 approved]. F1000Research 5:1 10. https://doi.org/10.12688/F1000RESEARCH.7678.1

Mubeen BR, Khan AH, Singh AK, Singh SP, Yadav RK, Singh U (2015) Response of mungbean (*Vigna radiata* L.) to rhizobium inoculation and growth regulators on biochemical changes and yield. Environ Ecol 33(1):105–109

Nadeem M, Li J, Yahya M, Sher A, Ma C, Wang X, Qiu L (2019) Research progress and perspective on drought stress in legumes: a review. Int J Mol Sci 20(10). https://doi.org/10.3390/ijms20102541

Nair RM, Pandey AK, War AR, Hanumantharao B, Shwe T, Alam A, Pratap A, Malik SR, Karimi R, Mbeyagala EK (2019) Biotic and abiotic constraints in mungbean production—progress in genetic improvement. Front Plant Sci 10:1340

Naresh RK, Singh SP, Dwivedi A, Kumar V (2013) Effects of water stress on physiological processes and yield attributes of different mungbean (L.) varieties. Afr J Biochem Res 7(5): 55–62

Nithila S, Venkatasalam K, Santhoshkumar D, Kavipriyan V (2019) Physiological characterization of green gram (*Vigna radiata* L.) genotypes for drought and salt stress tolerance. J Pharmacogn Phytochem 2:443–446

Niyogi KK (1999) Photoprotection revisited: genetic and molecular approaches. Annu Rev Plant Biol 50(1):333–359

Paul MJ, Foyer CH (2001) Sink regulation of photosynthesis. J Exp Bot 52(360):1383–1400

Pratap A, Gupta S, Nair RM, Gupta SK, Schafleitner R, Basu PS, Singh CM, Prajapati U, Gupta AK, Nayyar H (2019) Using plant phenomics to exploit the gains of genomics. Agronomy 9(3): 126

Rahman MM, Khan A, Hasan MM, Banu LA, Howlader MHK (2018) Effect of foliar application of gibberellic acid on different growth contributing characters of mungbean. Progress Agric 29(3): 233–238

Raina SK, Govindasamy V, Kumar M, Ajay Kumar S, Jagadesh R, Minhas PS (2016) Genetic variation in physiological responses of mungbeans (*Vigna radiata* (L.) Wilczek) to drought. Acta Physiol Plant 38:263. https://doi.org/10.1007/s11738-016-2280-x

Ratnasekera D, Subhashi AT (2015) Morpho-physiological response of selected mungbean (*Vigna radiata* L.) Sri Lankan genotypes to drought stress. J AgriSearch 2(3):162–166

Robertson MJ, Fukai S, Peoples MB (2004) The effect of timing and severity of water deficit on growth, development, yield accumulation and nitrogen fixation of mungbean. Field Crop Res 86(1):67–80

Rossi M, Bermudez L, Carrari F (2015) Crop yield: challenges from a metabolic perspective. Curr Opin Plant Biol 25:79–89

Sadasivan R, Natrajaratnam N, Dabu R, Muralidharan V, Rangasmay SR (1988) Response of mungbean cultivars to soil moisture stress at different growth phases. In: Mungbean proceeding of the second international symposium. AVRCD, pp 260–262

Sharma A, Kumar V, Shahzad B, Ramakrishnan M, Singh Sidhu GP, Bali AS, Handa N, Kapoor D, Yadav P, Khanna K, Bakshi P, Rehman A, Kohli SK, Khan EA, Parihar RD, Yuan H, Thukral AK, Bhardwaj R, Zheng B (2020) Photosynthetic response of plants under different abiotic stresses: a review. J Plant Growth Regul 39(2):509–531. https://doi.org/10.1007/s00344-019-10018-x

Sharp RE (2002) Interaction with ethylene: changing views on the role of abscisic acid in root and shoot growth responses to water stress. Plant Cell Environ 25(2):211–222. https://doi.org/10.1046/j.1365-3040.2002.00798.x

Sital JS, Malhotra JS, Sharma S, Singh S (2011) Comparative studies on biochemical components in mung bean [*Vigna radiata* (L.) Wilczek] varieties cultivated in summer and kharif seasons. Indian J Agric Biochem 24(1):68–72

Souframanien J, Dhole VJ, Reddy KS (2021) Breeding for low phytates and oligosaccharides in mungbean and blackgram. In: Breeding for enhanced nutrition and bio-active compounds in food legumes. Springer, New York, pp 99–130

Sowmiya Selvanayagi C (2020) Morphophysiology, biochemical and yield response of mungbean (*Vigna radiata* L. Wilczek) under drought stress. Annamalai University

Sunil Kumar B, Gokulakrishnan J, Sathiyanarayanan G, Prakash M (2016) Impact of osmotic stress on seed germination and seedling growth in mungbean (*Vigna radiata* L. Wilczek). Int J Trop Agric 34(3):645–652

Swathi L, Reddy DM, Sudhakar P, Vineela V (2017) Screening of mungbean (*Vigna radiata* L. Wilczek) genotypes against water stress mediated through polyethylene glycol. Int J Curr Microbiol App Sci 6(10):2524–2531

Tajoddin MD, Shinde M, Lalitha J (2011) In vivo reduction of the phytic acid content of mung bean (*Phaseolus aureus* L) cultivars during germination. Am Eurasian J Agric Environ Sci 10 (1):127–132

Tezara W, Mitchell VJ, Driscoll SD, Lawlor DW (1999) Water stress inhibits plant photosynthesis by decreasing coupling factor and ATP. Nature 401(6756):914–917

Tietjen B, Schlaepfer DR, Bradford JB, Lauenroth WK, Hall SA, Duniway MC, Hochstrasser T, Jia G, Munson SM, Pyke DA (2017) Climate change-induced vegetation shifts lead to more ecological droughts despite projected rainfall increases in many global temperate drylands. Glob Chang Biol 23(7):2743–2754

Tuberosa R (2012) Phenotyping for drought tolerance of crops in the genomics era. Front Physiol 3: 347

Turgeon R (1996) Phloem loading and plasmodesmata. Trends Plant Sci 1(12):418–423

Uddin S, Parvin S, Awal MA (2013) Morpho-physiological aspects of mungbean (*Vigna radiata* L.) in response to water stress. Int J Agric Sci Res 3(2):137–148

Varma A, Mishra SP, Tripathi A, Shukla UK (2018) Biochemical composition and storage protein profiling of mungbean (*Vigna radiata* L. Wilczek) cultivars. J Pharmacogn Phytochem 7 (4):708–713

Venkateswarlu S, Subramanian VB (1993) Photosynthesis during the reproductive phase of rainfed green gram. Indian J Plant Physiol XXXVI(3):174–177

Verslues PE, Agarwal M, Katiyar-Agarwal S, Zhu J, Zhu J (2006) Methods and concepts in quantifying resistance to drought, salt and freezing, abiotic stresses that affect plant water status. Plant J 45(4):523–539

White AC, Rogers A, Rees M, Osborne CP (2016) How can we make plants grow faster? A source-sink perspective on growth rate. J Exp Bot 67(1):31–45

Wilkinson S, Kudoyarova GR, Veselov DS, Arkhipova TN, Davies WJ (2012) Plant hormone interactions: innovative targets for crop breeding and management. J Exp Bot 63(9):3499–3509

Xiong L, Zhu JK (2002) Molecular and genetic aspects of plant responses to osmotic stress. Plant Cell Environ 25(2):131–139. https://doi.org/10.1046/j.1365-3040.2002.00782.x

Yordanov I, Velikova V, Tsonev T (2000) Plant responses to drought, acclimation, and stress tolerance. Photosynthetica 38:171–186

Genetic Engineering for Enhancing Abiotic Stress Tolerance in Pulses

14

Prateek Singh, Shallu Thakur, Sudhir Kumar, Biswajit Mondal, Meenal Rathore, and Alok Das

Abstract

Pulses are climate-smart grain legumes important to nutritional security and sustainable agriculture. Abiotic stresses take a heavy toll in pulse production, and genetic engineering offers a solution to add adaptive traits in the germplasm. Abiotic stresses being mostly polygenic are difficult to manipulate and require a thorough understanding of the underlying mechanism. Impact of abiotic stresses in eight different pulses, genetic mechanism involved, and transgenics approach adopted for enhancing the stress tolerance in those pulses are discussed. Traits engineered in chickpea (drought and salt tolerance), pigeon pea (salt tolerance), mung bean (salt and cold tolerance), urdbean (salt and drought tolerance and aluminum toxicity), cowpea (salt tolerance), field pea (salt, frost, and heat tolerance), common bean (drought tolerance), and lentil (cold and freezing tolerance) and resulting phenotypes are also discussed. Currently, only two transgenic pulses for biotic stress (insect resistance for cowpea and golden mosaic virus in common bean) are commercialized. Climate change poses various challenges, and genetic engineering and emerging genome editing techniques for abiotic stress-adaptive traits shall play a crucial role in abiotic stress management.

P. Singh · S. Thakur · S. Kumar · M. Rathore · A. Das (✉)
Division of Plant Biotechnology, ICAR-Indian Institute of Pulses Research, Kanpur, Uttar Pradesh, India
e-mail: alok.das@icar.gov.in

B. Mondal
Division of Crop Improvement, ICAR-Indian Institute of Pulses Research, Kanpur, Uttar Pradesh, India

P. Muthu Arjuna Samy et al. (eds.), *Legumes: Physiology and Molecular Biology of Abiotic Stress Tolerance*, https://doi.org/10.1007/978-981-19-5817-5_14

Keywords

Genetic Engineering · Pulses · Abiotic Stress · Drought Tolerance · Salt Tolerance · Heat Tolerance · Frost Tolerance

14.1 Introduction

The word pulse is of Latin origin derived from "*puls*" meaning "*porridge*." Pulses are edible seeds of the family Fabaceae/Leguminosae comprising several thousand species. Pulses are a cheap source of dietary protein among the vegetarian population living in developing countries. They play a critical role in ensuring global food and nutritional security accounting for one-third of the world's crop production. Besides their nutritional significance, pulses are also used as a fodder crop and green manure. Pulse crops tend to enhance soil fertility via biological nitrogen fixation. Pulses are members of the Papilionoideae subfamily that are further categorized into three subgroups Hologalegina (~5000 species), Millettioids, and Phaseoloids (combined ~4000 species) (Foyer et al. 2016; Lavin et al. 2005). The major pulses widely grown in India are chickpea (*Cicer arietinum* L.), pigeon pea (*Cajanus cajan* L.), mung bean (*Vigna radiata* L.), urdbean (*Vigna mungo* L. Hepper.), lentil (*Lens culinaris* Medik.), field pea (*Pisum sativum* L.), cowpea (*Vigna unguiculata* L.), and common bean (*Phaseolus* spp.). Despite their nutritional and agricultural significance, the productivity of the majority of pulses is stagnating at <1000 kg/ha (Joshi and Rao 2016), due to the various constraints encountered during the crop's life cycle. Pulses are considered as a hardy crop that is generally grown on marginal lands and are more often exposed to harsh unpredictable environments (Allard 1999). On the global scale, pulses are grown on 89.90 mha land with an annual production of 88.37 mt (FAOSTAT 2019). Development of high-yielding varieties resilient and adapted to specific niches resulted in 10.92% increase in yield with an average yield of 953 kg/ha during the last decade (2010–2019) (FAOSTAT 2019, Accessed on September, 2021).

The Intergovernmental Panel on Climate Change Report 2021 of the United Nations (IPCC 2021) suggests that there is an urgent need to address issues related to global climate variability. The occurrence of extreme weather conditions like uneven precipitation and drought has rapidly increased. These climatic changes have adversely affected global crop production and are likely to aggravate further. A temperature rise in the global temperature beyond 1.5 °C will result in more frequent heat extremes reaching the critical tolerance thresholds for agriculture and health. Revolution in traditional agricultural practice can limit the carbon release, but these changes are slow and inadequate to exhibit a prerequisite impact. Pulses are considered *climate-smart* crops that are the most promising component for nutritional security. They are extensively grown in arid and semiarid regions of the world, and India is its largest producer accounting for more than one-fourth of the global pulse production. Global climate change over the years has increased incidences of abiotic stresses, thereby adversely affecting the production of rainfed crops including pulses.

Abiotic stresses like drought, intense temperatures, salinity, and waterlogging have toxic effects on the plant physiology and biochemical landscape of the crop. The plants' physiological responses under abiotic stresses include *inter alia* photo-inhibition, loss of osmoregulation, and inhibition of biological nitrogen fixation. These physiological responses are the consequences of accumulated stress-induced osmolytes (proline, polyamines, glycine betaine, etc.) and reactive oxygen species (ROS) in the plant cells. ROS interfere with biochemical pathways of the plant cell causing photo-inhibition and DNA damage resulting in yield losses (Sultana et al. 2014). Drought and salinity stresses are the principal abiotic stresses that cause significant yield losses (up to 70% annually) in various crops (Wild 2003). Salt-induced stress in plants affects the growth of their roots and slows down the process of nodulation, thus interfering with the nutrient uptake and distribution mechanism (Naher and Alam 2010). Details of various abiotic stresses affecting pulses at various stages have been compiled (Table 14.1).

Abiotic stresses generally coexist that disrupt the normal physiology and morphology of the crop leading to significant yield losses. Plant's responses to these stresses are usually interconnected (Beck et al. 2007) involving activation of multiple stress-regulated genes that results in molecular, biochemical, physiological, and morphological modifications (Atkinson and Urwin 2012). These responses are genotype dependent and are based on the duration and intensity of abiotic stress (Daryanto et al. 2017) that involves a complex growth-specific phenomenon (Sehrawat et al. 2013a).

Stress response affects multiple overlapping pathways in plants, and various attempts to enhance abiotic stress-adaptive traits were reported. Strategies effectively employed are conventional breeding, molecular breeding, genetic engineering, and genome editing. The use of modern biotechnological tools for developing genetically enhanced germplasm/genotypes is imperative. Genetic transformation for introducing the desired trait into plants for better stress management targeting specific pathways/mechanisms is a potential option as demonstrated in commercially approved genetically engineered drought-tolerant maize and soybean. This chapter focuses on the impact of abiotic stresses in pulses, the genetic mechanism involved, and the transgenic approach adopted for enhancing the stress tolerance in pulses. We also discuss the use of genes derived from pulses for enhancing abiotic stress tolerance in model plants like *Arabidopsis* and tobacco and tree species like poplar.

14.1.1 Drought

The rapid change in the global climate has favored the occurrence of adverse and unpredictable weather phenomenon causing huge yield losses and crop failure throughout the world. Among the major events occurring as a result of global climate change, drought is considered the most limiting constraint. Pulses exhibit high

Table 14.1 List of abiotic stresses affecting pulses

Pulses	Prominent abiotic stresses	Stage(s)	References
Chickpea (*Cicer arietinum* L.)	Drought	Reproductive stages	Ahmad et al. (2005)
	Terminal heat	Reproductive stages	Basu et al. (2009)
Pigeon pea (*Cajanus cajan* L.)	Drought	Seedling and reproductive stages	Saxena (2008)
	Waterlogging	Germination and early vegetative stages	Singh et al. (1986)
Mung bean (*Vigna radiata* L.)	Salinity	Germination and seedling stage	Sehrawat et al. (2013b)
	Drought	Seedling and reproductive stages	Kumar et al. (2020)
Urd bean (*Vigna mungo* L.)	Salinity	Germination and seedling stage	Priyadharshini et al. (2019)
	Drought	Seedling stages	Sai and Chidambaranatham (2019)
	Waterlogging	Vegetative stage	Bansal et al. (2019)
Cowpea (*Vigna unguiculata* L.)	Waterlogging	Vegetative and seed production	Minchin et al. (1978)
	Salinity	Vegetative stage	Maas and Poss (1989)
Field pea (*Pisum sativum* L.)	Drought	Flowering and seed-filling stages	Martin and Jamieson (1996)
	Waterlogging	Germination	Crawford (1977)
	Heat	Flowering and pod formation	Mohapatra et al. (2020)
Common bean (*Phaseolus vulgaris* L.)	Drought	Reproductive stages	Beebe et al. (2013)
	Salinity	Germination stage	Bayuelo-Jiménez et al. (2002)
Lentil (*Lens culinaris* L.)	Drought	Vegetative and reproductive stage	Choukri et al. (2020)
	Heat	Flowering and seed-filling stages	Choukri et al. (2020)

sensitivity to drought stress, prominently during the reproductive as well as vegetative stages. Plants under drought stress suffer from stunted growth due to photoinhibition, and a reduction in nutrient uptake. Pulses under water-deficit conditions had a severely reduced rate of photosynthesis and nitrogen assimilation (Chaves 1991; Valentine et al. 2018). The low rate of nitrogen fixation is due to a reduction in the level of leghemoglobin and assimilation of 1-amino-cyclopropane 1-carboxylate (ACC) that reduces root nodulation (Glick et al. 2007). Furthermore, there is a reduction in transpiration efficiency of the plant resulting from the low stomatal conductance, leaf abscission, and stomatal closure under drought stress. These stress responses cause a drop in CO_2 levels, thereby reducing the photosynthesis potential

(Dutta et al. 2018). At the biochemical level, drought stress inhibits carbon metabolism and interferes with the cell signaling pathway.

14.1.2 Salinity

Pulses are vulnerable to salt stress, accounting for 50% yield loss in the arid and semiarid regions (Toker and Mutlu 2011). Salt stress disrupts the plant's osmoregulation pathway and reduces water availability creating a solute imbalance and ion toxicity in the cytosol (Conde et al. 2011). The physiological response of pulses under salt stress includes reduced seed germination rate, slow carbon metabolism, and photo-inhibition (Latef and Ahmad 2014). The rate of photosynthesis and nitrogen assimilation is greatly reduced under high salt stress. Plants suffering from salt stress exhibit an elevated level of anthocyanin accumulation in the stem and leaves that is reported to significantly lower the rate of germination and seedling formation. Furthermore, salt stress in cowpea inhibits α-amylase and β-amylase activity that substantially reduces germination potential (Eneas Filho et al. 1995). Plants when exposed to a high concentration of sodium (Na^+) and chlorine (Cl^-) ions disrupt nutritional balance and interfere with the absorption of other essential elements (Doering et al. 1984).

14.1.3 Waterlogging

Flooding is generally a situation arising from intensive rainfall occurring during a time interval that can cause damage and loss to crops. Soil flooding stress and its derivatives like submergence, waterlogging, hypoxia, and anoxia affect many biological and biochemical processes. Pulses like pigeon pea, urdbean, mung bean, and field pea exhibit high sensitivity to waterlogging (Sultana et al. 2014). Waterlogging is reported to inhibit the process of seed germination in several legume crops. Flooding causes hypoxia ($<2.0\%$ O_2) in the plant that increases the vulnerability to pathogen attack (Hsu et al. 2013). Furthermore, waterlogging can cause self-poisoning from the by-products of anaerobic metabolism (Rana et al. 2016).

14.1.4 Temperature Extremities

Pulses exhibit temperature sensitivity with an optimum temperature of 10–25 °C for *rabi* pulses and 15–30 °C for *kharif* pulses. High and chilling temperature stress induces photo-inhibition, disrupting respiration and membrane stability, and has a pronounced effect on pollen germination and fertility.

14.2 Genetically Engineered Pulses for Abiotic Stress Tolerance

Genetic engineering can effectively address abiotic stress-specific adaptive traits by targeting specific pathways or regulators governing the pathway. Reports on transgenic development in eight pulses, viz. chickpea, pigeon pea, mung bean, urdbean, cowpea, field pea, common bean, and lentil, have been discussed. Reports on the development and characterization of transgenic pulses are very few and have been enlisted (Table 14.2).

14.2.1 Chickpea

Chickpea or Bengal gram (*Cicer arietinum* L.) is the second most important food legume globally after dry beans in terms of annual production (FAOSTAT 2019). It is grown on 13.2 mha (mean of 2010–2019, FAOSTAT) producing 12.8 mt at an average yield of 971.5 kg/ha (mean of 2010–2019, FAOSTAT). Chickpea is generally grown in the arid and semiarid regions of the world. India, being its largest producer, accounts for approximately 70% of the world's production which is cultivated on 9.5 mha land (FAOSTAT 2019). However, the productivity of chickpea is still stagnating and is below the expected threshold due to the various stresses that significantly reduce their yield. Among the various abiotic stresses, waterlogging, drought, extreme temperatures, high salt, and freezing are the most adverse that severely reduce chickpea productivity. Drought accounts for 45–50% yield losses in chickpea (Varshney et al. 2014). Terminal drought in chickpea occurs during the post-flowering stage that is the most devastating, accounting for heavy yield losses of up to 50% (Das et al. 2021). Soil salinity is another major abiotic stress in chickpea that disturbs the plant's biochemical pathways resulting in huge yield losses in chickpea (Flowers et al. 2010). Several traditional breeding techniques have been attempted for crop improvement, but only limited success has been reported due to the multigenic nature of abiotic stresses and lack of desired trait in the available gene pool (Acharjee and Sarmah 2013; Mantri et al. 2012). Furthermore, plants' biochemical, molecular, and physiological responses to these abiotic stresses are complex involving multiple gene functions and lack proper screening technique. Alternatively, stress-tolerant traits from stress-tolerant species have been introduced in chickpea using transgenic and genome editing approaches.

The first abiotic stress tolerance in transgenic chickpea was reported in the year 2009, where the mutated osmoregulatory gene *P5CSF* derived from *Vigna aconitifolia* driven by *35S* promoter was introduced in *desi* variety C-235 (Bhatnagar-Mathur et al. 2009). The *P5CSF* gene encoding Δ^1-pyrroline-5-carboxylate synthase (P5CS) is responsible for the overproduction of proline in the transgenic chickpea lines under water-stress conditions. However, a relatively slight increase in the transpiration efficiency (TE) and moderate enhancement in the crop yield were reported in the transgenic events. Later, the *P5CS* gene was transformed into *desi* chickpea cv. Annigeri for developing salinity tolerance in transgenic chickpea. The transgenic chickpea lines exhibited a high level of proline

Table 14.2 Transgenic pulses for abiotic stress tolerance

Crop	Gene	Mechanism	Trait/phenotype	References
Chickpea	*P5CSF129A*	Proline accumulation	Drought tolerance	Bhatnagar-Mathur et al. (2009)
	P5CSF129A	Proline accumulation	Salt tolerance	Ghanti et al. (2011)
	Lec-RLK	Triggers saccharide signaling pathway	Salt tolerance	Singh (2018)
	OsRuvB	Better water retention, proline accumulation, and chlorophyll retention	Salt tolerance	Preeti (2018)
	AtDREB1A	Triggers activation of stress-induced genes	Drought tolerance	Anbazhagan et al. (2015), Das et al. (2021)
	CaCKX6	Higher root-to-shoot biomass	Drought tolerance	Khandal et al. (2020)
Pigeon pea	*P5CSF129A*	Proline accumulation	Salt tolerance	Surekha et al. (2014)
	OsRuvB	Better water retention, proline accumulation, and chlorophyll retention	Salt tolerance	Singh et al. (2020)
	Psp68	Triggers stress-induced signaling pathways	Salt tolerance	Neha (2019)
Mung bean	NHX1	Homeostasis, osmoregulation, and undeterred photosynthesis	Salt tolerance	Sahoo et al. (2016), Kumar et al. (2017)
	codA	Triggers COD pathway	Salt tolerance	Baloda et al. (2017)
	ICE1	Triggers cold-stress gene	Cold tolerance	Rout et al. (2020)
Urdbean	*Glyoxalase I*	Detoxifies cytotoxic methylglyoxal	Salt tolerance	Bhomkar et al. (2008)
	ALDRXV4	Osmoprotectants and neutralizes stress-induced toxins	Multiple stress tolerance—drought, salinity, undeterred photosynthesis	Singh et al. (2016)
	ALMT1	Higher exudation of malate in the rhizosphere	Aluminum (Al^{3+}) tolerance	Saha et al. (2020)
Cowpea	*NHX1*	Vacuolar Na^+/H^+ antiporter	Salt tolerance	Mishra et al. (2014)
Field pea	*NHX1*	Vacuolar Na^+/H^+ antiporter	Salt and frost tolerance	Ali et al. (2018)
	HsfA1d	Triggers heat-responsive genes	Heat tolerance	Shah et al. (2020)

(continued)

Table 14.2 (continued)

Crop	Gene	Mechanism	Trait/phenotype	References
Common bean	*pdf1.2, avp1*	Regulates proton pump; triggers drought tolerance responses	Drought tolerance	Espinosa-Huerta et al. (2013)
	HVA1	Induces LEA response	Drought tolerance	Kwapata et al. (2012)
Lentil	*codA*	Catalyzes synthesis of glycine betaine	Cold and freezing tolerance	Zaker et al. (2016)

accumulation with no yield losses under high salt stress conditions (250 mM NaCl) (Kiran Kumar et al. 2011).

A higher drought tolerance response with significantly higher TE value was reported in transgenic chickpea (cv. C-235) expressing dehydration-responsive element-binding protein 1A (*DREB1A*) gene driven by a stress-inducible promoter *rd29A* both isolated from *Arabidopsis thaliana* (Anbazhagan et al. 2015). The terminal water-stress response was evaluated using a lysimetric system that indicated elevated biomass partitioning in the shoot: denser, broader, and deeper rooting profile with higher TE than the untransformed control. Recently, in another study employing the *AtDREB1a* gene driven by stress-inducible promoter *rd29A*, transgenic chickpea lines exhibited better drought tolerance response in terms of relative water content, higher photo-adaptation along with chlorophyll retention, and osmotic adjustment as compared to control (Das et al. 2021).

Interestingly, in a recent report on developing drought tolerance in *Arabidopsis thaliana* and chickpea plants, a stress-induced chickpea gene *cytokinin oxidase/dehydrogenase 6* (*CaCKX6*) driven by a chickpea root-specific promoter *CaWRKY31* was demonstrated (Khandal et al. 2020). Enhanced root morphology along with elevated root-to-shoot biomass ratio was observed in *Arabidopsis* and chickpea, without compromising their vegetative and reproductive growth. Transgenic chickpea lines exhibited elevated CKX activity in roots without compromising nodulation in chickpea plants. Furthermore, there is a 25% increase in the chickpea yield, and the transgenic chickpea seeds exhibited higher mineral content.

Efficacy for salinity stress tolerance in chickpea was demonstrated in two different reports of transgenic chickpea (cv. HC-1) employing *Lec-RLK* gene (Singh 2018) and *OsRuvB* gene (Preeti 2018). The transgenic chickpea plants exhibited high proline content and chlorophyll retention resulting in minimal cell injury under high salt stress (100 mM). A novel strategy using miR408 was employed for inducing drought tolerance in chickpea (Hajyzadeh et al. 2015). The overexpression of miR408 causes repression of plantacyanin transcript and induces the expression of DREB and other drought-responsive genes in the transgenic lines that imparted tolerance in the tested transgenic lines.

In another study, an attempt to modulate the expression of stress-induced homeodomain-leucine zipper (I) (HD-Zip (I)) gene using chickpea transcription factor WRKY70 for inducing abiotic stress tolerance in transgenic tobacco and chickpea was reported (Sen et al. 2017). The molecular analysis of transgenic chickpea and tobacco lines indicated accumulation of HDZ12 transcripts with enhanced tolerance to osmotic stresses arising from drought and salinity compared to the wild type. Recently, a first report was demonstrated on understanding the role of two drought tolerance genes, viz. 4-coumarate ligase (4CL) and Reveille 7 (RVE7) employing CRISPR/Cas9 for editing in chickpea protoplast (Badhan et al. 2021).

The two stress tolerance genes *CBL-CIPKs* and *NAC* from chickpea have been transformed in transgenic tobacco (Meena et al. 2015) and poplars (Movahedi et al. 2015), respectively, for inducing abiotic stress tolerance in the transgenic lines. The *CBL-CIPK* gene encodes for calcineurin B-like interacting protein kinases that constitute signaling modules that relay calcium signals. The CBL-CIPK complexes provide stress tolerance by regulating membrane transport through available transporters and pores in the plasma membrane and tonoplast. The NAC transcription factor functions as a stress-induced regulator of plant immunity that modulates the hypersensitive responses and stomatal immunity.

In another report, for enhancing drought and salinity tolerance in transgenic tobacco, the *CAP2* gene derived from chickpea under 35S promoter was introduced. *CAP2* proteins belong to APETALA2-family transcription factor that binds to DNA and exhibits a key role in plant morphology, development, and stress response. The transgenic tobacco lines exhibited a drastic increase in the leaf cell size and enhanced root morphology (Shukla et al. 2006).

14.2.2 Pigeon Pea

Pigeon pea or red gram (*Cajanus cajan* L.) is an inherently sturdy crop widely grown in Asia and Africa. There are reports of imparting abiotic stress tolerance in transgenic pigeon pea using *P5CS* (*pyrroline-5-carboxylate synthase*), *p68* (*portable ATP-dependent helicase-DDX5*), and *ruvB* (ATP-dependent DNA helicase) genes. Interestingly, genes *HyPRP* (*hybrid-proline-rich protein*), *CDR* (*cold and drought regulators*), *CYP* (*cyclophilin*), and *CKS* (*cyclin-dependent kinases*) derived from pigeon pea have also been demonstrated for stress tolerance in the model plants like *Arabidopsis* and rice.

Pigeon pea is a rainfed water-intensive crop, which experiences several water stresses during the crop's life cycle. Apart from drought, pigeon pea also experiences salinity and waterlogging stress that causes severe yield losses (Banerjee et al. 2018). Pigeon pea exhibits phytotoxic symptoms (stunted growth, modified leaf morphology, hampered reproductive growth) when exposed to a high salt concentration (>50 mM) (Singh et al. 2020). Furthermore, the accumulation of sodium ions (Na^+) due to the salt stress can denature the cell's genetic material and proteins, thereby inhibiting the biochemical pathways (Tayyab et al. 2016). Several stress molecules like proline, glycine betaine, and polyamines that function as cellular osmolytes are produced in response to the stress. Abiotic stresses promote the production of cytotoxic molecules, viz. ROS and H_2O_2, inside the plant cell. Accumulated cytotoxic molecules can damage the photosynthetic apparatus and downregulate the production of photosynthetic pigments leading to photo-inhibition. High salt stress is found to inhibit chlorophyll production and promote its degradation (Yang et al. 2011). Similarly, high salt stress decreases β-carotene level, which is converted into zeaxanthin for protection against photo-inhibition (Banerjee and Roychoudhury 2016). To address the issues arising due to various abiotic stresses, several stress-tolerant varieties of pigeon pea have been developed.

In an attempt to enhance salinity tolerance in pigeon pea plants, a mutated Δ^1-pyrroline-5-carboxylate synthetase gene (P5CSF129A) derived from moth bean (*Vigna aconitifolia*) was successfully introduced in the three pigeon pea genotypes using the *Agrobacterium tumefaciens*-mediated transformation (ATMT) method (Surekha et al. 2014). Better growth response along with a higher level of chlorophyll and enhanced proline accumulation was reported in the transgenic pigeon pea lines compared to the control under induced high salt stress (up to 200 mM NaCl). Proline functions as an osmoprotectant that plays a significant role in maintaining osmotic balance to protect subcellular structures and enzymes and increase cellular osmolarity that provides the necessary turgidity for cell proliferation under stress conditions. P5CS is a rate-limiting enzyme in the proline biosynthetic pathway that confers enhanced tolerance due to feedback inhibition by proline.

In another report, *Psp68*, a DEAD-box helicase, has been introduced in transgenic pigeon pea using ATMT for imparting salinity tolerance (Neha 2019). Salt stress is reported to induce toxicity by disrupting the ionic concentration of the plant cell. p68 is a stress-induced molecule that exhibits helicase and ATPase activities. Furthermore, it interacts with Ca^{2+}-CaM and regulates signaling cascades linked with plant's response to salt stress (Wang et al. 2013). The transgenic pigeon pea lines exhibited enhanced salinity tolerance by modulating osmoregulation and higher photosynthesis along with enhanced catalase and peroxidase activity.

In planta transformation was attempted in pigeon pea for inducing salinity tolerance using *Os*RuvB gene derived from rice. A transformation efficiency of 35.7% was obtained, and the transgenic pigeon pea lines exhibited better photosynthetic and osmoregulatory responses than the control plants under induced salt stress (75 mM NaCl). The transgenic lines exhibited enhanced protection against membrane injury and lipid peroxidation due to the proline accumulation that triggers the stress tolerance pathway (Singh et al. 2020).

Pigeon pea plant has an inherent climate tolerance mechanism regulated by a multigene family. Various stress tolerance genes from pigeon pea have been identified for introducing abiotic stress tolerance in *Arabidopsis* and rice plants. The stress-induced *HyPRP* (Mellacheruvu et al. 2016) and *CDR* (Sunitha et al. 2017) genes of pigeon pea have been introduced in rice for developing stress tolerance. High expression of *Cajanus cajan*-derived hybrid-proline-rich protein-encoding gene (*CcHyPRP*) in the transgenic rice has provided enhanced tolerance against both abiotic and biotic stresses. In addition, transgenic lines exhibited enhanced growth response including better grain morphology in terms of size and number compared to control plants under various abiotic stress conditions. Furthermore, higher endochitinase activity and enhanced tolerance against the fungal pathogen,

Magnaporthe grisea, were reported in the transgenic lines of rice. Similarly, transgenic *indica* rice expressing pigeon pea *CcCDR* (cold and drought regulators) driven by stress-induced (*rd29A*) and constitutive (*CaMV35S*) promoters conferred high level of tolerance against several abiotic stresses (Sunitha et al. 2017). The stress responses are modulated by multiple regulatory genes of ABA-dependent and -independent signaling pathways. The transgenic rice lines expressing *CcCDR* gene exhibited enhanced physiological responses in terms of germination, survivability, and overall plant morphology under salinity, drought, and low-temperature stresses. Three stress-regulatory genes *CYP* (Sekhar et al. 2010), *CDR* (Tamirisa et al. 2014), and *CKS* (Tamirisa et al. 2017) of pigeon pea have been successfully introduced for developing multiple abiotic stress tolerance in transgenic *Arabidopsis*. The CcCYP gene encodes for cyclophilin, which is reported to enhance peptidyl-propyl *cis-trans* isomerase (PPIase) activity under stress conditions. Furthermore, CcCYP is involved in maintaining cellular homeostasis resulting in salinity tolerance in the transgenic plants. Similarly, under stress conditions, transgenic *Arabidopsis* lines expressing CcCKS exhibited better physiological response under stress conditions. However, under favorable conditions, the growth performance of the nontransgenic lines was better.

14.2.3 Mung Bean

Mung bean or green gram (*Vigna radiata* L.) is a short duration hardy pulse crop that is preferably cultivated as an intercrop on marginal land predominantly in Asian countries. Its seeds are a rich source of easily digestible protein (21–31%), antioxidants, dietary fiber, and essential phytonutrients (Burstin et al. 2011). The nonedible parts of the plant can serve as cattle feed and green manure. Mung bean is often grown as a short-duration intercrop with cereals and other pulses (mainly, pigeon pea) for improving soil fertility.

The average productivity of mung bean is stagnating low at 0.5 t/ha due to various abiotic and biotic constraints witnessed during the crop life cycle (Pratap et al. 2019). Drought, soil salinity, waterlogging, and heat stress, especially during the flowering and seed/pod development stages, are responsible for heavy productivity losses in mung bean (HanumanthaRao et al. 2016; Singh and Singh 2011). Premature sprouting due to rainfall during the reproductive stages is also reported to cause significant yield losses in mung bean (Sharma and Dhanda 2014). Drought stress in the flowering stages causes flower abscission and causes yield loss of up to 31–57% in mung bean (Nadeem et al. 2019). Furthermore, exposure to drought promotes the production of ROS giving rise to harmful superoxide molecules that damage the plant cell. Few successful studies have been reported for developing stress-tolerant short-duration varieties employing various breeding techniques. However, the slow development of stress-tolerant varieties and lack of genetic diversity in cultivated mung bean species are the major limiting factors in breeding. Enhancement of the existing stress tolerance trait and introgression and overexpression of novel tolerance genes through genetic engineering are potential approaches for imparting abiotic stress tolerance in mung bean. Three transgenes were employed, viz. *codA*, *NHX1*, and *ICE1*, which have been reported for inducing abiotic stress tolerance in mung bean.

In a report, inducing of salinity tolerance in genetically engineered mung bean expressing *Arabidopsis* antiporter *NHX1* gene driven by 35S constitutive promoter was demonstrated (Sahoo et al. 2016). The NHX regulators (Na^+/H^+ exchanger) play a critical role in cellular Na^+ ion homeostasis and regulate the movement of Na^+/H^+ across tonoplast membrane, thereby maintaining the Na^+ levels inside the cell cytoplasm (Blumwald et al. 2000; Deinlein et al. 2014). Efficient distribution of Na^+ ions via overexpression of vacuolar antiporter gene, NHX1, is employed for developing salinity tolerance in the transgenic mung bean lines. The transgenic line expressing *AtNHX1* gene exhibited improved intracellular ion homeostasis with higher level antioxidant and proline production than the control plants under salinity stress. Furthermore, the transgenic mung bean lines exhibited higher photosynthesis efficiency with reduced levels of reactive oxygen species (H_2O_2 and O_2^-). However, the later transgenic mung bean progenies exhibited reduced height with low productivity resulting from metabolic payoffs from the constitutive expression of the transgene. A similar report, on transgenic mung bean co-expressing *AtNHX1* and *bar* genes for inducing improved tolerance against salt and oxidative stresses along with herbicide tolerance, was demonstrated using a stress-induced *rd29A* promoter (Kumar et al. 2017). The use of stress-inducible promoter *rd29A* was preferentially achieved for optimal level of stress tolerance at the desired time, thereby reducing the extra load due to the unintended metabolic payoffs. Another gene *codA* encoding for choline oxidase was introduced in transgenic mung bean for inducing abiotic stress tolerance via ATMT using cotyledonary node explants (Baloda and Madanpotra 2017). Choline oxidase targets the COD pathway changing choline directly into glycine betaine (GB). GB functions as an osmoprotectant that regulates the osmotic pressure of cytoplasm and protects the photosynthesis apparatus of the chloroplast under stress conditions. The leaf disc test of the transgenic mung bean lines exhibited high salinity (50–200 mM NaCl) tolerance.

Recently, enhanced expression of ICE1 gene in transgenic mung bean (cv. OBGG-52) for successfully inducing cold tolerance at the seedling stage has been reported (Rout et al. 2020). The cold-stress response in a plant is a complex process that is induced by the CCAATT motif-binding factor (CBF) that regulates the gene responsible for cold tolerance. ICE1 is a transcription factor that regulates CBF and confers cold tolerance, making it a potential candidate for inducing cold tolerance in transgenic mung bean. The transgenic lines exhibited well-developed plant morphology and had a significantly higher rate of photosynthesis along with proline, chlorophyll, and lipid accumulation as compared to control under the low-temperature stress. Similarly, the transgenic seeds produced exhibited significantly higher germination responses and growth on artificial medium at 10–14 °C.

14.2.4 Urdbean

Urdbean or black gram (*Vigna mungo* L. Hepper.) is a nutritious grain legume mostly cultivated in tropical and subtropical regions of South and Southeast Asia for a short-duration (90–120 days) intercrop with rice. India is the largest producer of urdbean accounting for about 70% of the world's urdbean production. India produces an average 1.5 mt of urdbean annually cultivated on 3.08 mha land, with an average productivity of 486.1 kg/ha (2004–2014). The crop witnesses various abiotic and biotic constraints that severely reduce crop productivity. Among the various abiotic stresses, salinity and drought are the most devastating. Like other legume crops, only limited genetic variability is present in the urdbean germplasm. So, developing transgenic urdbean varieties employing genetic engineering seems to be a viable option to overcome the limitations associated with current crop improvement strategies.

Only a few reports on developing transgenic urdbean for inducing abiotic stress tolerance are available. The first successful development of salt tolerance transgenic urdbean (cv. LBG-20) overexpressing *glyoxalase I* gene driven by Cestrum yellow leaf curling virus (CmYLCV) promoter using ATMT was demonstrated (Bhomkar et al. 2008). The nodal region of the embryonal axis of urdbean was used for in vitro regeneration and transformation experiments. The *glyoxalase I* gene used for inducing salt tolerance was isolated from mustard (Veena et al. 1999). The enzyme belongs to the metallo-glutathione (GSH) transferase superfamily that has a key role in detoxification of the cytotoxic methylglyoxal to *S*-D-lactoylglutathione (Aronson et al. 1978). The transgenic urdbean lines exhibited relatively high salt tolerance at 100 mM NaCl over the control plants.

In a similar study, *Glyoxalase 1* (*Gly1*) gene driven by two distinct promoters, constitutive Cestrum viral promoter (Stavolone et al. 2003) and a stress-inducible *rd29A* promoter (Kasuga et al. 1999), in two different expression cassettes was employed for inducing salinity tolerance in urdbean using cotyledonary node explants (Bhalla-Sarin et al. 2004). The expression vector has a kanamycin resistance (*nptII*) gene flanked by *lox* sites for employing marker-free approach in the transgenic lines. The efficacy of salt tolerance in the transgenic urdbean lines was tested by leaf disc assay. The leaves of the transgenic line exhibited relatively higher chlorophyll content at 600 mM NaCl than the non-transformed control lines.

In another report on inducing multiple abiotic stress tolerance in the *ALDRXV4* gene isolated from desiccation-tolerant plant, *Xerophyta viscose* was introduced in urdbean (cv. LBG-20) using ATMT (Singh et al. 2016). The ALDRXV4 is a member of the aldo-keto reductase superfamily, is involved in the reduction of specific metabolites in the cells, functions as an osmoprotectant, and neutralizes reactive carbonyl species that are produced during stress conditions. The transgenic urdbean lines expressing the *ALDRXV4* gene exhibited broad-spectrum tolerance against various environmental stresses, including salinity, drought, methyl viologen, and H_2O_2-induced oxidative stress.

Recently, for inducing enhanced aluminum tolerance, the *Arabidopsis malate transporter 1* (*ALMT1*) gene was transferred to urdbean (cv. PU 30) (Saha et al.

2020). The cotyledon node explants were used for regeneration and transformation of urdbean using ATMT. The transgenic lines exhibited enhanced adaptability in acidic soil due to enhanced exudation of malate in the rhizosphere during the Al^{3+} stresses. Furthermore, the transgenic urdbean lines have a better root profile and higher photosynthesis efficiency under Al^{3+} stress than the wild type.

14.2.5 Cowpea

Cowpea or black-eyed pea (*Vigna unguiculata* L.) is an important grain legume crop well adapted to arid and semiarid tropics in Asia, Africa, North America, and South America. It is globally cultivated on 14.4 mha of land with an annual production of 8.9 mt accounting for a yield of 616.3 kg/ha (FAOSTAT 2019). Nigeria with an annual production of 3.58 mt is the largest producer of cowpea accounting for 40.17% of the annual cowpea production worldwide (FAOSTAT 2019). Cowpea is consumed as a dry grain and as a fresh vegetable. It is a cheap source of vegetable protein, carbohydrate, vitamins, and fiber and is reported to provide various health benefits, e.g., weight management and managing cardiovascular diseases (Gonçalves et al. 2016). Besides its nutritional benefits, cowpea serves as a fodder for livestock, enhances soil fertility, and suppresses weed production in the field. Recently, transgenic cowpea for insect-resistant traits has been approved for commercial cultivation in Nigeria.

Salinity and waterlogging are the major abiotic stresses affecting the yield potential of cowpea. Furthermore, the lack of enhanced varieties that can withstand abiotic stresses adds to low production of cowpea. There is only a single report of transgenic cowpea exhibiting tolerance against abiotic stress. In the report, vacuolar Na^+/H^+ antiporter gene (*NHX1*) isolated from mung bean driven by a constitutive promoter was introgressed in cowpea cv. Pusa Komal using ATMT for inducing salinity tolerance in the transgenic cowpea lines (Mishra et al. 2014). The transgenic cowpea lines in the leaf disc assay exhibited higher salinity tolerance to 200 mM NaCl compared to the control.

Interestingly, *DREB2A* isolated from cowpea was successfully employed for developing drought tolerance in transgenic *Arabidopsis* (Sadhukhan et al. 2014). The expression of *Vu*DREB2A in the transgenic *Arabidopsis* plant significantly enhanced its drought tolerance capabilities. Furthermore, overexpression of a modified *Vu*DREB2A, after removal of a putative negative regulatory domain, resulted in a dwarf phenotype in the transgenic plants.

14.2.6 Field Pea

Field pea or garden pea (*Pisum sativum* L.) is one of the prominent cool-season pulse crops grown in semiarid regions throughout the world. Dry peas are the third most widely grown legume in the world that is cultivated on about 7.17 mha of land with an average annual production of 14.18 mt (FAOSTAT 2019). China is the largest producer of field pea with an annual production of 1.45 mt followed by India (0.81 mt) (FAOSTAT 2019). Pea seeds are consumed as fresh vegetables and also in dried form. Pea seeds are a rich source of dietary protein, fibers, and other essential phytonutrients that exhibit antioxidant and anti-inflammatory properties. The non-edible parts of the pea plant can be used as cattle feed and green manure. Among abiotic stresses, heat stress predominantly reduces global pea production mainly during the flowering stages of the crop. A one-degree rise in the average atmospheric temperature during the flowering stage is reported to reduce pea productivity by 600 kg/ha (Ridge and Pye 1985).

Three reports of transgenic field pea harboring *NHX1* and *HsfA1d* genes isolated from *Arabidopsis* exhibited tolerance against salinity and extreme temperature stresses. In the first transgenic study, the *NHX1* gene was used as a physiological trait governing salinity tolerance for establishing the expression capability of the di-cistronic binary vector system in the transgenic pea plant (Ali et al. 2010). In further study, the progenies of the transgenic pea lines were evaluated for five subsequent generations (Zahid et al. 2018). The progenies exhibited stability integration of *AtNHX1* along with persistent morphological features. Interestingly, further analysis of the obtained progeny indicated the development of an additional trait for frost tolerance (however, further investigation was needed for validation). Recently, heat-shock factor *HsfA1d* isolated from *Arabidopsis* was employed for enhancing ROS scavenging system for imparting tolerance against thermal stress in transgenic pea (Shah et al. 2020). The transgenic lines exhibited better heat tolerance at 42 °C due to the accumulation of stress-induced molecules like proline and ascorbate peroxidase that reduced the toxicity caused by ROS.

Three stress-tolerant genes *MnSOD*, *p68*, and *cAPX* isolated from pea have been demonstrated to impart abiotic stress tolerance in rice, tobacco, and tomato, respectively. MnSOD is a manganese-dependent enzyme that triggers the plant's ROS scavenging systems and protects the transgenic rice lines against drought stresses (Wang et al. 2005a). The p68 gene functions as a DEAD-box helicase that controls the generation of stress-induced ROS and modulates the antioxidative defense machinery in transgenic tobacco against salinity stress (Tuteja et al. 2014). The cytosolic ascorbate peroxidase cAPX plays a key role in H_2O_2 metabolism and confers protection against oxidative stress arising due to various environmental constraints. The transgenic tomato plants expressing *cAPX* exhibited tolerance against chilling and salinity stresses in transgenic tomatoes (Wang et al. 2005b).

14.2.7 Common Bean

Common bean or French bean (*Phaseolus vulgaris* L.) or green bean is the most consumed food legume worldwide. The crop has been reported to be originated from two centers in the modern world, namely, Andes and Mesoamerica that represent the two major gene pools. It is globally cultivated on ~33.06 mha of land with an annual production of ~28.90 mt. Myanmar with an annual production of 5.84 mt is the largest producer of dry bean, accounting for 20.21% of the world's production

(FAOSTAT 2019). It is considered as the most economical source of dietary protein, with a commercial value higher than that of all other legume crops combined (Porch et al. 2013). The immature pods, also known as green bean/snap bean, are consumed as a vegetable, and the dried mature seeds are consumed either in the form of soup (dal) or as roasted beans. The dried seeds are rich in lysine and tryptophan and provide beneficial phytochemicals, antioxidants, and flavonoids.

The productivity of common bean is severely hampered by various abiotic stresses like extreme temperature, water availability, high salt, ultraviolet (UV) radiation, and heavy metal accumulation. Global production loss of more than 60% has been reported due to terminal or intermittent drought stress in common bean (Rao et al. 2013). Various trait improvements of common bean have been reported via conventional breeding methods. However, these methods are time consuming and depend on the traits available in the *P. vulgaris* gene pool. To counter these issues, genetic engineering has evolved as a successful strategy complementing the existing technology for the introduction of a new trait in the susceptible species of common bean. Golden mosaic virus-resistant transgenic common bean has been approved for commercial cultivation in Brazil.

Three successful reports for drought tolerance in transgenic common bean using *HVA1* (encoding for type III, late embryogenesis abundant (LEA) proteins) isolated from barley and *avp1* (triggers proton pump pyrophosphatase) isolated from *A. thaliana* are available. In the first report, efficacy of drought tolerance was tested using *HVA1* gene driven by a constitutive rice actin 1 promoter in five common bean cultivars, namely, Condor, Matterhorn, Sedona, Olathe, and Montcalm, using the direct transformation (Kwapata et al. 2012). The apical shoot meristem primordium derived from the mature seeds was used for the direct transformation of common bean. The transgenic lines harboring the HVA1 gene exhibited better drought tolerance due to the higher level of LEA protein that plays a key role in the protection against protein aggregation under osmotic stresses. In another report for demonstrating transformation efficacy of *Agrobacterium tumefaciens* in common bean, two genes *pdf1:2* and *avp1* isolated from *A. thaliana* driven by a double enhancer-based *CaMV* promoter were introduced in hypocotyl explants derived from two common bean cultivars, viz. Flor de Mayo Anita (FMA) and Pinto Saltillo (Espinosa-Huerta et al. 2013). The *pdf1:2* encodes for defensins that are cystine-rich peptides that impart innate immunity in plants against fungal pathogens, and the *avp1* encodes for proton pump pyrophosphatase that provides drought tolerance in plants. Similarly, the transgenic common bean lines expressing the *pdf1:2* and *avp1* genes depicted enhanced tolerance against both biotic and abiotic stresses. In another report, the *vacuolar pyrophosphatase-1* (*avp1*) gene of *Arabidopsis thaliana* was introduced in hypocotyl explants of beans cv. Pinto Saltillo via ATMT for inducing drought tolerance in the transgenic lines (Cadena-Hernández et al. 2019). The result indicated better adaptation along with a higher photosynthetic rate in all PS-avp1 lines as compared to the control under drought stress.

Two stress tolerance genes, viz. P5CS (Chen et al. 2013) and ethylene-responsive element-binding transcriptional factor (ERF) (Kavas et al. 2020), were derived from the common bean for inducing salinity tolerance in *Arabidopsis* and tobacco plants, respectively. The P5CS gene encodes Δ^1-pyrroline-5-carboxylate synthase that is a crucial enzyme in proline biosynthesis in plants. The transgenic *Arabidopsis* lines exhibited significantly higher levels of P5CS transcripts under salt-stress conditions (up to 200 mM NaCl) than the control plant. Similarly, the ERF transcriptional factor regulates plant's response by triggering the stress hormones and improves its survival under various stress conditions.

14.2.8 Lentil

Lentil (*Lens culinaris* Medik.) is among the oldest domesticated grain legume crop in the world. It is cultivated on ~4.8 mha of land worldwide with an annual production of ~5.73 mt. Canada (~2.16 mt) is the largest producer of lentil followed by India (~1.22 mt), Australia (~0.53 mt), and Turkey (~0.35 mt). Lentil is well adapted to diverse soil terrains. Despite being a hardy crop, it witnesses severe yield losses due to the terminal heat and drought stresses particularly occurring during the reproductive and seed-filling stages of the crop. Only a single report is available on abiotic tolerance in transgenic lentil (cv. Gachsaran) expressing bacterial *codA* gene that was developing tolerance against cold and freezing (Zaker Tavallaie et al. 2017). The genetic transformation was done employing ATMT using cotyledon with a slight part of embryo axes explants. The *codA* gene used in the study encodes for choline oxidase enzyme that catalyzes the synthesis of glycine betaine from choline. Glycine betaine is a stress-induced molecule that stabilizes and regulates activities of stress enzymes and protein complexes and also maintains the membrane integrity under salt stress.

14.3 Conclusions

Pulses have narrow genetic diversity, and hence the addition of traits through transgenic technology is imperative. There are several abiotic stresses encountered by pulses throughout the growth phases, and the situation is predicted to aggravate with changing climate conditions. Genetic engineering can add precise gene/locus that can assist certain adaptive traits for imparting abiotic stress tolerance in plants. An important feature is a "*regulon technology*" that can regulate the master regulator (mostly transcription factors), which in turn can activate several downstream genes to affect the abiotic stress-adaptive traits. Currently, only a few abiotic stress-tolerant traits were engineered in pulses; however, there is a large scope of introducing additional traits. Though the technology is promising, regulatory issues and public acceptability are the key areas of concern. Genome editing technologies can address some of the perceived issues associated with transgenic crops, and hopefully, in the near future, engineered/edited pulses with *inbuilt* abiotic stress tolerance shall be considered for cultivation.

References

Acharjee S, Sarmah BK (2013) Biotechnologically generating 'super chickpea' for food and nutritional security. Plant Sci 207:108–116

Ahmad F, Gaur PM, Croser J (2005) Chickpea (Cicer arietinum L.). In: Singh RJ, Jauhar PP (eds) Genetic resources, chromosome engineering, and crop improvement, (Genetic resources, chromosome engineering, and crop improvement series), vol 1. CRC Press, pp 187–217

Ali Z, Schumacher HM, Heine-Dobbernack E, El-Banna A, Hafeez FY, Jacobsen H-J, Kiesecker H (2010) Dicistronic binary vector system—a versatile tool for gene expression studies in cell cultures and plants. J Biotechnol 145(1):9–16

Ali Z, Saeed W, Naseem S, Ahmad F, Akrem A et al (2018) Phenotypic evaluation of transgenic peas (*Pisum sativum* L.) harboring AtNHX1 demonstrates stable gene expression and conserved morphology in subsequent generations. Turk J Bot 42(2):1–9

Allard RW (1999) Principles of plant breeding. Wiley

Anbazhagan K, Bhatnagar-Mathur P, Vadez V, Dumbala SR, Kishor PB, Sharma KK (2015) DREB1A overexpression in transgenic chickpea alters key traits influencing plant water budget across water regimes. Plant Cell Rep 34(2):199–210

Aronson AC, Marmstal E, Mannervik B (1978) Glyoxalase I, a zinc containing metalloenzyme of mammals and yeast. Biochem Biophys Res Commun 81:4235–4250

Atkinson NJ, Urwin PE (2012) The interaction of plant biotic and abiotic stresses: from genes to the field. J Exp Bot 63(10):3523–3543

Badhan S, Ball AS, Mantri N (2021) First report of CRISPR/Cas9 mediated DNA-free editing of 4CL and RVE7 genes in chickpea protoplasts. Int J Mol Sci 22(1):396

Baloda A, Madanpotra S (2017) Transformation of mungbean plants for salt and drought tolerance by introducing a gene for an osmoprotectant glycine betaine. J Plant Stress Physiol 3:5–11

Baloda A, Madanpotra S, Jaiwal PK (2017) Transformation of mungbean plants for abiotic stress tolerance by introducing codA gene, for an osmoprotectant glycine betaine. J Plant Stress Physiol 3:5–11

Banerjee A, Roychoudhury A (2016) Plant responses to light stress: oxidative damages, photoprotection, and role of phytohormones. In: Plant hormones under challenging environmental factors. Springer, pp 181–213

Banerjee A, Ghosh P, Roychoudhury A (2018) Salt stress responses in pigeon pea (*Cajanus cajan* L.). In: Pulse improvement. Springer, pp 99–108

Bansal R, Sharma SS, Tripathi K, Kumar A (2019) Waterlogging tolerance in blackgram [*Vigna mungo* (L.) Hepper] is associated with chlorophyll content and membrane integrity. Ind J Biochem Biophysics 56(1):81–85

Basu PS, Ali M, Chaturvedi SK (2009) Terminal heat stress adversely affects chickpea productivity in northern India—Strategies to improve thermotolerance in the crop under climate change. In: Challenges and opportunities in agrometeorology, New Delhi, February, 23–25, pp 189–193

Bayuelo-Jiménez JS, Craig R, Lynch JP (2002) Salinity tolerance of Phaseolus species during germination and early seedling growth. Crop Sci 42(5):1584–1594

Beck EH, Fettig S, Knake C, Hartig K, Bhattarai T (2007) Specific and unspecific responses of plants to cold and drought stress. J Biosci 32(3):501–510

Beebe S, Rao I, Blair M, Acosta J (2013) Phenotyping common beans for adaptation to drought. Front Physiol 4(35):1–20

Bhalla-Sarin N, Bhomkar P, Debroy S, Sharma N, Saxena M, Upadhyaya CP, Muthusamy A, Pooggin M, Hohn T (2004) Transformation of *Vigna mungo* (blackgram) for abiotic stress tolerance using marker free approach. In: New directions for a diverse planet: proceedings of the 4th International Crop Congress Brisbane, Australia, Oct, p 454

Bhatnagar-Mathur P, Vadez V, Jyostna Devi M, Lavanya M, Vani G, Sharma KK (2009) Genetic engineering of chickpea (*Cicer arietinum* L.) with the P5CSF129A gene for osmoregulation with implications on drought tolerance. Mol Breed 23(4):591–606

Bhomkar P, Upadhyay CP, Saxena M, Muthusamy A, Shiva Prakash N, Pooggin M, Hohn T, Sarin NB (2008) Salt stress alleviation in transgenic *Vigna mungo* L. Hepper (blackgram) by overexpression of the glyoxalase I gene using a novel Cestrum yellow leaf curling virus (CmYLCV) promoter. Mol Breed 22(2):169–181

Blumwald E, Aharon GS, Apse MP (2000) Sodium transport in plant cells. Biochim Biophys Acta Biomembranes 1465(1–2):140–151

Burstin J, Gallardo K, Mir RR, Varshney RK, Duc G (2011) Chapter 20: Improving protein content and nutrition quality. In: Biology and breeding of food legumes. CAB International, p 314

Cadena-Hernández D, Gaxiola-Ariza R, Mora-Avilés MA (2019) Water stress tolerance in beans cv. Pinto Saltillo modified with the vacuolar pyrophosphatase-1 gene of Arabidopsis thaliana. Revista Mexicana de Ciencias Agrícolas 10(6):1231–1245

Chaves MM (1991) Effects of water deficits on carbon assimilation. J Exp Bot 42(1):1–16

Chen JB, Yang JW, Zhang ZY, Feng XF, Wang SM (2013) Two P5CS genes from common bean exhibiting different tolerance to salt stress in transgenic Arabidopsis. J Genet 92(3):461–469

Choukri H, Hejjaoui K, El-Baouchi A, El-haddad N, Smouni A, Maalouf F, Thavarajah D, Kumar S (2020) Heat and drought stress impact on phenology, grain yield, and nutritional quality of lentil (*Lens culinaris* Medikus). Front Nutr 7:596307

Conde A, Chaves MM, Gerós H (2011) Membrane transport, sensing and signaling in plant adaptation to environmental stress. Plant Cell Physiol 52(9):1583–1602

Crawford RMM (1977) Tolerance of anoxia and ethanol metabolism in germinating seeds. New Phytol 79:511–517

Daryanto S, Wang L, Jacinthe P-A (2017) Global synthesis of drought effects on cereal, legume, tuber and root crops production: a review. Agric Water Manag 179:18–33

Das A, Basu PS, Kumar M, Ansari J, Shukla A, Thakur S, Singh P, Datta S, Chaturvedi SK, Sheshshayee MS (2021) Transgenic chickpea (*Cicer arietinum* L.) harbouring AtDREB1a are physiologically better adapted to water deficit. BMC Plant Biol 21(1):1–17

Deinlein U, Stephan AB, Horie T, Luo W, Xu G, Schroeder JI (2014) Plant salt-tolerance mechanisms. Trends Plant Sci 19(6):371–379

Doering HW, Schulze G, Roscher P (1984) Salinity effects on the micronutrient supply of plants differing in salt resistance. In: 6. Colloque International Pour l'optimisation de La Nutrition Des Plantes, Montpellier (France), 2–8 Sept 1984

Dutta S, Mitra M, Agarwal P, Mahapatra K, De S, Sett U, Roy S (2018) Oxidative and genotoxic damages in plants in response to heavy metal stress and maintenance of genome stability. Plant Signal Behav 13(8):e1460048

Eneas Filho J, Oliveira NOB, Prisco JT, Gomes Filho E, Nogueira CM (1995) Effects of salinity in vivo and in vitro on cotyledonary galactosidases from *Vigna unguiculata* (L.) Walp. during seed germination and seedling establishment. Rev Bras Fisiol Veg 7:135–142

Espinosa-Huerta E, Quintero-Jiménez A, Cabrera-Becerra K, Mora-Avilés MA (2013) Stable and efficient Agrobacterium tumefaciens mediated transformation of *Phaseolus vulgaris*. Agrociencia 47(4):319–333

FAOSTAT (2019) (Food and Agriculture Organization of the United Nations), 2019. Production: crops. http://faostat.fao.org

Flowers TJ, Gaur PM, Gowda CLL, Krishnamurthy L, Samineni S, Siddique KHM, Turner NC, Vadez V, Varshney RK, Colmer TD (2010) Salt sensitivity in chickpea. Plant Cell Environ 33(4):490–509

Foyer CH, Lam H-M, Nguyen HT, Siddique KHM, Varshney RK, Colmer TD, Cowling W, Bramley H, Mori TA, Hodgson JM (2016) Neglecting legumes has compromised human health and sustainable food production. Nat Plants 2(8):1–10

Ghanti KKS, Sujata KG, Vijay K et al (2011) Heterologous expression of P5CS gene in chickpea enhances salt tolerance without affecting yield. Biol Plant 55:634

Glick BR, Cheng Z, Czarny J, Duan J (2007) Promotion of plant growth by ACC deaminase-producing soil bacteria. In: New perspectives and approaches in plant growth-promoting rhizobacteria research, pp 329–339

Gonçalves A, Goufo P, Barros A, Domínguez-Perles R, Trindade H, Rosa EAS, Ferreira L, Rodrigues M (2016) Cowpea (*Vigna unguiculata* L. Walp), a renewed multipurpose crop for a more sustainable agri-food system: nutritional advantages and constraints. J Sci Food Agric 96(9):2941–2951

Hajyzadeh M, Turktas M, Khawar KM, Unver T (2015) miR408 overexpression causes increased drought tolerance in chickpea. Gene 555(2):186–193

HanumanthaRao B, Nair RM, Nayyar H (2016) Salinity and high temperature tolerance in mungbean [*Vigna radiata* (L.) Wilczek] from a physiological perspective. Front Plant Sci 7:957

Hsu FC, Chou MY, Chou SJ, Li YR, Peng HP, Shih MC (2013) Submergence confers immunity mediated by the WRKY22 transcription factor in *Arabidopsis*. Plant Cell 25:2699–2713

IPCC (2021) Climate change 2021: the physical science basis. Contribution of Working Group I to the sixth assessment report of the Intergovernmental Panel on Climate Change. Cambridge University Press

Joshi PK, Rao PP (2016) Global and regional pulse economies: current trends and outlook, vol 144. Intl Food Policy Res Inst

Kasuga M, Liu Q, Miura S, Yamaguchi-Shinozaki K, Shinozaki K (1999) Improving plant drought, salt, and freezing tolerance by gene transfer of a single stress-inducible transcription factor. Nat Biotechnol 17(3):287–291

Kavas M, Gökdemir G, Seçgin Z, Bakhsh A (2020) Ectopic expression of common bean ERF transcription factor PvERF35 promotes salt stress tolerance in tobacco. Plant Biol 22(6): 1102–1112

Khandal H, Gupta SK, Dwivedi V, Mandal D, Sharma NK, Vishwakarma NK, Pal L, Choudhary M, Francis A, Malakar P (2020) Root-specific expression of chickpea cytokinin oxidase/dehydrogenase 6 leads to enhanced root growth, drought tolerance and yield without compromising nodulation. Plant Biotechnol J 18(11):2225–2240

Kiran Kumar Ghanti S, Sujata KG, Kumar V, Nataraja Karba N, Srinath Rao M, Kavi Kishor PB (2011) Heterologous expression of P5CS gene in chickpea enhances salt tolerance without affecting yield. Biol Plant 55(4):634–640

Kumar S, Kalita A, Srivastava R, Sahoo L (2017) Co-expression of Arabidopsis NHX1 and bar improves the tolerance to salinity, oxidative stress, and herbicide in transgenic mungbean. Front Plant Sci 8:1896

Kumar S, Ayachit G, Sahoo L (2020) Screening of mungbean for drought tolerance and transcriptome profiling between drought-tolerant and susceptible genotype in response to drought stress. Plant Physiol Biochem 157:229–238

Kwapata K, Nguyen T, Sticklen M (2012) Genetic transformation of common bean (*Phaseolus vulgaris* L.) with the gus color marker, the bar herbicide resistance, and the barley (Hordeum vulgare) HVA1 drought tolerance genes. Int J Agron 2012:198960

Latef A, Ahmad P (2014) Legumes and breeding under abiotic stress: an overview. In: Legumes under environmental stress: yield, improvement and adaptations. Wiley, pp 1–20

Lavin M, Herendeen PS, Wojciechowski MF (2005) Evolutionary rates analysis of Leguminosae implicates a rapid diversification of lineages during the tertiary. Syst Biol 54(4):575–594

Maas EV, Poss JA (1989) Salt sensitivity of cowpea at various growth stages. Irrig Sci 10(2): 313–320

Mantri N, Patade V, Penna S, Ford R, Pang E (2012) Abiotic stress responses in plants: present and future. In: Abiotic stress responses in plants. Springer, pp 1–19

Martin RJ, Jamieson PD (1996) Effect of timing and intensity of drought on the growth and yield of field peas (Pisum sativum L.). N Z J Crop Hortic Sci 24:167–174

Meena MK, Ghawana S, Dwivedi V, Roy A, Chattopadhyay D (2015) Expression of chickpea CIPK25 enhances root growth and tolerance to dehydration and salt stress in transgenic tobacco. Front Plant Sci 6:683

Mellacheruvu S, Tamirisa S, Vudem DR, Khareedu VR (2016) Pigeonpea hybrid-proline-rich protein (CcHyPRP) confers biotic and abiotic stress tolerance in transgenic rice. Front Plant Sci 6:1167

Minchin F, Summerfield R, Eaglesham A, Stewart K (1978) Effects of short-term waterlogging on growth and yield of cowpea (*Vigna unguiculata*). J Agric Sci 90(2):355–366

Mishra S, Behura R, Awasthi JP, Dey M, Sahoo D, Das Bhowmik SS, Panda SK, Sahoo L (2014) Ectopic overexpression of a mungbean vacuolar Na^+/H^+ antiporter gene (VrNHX1) leads to increased salinity stress tolerance in transgenic *Vigna unguiculata* L. Walp. Mol Breed 34 (3):1345–1359

Mohapatra C, Chand R, Tiwari JK, Singh AK (2020) Effect of heat stress during flowering and pod formation in pea (*Pisum sativum* L.). Physiol Mol Biol Plants 26(6):1119–1125

Movahedi A, Zhang J, Gao P, Yang Y, Wang L, Yin T, Kadkhodaei S, Ebrahimi M, Zhuge Q (2015) Expression of the chickpea CarNAC3 gene enhances salinity and drought tolerance in transgenic poplars. Plant Cell Tissue Organ Cult 120(1):141–154

Nadeem M, Li J, Yahya M, Sher A, Ma C, Wang X, Qiu L (2019) Research progress and perspective on drought stress in legumes: a review. Int J Mol Sci 20(10):2541. https://doi.org/10.3390/ijms20102541

Naher N, Alam AK (2010) Germination, growth and nodulation of mungbean (*Vigna radiata* L.) as affected by sodium chloride. Int J Sustain Crop Prod 5(2):8–11

Neha KP (2019) Psp68, a dead box helicase confers salinity tolerance in transgenic pigeon pea. Int J Curr Microbiol App Sci 8(4):309–324

Porch TG, Beaver JS, Debouck DG, Jackson SA, Kelly JD, Dempewolf H (2013) Use of wild relatives and closely related species to adapt common bean to climate change. Agronomy 3(2): 433–461

Pratap A, Gupta S, Basu PS, Tomar R, Dubey S, Rathore M, Prajapati US, Singh P, Kumari G (2019) Towards development of climate smart mungbean: challenges and opportunities. In: Genomic designing of climate-smart pulse crops. Springer, pp 235–264

Preeti (2018) Development and characterization of transgenic chickpea (*Cicer arietinum* L.) plants with OsRuvB gene for salt stress tolerance [With CD Copy], Molecular biology & biotechnology. CCSHAU, Hisar

Priyadharshini B, Vignesh M, Prakash M, Anandan R (2019) Evaluation of blackgram genotypes for saline tolerance at seedling stage. Indian J Agric Res 53(1):83–87

Rana D, Dass A, Rajanna G, Kaur R (2016) Biotic and abiotic stress management in pulses. Indian J Agron 61:238–248. https://doi.org/10.1007/978-981-13-6986-5_12

Rao I, Beebe S, Polania J, Ricaurte J, Cajiao C, Garcia R, Rivera M (2013) Can tepary bean be a model for improvement of drought resistance in common bean? Afr Crop Sci J 21(4):265–281

Ridge PE, Pye DL (1985) The effects of temperature and frost at flowering on the yield of peas grown in a Mediterranean environment. Field Crop Res 12:339–346

Rout GR, Bansal A, Swain D, Jadhao KR, Shelke RG, Panda SK (2020) Overexpression of ICE1 gene in mungbean (*Vigna radiata* L.) for cold tolerance. Plant Cell Tissue Organ Cult 143(3): 593–608

Sadhukhan A, Kobayashi Y, Kobayashi Y, Tokizawa M, Yamamoto YY, Iuchi S, Koyama H, Panda SK, Sahoo L (2014) VuDREB2A, a novel DREB2-type transcription factor in the drought-tolerant legume cowpea, mediates DRE-dependent expression of stress-responsive genes and confers enhanced drought resistance in transgenic Arabidopsis. Planta 240(3): 645–664

Saha B, Swain D, Borgohain P, Rout GR, Koyama H, Panda SK (2020) Enhanced exudation of malate in the rhizosphere due to AtALMT1 overexpression in blackgram (*Vigna mungo* L.) confers increased aluminium tolerance. Plant Biol 22(4):701–708

Sahoo DP, Kumar S, Mishra S, Kobayashi Y, Panda SK, Sahoo L (2016) Enhanced salinity tolerance in transgenic mungbean overexpressing Arabidopsis antiporter (NHX1) gene. Mol Breed 36(10):1–15

Sai CB, Chidambaranatham P (2019) Reproductive stage droughttolerance in blackgram is associated with role of antioxidants on membrane stability. Plant Physiol Rep 24(3):399–409

Saxena KB (2008) Genetic improvement of pigeon pea — A review. Trop Plant Biol 1(2):159–178

Sehrawat A, Gupta R, Deswal R (2013a) Nitric oxide-cold stress signalling cross-talk, evolution of a novel regulatory mechanism. Proteomics 13(12–13):1816–1835

Sehrawat N, Jaiwal PK, Yadav M, Bhat KV, Sairam RK (2013b) Salinity stress restraining mungbean (*Vigna radiata* (L.) Wilczek) production: gateway for genetic improvement. Int J Agric Crop Sci 6(9):505

Sekhar K, Priyanka B, Reddy VD, Rao KV (2010) Isolation and characterization of a pigeonpea cyclophilin (CcCYP) gene, and its over-expression in Arabidopsis confers multiple abiotic stress tolerance. Plant Cell Environ 33(8):1324–1338

Sen S, Chakraborty J, Ghosh P, Basu D, Das S (2017) Chickpea WRKY70 regulates the expression of a homeodomain-leucine zipper (HD-Zip) I transcription factor CaHDZ12, which confers abiotic stress tolerance in transgenic tobacco and chickpea. Plant Cell Physiol 58(11): 1934–1952

Shah Z, Iqbal A, Khan FU, Khan HU, Durrani F, Ahmad MZ (2020) Genetic manipulation of pea (*Pisum sativum* L.) with Arabidopsis's heat shock factor HsfA1d improves ROS scavenging system to confront thermal stress. Genet Resour Crop Evol 67(8):2119–2127

Sharma A, Dhanda S (2014) Abiotic stress response in *Vigna radiata* L. (Mungbean). Int J Life Sci Biotechnol Pharma Res 3(4):14

Shukla RK, Raha S, Tripathi V, Chattopadhyay D (2006) Expression of CAP2, an APETALA2-family transcription factor from chickpea, enhances growth and tolerance to dehydration and salt stress in transgenic tobacco. Plant Physiol 142(1):113–123

Singh G (2018) Development of transgenic chickpea (*Cicer arietinum* L.) plants with lectin receptor-like kinase (Lec-RLK) gene for improving salt tolerance [With CD Copy], Molecular biology and biotyechnology. CCSHAU, Hisar

Singh DP, Singh BB (2011) Breeding for tolerance to abiotic stresses in mungbean. J Food Legum 24(2):83–90

Singh UP, Zargar GH, Fazili MA (1986) Growth of foliage of pigeon pea (*Cajanus cajan* [L.] Millsp.) and disease progress of Alternaria blight. J Phytopathol 115(2):147–151

Singh P, Kumar D, Sarin NB (2016) Multiple abiotic stress tolerance in *Vigna mungo* is altered by overexpression of ALDRXV4 gene via reactive carbonyl detoxification. Plant Mol Biol 91 (3):257–273

Singh R, Sharma S, Kharb P, Saifi S, Tuteja N (2020) OsRuvB transgene induces salt tolerance in pigeon pea. J Plant Interact 15(1):17–26

Stavolone L, Ragozzino A, Hohn T (2003) Characterization of Cestrum yellow leaf curling virus: a new member of the family Caulimoviridae. J Gen Virol 84(12):3459–3464

Sultana R, Choudhary AK, Pal AK, Saxena KB, Prasad BD, Singh R (2014) Abiotic stresses in major pulses: current status and strategies. In: Approaches to plant stress and their management. Springer, pp 173–190

Sunitha M, Srinath T, Reddy VD, Rao KV (2017) Expression of cold and drought regulatory protein (CcCDR) of pigeonpea imparts enhanced tolerance to major abiotic stresses in transgenic rice plants. Planta 245(6):1137–1148

Surekha CH, Kumari KN, Aruna LV, Suneetha G, Arundhati A, Kavi Kishor PB (2014) Expression of the *Vigna aconitifolia* P5CSF129A gene in transgenic pigeonpea enhances proline accumulation and salt tolerance. Plant Cell Tissue Organ Cult 116(1):27–36

Tamirisa S, Vudem DR, Khareedu VR (2014) Overexpression of pigeonpea stress-induced cold and drought regulatory gene (CcCDR) confers drought, salt, and cold tolerance in Arabidopsis. J Exp Bot 65(17):4769–4781

Tamirisa S, Vudem DR, Khareedu VR (2017) A cyclin dependent kinase regulatory subunit (CKS) gene of pigeonpea imparts abiotic stress tolerance and regulates plant growth and development in Arabidopsis. Front Plant Sci 8:165

Tayyab A, Azeem M, Qasim M, Ahmed N, Ahmad R (2016) Salt stress responses of pigeon pea (*Cajanus cajan*) on growth, yield and some biochemical attributes. Pak J Bot 48(4):1353–1360

Toker C, Mutlu N (2011) Chapter 16: Breeding for abiotic stresses. In: Biology and breeding of food legumes. CAB International, p 241

Tuteja N, Banu MSA, Huda KMK, Gill SS, Jain P, Pham XH, Tuteja R (2014) Pea p68, a DEAD-box helicase, provides salinity stress tolerance in transgenic tobacco by reducing oxidative stress and improving photosynthesis machinery. PLoS One 9(5):e98287

Valentine AJ, Benedito VA, Kang Y (2018) Legume nitrogen fixation and soil abiotic stress: from physiology to genomics and beyond. Annu Plant Rev Online:207–248

Varshney RK, Thudi M, Nayak SN, Gaur PM, Kashiwagi J, Krishnamurthy L, Jaganathan D, Koppolu J, Bohra A, Tripathi S, Rathore A, Jukanti AK, Jayalakshmi V, Vemula A, Singh SJ, Yasin M, Sheshshayee MS, Viswanatha KP (2014) Genetic dissection of drought tolerance in chickpea (*Cicer arietinum* L.). Theor Appl Genet 127(2):445–462. https://doi.org/10.1007/s00122-013-2230-6

Veena, Reddy VS, Sopory SK (1999) Glyoxalase I from Brassica juncea: molecular cloning, regulation and its over-expression confer tolerance in transgenic tobacco under stress. Plant J 17(4):385–395

Wang F-Z, Wang Q-B, Kwon S-Y, Kwak S-S, Su W-A (2005a) Enhanced drought tolerance of transgenic rice plants expressing a pea manganese superoxide dismutase. J Plant Physiol 162(4): 465–472

Wang Y, Wisniewski M, Meilan R, Cui M, Webb R, Fuchigami L (2005b) Overexpression of cytosolic ascorbate peroxidase in tomato confers tolerance to chilling and salt stress. J Am Soc Hortic Sci 130(2):167–173

Wang H, Gao X, Yang JJ, Liu Z-R (2013) Interaction between p68 RNA helicase and Ca^{2+}-calmodulin promotes cell migration and metastasis. Nat Commun 4(1):1–12

Wild A (2003) Soils, land and food: managing the land during the twenty-first century. Cambridge University Press

Yang JY, Zheng W, Tian Y, Wu Y, Zhou DW (2011) Effects of various mixed salt-alkaline stresses on growth, photosynthesis, and photosynthetic pigment concentrations of Medicago ruthenica seedlings. Photosynthetica 49(2):275–284

Zahid ALI, Saeed W, Naseem S, Ahmad F, Akrem A, Yasmeen N, Jacobsen HJ (2018) Phenotypic evaluation of transgenic peas (*Pisum sativum* L.) harboring AtNHX1 demonstrates stable gene expression and conserved morphology in subsequent generations. Turk J Bot 42(2):150–158

Zaker F, Ghareyazie B, Bagheri A, Sharma KK (2016) Genetic transformation of Lentil (*Lens culinaris* M.) and production of transgenic fertile plants. Iranian J Pulses Res 7(2):215–229

Zaker Tavallaie F, Ghareyazie B, Bagheri A, Sharma KK (2017) Genetic transformation of Lentil (*Lens culinaris* M.) and production of transgenic fertile plants. Iran J Pulses Res 7(2):215–229

Aluminum Toxicity Tolerance in Food Legumes: Mechanisms, Screening, and Inheritance

15

Jyoti Taunk, Chandan Kumar Singh, Deepti Singh, Ram Sewak Singh Tomar, Dharmendra Singh, and Madan Pal

Abstract

Legumes are a vital food source next to cereals. Their productivity is restricted in acidic soils as most of them are sensitive to aluminum (Al) stress. Al can quickly inhibit cell division, disrupt cell structure, diminish water and nutrient uptake, and hinder root elongation in food legumes. An increase in rhizospheric pH, Al avoidance, and shift of nutrient element circulation pay to Al tolerance in food legumes. Also, exudation of organic acids and induction of antioxidant activities portray a significant part in Al stress tolerance of leguminous species. Molecular mapping of Al-tolerant genes helps in designing breeding strategies to improve crop production on Al toxic soils. This chapter focuses on various aspects of Al toxicity tolerance covering mechanisms, screening techniques, genetic control, mapping, and molecular advancements in legumes.

J. Taunk
Division of Plant Physiology, ICAR-Indian Agricultural Research Institute, New Delhi, India

Present Address: Department of Biotechnology, University Centre for Research and Development (UCRD), Chandigarh University, Mohali, India

C. K. Singh · D. Singh (✉)
Division of Genetics, ICAR-Indian Agricultural Research Institute, New Delhi, India

D. Singh
Department of Botany, Meerut College, Meerut, India

R. S. S. Tomar
College of Horticulture and Forestry, Rani Lakshmi Bai Central Agricultural University, Jhansi, India

M. Pal
Division of Plant Physiology, ICAR-Indian Agricultural Research Institute, New Delhi, India

P. Muthu Arjuna Samy et al. (eds.), *Legumes: Physiology and Molecular Biology of Abiotic Stress Tolerance*, https://doi.org/10.1007/978-981-19-5817-5_15

Keywords

Aluminum · Legumes · Molecular mapping · Plant stress · Screening · QTL

15.1 Introduction

Aluminum (Al) is a plenteous element in the earth's crust and the most commonly expended nonferrous post-transition metal for human purposes. However, in soil, its large quantity is subsumed as aluminosilicates, and only a small quantity is biologically available (Bhalerao and Prabhu 2013). Sometimes in small intensities, Al can benefit plant development or induce some advantageous outcomes (Rout et al. 2001). However, mostly it is considered noxious to plants. Al toxicity depends upon several components like its concentration in the soil, pH of the soil, its chemical form in the soil, and plant species (Bojórquez-Quintal et al. 2017). Al becomes poisonous to plants once its concentration becomes more than 2–3 ppm (Rout et al. 2001). Soluble Al turns out to be accessible to plants only when soil pH falls below 5.5, making it the most notorious stress condition for acidic soils. Soil acidification has been intensified by continuous intensive agriculture and climate change. Around 50% of the cultivable area is adversely affected by Al stress in acidic soils (Singh et al. 2017). When soil pH declines under 5.5, aluminosilicate clays and aluminum hydroxide minerals start to melt, letting out aluminum-hydroxy cations and $Al(H_2O)_6{}^{3+}$ (Al^{3+}), which later interchange with other cations. The toxicity of various Al species on plants is in the following sequence: $Al_{13} > Al^{3+} > Al(OH)^{2+-} > Al(OH)^{2+} > Al\ (OH)^{4-}$. Therefore, Al^{3+} and Al_{13} are contemplated as the most noxious types (Kinraide 1997). There is variability in different plant species for sensitivity towards Al stress. Most plant genera are susceptible to even micromolar concentrations of Al. Few (tea, pineapple) are considered tolerant to elevated levels of Al (Delhaize and Ryan 1995; Silva 2012).

Al alters plant functioning at very early stages. Toxic Al ions mostly target root tips and inhibit cell elongation and division in this zone. The latter results in root arresting supplemented by decreased water and nutrient uptake. Root tips become swollen and damaged. Sometimes, root lesions also occur. Plants have numerous Al-binding sites, which include cell walls, plasma membranes, cytoskeleton, and nucleus. Al expeditiously amasses in the plasma membrane and symplasm and affects various cellular processes (Panda et al. 2009). It interacts with lipids inducing lipid peroxidation; causes an increase in reactive oxygen free radicals; disrupts cytoplasmic Ca^{2+} homeostasis; accumulates callose in the plasmodesmata; and disrupts cytoskeleton (Panda et al. 2009). These changes ultimately affect several signalling cascades and processes operating inside the cell directly or indirectly. Al toxicity also causes accumulation of certain metabolites, induces behavioural changes in many enzymes, lowers P availability to plants, increases plant's susceptibility towards drought stress, and causes lodging (Arunakumara et al. 2013).

Legumes are the sustainable and nutritionally valuable food sources that are ranked after cereals in the food pyramid. Chickpea, pigeon pea, lentil, mung bean, urdbean, and soybean are the most important food legume plants. They provide 20–45% proteins in the form of essential amino acids, along with complex carbohydrates, dietary fibers, and a significant amount of minerals and vitamins. They lack cholesterol and are low in fat (Maphosa and Jideani 2017). In the current climate change scenario, legume crops can play a crucial role in future food security by carrying out numerous actions in line with sustainability percept. Legumes produce 5–7 times less greenhouse gases than other crops, allow carbon sequestration, improve soil fertility, and can be utilized in the form of green manure (Stagnari et al. 2017). They are good for intercropping or relay intercropping. However, legume productivity is severely constrained by Al toxicity. Apart from other effects of Al toxicity, it also causes a fatal effect on legume/rhizobia symbiosis and ultimately on the nitrogen-fixation process. Al toxicity affects various stratagems of nitrogen fixation, either being root hair formation, rhizobial population, nitrogen metabolism, nitrogenase activity, or uptake of hydrogenases (Jaiswal et al. 2018). In a study by Scheffer-Basso and Prior (2015), it was found that the topological attributes were considerably diminished by Al in all the studied leguminous plants. Nodulated legumes were registered to be extra sensitive towards Al toxicity in comparison to legumes that take mineral nitrogen (Hungria and Vargas 2000).

Legumes respond to Al stress by an increase in rhizospheric pH, Al avoidance, and alteration of nutrient element distribution conditioned through signal transduction, metabolite production, and Al-induced gene expressions. Efflux of organic anions as that of citrate, malate, and oxalate from roots has been recognized as one of the climacteric mechanisms for Al resistance in a variety of plants (Delhaize et al. 2007). This attribute is operated by genes belonging to two separate gene families ALMT and MATE that encrypt membrane proteins, which expedite the secretion of organic anions through the plasma membrane (Zhou et al. 2011). Identification and validation of such counter-susceptive genes provide ample opportunities for improving Al^{3+} resistance in plants via classical breeding and/or biotechnological schema. However, classical breeding for improving Al tolerance has showcased limited success stories, especially in legumes. Therefore, the inclusion of advanced techniques like molecular breeding is essential to increase legume production under Al stress. Further, for fast and accurate identification of tolerant genotypes, rapid screening techniques are needed which can easily select a large number of genotypes for breeding purpose. Besides, identification of reliable traits, mechanisms, and genetic dissection of Al tolerance are also required for improving Al tolerance in crop plants including legumes. Further, appropriate screening techniques are essential for screening Al-tolerant genotypes. Many studies have reported few screening techniques in different crop plants such as barley (Hossain et al. 2005; Wang et al. 2006), spring rye (Hede et al. 2002), maize (Giaveno and Miranda Filho 2000; Mendes et al. 1984), and wheat (Baier et al. 1995). Compilations of different screening techniques for assessment of Al tolerance in plants including legumes are limited. Keeping in view the above facts, this chapter discusses varied facets of Al toxicity tolerance encompassing screening approaches, mechanisms, genetics, and molecular improvements in legumes.

15.2 Genotypic Differences in Al Tolerance Among Legumes

There are amazing disparities in the adjustability of various legumes towards Al stress. Narayanan and Sayamala (1989) stated that plant height was considerably decreased above 20 ppm Al in pigeon pea, while in case of soybean, Sapra et al. (1982) found that even 8 ppm Al was sufficient to reduce the plant height. There was a drop in leaf number and size in pigeon pea only at very extreme concentrations of 40 and 60 ppm Al. Singh et al. (2012) reported that root regrowth after hematoxylin staining, root and shoot lengths and their dry weights, and pods/plant reduced notably at 148 μm Al concentration. Dessureaux (1969) testified that at 20 ppm, leaf size was considerably condensed in alfalfa seedlings. The tap root length was considerably impeded at 40 ppm Al, although at 10 ppm, root length was stimulated. Klimashevskii et al. (1970) noted in field pea plants that Al-tolerant cultivar exhibited only 32% diminution in growth at 11 mg Al^{3+}/L; however, this concentration was totally injurious to Al-sensitive one. Root elongation was dwindled by approximately 50% under 9.3 mM $AlCl_3$ mM/m^3 in the rooting medium in the case of faba bean (Grauer and Horst 1990). On the other hand, root elongation was wholly inhibited by 100 mM/m^3 $AlCl_3$ in case of field pea (Matsumoto 1991). The level of 20 ppm Al separated sensitive from tolerant chickpea genotypes via hematoxylin staining and root regrowth under short-term Al exposure (Singh et al. 2011). However, at a level of 5 ppm Al^{3+}, chickpea shoot dry weight was decreased by 70% in sensitive cultivars while intolerant cultivars decreased only by 27% (Rai 1991). Al is mostly accumulated in the root apex of crop plants including *Fabaceae*. Al accumulation in these food legumes influences plant growth as well as yield. This inhibition of growth caused due to Al in lentil and mung bean cells was found to be well associated with the deposition of callose (Singh et al. 2016).

15.3 Symptoms of Al Toxicity in Legumes

Aluminum can quickly inhibit cell division, damage cell structure, diminish water as well as nutrient uptake, and hinder root elongation in leguminous plants (Arunakumara et al. 2013). The influence of Al stress is more prominent on roots. The most visible symptom of Al stress is root growth inhibition. The influenced roots become short, stubby, and lateral roots converting into peg-like or weaken to grow, and thus the entire root systems stop elongating and acquire brownish coloration as reported in pea plants (Singh and Choudhary 2010).

Shoot growth is often considered a secondary perceptible indication of Al noxiousness and often similar to deficits of phosphorus, calcium, magnesium, and iron (Foy 1984). Generally, the plant canopy of Al-toxic plants surfaces as phosphorus becomes deficient. This imitates Al displacement of the plant's phosphorus metabolic process. Foliar symptoms resembling phosphorus deficiency have been reported in legumes like chlorosis in soybean (Foy et al. 1973) and purple coloration in leaves and stems of lentil (Singh et al. 2012).

The inhibition of root elongation due to Al toxicity has been highly utilized as an attribute for the assessment of Al-tolerant cultivars in lentil (Singh et al. 2021a). Various mechanisms triggered the decline of root growth, nutrient deficiencies, and yield damages (Kochian 1995). Under Al treatment, activities of various antioxidant enzymes like superoxide dismutase (SOD), ascorbate peroxidase (APX), and guaiacol peroxidase (GPX) also increased in legumes (Arunakumara et al. 2013). Al stress caused an increase in callose and reactive oxygen species (ROS) production in the roots of lentil and mung bean (Singh et al. 2016). Production of ROS promotes inhibition of root elongation triggered by Al (Wang et al. 2019). Reports on Al toxicity in legumes reflecting species type, treatment levels, and duration, together with the effects on plants, are listed in Table 15.1.

15.4 Physiological and Biochemical Mechanisms of Al Tolerance in Food Legumes

With an aim to alleviate Al stress, plants have acquired some Al tolerance mechanisms which are based on the detoxification sites. These mechanisms can be classified into exclusion (apoplast) mechanism externally and tolerance (symplast) mechanism internally (Kochian 1995). In respect of exclusion mechanism, secretion of organic acids and rise of rhizospheric pH supply Al tolerance in *Fabaceae*. Al exclusion from the root zone was found to be the chief mechanism in case of pea plants (Kichigina et al. 2017). Detoxification of Al externally by the exudation of organic acids such as malate and citrate seems to be another mechanism for Al tolerance in food legumes (Miyasaka et al. 1991; Yang et al. 2000). Citrate and malate were released from roots of Al-tolerant cultivars and wild accession of lentil (Singh et al. 2016) and soybean (Yang et al. 2000). Miyasaka et al. (1991) stated that Al-tolerant snap bean cultivar grown in the presence of Al secreted 70 times more citrate in the presence of Al, whereas Al-sensitive cultivar secreted it up to 10 times only.

Two types of exudations of organic acids have been suggested in plants as per the required time (Ma et al. 1997). The secretion of organic acids is swift in pattern I, while it is postponed for several hours in case of pattern II once Al is added in the nutrient solution. Stimulation of an anion channel positioned on the plasma membrane by Al is the likely mechanism accountable for swift exudation in pattern I (Delhaize et al. 2007; Delhaize and Ryan 1995; Yang and Zhang 1998). Further, induction of novel protein is not required (Ma et al. 1997). In pattern II, protein induction is required; therefore, organic acid secretion is postponed for several hours following Al exposure (Yang et al. 2005). For instance, in lentil, maximal efflux of malate occurred after 3-h exposure to Al (Singh et al. 2016). Rice bean roots secreted citrate to alleviate Al stress, but the exudation was belated by approximately 3 h (Yang et al. 2006). At adequate strengths, these organic acids can make complexes with Al ions, avert them from attaching to the fixed negative sites of the cell wall and plasma membrane, and bestow Al tolerance to the plants.

Table 15.1 Aluminum toxicity tolerance studies in different legume crops

Legume species	Al-tolerant genotype	Al treatment level	Duration of treatment	Major finding	Reference
Chickpea	ICC14880 IPC92-39	5, 10, and 20 ppm Al	24–48 h	Al depressed root regrowth and increased root staining	Singh et al. (2011)
Pigeon pea	IPA7-10 and T-7	30 ppm Al	24–48 h	Al reduced root regrowth and increased staining	Singh et al. (2011)
Lentil	L-7903, L-4602, and ILL-6002	74, 148, 222, and 296 μM	24 h to 65 days	Al depressed root and shoot growth and pods/ plant	Singh et al. (2012)
	L-7903, L-4602, and ILWL-185	148 μM	24 h to 130 days	Al reduced root and shoot growth, increased callose, ROS production, triggered exudation of organic acids and antioxidant activities in tolerant genotypes	Singh et al. (2017)
Mung bean	Pusa-672	74 and 185 μM	48 h	Al inhibited root elongation rate and root regrowth and augmented buildup of Al, callose, H_2O_2, and lipid peroxidation. It triggered an antioxidant response in the tolerant genotype	Singh et al. (2015)
Urdbean	Mash-114	74 and 185 μM	48 h	Al treatment increased callose and ROS production and triggered antioxidant activities	Singh et al. (2015)
Rice bean	RBL-6	74 and 185 μM	48 h	Al treatment increased callose and ROS production and triggered antioxidant activities	Singh et al. (2015)
Soybean	PI417021, PI416937, and Biloxi	2 and 5 μM	3 days	Reduced taproot elongation	Villagarcia et al. (2001)
Pea	PC-55-11-1-2	10, 20, 30, and 40 ppm Al	24 h to 24 days	Al stress reduced relative root growth and increased root staining	Singh et al. (2007)
	Azad P1 and PC-55-11-1-2	30 ppm Al	24–72 h	Al stress increased root staining and decreased root growth	

Al aluminum, *ROS* reactive oxygen species

In internal tolerance mechanism, Al ions absorbed by cells are accumulated and chelating of Al takes place in the cytosol. This occurs with the help of organic acids, Al-binding proteins, localization of Al into the vacuole, and induction of protein synthesis that chelates Al in the symplast. Physiological and biochemical components of *Vigna* species for evaluating Al stress were reported by Singh et al. (2015).

15.5 Physiological and Biochemical Parameters Associated with Al Tolerance in Legumes

15.5.1 Organic Acid Exudation

Organic acid secretion is involved in providing tolerance to Al stress. The tolerance level of legume can be detected by measuring the level of organic acid exudation from the root of stressed and non-stressed plants. Estimation of malate and citrate was realized to be greater in tolerant genotypes of lentil as compared to sensitive ones (Singh et al. 2016, 2018, 2021a). These reports also suggested that malate was secreted in higher amount than that of citrate. Level of malate was estimated in F_6 population to identify the QTLs associated with malate secretion in lentil (Singh et al. 2021b). Al-tolerant genotype of soybean was reported to release higher citrate level from roots than the sensitive genotypes (Dong et al. 2004; Silva 2012).

15.5.2 Callose Accumulation

Callose is a polysaccharide (β-1,3-glucan) found in the cell walls of higher plants (Chen and Kim 2009). Callose accumulation in exposed root tips is strongly connected to Al-induced root inhibition. Thus, Al-induced callose development in root tips is considered as a mark of Al susceptibility and often used as a selection criterion to detect Al tolerance (Horst et al. 1997). Al-influenced callose production in roots of bean cultivars has been reported as an indicator of Al toxicity (Massot et al. 1999). Al-induced callose formation has also been testified in soybean within 30 min of exposure to Al stress (Wissemeier et al. 1987).

15.5.3 Mucilage Secretion

Plants showing tolerance to Al stress release mucilage from the roots that combines with Al and lowers toxicity. Secretion of Al-induced mucilage was reported in legumes such as cowpea (Horst et al. 1982), pea (Brigham et al. 2001), and *Glycine max* (Cai et al. 2013).

15.5.4 Al-Induced Antioxidant Enzyme Production

Antioxidant enzymes are the most important enzymes used against oxidative stress caused due to Al toxicity in plant cells. The activity of these enzymes increases under Al stress; therefore, measurement of their activities can be utilized to distinguish Al-tolerant and Al-sensitive genotypes under Al-stress conditions. Root and shoot samples exposed to control and Al-stress condition can be collected for the estimation of antioxidant enzyme activities like those of SOD, APX, GPX, and CAT.

CAT activity can be estimated by tracking the method of Aebi (1984). SOD activity can be estimated by documenting the enzyme-induced drop in the absorbance of formazan built by reaction of *p*-nitroblue tetrazolium chloride (NBT) and superoxide radicals tracking the method of Dhindsa et al. (1981). One unit of SOD activity is represented as the quantity of enzyme that inhibited 50% NBT photoreduction. Shamsi et al. (2008) evaluated antioxidant enzyme activities of two soybean cultivars with variable Al tolerance under hydroponic conditions. Higher SOD and peroxidase activities were noticed in plants subjected to Al stress. In a study by Singh et al. (2016), higher SOD and APX activities were observed in the roots and shoots of lentil genotypes exposed to Al stress. Their activity was observed to be extra prominent in roots than in shoots, whereas catalase activity was observed to be decreased significantly in both the resistant and sensitive genotypes when matched to their corresponding controls. Al stress also triggered a considerable upsurge in GPX activity in all the plants. Al-tolerant broad bean cultivars exhibited higher levels of antioxidants like SOD, POD, catalase, and GST in reaction to Al stress. Also, in case of common bean cultivars, activity of SOD and POD was observed to be significantly greater under Al treatment (Tóth et al. 2021).

15.5.5 Lipid Peroxidation

Lipid peroxidation can be articulated as Malondialdehyde (MDA) content, and its concentration in roots can be estimated subsequent to reaction with thiobarbituric acid (TBA) (Heath and Packer 1968). Nonspecific absorbance is calculated at 600 nm which is deducted from the absorbance at 530 nm, and extinction coefficient of 155/mm/cm is used to determine MDA content which is stated as μMol/g of fresh weight (Ribeiro et al. 2012). Panda et al. (2003) described enhanced lipid peroxidation due to increase in Al concentration in green gram leaves that was calculated in the form of TBARS. They concluded that it was due to excessive generalization of hydroxyl radicals (Panda et al. 2003). Lipid peroxidation is a portion of the hole production expression of Al toxicity in roots. Boosted lipid peroxidation by oxygen free radicals is a result of the prime effects of Al on membrane structure. They found that lipid peroxidation in the root tip was enhanced in response to Al toxicity only after long periods of treatment. Lipid peroxidation has been reported to be a comparatively prompt symptom of Al toxicity persuaded by the build-up of Al and seems to partially instigate callose production, but not the root elongation inhibition (Yamamoto et al. 2001). The enhancement of lipid peroxidation displayed Al dose dependency in the case of pea plants (Yamamoto et al. 2001).

15.5.6 Nutritional Interaction

Many reports have stated that Al intervenes with the uptake, transport, and utilization efficiency of elements such as phosphorus (Liao et al. 2006), potassium, calcium (Rengel 1992), manganese (Foy 1984), boron (Stass et al. 2007; Yang and Zhang 1998), copper, and zinc (Yang et al. 2021) in legumes.

15.5.7 Visual Detection of Al Contents

Morin is a fluorescence-emitting dye that develops a highly specific complex with Al. Therefore, it is utilized in the detection of Al accumulation as well as its localization within the cells. The Al contents in roots as well as shoots amplified with increment in Al concentration in the nutrient solution. Further, it was observed greater in roots as compared to shoots in the case of lentil (Singh et al. 2016). Choudhary and Singh (2011) found that root and shoot Al contents were considerably lesser in the tolerant genotypes than the sensitive ones. On the basis of this observation, they stated that the Al tolerance mechanism involves Al exclusion and perhaps internal detoxification. Al content and collection were found to be greater in the roots as compared to the shoots in case of castor beans (de Freitas et al. 2019).

15.6 Screening Techniques for Al Tolerance

The screening methods must be potent enough to distinguish the genotypes and constitute the focused production environment. The preliminary screening pursuits are typically accomplished on seedlings under commanded conditions with controlled Al treatment, and the prominence is provided to the phenotype tolerance to select tolerant genotypes. Numerous techniques are employed to judge tolerance with the main focus on root and shoot growth and development (Abate et al. 2013). Comparative studies of screening methods for tolerance towards Al toxicity have been conducted in pigeon pea (Choudhary et al. 2011a, b; Singh et al. 2011), chickpea (Singh et al. 2011), lentil (Singh et al. 2012, 2016, 2021c), mung bean (Singh et al. 2015), urdbean (Singh et al. 2015), pea (Singh et al. 2007), and soybean (Villagarcia et al. 2001) based on short- and long-term techniques.

15.6.1 Short-Term Screening Techniques

Short-term screening techniques involve many staining and nonstaining methods for evaluation of Al toxicity tolerance.

15.6.1.1 Hematoxylin Staining Method

Bioaccumulation of Al in crops is a well-documented phenomenon. Hoffer and Carr (1923) were the first to demonstrate the use of hematoxylin dye in visualizing Al accumulation in plants. Polle et al. (1978) improvised their protocol into a non-destructive form of hematoxylin staining that allowed rapid visual detection of Al. Hematoxylin staining is a primitive indicator of Al toxicity and is the most reliable technique among staining techniques (Polle et al. 1978; Roy and Bhadra 2013). Therefore, it is utilized as an assenting test for tolerance towards Al stress. It depicts a visual signal of Al uptake by sensitive plants where Al binds to form a purple-bluish complex in the root tips. The greater the concentration of Al, the darker the purple color perceived in the root tips (Miftahudin et al. 2007). Lack of color in root tips of tolerant genotypes exhibited that they have either excluded Al or bound it in complexes that are unapproachable to hematoxylin. This dye creates complexes with Al that precipitate with phosphate as $AlPO_4$ in the intercellular spaces (Ownby 1993). This method was commonly used for the evaluation of Al tolerance in pea (Singh et al. 2007; Singh and Choudhary 2010), chickpea (Singh et al. 2011) and pigeon pea (Singh et al. 2011).

15.6.1.2 Eriochrome Cyanine R Staining

Similar to hematoxylin, eriochrome cyanine also distinguishes genotypes on the basis of staining pattern (Hede et al. 2002). The degree of damage due to Al stress was determined by the intensity of uptake of the stain (Vishnyakova et al. 2015), while root re-elongation after eriochrome cyanine staining was determined in case of pea (Kichigina et al. 2017).

15.6.1.3 Root Regrowth After Staining

Hematoxylin staining is also employed as a means of measuring root regrowth (RRG). Singh et al. (2012) reported that hematoxylin with tailored pulse technique assesses Al tolerance on the basis of the capability of Al-tolerant seedlings to maintain root growth after a brief pulse treatment with high Al concentration in lentil. Al-sensitive seedlings did not show RRG because their apical meristem was damaged, whereas tolerant genotypes showed continued root growth. Singh et al. (2012) examined variation of Al tolerance in lentil and found that RRG after staining had significant correlation with root and shoot length, dry weight of roots and shoots, and pods/plant. Later, Singh et al. (2016) also evaluated Al resistance in 285 wild and cultivated lentil genotypes in a nutrient solution by measuring RRG after hematoxylin staining of root apices. On the basis of this parameter, they were able to distinguish genotypes into different groups. Genotypes that had mean primary RRG <0.5 cm were categorized as Al sensitive. On the other hand, genotypes with mean primary RRG significantly >1.0 cm were counted as resistant. Seedlings exhibiting intermediate RRG (0.50–1.00 cm) were considered as moderately resistant. They also found that RRG was correlated with seed yield under Al toxic field conditions. Further, quantitative trait loci (QTLs) for the trait RRG after hematoxylin staining were also mapped in F_2 and F_3 mapping population of lentil (Singh et al. 2018). Screening techniques involving the use of staining dyes are represented in Fig. 15.1a, b.

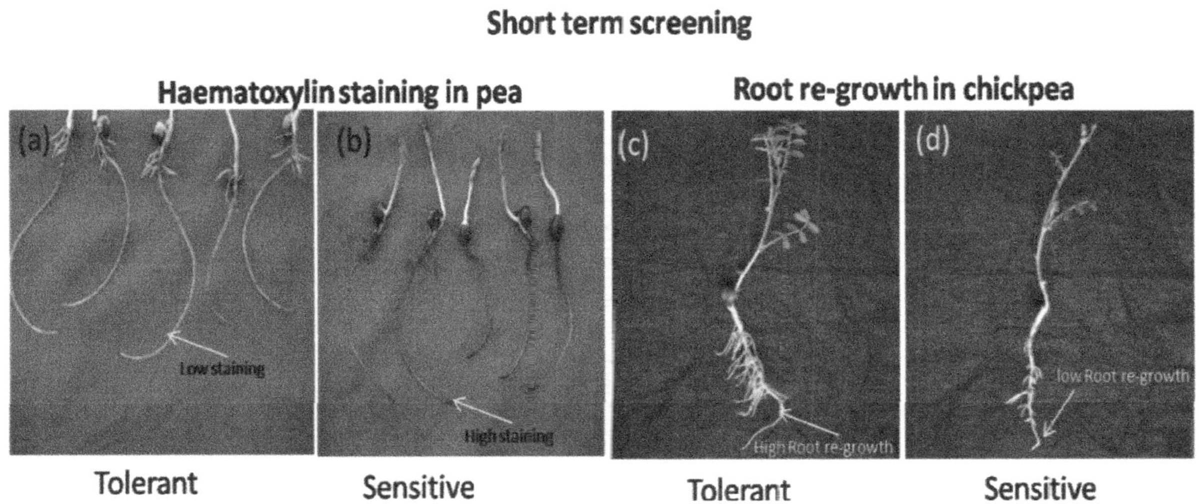

Fig. 15.1 Short-term screening technique for Al tolerance (**a**, **b**) using hematoxylin staining and (**c**, **d**) root regrowth

15.6.1.4 Root Regrowth Without Staining

Root regrowth without staining has been used as an indispensable morphological marker for testing Al tolerance in plants. Choudhary and Singh (2011) efficiently screened 32 genotypes of pigeon pea under Al toxic conditions using RRG as parameter. This screening method has also been used in chickpea (Singh et al. 2011) and pea (Singh and Choudhary 2010) (Fig. 15.1c, d).

15.6.1.5 Fluorescence Staining Methods

Use of fluorescence dyes such as aniline blue, morin, and fluorescein diacetate (FDA) to differentiate Al-tolerant genotypes from sensitive ones has been testified in many legumes (Singh et al. 2016). These dyes can be used to detect callose deposition, Al-induced H_2O_2 production, and presence and estimation of Al contents in roots and shoots.

15.6.1.6 Callose Deposition

The higher the Al-induced injury to the root, the higher is the Al-induced callose deposition. Due to higher affinity of aniline blue dye with callose, higher accumulation of callose can be denoted by the level of fluorescence due to Al-morin complex (Singh et al. 2015). Callose synthesis was found to be positively associated with internal Al concentration and negatively associated with root elongation rate in the case of bean cultivars under Al toxic condition (Massot et al. 1999). Singh et al. (2021a, b, c) exhibited that callose formation is induced by Al as a mark of injury, markedly in the root apex. Singh et al. (2018) have mapped Al resistance loci in lentil using RRG after hematoxylin staining and callose accumulation as markers. Al stress also triggered callose production in the root tips of alfalfa (An et al. 2020).

15.6.1.7 Detection of Al-Induced H_2O_2 Level

Level of DCF-DA fluorescence depicts the level of Al-induced injury caused due to production of H_2O_2. Higher injury corresponds to higher damage due to Al ion, while lower fluoresce depicts less Al-induced injury to the roots. Evans blue (0.025%, w/v) is used for localizing the loss of plasma membrane integrity (Yamamoto et al. 2001). Hydrogen peroxide- and H_2O_2-generated apoplast diamine oxidase (DAO) activities were received chemically via transmission electron microscopy in pea (Sujkowska-Rybkowska and Borucki 2014). They found the participation of DAO in the production of a huge quantity of H_2O_2 in the nodule apoplast under Al toxicity. Hydrogen peroxide production was visualized in lentil roots by DCF-DA, which produced green fluorescence (Singh et al. 2016). The DCF-DA fluorescence in the root tips of control plants was insignificant, while it amplified significantly under Al stress. The level of H_2O_2 was found to be increased in both the resistant and sensitive genotypes although low signals were observed in resistant breeding lines while intense green fluorescence was observed in the root's tips of sensitive cultivars. H_2O_2 was determined in both roots and shoots by the method of Sagisaka (1976) in the case of black gram. The H_2O_2 content was observed to increase progressively in all the treated samples with the rising period of stress and concentration of Al^{3+} (Awasthi et al. 2017). Under Al stress, H_2O_2 production was found to be more in *Vigna radiata* than in *V. mungo* and *V. umbellate*.

15.6.2 Long-Term Screening Techniques

Al tolerance in legumes can be screened by three main long-term phenotyping techniques including nutrient solution culture with or without staining and soil and sand culture together with field screening.

15.6.2.1 Nutrient Solution Culture Without Staining

Solution culture technique is based on the inhibition of root growth under Al toxic conditions. In solution culture technique without staining, ratio of root growth in the presence of Al to its absence is determined. This technique is repeatable, non-destructive, rapid, cost effective, and independent of soil nutritional factors. Moreover, a huge quantity of plants can be accommodated in a brief period of time. However, it is not effective for the evaluation of Al tolerance in vegetatively propagated plants and at adult plant stages.

Al toxicity also causes morphological damage to plant parts. Therefore, many root- and shoot-based morphological features are used for the evaluation of Al tolerance in legumes. These include traits like relative root elongation, root regrowth, root and shoot length and their dry weights, and root system architecture.

15.6.2.2 Relative Root Length (RRL)

RRL is described as the ratio of the maximum root length under Al stress to that of the maximum root length under control condition. Long-term screening technique for Al tolerance using relative root length as an attribute in legumes is represented in Fig. 15.2. This type of screening strategy can be adopted under either hydroponic or sand assays.

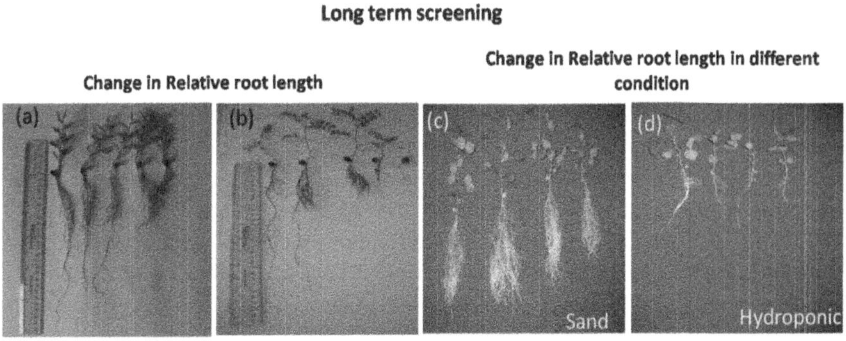

Fig. 15.2 Long-term screening technique for Al tolerance using relative root length. (**a**, **b**) Change in relative root lengths of chickpea genotypes, (**c**, **d**) relative root length as parameter to differentiate Al-tolerant and -sensitive genotypes of pea under hydroponic and sand condition

15.6.2.3 Root System Architecture

Another major morphological character is root system architecture (RSA), which represents geometric organization of the discrete roots within a root system in the soil volume the root system occupies. There are five main components of RSA: branch magnitude, topology, link lengths, root angles, and link radius (Jung and McCouch 2013). RSA is dynamic and is modulated by the external environment. Various root characteristics qualify plants to respond, adapt, and survive in varied environments (Paez-Garcia et al. 2015). Legumes have wide diversity of RSA among different species. Every type of RSA is supervised by a genetically regulated "post-embryonary root developmental program," which is multidimensional and allows phenotypic plasticity in reply towards stress including Al toxicity. RSA qualities like anchorage, soil nutrient exploitation, and developmental plasticity have profound effects on yield, more specifically under stress conditions (Jung and McCouch 2013). The development of nonintrusive techniques to actively study RSA may help in designing cultivars with optimum root systems for soils with Al toxicity (Rao et al. 2016). Usually, hydroponics screening to denote RSA is preferred over the soil-based screening due to non-destructive approaches followed under hydroponic culture. Evaluation of root architecture at the seedling stage, i.e., seedling root architecture (SRA) under Al-stress conditions, has also helped to deduce Al tolerance within a large number of genotypes in one go. It also helps in the early detection of Al tolerance within the genotypes and allows breeders to develop Al-tolerant varieties (Singh et al. 2021a, b, c). The results for Al tolerance under hydroponics and field conditions are also found to be significantly correlated as evident in many reports (Singh et al. 2016, 2021c).

15.6.2.4 Sand Culture

Acidic soils with toxic amounts of exchangeable Al and sand assays have been exploited to detect tolerance in plants based on the growth of crop plants. However, results of sand assay were comparable with solution culture assay and more closely reflect Al tolerance in the field. However, the major demerit of this technique is that plants are exposed two times a day, firstly with an acidic Al solution and secondly to an acidic nutrient solution. In sand culture, Al and nutrients are supplied in solution form. This is because sand is nearly inert, and the dose of Al applied to plants can be controlled and replicated with precision. Previous results showed that sand culture provides more accurate results. In a study on pigeon pea, where the hydroponic and sand assays were compared for Al tolerance study, it was found that both the studies consistently differentiated tolerant and sensitive genotypes. These two approaches interrelated well and were comparable over time and place (Choudhary et al. 2011a, b). In contrast, the results of sand culture were not well correlated with solution culture as per Villagarcia et al. (2001). They observed that sand culture was required in ten times higher proportion to inhibit root elongation as compared to the hydroponic system. Grauer and Horst (1990) described a weak association among Al tolerance of 31 soybean genotypes in solution and sand culture. However, the precise basis for greater Al concentration in sand culture is still uncertain.

15.6.2.5 Soil Culture

Evaluation of crop plants is usually conducted in Al toxic fields as this is the most direct screening method to measure agronomic traits and yield components. Selection on acidic soil is an intermediary phase earlier to field testing to assess genotypes under an environment closer to the field condition. Among the soil-based short-term technique, pot culture has been successfully applied based on the root development of seedlings (Ahlrichs et al. 1990). Singh et al. (2016, 2021a, b, c) conducted field experiments in acidic soils of North Eastern states of India following a randomized block design with three replications to detect the performance of lentil genotypes. Soil samples were first detected for nutritional value such as organic carbon, exchangeable Al, and available N, P, K, Ca, and Mg along with soil pH. Both the field experiments showed higher seed yield per plant of tolerant genotypes over the sensitive and check varieties. Overall parameters and traits associated with aluminum toxicity have been summarized in Fig. 15.3.

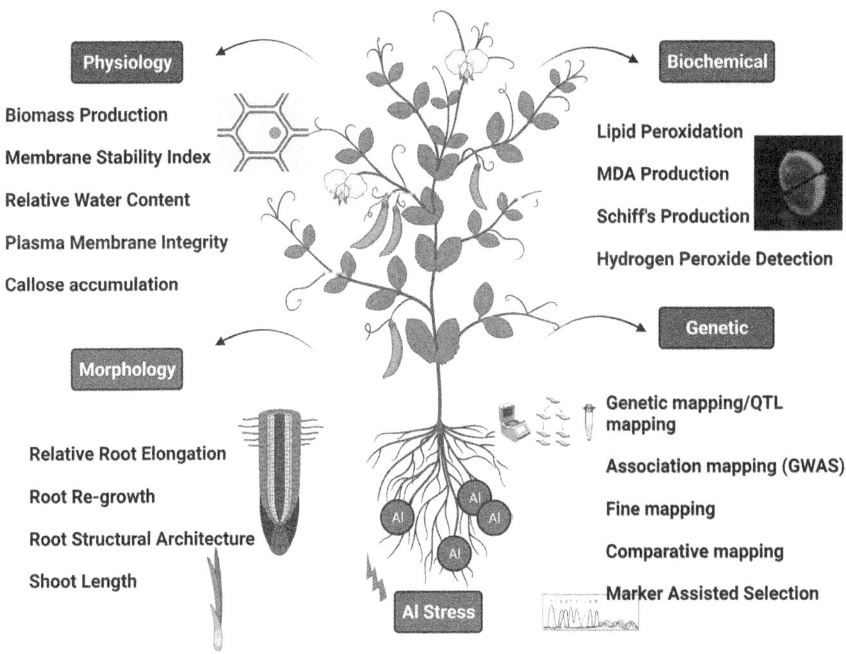

Fig. 15.3 Summary of different parameters and screening techniques used for the evaluation of Al tolerance

15.7 Genetics and Molecular Aspects of Al Tolerance in Legumes

Al toxicity tolerance appeared to be controlled by single or multiple genes in crop plants. In pea and chickpea, tolerance to Al toxicity was found to be regulated by a single dominant gene (Singh and Choudhary 2010; Singh et al. 2011). Al tolerance gene(s) and their loci can be identified by utilizing molecular mapping techniques like QTL mapping, association, and fine mapping. QTL mapping basically determines the total number of genes, their location, and action pattern. Modern breeding strategies relying on QTL mapping have not yet been extensively utilized for improving Al-stress tolerance, though various studies revealing these loci have been accomplished. Further, unlike the cereal crops, limited progress has been made towards the identification of QTL(s) related to Al tolerance in lesser studied legumes.

Root system is complex, and its progress is absolutely affected by Al stress. Custody of a deep and thick root system which permits way to deep water in the soil has been regarded crucial for Al tolerance. Root traits such as root tolerance index, relative root length, relative root elongation, and net root length have often been used as criteria to evaluate Al tolerance. Numerous QTLs for root system architecture traits and their genotypic variation have been characterized. Two major QTLs were identified in lentil, one for fluorescent signals, and another for RRG. Both were mapped on linkage group (LG) 1 under Al-stress conditions when checked on F_2 mapping population derived from a cross between BM-4 and L-4602 (Singh et al. 2018). In soybean, a RIL population was tested to evaluate RRE and apical Al^{3+} content (AAC) where a significant negative correlation was noticed between them. Five QTLs explained 39.65% of RRE and AAC variation. These five QTLs were detected on chromosomes Gm04, Gm16, Gm17, and Gm19 (Wang et al. 2019). In another study, multiple regression analysis divulged five QTLs from autonomous linkage groups in soybean, which conditioned root extension under HIAL (2 mM Al^{3+}) stress (Bianchi Hall et al. 2000).

In the case of alfalfa, three putative Al tolerance QTLs were identified on LG I, LG II, and LG III enlightening 38%, 16%, and 27% phenotypic variation, respectively (Narasimhamoorthy et al. 2007).

Later, three more QTLs were identified on LG 1, 4, and 7, which explicated 20.8%, 15.2%, and 21.7% of the variation, respectively (Khu et al. 2013). Tesfaye et al. (2001) used *Agrobacterium*-mediated gene transfer to enhance Al-stress tolerance in case of transgenic alfalfa by overexpression of malate dehydrogenase gene, which boosted organic acid synthesis.

Five soybean QTLs based on average membership index as a sign of Al tolerance which is represented by four traits, namely plant height, number of leaves, shoot dry weight, and root dry weight at seedling stage, were identified in a RIL population derived from a cross of Kefeng No. 1 × Nannong 1138-2 (Qi et al. 2008).

Major documented physiological mechanism of Al tolerance involves Al-activated secretion of Al-binding organic acids from the root tip, averting uptake into the chief site of toxicity. In case of soybean, a previously constructed high-density genetic linkage map using RAD-seq technology was used to fine-map Al tolerance QTLs. A major QTL (*qAl-HC2*) was fine-mapped onto a narrowed interval of 69.95 kb (Cai et al. 2019). Another study utilized high-density SNP genetic map to fine-map gene GmGSTU9 encrypting GST in qAl06 and three genes (GmPrx143, GmPrx144, and GmPrx145) encrypting PRX in qAl-HC2 with the intervals of 36.89 kb and 69.95 kb, respectively (Cai et al. 2019). Zhang et al. (2020) executed genome-wide documentation of MATE-encoding genes in the *Cicer* genome. In totality, 56 annotated MATE genes were recognized, which were partitioned into four main phylogeny clusters.

Transcriptomic analysis in contrasting genotypes for Al tolerance has been performed to identify candidate genes and pathways along with large number of SNP, SSR, and indel markers for Al tolerance in many legumes such as common bean (Eticha et al. 2010), pigeon pea (Gao et al. 2020), soybean (Huang et al. 2017; You et al. 2011; Zhang et al. 2020), and lentil (Singh et al. 2021a, b, c). Metabolomics has been performed in *Vigna mungo* under Al stress that showed metabolic shifts and excessive ROS triggering detoxification defense mechanism (Chowra et al. 2017).

15.8 Conclusion and Future Perspectives

Al stress inhibits the growth of food legumes severely on acidic soil. Until now, research on Al toxicity has been mostly focused on the evaluation of morphophysiological and anatomical traits in legumes. There are few reports avail able on the efflux of organic acids and signal transduction under Al stress in these plants. In view of the earlier studies, physiological mechanisms of Al tolerance in few legume plants are suggested. There are limited reports about the resistance and physio-biochemical effects of Al toxicity in *Fabaceae* plants. The comprehension about uptake, transport, and distribution of Al in legumes is still missing. With the advances in molecular and biochemical sciences, our knowledge of different molecular and biochemical approaches for Al resistance in food legume crops can be used in legume improvement programs to minimize the economic loss. The QTLs controlling Al tolerance-related traits could be immediately deployed in breeding schemes through marker-assisted selection. Equally important will be to invest on legume germplasm collection programs for improving Al tolerance. Molecular breeding based on "omics" has better advantage and renders different opportunities over conventional breeding. These techniques can be used for screening a large, diversified germplasm in a limited time and space resulting in an early and precise detection of candidate gene(s). Application of machine learning (ML) in quantitative trait locus (QTL) mining and artificial intelligence can further help in determining the genetic determinants of Al tolerance in pulses.

References

Abate E, Hussien S, Laing M, Mengistu F (2013) Aluminium toxicity tolerance in cereals: mechanisms, genetic control and breeding methods. Afr J Agric Res 8(9):711–722
Aebi H (1984) Catalase in vitro. In: Methods in enzymology, vol 105. Academic, pp 121–126
Ahlrichs JL, Karr MC, Baligar VC, Wright RJ (1990) Rapid bioassay of aluminum toxicity in soil. Plant Soil 122(2):279–285
An Y, Wang S, Zhou P (2020) Effects of abscisic acid on alleviation of aluminum toxicity in alfalfa. International Grassland Congress Proceedings
Arunakumara KKIU, Walpola BC, Yoon M-H (2013) Aluminum toxicity and tolerance mechanism in cereals and legumes—a review. J Korean Soc Appl Biol Chem 56(1):1–9
Awasthi JP, Saha B, Regon P, Sahoo S, Chowra U, Pradhan A, Roy A, Panda SK (2017) Morpho-physiological analysis of tolerance to aluminum toxicity in rice varieties of North East India. PLoS One 12(4):e0176357
Baier AC, Somers DJ, Gusiafson JP (1995) Aluminium tolerance in wheat: correlating hydroponic evaluations with field and soil performances. Plant Breed 114(4):291–296
Bhalerao S, Prabhu D (2013) Aluminium toxicity in plants—a review. J Appl Chem 2 (May):447–474
Bianchi Hall CM, Carter TE, Bailey MA, Mian MAR, Rufty TW, Ashley DA, Boerma HR, Arellano C, Hussey RS, Parrott WA (2000) Aluminum tolerance associated with quantitative trait loci derived from soybean PI 416937 in hydroponics. Crop Sci 40(2):538–545
Bojórquez-Quintal E, Escalante-Magaña C, Echevarría-Machado I, Martínez-Estévez M (2017) Aluminum, a friend or foe of higher plants in acid soils. Front Plant Sci 8:1767
Brigham LA, Hawes MC, Miyasaka SC (2001) Avoidance of aluminium toxicity: role of root border cells. In: Plant nutrition. Springer, pp 452–453
Cai M, Wang N, Xing C, Wang F, Wu K, Du X (2013) Immobilization of aluminum with mucilage secreted by root cap and root border cells is related to aluminum resistance in Glycine max L. Environ Sci Pollut Res 20(12):8924–8933
Cai Z, Cheng Y, Xian P, Lin R, Xia Q, He X, Liang Q, Lian T, Ma Q, Nian H (2019) Fine-mapping QTLs and the validation of candidate genes for aluminum tolerance using a high-density genetic map. Plant Soil 444(1):119–137
Chen X-Y, Kim JY (2009) Callose synthesis in higher plants. Plant Signal Behav 4(6):489–492
Choudhary AK, Singh D (2011) Screening of pigeonpea genotypes for nutrient uptake efficiency under aluminium toxicity. Physiol Mol Biol Plants 17(2):145–152
Choudhary AK, Singh D, Iquebal MA (2011a) Selection of pigeonpea genotypes for tolerance to aluminium toxicity. Plant Breed 130(4):492–495
Choudhary AK, Singh D, Kumar J (2011b) A comparative study of screening methods for tolerance to aluminum toxicity in pigeonpea ['*Cajanus cajan*' (L.) Millspaugh]. Aust J Crop Sci 5(11): 1419–1426
Chowra U, Yanase E, Koyama H, Panda SK (2017) Aluminium-induced excessive ROS causes cellular damage and metabolic shifts in black gram *Vigna mungo* (L.) Hepper. Protoplasma 254(1):293–302
de Freitas PAF, de Carvalho HH, Costa JH, Miranda RS, Saraiva KDC, de Oliveira FDB, Coelho DG, Prisco JT, Gomes-Filho E (2019) Salt acclimation in sorghum plants by exogenous proline: physiological and biochemical changes and regulation of proline metabolism. Plant Cell Rep 38(3):403–416
Delhaize E, Ryan PR (1995) Aluminum toxicity and tolerance in plants. Plant Physiol 107(2):315
Delhaize E, Gruber BD, Ryan PR (2007) The roles of organic anion permeases in aluminium resistance and mineral nutrition. FEBS Lett 581(12):2255–2262
Dessureaux L (1969) Effect of aluminium on alfalfa seedlings. Plant Soil 30(1):93–98
Dhindsa RS, Plumb-Dhindsa P, Thorpe TA (1981) Leaf senescence: correlated with increased levels of membrane permeability and lipid peroxidation, and decreased levels of superoxide dismutase and catalase. J Exp Bot 32(1):93–101

Dong D, Peng X, Yan X (2004) Organic acid exudation induced by phosphorus deficiency and/or aluminium toxicity in two contrasting soybean genotypes. Physiol Plant 122(2):190–199

Eticha D, Zahn M, Bremer M, Yang Z, Rangel AF, Rao IM, Horst WJ (2010) Transcriptomic analysis reveals differential gene expression in response to aluminium in common bean (*Phaseolus vulgaris*) genotypes. Ann Bot 105(7):1119–1128

Foy CD (1984) Physiological effects of hydrogen, aluminum, and manganese toxicities in acid soil. Soil Acid Liming 12:57–97

Foy CD, Gerloff GC, Gabelman WH (1973) Differential effects of aluminum on the vegetative growth of tomato cultivars in acid soil and nutrient solutions. J Am Soc Hortic Sci 98(5)

Gao Z, Dong B, Cao H, He H, Yang Q, Meng D, Fu Y (2020) Time series RNA-seq in pigeonpea revealed the core genes in metabolic pathways under aluminum stress. Genes 11(4):380

Giaveno CD, Miranda Filho JB (2000) Rapid screening for aluminum tolerance in maize (*Zea mays* L.). Genet Mol Biol 23(4):847–850. https://doi.org/10.1590/S1415-47572000000400024

Grauer UE, Horst WJ (1990) Effect of pH and nitrogen source on aluminium tolerance of rye (*Secale cereale* L.) and yellow lupin (*Lupinus luteus* L.). Plant Soil 127(1):13–21

Heath RL, Packer L (1968) Photoperoxidation in isolated chloroplasts: I. Kinetics and stoichiometry of fatty acid peroxidation. Arch Biochem Biophys 125(1):189–198

Hede AR, Skovmand B, Ribaut J, González de León D, Stølen O (2002) Evaluation of aluminium tolerance in a spring rye collection by hydroponic screening. Plant Breed 121(3):241–248

Hoffer GN, Carr RH (1923) Accumulation of aluminum and iron compounds in corn plants and its probable relation to root rots. J Agric Res 3(10):323–331

Horst WJ, Wagner A, Marschner H (1982) Mucilage protects root meristems from aluminium injury. Z Pflanzenphysiol 105(5):435–444

Horst WJ, Püschel AK, Schmohl N (1997) Induction of callose formation is a sensitive marker for genotypic aluminium sensitivity in maize. Plant Soil 192(1):23–30

Hossain M, Zhou M, Mendham N (2005) A reliable screening system for aluminium tolerance in barley cultivars. Aust J Agric Res 56(5):475–482

Huang SC, Chu SJ, Guo Y-M, Ji Y-J, Hu D-Q, Cheng J, Lu G-H, Yang R-W, Tang C-Y, Qi J-L (2017) Novel mechanisms for organic acid-mediated aluminium tolerance in roots and leaves of two contrasting soybean genotypes. AoB Plants 9(6):plx064

Hungria M, Vargas MA (2000) Environmental factors affecting N_2 fixation in grain legumes in the tropics, with an emphasis on Brazil. Field Crop Res 65(2–3):151–164

Jaiswal SK, Naamala J, Dakora FD (2018) Nature and mechanisms of aluminium toxicity, tolerance and amelioration in symbiotic legumes and rhizobia. Biol Fertil Soils 54(3):309–318

Jung JKHM, McCouch SRM (2013) Getting to the roots of it: genetic and hormonal control of root architecture. Front Plant Sci 4:186

Khu DM, Reyno R, Han Y, Zhao PX, Bouton JH, Brummer EC, Monteros MJ (2013) Identification of aluminum tolerance quantitative trait loci in tetraploid alfalfa. Crop Sci 53(1):148–163

Kichigina NE, Puhalsky JV, Shaposhnikov AI, Azarova TS, Makarova NM, Loskutov SI, Safronova VI, Tikhonovich IA, Vishnyakova MA, Semenova EV (2017) Aluminum exclusion from root zone and maintenance of nutrient uptake are principal mechanisms of Al tolerance in *Pisum sativum* L. Physiol Mol Biol Plants 23(4):851–863

Kinraide TB (1997) Reconsidering the rhizotoxicity of hydroxyl, sulphate, and fluoride complexes of aluminium. J Exp Bot 48(5):1115–1124

Klimashevskii EL, Markova YA, Seregina ML, Grodzinskii DM, Kozarenko TD (1970) Specifics of the physiological activity of pea plants in connection with unequal resistance of different varieties to mobile aluminum. Soviet Plant Physiol 17(3):372–378

Kochian LV (1995) Cellular mechanisms of aluminum toxicity and resistance in plants. Annu Rev Plant Biol 46(1):237–260

Liao H, Wan H, Shaff J, Wang X, Yan X, Kochian LV (2006) Phosphorus and aluminum interactions in soybean in relation to aluminum tolerance. Exudation of specific organic acids. from different regions of the intact root system. Plant Physiol 141(2):674–684

Ma JF, Zheng SJ, Li XF, Takeda K, Matsumoto H (1997) A rapid hydroponic screening for aluminium tolerance in barley. Plant Soil 191(1):133–137

Maphosa Y, Jideani VA (2017) The role of legumes in human nutrition. In: Functional food-improve health through adequate food, vol 1. IntechOpen, p 13

Massot N, Llugany M, Poschenrieder C, Barceló J (1999) Callose production as indicator of aluminum toxicity in bean cultivars. J Plant Nutr 22(1):1–10

Matsumoto H (1991) Biochemical mechanism of the toxicity of aluminium and the sequestration of aluminium in plant cells. In: Plant-soil interactions at low pH. Springer, pp 825–838

Mendes P, Farina MPW, Channon P (1984) Assessment of aluminium tolerance in maize using a rapid screening procedure. South Afr J Plant Soil 1(3):83–86

Miftahudin M, Nurlaela N, Juliarni J (2007) Uptake and distribution of aluminum in root apices of two rice varieties under aluminum stress. HAYATI J Biosci 14(3):110

Miyasaka SC, Buta JG, Howell RK, Foy CD (1991) Mechanism of aluminium tolerance in snapbeans: root exudation of citric acid. Plant Physiol 96(3):737–743

Narasimhamoorthy B, Bouton JH, Olsen KM, Sledge MK (2007) Quantitative trait loci and candidate gene mapping of aluminum tolerance in diploid alfalfa. Theor Appl Genet 114(5): 901–913

Narayanan A, Sayamala R (1989) Response of pigeon pea (*Cajanus cajan* L.) genotypes to aluminium toxicity. Indian J Plant Physiol 32:17–24

Ownby JD (1993) Mechanisms of reaction of hematoxylin with aluminium treated wheat roots. Physiol Plant 87(3):371–380

Paez-Garcia A, Motes CM, Scheible W-R, Chen R, Blancaflor EB, Monteros MJ (2015) Root traits and phenotyping strategies for plant improvement. Plants 4(2):334–355

Panda SK, Chaudhury I, Khan MH (2003) Heavy metals induce lipid peroxidation and affect antioxidants in wheat leaves. Biol Plant 46(2):289–294

Panda SK, Baluška F, Matsumoto H (2009) Aluminum stress signaling in plants. Plant Signal Behav 4(7):592–597

Polle EKCF, Konzak CF, Kattrick JA (1978) Visual detection of aluminum tolerance levels in wheat by hematoxylin staining of seedling roots 1. Crop Sci 18(5):823–827

Qi B, Korir P, Zhao T, Yu D, Chen S, Gai J (2008) Mapping quantitative trait loci associated with aluminum toxin tolerance in NJRIKY recombinant inbred line population of soybean (*Glycine max*). J Integr Plant Biol 50(9):1089–1095

Rai R (1991) Effects of soil acidity factors on interaction of chickpea (*Cicer arietinum* L.) genotypes and Rhizobium strains: symbiotic N-fixation, grain quality and grain yield in acid soils. In: Plant-soil interactions at low pH. Springer, pp 619–631

Rao IM, Miles JW, Beebe SE, Horst WJ (2016) Root adaptations to soils with low fertility and aluminium toxicity. Ann Bot 118(4):593–605

Rengel Z (1992) Role of calcium in aluminium toxicity. New Phytol 121(4):499–513

Ribeiro C, Cambraia J, Peixoto PHP, Fonseca Júnior ÉMD (2012) Antioxidant system response induced by aluminum in two rice cultivars. Braz J Plant Physiol 24(2):107–116

Rout G, Samantaray S, Das P (2001) Aluminium toxicity in plants: a review. Agronomie 21(1): 3–21

Roy B, Bhadra S (2013) Hematoxylin staining of seedling roots is a potential phenotypic index for screening of aluminium tolerance in rice (*Oryza sativa* L.). Indian J Genet Plant Breed 73(2): 194–198

Sagisaka S (1976) The occurrence of peroxide in perennial plant of *Populas gelrica*. Plant Physiol 57:308–309

Sapra VT, Mebrahtu T, Mugwira LM (1982) Soybean germplasm and cultivar aluminum tolerance in nutrient solution and Bladen clay loam soil 1. Agron J 74(4):687–690

Scheffer-Basso SM, Prior BC (2015) Aluminum toxicity in roots of legume seedlings assessed by topological analysis. Acta Scientiarum Agronomy 37:61–68

Shamsi IH, Wei K, Zhang GP, Jilani GH, Hassan MJ (2008) Interactive effects of cadmium and aluminium on growth and antioxidative enzymes in soybean. Biol Plant 52(1):165–169

Silva S (2012) Aluminium toxicity targets in plants. J Bot 2012:219462., 8 pages. https://doi.org/10.1155/2012/219462
Singh D, Choudhary AK (2010) Inheritance pattern of aluminum tolerance in pea. Plant Breed 129(6):688–692
Singh D, Rai AK, Sureja AK, Bhardwaj R (2007) Hematoxylin staining: a rapid method for assessment of aluminum tolerance in pea genotypes. In: Paper presented in second Indian Horticulture Congress held in ICAR Research Complex Barapani, Meghalaya, India from 18–21 April, 2007
Singh D, Raje RS, Choudhary AK (2011) Genetic control of aluminium tolerance in pigeonpea (*Cajanus cajan* L.). Crop Pasture Sci 62(9):761–764
Singh D, Dikshit HK, Singh R (2012) Variation of aluminium tolerance in lentil (*Lens culinaris* Medik.). Plant Breed 131(6):751–761
Singh D, Pal M, Singh R, Singh CK, Chaturvedi AK (2015) Physiological and biochemical characteristics of *Vigna* species for Al stress tolerance. Acta Physiol Plant 37(4):1–13
Singh D, Pal M, Singh CK, Taunk J, Jain P, Chaturvedi AK, Maurya S, Karwa S, Singh R, Tomar RSS (2016) Molecular scanning and morpho-physiological dissection of component mechanism in *Lens* species in response to aluminium stress. PLoS One 11(7):e0160073
Singh D, Singh CK, Taunk J, Tomar RSS, Chaturvedi AK, Gaikwad K, Pal M (2017) Transcriptome analysis of lentil (*Lens culinaris* Medikus) in response to seedling drought stress. BMC Genomics 18(1):1–20
Singh CK, Singh D, Tomar RSS, Karwa S, Upadhyaya KC, Pal M (2018) Molecular mapping of aluminium resistance loci based on root re-growth and Al-induced fluorescent signals (callose accumulation) in lentil (*Lens culinaris* Medikus). Mol Biol Rep 45(6):2103–2113
Singh CK, Singh D, Sharma S, Chandra S, Taunk J, Konjengbam NS, Singh D, Kumar A, Upadhyaya KC, Pal M (2021a) Morpho-physiological characterization coupled with expressional accord of exclusion mechanism in wild and cultivated lentil under aluminum stress. Protoplasma 258(5):1029–1045
Singh CK, Singh D, Sharma S, Chandra S, Tomar RSS, Kumar A, Upadhyaya KC, Pal M (2021b) Mechanistic association of quantitative trait locus with malate secretion in lentil (*Lens culinaris* Medikus) seedlings under aluminium stress. Plants 10(8):1541
Singh CK, Singh D, Taunk J, Chaudhary P, Chandra S, Singh D, Singh MP, Konjengbam NS, Singh MP, Singh Sengar R (2021c) Comparative inter and intra species transcriptomics revealed key differential pathways associated with aluminium stress tolerance in lentil. Front Plant Sci 12:693630
Stagnari F, Maggio A, Galieni A, Pisante M (2017) Multiple benefits of legumes for agriculture sustainability: an overview. Chem Biol Technol Agric 4(1):1–13
Stass A, Kotur Z, Horst WJ (2007) Effect of boron on the expression of aluminium toxicity in *Phaseolus vulgaris*. Physiol Plant 131(2):283–290
Sujkowska-Rybkowska M, Borucki W (2014) Localization of hydrogen peroxide accumulation and diamine oxidase activity in pea root nodules under aluminum stress. Micron 57:13–22
Tesfaye M, Temple SJ, Allan DL, Vance CP, Samac DA (2001) Overexpression of malate dehydrogenase in transgenic alfalfa enhances organic acid synthesis and confers tolerance to aluminum. Plant Physiol 127(4):1836–1844
Tóth B, Moloi MJ, Szőke L, Danter M, Grusak MA (2021) Cultivar differences in the biochemical and physiological responses of common beans to aluminum stress. Plants 10(10):2097
Villagarcia MR, Carter TE, Rufty TW, Niewoehner AS, Jennette MW, Arrellano C (2001) Genotypic rankings for aluminum tolerance of soybean roots grown in hydroponics and sand culture. Crop Sci 41(5):1499–1507
Vishnyakova MA, Semenova EV, Kosareva IA, Kravchuk ND, Loskutov SI, Pukhalskii IV, Shaposhnikov AI, Sazanova AL, Belimov AA (2015) Method for rapid assessment of aluminum tolerance of pea (*Pisum sativum* L.). Сельскохозяйственная Биология 3(eng):353–360

Wang J, Raman H, Zhang G, Mendham N, Zhou M (2006) Aluminium tolerance in barley (*Hordeum vulgare* L.): physiological mechanisms, genetics and screening methods. J Zhejiang Univ Sci B 7(10):769–787

Wang X, Cheng Y, Yang C, Yang C, Mu Y, Xia Q, Ma Q (2019) QTL mapping for aluminum tolerance in RIL population of soybean (*Glycine max* L.) by RAD sequencing. PLoS One 14(10):e0223674

Wissemeier AH, Klotz F, Horst WJ (1987) Aluminium induced callose synthesis in roots of soybean (*Glycine max* L.). J Plant Physiol 129(5):487–492

Yamamoto Y, Kobayashi Y, Matsumoto H (2001) Lipid peroxidation is an early symptom triggered by aluminum, but not the primary cause of elongation inhibition in pea roots. Plant Physiol 125(1):199–208

Yang Y, Zhang H (1998) Boron amelioration of aluminum toxicity in mungbean seedlings. J Plant Nutr 21(5):1045–1054

Yang JL, Zheng SJ, He YF, Tang CX, Zhou GD (2005) Genotypic differences among plant species in response to aluminum stress. J Plant Nutr 28(6):949–961

Yang ZM, Sivaguru M, Horst WJ, Matsumoto H (2000) Aluminium tolerance is achieved by exudation of citric acid from roots of soybean (Glycine max). Physiol Plant 110(1):72–77

Yang JL, Zhang LEI, Li YY, You JF, Wu P, Zheng SJ (2006) Citrate transporters play a critical role in aluminium-stimulated citrate efflux in rice bean (Vigna umbellata) roots. Ann Bot 97(4): 579–584

Yang Q, Lin Y, Jin L, Ren X, He C, Liu Q (2021) Responses of mineral nutrient contents and transport in red clover under aluminum stress. Legum Sci 3(4):e94

You J, Zhang H, Liu N, Gao L, Kong L, Yang Z (2011) Transcriptomic responses to aluminum stress in soybean roots. Genome 54(11):923–933

Zhang X, Weir B, Wei H, Deng Z, Zhang X, Zhang Y, Xu X, Zhao C, Berger JD, Vance W (2020) Genome-wide identification and transcriptional analyses of MATE transporter genes in root tips of wild Cicer spp. under aluminium stress. BioRxiv

Zhou G, Delhaize E, Zhou M, Ryan PR (2011) Biotechnological solutions for enhancing the aluminium resistance of crop plants. Abiotic stress in plants—mechanisms and adaptations. InTechOpen, Brisbane, pp 119–142

The manufacturer's authorised representative in the EU is Springer Nature Customer Service Centre GmbH, Europaplatz 3, 69115 Heidelberg, Germany. If you have any concerns regarding our products, please contact ProductSafety@springernature.com

Printed and bound by CPI Group (UK) Ltd, Croydon, CR0 4YY
30/09/2026
02225237-0015